PARTIAL INTEGRAL OPERATORS AND INTEGRO-DIFFERENTIAL EQUATIONS

PURE AND APPLIED MATHEMATICS

A Program of Monographs, Textbooks, and Lecture Notes

MONOGRAPHS AND TEXTBOOKS IN
PURE AND APPLIED MATHEMATICS

1. *K. Yano*, Integral Formulas in Riemannian Geometry (1970)
2. *S. Kobayashi*, Hyperbolic Manifolds and Holomorphic Mappings (1970)
3. *V. S. Vladimirov*, Equations of Mathematical Physics (A. Jeffrey, ed.; A. Littlewood, trans.) (1970)
4. *B. N. Pshenichnyi*, Necessary Conditions for an Extremum (L. Neustadt, translation ed.; K. Makowski, trans.) (1971)
5. *L. Narici et al.*, Functional Analysis and Valuation Theory (1971)
6. *S. S. Passman*, Infinite Group Rings (1971)
7. *L. Dornhoff*, Group Representation Theory. Part A: Ordinary Representation Theory. Part B: Modular Representation Theory (1971, 1972)
8. *W. Boothby and G. L. Weiss, eds.*, Symmetric Spaces (1972)
9. *Y. Matsushima*, Differentiable Manifolds (E. T. Kobayashi, trans.) (1972)
10. *L. E. Ward, Jr.*, Topology (1972)
11. *A. Babakhanian*, Cohomological Methods in Group Theory (1972)
12. *R. Gilmer*, Multiplicative Ideal Theory (1972)
13. *J. Yeh*, Stochastic Processes and the Wiener Integral (1973)
14. *J. Barros-Neto*, Introduction to the Theory of Distributions (1973)
15. *R. Larsen*, Functional Analysis (1973)
16. *K. Yano and S. Ishihara*, Tangent and Cotangent Bundles (1973)
17. *C. Procesi*, Rings with Polynomial Identities (1973)
18. *R. Hermann*, Geometry, Physics, and Systems (1973)
19. *N. R. Wallach*, Harmonic Analysis on Homogeneous Spaces (1973)
20. *J. Dieudonné*, Introduction to the Theory of Formal Groups (1973)
21. *I. Vaisman*, Cohomology and Differential Forms (1973)
22. *B.-Y. Chen*, Geometry of Submanifolds (1973)
23. *M. Marcus*, Finite Dimensional Multilinear Algebra (in two parts) (1973, 1975)
24. *R. Larsen*, Banach Algebras (1973)
25. *R. O. Kujala and A. L. Vitter, eds.*, Value Distribution Theory: Part A; Part B: Deficit and Bezout Estimates by Wilhelm Stoll (1973)
26. *K. B. Stolarsky*, Algebraic Numbers and Diophantine Approximation (1974)
27. *A. R. Magid*, The Separable Galois Theory of Commutative Rings (1974)
28. *B. R. McDonald*, Finite Rings with Identity (1974)
29. *J. Satake*, Linear Algebra (S. Koh et al., trans.) (1975)
30. *J. S. Golan*, Localization of Noncommutative Rings (1975)
31. *G. Klambauer*, Mathematical Analysis (1975)
32. *M. K. Agoston*, Algebraic Topology (1976)
33. *K. R. Goodearl*, Ring Theory (1976)
34. *L. E. Mansfield*, Linear Algebra with Geometric Applications (1976)
35. *N. J. Pullman*, Matrix Theory and Its Applications (1976)
36. *B. R. McDonald*, Geometric Algebra Over Local Rings (1976)
37. *C. W. Groetsch*, Generalized Inverses of Linear Operators (1977)
38. *J. E. Kuczkowski and J. L. Gersting*, Abstract Algebra (1977)
39. *C. O. Christenson and W. L. Voxman*, Aspects of Topology (1977)
40. *M. Nagata*, Field Theory (1977)
41. *R. L. Long*, Algebraic Number Theory (1977)
42. *W. F. Pfeffer*, Integrals and Measures (1977)
43. *R. L. Wheeden and A. Zygmund*, Measure and Integral (1977)
44. *J. H. Curtiss*, Introduction to Functions of a Complex Variable (1978)
45. *K. Hrbacek and T. Jech*, Introduction to Set Theory (1978)
46. *W. S. Massey*, Homology and Cohomology Theory (1978)
47. *M. Marcus*, Introduction to Modern Algebra (1978)
48. *E. C. Young*, Vector and Tensor Analysis (1978)
49. *S. B. Nadler, Jr.*, Hyperspaces of Sets (1978)
50. *S. K. Segal*, Topics in Group Kings (1978)
51. *A. C. M. van Rooij*, Non-Archimedean Functional Analysis (1978)
52. *L. Corwin and R. Szczarba*, Calculus in Vector Spaces (1979)
53. *C. Sadosky*, Interpolation of Operators and Singular Integrals (1979)
54. *J. Cronin*, Differential Equations (1980)
55. *C. W. Groetsch*, Elements of Applicable Functional Analysis (1980)

56. *I. Vaisman*, Foundations of Three-Dimensional Euclidean Geometry (1980)
57. *H. I. Freedan*, Deterministic Mathematical Models in Population Ecology (1980)
58. *S. B. Chae*, Lebesgue Integration (1980)
59. *C. S. Rees et al.*, Theory and Applications of Fourier Analysis (1981)
60. *L. Nachbin*, Introduction to Functional Analysis (R. M. Aron, trans.) (1981)
61. *G. Orzech and M. Orzech*, Plane Algebraic Curves (1981)
62. *R. Johnsonbaugh and W. E. Pfaffenberger*, Foundations of Mathematical Analysis (1981)
63. *W. L. Voxman and R. H. Goetschel*, Advanced Calculus (1981)
64. *L. J. Corwin and R. H. Szczarba*, Multivariable Calculus (1982)
65. *V. I. Istrătescu*, Introduction to Linear Operator Theory (1981)
66. *R. D. Järvinen*, Finite and Infinite Dimensional Linear Spaces (1981)
67. *J. K. Beem and P. E. Ehrlich*, Global Lorentzian Geometry (1981)
68. *D. L. Armacost*, The Structure of Locally Compact Abelian Groups (1981)
69. *J. W. Brewer and M. K. Smith, eds.*, Emmy Noether: A Tribute (1981)
70. *K. H. Kim*, Boolean Matrix Theory and Applications (1982)
71. *T. W. Wieting*, The Mathematical Theory of Chromatic Plane Ornaments (1982)
72. *D. B. Gauld*, Differential Topology (1982)
73. *R. L. Faber*, Foundations of Euclidean and Non-Euclidean Geometry (1983)
74. *M. Carmeli*, Statistical Theory and Random Matrices (1983)
75. *J. H. Carruth et al.*, The Theory of Topological Semigroups (1983)
76. *R. L. Faber*, Differential Geometry and Relativity Theory (1983)
77. *S. Barnett*, Polynomials and Linear Control Systems (1983)
78. *G. Karpilovsky*, Commutative Group Algebras (1983)
79. *F. Van Oystaeyen and A. Verschoren*, Relative Invariants of Rings (1983)
80. *I. Vaisman*, A First Course in Differential Geometry (1984)
81. *G. W. Swan*, Applications of Optimal Control Theory in Biomedicine (1984)
82. *T. Petrie and J. D. Randall*, Transformation Groups on Manifolds (1984)
83. *K. Goebel and S. Reich*, Uniform Convexity, Hyperbolic Geometry, and Nonexpansive Mappings (1984)
84. *T. Albu and C. Năstăsescu*, Relative Finiteness in Module Theory (1984)
85. *K. Hrbacek and T. Jech*, Introduction to Set Theory: Second Edition (1984)
86. *F. Van Oystaeyen and A. Verschoren*, Relative Invariants of Rings (1984)
87. *B. R. McDonald*, Linear Algebra Over Commutative Rings (1984)
88. *M. Namba*, Geometry of Projective Algebraic Curves (1984)
89. *G. F. Webb*, Theory of Nonlinear Age-Dependent Population Dynamics (1985)
90. *M. R. Bremner et al.*, Tables of Dominant Weight Multiplicities for Representations of Simple Lie Algebras (1985)
91. *A. E. Fekete*, Real Linear Algebra (1985)
92. *S. B. Chae*, Holomorphy and Calculus in Normed Spaces (1985)
93. *A. J. Jerri*, Introduction to Integral Equations with Applications (1985)
94. *G. Karpilovsky*, Projective Representations of Finite Groups (1985)
95. *L. Narici and E. Beckenstein*, Topological Vector Spaces (1985)
96. *J. Weeks*, The Shape of Space (1985)
97. *P. R. Gribik and K. O. Kortanek*, Extremal Methods of Operations Research (1985)
98. *J.-A. Chao and W. A. Woyczynski, eds.*, Probability Theory and Harmonic Analysis (1986)
99. *G. D. Crown et al.*, Abstract Algebra (1986)
100. *J. H. Carruth et al.*, The Theory of Topological Semigroups, Volume 2 (1986)
101. *R. S. Doran and V. A. Belfi*, Characterizations of C*-Algebras (1986)
102. *M. W. Jeter*, Mathematical Programming (1986)
103. *M. Altman*, A Unified Theory of Nonlinear Operator and Evolution Equations with Applications (1986)
104. *A. Verschoren*, Relative Invariants of Sheaves (1987)
105. *R. A. Usmani*, Applied Linear Algebra (1987)
106. *P. Blass and J. Lang*, Zariski Surfaces and Differential Equations in Characteristic *p* > 0 (1987)
107. *J. A. Reneke et al.*, Structured Hereditary Systems (1987)
108. *H. Busemann and B. B. Phadke*, Spaces with Distinguished Geodesics (1987)
109. *R. Harte*, Invertibility and Singularity for Bounded Linear Operators (1988)
110. *G. S. Ladde et al.*, Oscillation Theory of Differential Equations with Deviating Arguments (1987)
111. *L. Dudkin et al.*, Iterative Aggregation Theory (1987)
112. *T. Okubo*, Differential Geometry (1987)

113. *D. L. Stancl and M. L. Stancl,* Real Analysis with Point-Set Topology (1987)
114. *T. C. Gard,* Introduction to Stochastic Differential Equations (1988)
115. *S. S. Abhyankar,* Enumerative Combinatorics of Young Tableaux (1988)
116. *H. Strade and R. Farnsteiner,* Modular Lie Algebras and Their Representations (1988)
117. *J. A. Huckaba,* Commutative Rings with Zero Divisors (1988)
118. *W. D. Wallis,* Combinatorial Designs (1988)
119. *W. Wiesław,* Topological Fields (1988)
120. *G. Karpilovsky,* Field Theory (1988)
121. *S. Caenepeel and F. Van Oystaeyen,* Brauer Groups and the Cohomology of Graded Rings (1989)
122. *W. Kozlowski,* Modular Function Spaces (1988)
123. *E. Lowen-Colebunders,* Function Classes of Cauchy Continuous Maps (1989)
124. *M. Pavel,* Fundamentals of Pattern Recognition (1989)
125. *V. Lakshmikantham et al.,* Stability Analysis of Nonlinear Systems (1989)
126. *R. Sivaramakrishnan,* The Classical Theory of Arithmetic Functions (1989)
127. *N. A. Watson,* Parabolic Equations on an Infinite Strip (1989)
128. *K. J. Hastings,* Introduction to the Mathematics of Operations Research (1989)
129. *B. Fine,* Algebraic Theory of the Bianchi Groups (1989)
130. *D. N. Dikranjan et al.,* Topological Groups (1989)
131. *J. C. Morgan II,* Point Set Theory (1990)
132. *P. Biler and A. Witkowski,* Problems in Mathematical Analysis (1990)
133. *H. J. Sussmann,* Nonlinear Controllability and Optimal Control (1990)
134. *J.-P. Florens et al.,* Elements of Bayesian Statistics (1990)
135. *N. Shell,* Topological Fields and Near Valuations (1990)
136. *B. F. Doolin and C. F. Martin,* Introduction to Differential Geometry for Engineers (1990)
137. *S. S. Holland, Jr.,* Applied Analysis by the Hilbert Space Method (1990)
138. *J. Oknínski,* Semigroup Algebras (1990)
139. *K. Zhu,* Operator Theory in Function Spaces (1990)
140. *G. B. Price,* An Introduction to Multicomplex Spaces and Functions (1991)
141. *R. B. Darst,* Introduction to Linear Programming (1991)
142. *P. L. Sachdev,* Nonlinear Ordinary Differential Equations and Their Applications (1991)
143. *T. Husain,* Orthogonal Schauder Bases (1991)
144. *J. Foran,* Fundamentals of Real Analysis (1991)
145. *W. C. Brown,* Matrices and Vector Spaces (1991)
146. *M. M. Rao and Z. D. Ren,* Theory of Orlicz Spaces (1991)
147. *J. S. Golan and T. Head,* Modules and the Structures of Rings (1991)
148. *C. Small,* Arithmetic of Finite Fields (1991)
149. *K. Yang,* Complex Algebraic Geometry (1991)
150. *D. G. Hoffman et al.,* Coding Theory (1991)
151. *M. O. González,* Classical Complex Analysis (1992)
152. *M. O. González,* Complex Analysis (1992)
153. *L. W. Baggett,* Functional Analysis (1992)
154. *M. Sniedovich,* Dynamic Programming (1992)
155. *R. P. Agarwal,* Difference Equations and Inequalities (1992)
156. *C. Brezinski,* Biorthogonality and Its Applications to Numerical Analysis (1992)
157. *C. Swartz,* An Introduction to Functional Analysis (1992)
158. *S. B. Nadler, Jr.,* Continuum Theory (1992)
159. *M. A. Al-Gwaiz,* Theory of Distributions (1992)
160. *E. Perry,* Geometry: Axiomatic Developments with Problem Solving (1992)
161. *E. Castillo and M. R. Ruiz-Cobo,* Functional Equations and Modelling in Science and Engineering (1992)
162. *A. J. Jerri,* Integral and Discrete Transforms with Applications and Error Analysis (1992)
163. *A. Charlier et al.,* Tensors and the Clifford Algebra (1992)
164. *P. Biler and T. Nadzieja,* Problems and Examples in Differential Equations (1992)
165. *E. Hansen,* Global Optimization Using Interval Analysis (1992)
166. *S. Guerre-Delabrière,* Classical Sequences in Banach Spaces (1992)
167. *Y. C. Wong,* Introductory Theory of Topological Vector Spaces (1992)
168. *S. H. Kulkarni and B. V. Limaye,* Real Function Algebras (1992)
169. *W. C. Brown,* Matrices Over Commutative Rings (1993)
170. *J. Loustau and M. Dillon,* Linear Geometry with Computer Graphics (1993)
171. *W. V. Petryshyn,* Approximation-Solvability of Nonlinear Functional and Differential Equations (1993)

172. *E. C. Young*, Vector and Tensor Analysis: Second Edition (1993)
173. *T. A. Bick*, Elementary Boundary Value Problems (1993)
174. *M. Pavel*, Fundamentals of Pattern Recognition: Second Edition (1993)
175. *S. A. Albeverio et al.*, Noncommutative Distributions (1993)
176. *W. Fulks*, Complex Variables (1993)
177. *M. M. Rao*, Conditional Measures and Applications (1993)
178. *A. Janicki and A. Weron*, Simulation and Chaotic Behavior of α-Stable Stochastic Processes (1994)
179. *P. Neittaanmäki and D. Tiba*, Optimal Control of Nonlinear Parabolic Systems (1994)
180. *J. Cronin*, Differential Equations: Introduction and Qualitative Theory, Second Edition (1994)
181. *S. Heikkilä and V. Lakshmikantham*, Monotone Iterative Techniques for Discontinuous Nonlinear Differential Equations (1994)
182. *X. Mao*, Exponential Stability of Stochastic Differential Equations (1994)
183. *B. S. Thomson*, Symmetric Properties of Real Functions (1994)
184. *J. E. Rubio*, Optimization and Nonstandard Analysis (1994)
185. *J. L. Bueso et al.*, Compatibility, Stability, and Sheaves (1995)
186. *A. N. Michel and K. Wang*, Qualitative Theory of Dynamical Systems (1995)
187. *M. R. Darnel*, Theory of Lattice-Ordered Groups (1995)
188. *Z. Naniewicz and P. D. Panagiotopoulos*, Mathematical Theory of Hemivariational Inequalities and Applications (1995)
189. *L. J. Corwin and R. H. Szczarba*, Calculus in Vector Spaces: Second Edition (1995)
190. *L. H. Erbe et al.*, Oscillation Theory for Functional Differential Equations (1995)
191. *S. Agaian et al.*, Binary Polynomial Transforms and Nonlinear Digital Filters (1995)
192. *M. I. Gil'*, Norm Estimations for Operation-Valued Functions and Applications (1995)
193. *P. A. Grillet*, Semigroups: An Introduction to the Structure Theory (1995)
194. *S. Kichenassamy*, Nonlinear Wave Equations (1996)
195. *V. F. Krotov*, Global Methods in Optimal Control Theory (1996)
196. *K. I. Beidar et al.*, Rings with Generalized Identities (1996)
197. *V. I. Arnautov et al.*, Introduction to the Theory of Topological Rings and Modules (1996)
198. *G. Sierksma*, Linear and Integer Programming (1996)
199. *R. Lasser*, Introduction to Fourier Series (1996)
200. *V. Sima*, Algorithms for Linear-Quadratic Optimization (1996)
201. *D. Redmond*, Number Theory (1996)
202. *J. K. Beem et al.*, Global Lorentzian Geometry: Second Edition (1996)
203. *M. Fontana et al.*, Prüfer Domains (1997)
204. *H. Tanabe*, Functional Analytic Methods for Partial Differential Equations (1997)
205. *C. Q. Zhang*, Integer Flows and Cycle Covers of Graphs (1997)
206. *E. Spiegel and C. J. O'Donnell*, Incidence Algebras (1997)
207. *B. Jakubczyk and W. Respondek*, Geometry of Feedback and Optimal Control (1998)
208. *T. W. Haynes et al.*, Fundamentals of Domination in Graphs (1998)
209. *T. W. Haynes et al.*, Domination in Graphs: Advanced Topics (1998)
210. *L. A. D'Alotto et al.*, A Unified Signal Algebra Approach to Two-Dimensional Parallel Digital Signal Processing (1998)
211. *F. Halter-Koch*, Ideal Systems (1998)
212. *N. K. Govil et al.*, Approximation Theory (1998)
213. *R. Cross*, Multivalued Linear Operators (1998)
214. *A. A. Martynyuk*, Stability by Liapunov's Matrix Function Method with Applications (1998)
215. *A. Favini and A. Yagi*, Degenerate Differential Equations in Banach Spaces (1999)
216. *A. Illanes and S. Nadler, Jr.*, Hyperspaces: Fundamentals and Recent Advances (1999)
217. *G. Kato and D. Struppa*, Fundamentals of Algebraic Microlocal Analysis (1999)
218. *G. X.-Z. Yuan*, KKM Theory and Applications in Nonlinear Analysis (1999)
219. *D. Motreanu and N. H. Pavel*, Tangency, Flow Invariance for Differential Equations, and Optimization Problems (1999)
220. *K. Hrbacek and T. Jech*, Introduction to Set Theory, Third Edition (1999)
221. *G. E. Kolosov*, Optimal Design of Control Systems (1999)
222. *N. L. Johnson*, Subplane Covered Nets (2000)
223. *B. Fine and G. Rosenberger*, Algebraic Generalizations of Discrete Groups (1999)
224. *M. Väth*, Volterra and Integral Equations of Vector Functions (2000)
225. *S. S. Miller and P. T. Mocanu*, Differential Subordinations (2000)

226. *R. Li et al.*, Generalized Difference Methods for Differential Equations: Numerical Analysis of Finite Volume Methods (2000)
227. *H. Li and F. Van Oystaeyen*, A Primer of Algebraic Geometry (2000)
228. *R. P. Agarwal*, Difference Equations and Inequalities: Theory, Methods, and Applications, Second Edition (2000)
229. *A. B. Kharazishvili*, Strange Functions in Real Analysis (2000)
230. *J. M. Appell et al.*, Partial Integral Operators and Integro-Differential Equations (2000)

Additional Volumes in Preparation

PARTIAL INTEGRAL OPERATORS AND INTEGRO-DIFFERENTIAL EQUATIONS

Jürgen M. Appell
University of Würzburg
Würzburg, Germany

Anatolij S. Kalitvin
Pedagogical Institute of Lipetsk
Lipetsk, Russia

Petr P. Zabrejko
Belgo University
Minsk, Belorussia

MARCEL DEKKER, INC.   NEW YORK · BASEL

ISBN: 0-8247-0396-0

This book is printed on acid-free paper.

Headquarters
Marcel Dekker, Inc.
270 Madison Avenue, New York, NY 10016
tel: 212-696-9000; fax: 212-685-4540

Eastern Hemisphere Distribution
Marcel Dekker AG
Hutgasse 4, Postfach 812, CH-4001 Basel, Switzerland
tel: 41-61-261-8482; fax: 41-61-261-8896

World Wide Web
http://www.dekker.com

The publisher offers discounts on this book when ordered in bulk quantities. For more information, write to Special Sales/Professional Marketing at the headquarters address above.

Current printing (last digit):
10 9 8 7 6 5 4 3 2 1

PRINTED IN THE UNITED STATES OF AMERICA

Preface

This monograph is concerned with partial integral operators and so-called integro-differential equations of Barbashin type. Partial integral operators are, loosely speaking, linear or nonlinear operators with integrals which depend on a parameter. Barbashin equations are special integro-differential equations whose right-hand sides contain both a multiplication operator and an integral operator. Both notions are closely related; for example, an initial value problem for a Barbashin integro-differential equation, linear or nonlinear, leads in a natural way to a fixed point problem involving a partial integral operator.

Although partial integral equations and integro-differential equations arise in many applied problems of mechanics, physics, engineering, and even biology, there is no systematic **treatise on their theory,** methods, and applications so far. Of course, many special results are scattered over a large number of research papers, as may be seen from the impressive list of references at the end. The authors therefore think that there should be some interest in collecting the basic facts of the corresponding theory, presenting some of the methods which have turned out to be useful in the field, and illustrating the abstract results by means of some selected, though typical and important, examples and applications. This is the purpose of the present monograph.

As a matter of fact, many of the research papers contained in the list of references appeared in the former Soviet Union and are referred to here for the first time in book form. Thus, this monograph may also serve as a guide to the "state-of-the-art" in the Russian literature.

This book is the outcome of numerous meetings of the authors and some of their colleagues in Germany, Russia, Belorussia, Italy, and in the United States. Travel expenses and living costs were generously supported by the German Science Foundation (DFG), the German Academic Exchange Service (DAAD), the Belorussian Foundation for Fundamental Scientific Research, the International Soros

Science Education Program, and the Italian National Research Council (CNR). To all of them the authors express their deep and sincere gratitude. The first author acknowledges with particular thankfulness a DFG grant (Ap 40/6-1 and Ap 40/6-2) over several years which considerably improved our research activity at the University of Würzburg.

Jurgen M. Appell
Anatolij S. Kalitvin
Petr P. Zabrejko

Contents

§ 19. Applications of Barbashin equations   421

§ 20. Applications of partial integral equations   450

Introduction

This monograph consists of four chapters. The first two chapters are concerned with integro-differential equation of the form

$$(1) \qquad \frac{\partial x(t,s)}{\partial t} = c(t,s)x(t,s) + \int_a^b k(t,s,\sigma)x(t,\sigma)\,d\sigma + f(t,s).$$

Here $c: J \times [a,b] \to \mathbb{R}$, $k: J \times [a,b] \times [a,b] \to \mathbb{R}$, and $f: J \times [a,b] \to \mathbb{R}$ are given functions, where J is a bounded or unbounded interval; the function x is unknown.

The equation (1) has been studied first by Barbashin and his pupils (see BARBASHIN [1957, 1958], BARBASHIN-BISJARINA [1963], BAR-BASHIN-LIBERMAN [1958], BISJARINA [1963, 1964, 1964a, 1967], BY-KOV [1953, 1953a, 1961, 1962, 1962a, 1962b, 1962c, 1983], BYKOV-KULTAEV [1984], BYKOV-TANKIEV [1982], BYKOVA [1967, 1968, 1971], BYKOVA-BYKOVA [1969], KAMYTOV [1976], KHOTEEV [1976, 1984], KRIVOSHEJN-BYKOVA [1974], LIBERMAN [1958], MAMYTOV [1985], SALGAPAROV [1962], ATAMANOV [1977], BOJKOV [1985], IMANDIEV-DZHURAEV-PANKOV [1974], KALIEV [1974], and SAMEDOVA [1958]). For this reason this equation is nowadays called *integro-differential equation of Barbashin type* or simply *Barbashin equation*.

Identifying a real function $z = z(t,s)$ of two variables (t,s) with the abstract function $z = z(t)$ of one variable $t \in J$ which takes its values in some Banach space X of real functions on $[a,b]$ and is defined by $z(t)(s) = z(t,s)$, one may write equation (1) as an *ordinary differential equation*

$$(2) \qquad \frac{dx}{dt} = A(t)x + f(t)$$

in X. This identification which is a common device in the theory of partial differential equations when passing from a parabolic equation to an abstract *evolution equation* turns out to be useful also here.

Observe that the operator function $A(t)$ in (2) has a very special form: it is the sum

$$(3) \qquad A(t) = C(t) + K(t)$$

1

of a *multiplication operator*

(4) $$C(t)x(s) = c(t,s)x(s)$$

generated by the multiplier c, and an *integral operator*

(5) $$K(t)x(s) = \int_a^b k(t,s,\sigma)x(\sigma)\,d\sigma$$

generated by the kernel k; we shall call operators of this form *Barbashin operators* in what follows. This pleasant fact allows us to combine methods of the *theory of differential equations in Banach spaces* with methods of the *theory of linear integral operators*. As a consequence, we obtain a theory which is far more complete and richer than the theory of differential equations (2) containing an arbitrary operator function $A(t)$.

In particular, in many cases it is useful to study first the "reduced" equation

(6) $$\frac{dx}{dt} = C(t)x + f(t)$$

with $C(t)$ being defined by (4), and then to consider the "full" equation (2) as an *integral perturbation* of (6). *This philosophy will be used over and over again in our study of the Barbashin equation* (1).

Let us illustrate this by means of a simple example. It is a well-known fact that, under some natural hypotheses, the differential equation (2) defines a continuous operator function $U = U(t,\tau)$ of two variables, called the *evolution operator* (or *Cauchy operator*) of (2). This operator may be defined as the unique solution of the integral equation

$$U(t,\tau) = I + \int_\tau^t A(\tau)U(t,\tau)\,d\tau \quad (t,\tau \in J).$$

The integral operator with kernel U may then be considered as the right inverse of the differential operator $\frac{d}{dt} - A(t)$; in particular, the unique solution x of (2) satisfying the *initial condition*

(7) $$x(a) = x_a$$

is given by

$$x(t) = U(t,a)x_a + \int_a^t U(t,\tau)f(\tau)\,d\tau.$$

Now, the evolution operator $U_0 = U_0(t,\tau)$ of the reduced equation (6) may be calculated explicitly: as one expects, it is just the multiplication operator

$$(8) \qquad U_0(t,\tau)x(s) = e(t,\tau,s)x(s)$$

generated by the multiplier

$$(9) \qquad e(t,\tau,s) = \exp\left\{\int_\tau^t c(\xi,s)\,d\xi\right\}.$$

On the other hand, since the difference $K(t) = A(t) - C(t)$ is an integral operator, it is natural to expect that also the difference $H(t,\tau) = U(t,\tau) - U_0(t,\tau)$ of the corresponding evolution operators is an integral operator

$$H(t,\tau)x(s) = \int_a^b h(t,\tau,s,\sigma)x(\sigma)\,d\sigma$$

containing some kernel h which may be evaluated in terms of the kernel k. This is in fact true in many Banach spaces X which occur in applications.

A particularly simple, though important, special case is that of the *stationary integro-differential equation of Barbashin type*

$$(10) \qquad \frac{\partial x(t,s)}{\partial t} = c(s)x(t,s) + \int_a^b k(s,\sigma)x(t,\sigma)\,d\sigma + f(t,s),$$

i.e. the multiplier c and the kernel k do not depend on t. In this case the differential equation (2) reduces to

$$(11) \qquad \frac{dx}{dt} = Ax + f(t),$$

where $A = C + K$ is a *stationary Barbashin operator* with

$$(12) \qquad Cx(s) = c(s)x(s)$$

and

(13) $$K x(s) = \int_a^b k(s, \sigma) x(\sigma) \, d\sigma.$$

Of course, much more can be said in this case; for example, the evolution operator of the full equation (14) is then simply the *exponential operator* $U(t, \tau) = e^{A(t-\tau)}$.

We point out that the operator K in (5) is not a usual integral operator, since the kernel k depends on a parameter t, but the unknown solution x does not. Such operators are called *partial integral operators*, inasmuch as the integration is carried out only with respect to some arguments, while the other arguments of the integrand are "frozen". This is of course analogous to partial differential operators.

Apart from Barbashin equations, partial integral operators constitute the second topic of this monograph. In the most general form, a partial integral operator may be written in the form

(14)
$$P x(t, s) = c(t, s) x(t, s) + \int_T l(t, s, \tau) x(\tau, s) \, d\tau$$
$$+ \int_S m(t, s, \sigma) x(t, \sigma) \, d\sigma + \int_T \int_S n(t, s, \tau, \sigma) x(\tau, \sigma) \, d\sigma \, d\tau,$$

where T and S are arbitrary measure spaces (for example, bounded domains in Euclidean space). Introducing the operators

(15) $$L x(t, s) = \int_T l(t, s, \tau) x(\tau, s) \, d\tau,$$

(16) $$M x(t, s) = \int_S m(t, s, \sigma) x(t, \sigma) \, d\sigma,$$

and

(17) $$N x(t, s) = \int_T \int_S n(t, s, \tau, \sigma) x(\tau, \sigma) \, d\sigma \, d\tau,$$

we may write the operator (14) again as a sum $P = C + K$ of the operators

$$(18) \qquad\qquad Cx(t, s) = c(t, s)x(t, s)$$

and the operator

$$(19) \qquad\qquad K = L + M + N.$$

Observe that the operators L and M are partial integral operators, while the operator N is an ordinary integral operator acting on spaces of functions of two variables.

In contrast to ordinary integral operators, partial integral operators may have rather unpleasant properties. For instance, the operators (15) and (16) are in general not compact. This is probably the reason why partial integral operators (and partial integral equations) have not been studied yet as systematically as ordinary integral operators and equations, although they arise in many problems. For example, the partial integral equation

$$(20) \qquad\qquad x(t, s) = Kx(t, s) + g(t, s),$$

where K is the operator (19), occurs in various problems of mechanics and physics.

Integro-differential equations of Barbashin type and partial integral equations are connected to each other in several ways. To give an example, suppose we are interested in finding a solution of equation (1) which satisfies the initial condition

$$x(a, s) = x_a(s)$$

for some given function $x_a \in X$ (see (7)). Putting $y(t, s) = \partial x(t, s)/\partial t$ we arrive at the equation

$$y(t, s) = \int_a^t c(t, s)y(\tau, s)\, d\tau + \int_a^t \int_a^b k(t, s, \sigma)y(\tau, \sigma)\, d\sigma\, d\tau + g(t, s),$$

where

$$g(t, s) = f(t, s) + c(t, s)x_a(s) + \int_a^b k(t, s, \sigma)x_a(\sigma)\, d\sigma.$$

This is a partial integral equation of type (20) with $m(t, s, \sigma) \equiv 0$, $l(t, s, \tau) \equiv c(t, s)$, and $n(t, s, \tau, \sigma) \equiv k(t, s, \sigma)$. Consequently, all abstract results on the existence and uniqueness of solutions to the partial integral equation (20) may be used to obtain existence and uniqueness theorems for the integro-differential equation (1).

Let us now describe the contents of the four chapters of this monograph more in detail. Chapter I is devoted to the *general theory* of the integro-differential equations of Barbashin type (1) and (10). In § 1 we discuss the connections between the integro-differential equations (1) and (10) with the corresponding differential equations (2) and (11) in a Banach space X, respectively, with a particular emphasis on the important cases $X = C([a, b])$ and $X = L_p([a, b])$ ($1 \leq p \leq \infty$). Here we study also relations between the *evolution operators* of the equations (2) and (6) sketched above.

The following § 2 and § 3 are devoted to a thorough study of the operator (3) in the *Chebyshev space* $X = C([a, b])$ and the *Lebesgue space* $X = L_p([a, b])$, respectively. For instance, we give sufficient (sometimes also necessary) conditions for the *strong continuity* and the *norm continuity* of the operator function (3) in X, as well as *representation theorems* for the corresponding evolution operator. We are also interested in *algebraic properties* of the system $\mathfrak{L}_b(X)$ of all Barbashin operators, considered as a subset of the algebra $\mathfrak{L}(X)$ of bounded linear operators on X.

The Lebesgue spaces L_p are important examples of so-called *ideal spaces* (or L_∞-Banach lattices) of measurable functions we consider in § 4. Other typical examples of ideal spaces are *Orlicz spaces* (which are useful in problems involving strong nonlinearities), or *Lorentz* and *Marcinkiewicz spaces* (which naturally arise in interpolation theory for linear operators). It turns out that many results established in § 3 for Lebesgue spaces carry over to general ideal spaces; this considerably enlarges the applicability of the integro-differential equation (1).

In Chapter II we study *basic analytic properties of Barbashin operators and equations*, again mainly in the spaces $C([a, b])$ and $L_p([a, b])$.

First we consider the *Ljapunov exponent*

$$\omega = \varlimsup_{t \to \infty} \log \frac{\|U(t,0)\|}{t}$$

and the *Ljapunov-Bohl exponent*

$$\omega^* = \varlimsup_{\substack{\tau \to \infty \\ t-\tau \to \infty}} \log \frac{\|U(t,\tau)\|}{t-\tau}$$

of equation (2); as is well known, these exponents are of great importance in *stability theory*. The basic idea in § 5 is again to consider the operator (3) as an integral perturbation of the operator (4): in fact, we may use the explicit formula for the Ljapunov-Bohl exponent ω_0^* of the reduced equation (6), namely

$$\omega_0^* = \varlimsup_{\substack{\tau \to \infty \\ t-\tau \to \infty}} \operatorname*{ess\,sup}_{a \le s \le b} \frac{1}{t-\tau} \int_\tau^t c(\xi,s)\, d\xi$$

(see (8)/(9)), and the fact that the Ljapunov-Bohl exponent is *stable* under small perturbations, to obtain information on the Ljapunov-Bohl exponent of the full equation (2).

While large parts of Chapter I are concerned with *existence* and *uniqueness results*, we study the *continuous dependence* of solutions on a (small) scalar parameter in § 6. We give conditions for the continuous dependence of the evolution operator on a parameter; more regular (e.g. smooth) dependence is also studied.

Bounded and *periodic solutions* of the Barbashin equation (1) are studied in § 7. In this connection, the *Green function* of the Barbashin operator (3) plays a crucial role. Again, we obtain information on the Green function of equation (2) from analogous information on the Green function of the reduced equation (6). Moreover, we prove a result on the continuous dependence of Green's function for (2) on a small scalar parameter ε. In particular, if the dependence becomes "singular" as $\varepsilon \to 0$, this may be considered as an analogue to Bogoljubov's famous *averaging principle* for the differential equation

$$\frac{dx}{dt} = A(\frac{t}{\varepsilon})x.$$

Very little is known about periodic solutions of equation (1); we restrict ourselves to some elementary remarks in § 7, reducing the existence of periodic solutions, as usual, to a fixed point problem for the corresponding *Poincaré operator*.

§ 8 is concerned with Barbashin equations containing *degenerate kernels*. Here one may, at least in some special cases, construct solutions explicitly.

The following two sections are devoted to *boundary value problems* for the Barbashin equation (1). More precisely, we study the stationary equation (10) in § 9 and the nonstationary equation (1) in § 10, both for $0 \le t \le T$ and $-1 \le s \le 1$. The imposed *two-point boundary conditions* are

$$x(0, s) = \phi(s) \ (0 < s \le 1), \qquad x(T, s) = \psi(s) \ (-1 \le s < 0),$$

where $\phi : (0, 1] \to \mathbb{R}$ and $\psi : [-1, 0) \to \mathbb{R}$ are given. To this end, we first reduce this boundary value problem to an operator equation in a suitable function space in such a way that we may apply the classical *Fredholm theory*. In the second part we apply more sophisticated methods, such as (nonstandard) *fixed point principles* and *spectral estimates*.

Let us now pass to the description of Chapter III which is concerned with *partial integral operators* and *partial integral equations*. Apart from *continuity properties* of the operator (19), we are also interested in *algebraic properties* of certain subclasses of partial integral operators, e.g. those operators for which one of the kernels l, m, or n is zero. This will be the subject of § 11. Of course, the operator (19) has to be studied in spaces of functions of two variables. An important class of such spaces are the so-called *spaces with mixed norm* which are discussed, both from the theoretical viewpoint and in view of examples, in § 12. The most prominent example of a space with mixed norm is of course the space $U = [L_p(T) \to L_q(S)]$ or $V = [L_p(T) \leftarrow L_q(S)]$ $(1 \le p, q < \infty)$ of all measurable functions $x : T \times S \to \mathbb{R}$ with norms

$$\|x\|_U = \left\{ \int_S \left[\int_T |x(t, s)|^p \, dt \right]^{q/p} ds \right\}^{1/q}$$

and

$$||x||_V = \left\{ \int_S \left[\int_T |x(t,s)|^q \, ds \right]^{p/q} dt \right\}^{1/p},$$

respectively. These spaces arise naturally in the description of linear integral operators between L_p and L_q. For instance, kernels belonging to these spaces generate bounded linear integral operators between L_p and L_q.

The following § 13 is concerned with a systematic study of the partial integral operator (14) in the space $X = C(T \times S)$ of continuous functions on $T \times S$. In particular, we give conditions (both necessary and sufficient) for the *boundedness* and *continuity* of this operator. Moreover, we study again some *algebraic properties* of various subclasses of such operators in the algebra $\mathfrak{L}(X)$.

Spectral properties of partial integral operators are studied in detail in § 14, mainly for the (stationary) operator

$$(21) \qquad Kx(t,s) = \int_T l(t,\tau)x(\tau,s)\,d\tau + \int_S m(s,\sigma)x(t,\sigma)\,d\sigma.$$

Since this operator may be written as a *tensor product* $K = \tilde{L} \otimes I + I \otimes \tilde{M}$ with

$$(22) \qquad \tilde{L}u(t) = \int_T l(t,\tau)u(\tau)\,d\tau$$

and

$$(23) \qquad \tilde{M}v(s) = \int_S m(s,\sigma)v(\sigma)\,d\sigma,$$

it is not surprising that one may get information on the spectrum of the operator (21) in terms of the spectra of the operators (22) and (23). Some estimates for the *spectral radius* of these operators and an *index formula* for (21) are given in § 14, too.

The partial integral equation (20) is studied in § 15; in particular, we are interested in the *existence* and *uniqueness* of solutions to this equation. Although such equations occur frequently in applications, they have not yet been studied systematically.

The final Chapter IV is concerned with *generalizations* and *applications*. In § 16 we pass to functions of three variables and study the *generalized integro-differential equation*

$$\frac{\partial u(\varphi, t, s)}{\partial \varphi} = c(\varphi, t, s)u(\varphi, t, s) +$$

(24) $$\int_T l(\varphi, t, s, \tau)u(\varphi, \tau, s)\, d\tau + \int_S m(\varphi, t, s, \sigma)u(\varphi, t, \sigma)\, d\sigma$$

$$+ \int_T \int_S n(\varphi, t, s, \tau, \sigma)u(\varphi, \tau, \sigma)\, d\sigma\, d\tau.$$

Equations of this form occur, for example, when applying Fourier transforms to a stationary Schrödinger equation.

While all operators and equations considered so far are linear, in the following § 17 we study *nonlinear operators* and *equations*. For instance, the nonlinear analogue to (1) is the Barbashin equation

(25) $$\frac{\partial x(t, s)}{\partial t} = c(t, s)x(t, s) + \int_a^b k(t, s, \sigma, x(t, \sigma))\, d\sigma + f(t, s)$$

which contains a nonlinear *integral operator of Uryson type.* Similarly, the nonlinear analogue to (14) is

(26)
$$Px(t, s) = c(t, s)x(t, s) + \int_T l(t, s, \tau, x(\tau, s))\, d\tau$$

$$+ \int_S m(s, \sigma, x(t, \sigma))\, d\sigma + \int_T \int_S n(t, s, \tau, \sigma, x(\tau, \sigma))\, d\sigma\, d\tau,$$

which contains nonlinear *partial integral operators of Uryson type.* A particularly important special case is when the kernels of these operators are of *Hammerstein type,* i.e.

(27)
$$\frac{\partial x(t, s)}{\partial t} = c(t, s)x(t, s)$$

$$+ \int_a^b k(t, s, \sigma)h(t, \sigma, x(t, \sigma))\, d\sigma + f(t, s)$$

and

$$Px(t,s) = c(t,s)x(t,s) + \int_T l(t,s,\tau)h(\tau,s,x(\tau,s))\, d\tau$$

(28)
$$+ \int_S m(t,s,\sigma)h(t,\sigma,x(t,\sigma))\, d\sigma$$

$$+ \int_T \int_S n(t,s,\tau,\sigma)h(\tau,\sigma,x(\tau,\sigma))\, d\sigma\, d\tau,$$

respectively. In this case, the study of (27) and (28) reduces to that of the linear equations (1) and (14), respectively, and well-known properties of the *Nemytskij operator*

(29)
$$Hy(t,s) = h(t,s,y(t,s))$$

generated by the nonlinear function h. We will show in § 17 that the operator (28) may be studied by *topological methods, monotonicity methods,* and *variational methods.*

As a matter of fact, another useful method for solving nonlinear operator equations is the *Newton-Kantorovich method.* In § 18 we discuss the applicability of this method to the nonlinear partial integral equation $x = Px$, with P given by (28). The abstract results are illustrated in the spaces C and L_p for $1 \leq p \leq \infty$. Here it turns out that the hypotheses of the classical Newton-Kantorovich method lead to a strong degeneracy for the kernels l, m, and n in (28).

We remark that there are other topics which we did not include in this monograph to keep in within a reasonable size. For example, we do not discuss Barbashin equations containing *partial derivatives.* To the best of our knowledge, the forthcoming paper KOHL [2000] is the only reference about such equations. Of course, this part of the theory, let alone possible applications, is far from being complete.

Finally, § 19 and § 20 are concerned with *selected applications.* As mentioned before, both Barbashin equations and partial integral equations apply to very different fields of mathematics, mechanics, physics, engineering, and natural sciences. In § 19 we discuss applications of Barbashin equations to *probability theory, evolution equations,* and *systems with substantially distributed parameters.* We also give typical applications to *mathematical biology* and *transport problems* in § 19.

Partial integral equations occur in a large number of "real life" problems. Let us just give a sample list of such applications with corresponding references:

Random integral equations (ARINO-KIMMEL [1987]):

$$x(t,\omega) = h(t,\omega) + \int_0^{t+\delta(t)} h(t,\tau,\omega)f(\tau,x(\tau,\omega))\,d\tau.$$

Random integral equations (DING [1984]):

$$x(\omega,t) = x_0(\omega,t) + \int_{t_0}^t k(\omega,t,s,x(\omega,s))\,ds.$$

Random integral equations (JOSHI [1987]):

$$x(s,\omega) = \int_0^1 k(s,t,\omega)f(t,x(t,\omega),\omega)\,dt = y(s,\omega).$$

Random integral equations (SZYNAL-WEDRYCHOWICZ [1987]):

$$x(t,\omega) = x_0(t,\omega) + \int_{t_0}^t k(t,\omega,s,x(s,\omega))\,ds.$$

Biology and physics (PACHPATTE [1986]):

$$u(t,x) = f(t,x) + \int_0^t P[t,x,u(s,x)]\,ds$$

$$+ \int_\Omega Q[t,x,y,u(t,y)]\,dy + \int_0^t \int_\Omega F[t,x,s,y,u(s,y)]\,dy\,ds.$$

Differential equations (PACHPATTE [1983]):

$$z(x,y) = F\left(x,y,z(x,y),\int_0^y \int_0^y f_1[x,y,s,t,z(s,t)]\,ds\,dt,\right.$$

$$\left.\int_0^x f_2[x,y,s,z(s,y)]\,ds,\int_0^y f_3[x,y,t,z(x,t)]\,dt\right).$$

Differential equations (PACHPATTE [1984]):

$$z(x,y) = F(x,y,z(x,y), \int_0^{a(x,y)} \int_0^{b(x,y)} f_1[x,y,s,t,z(s,t)]\,ds\,dt,$$

$$\int_0^{c(x,y)} f_2[x,y,s,z(s,p(x,y))]\,ds, \int_0^{d(x,y)} f_3[x,y,t,z(q(x,y),t)]\,dt).$$

Stochastic equations (RAO [1983]):

$$x(t,\omega) = f(t,\omega) + \int_0^t a(t,s,\omega)g(s,x(s,\omega))\,ds$$

$$+ \int_0^\infty b(t,s,\omega)h(s,x(s,\omega))\,ds.$$

Quasiconformal mappings (SARVAS [1982]):

$$\psi(t_0,t,x) = x + \int_{t_0}^t F(s,\psi(t_0,s,x))\,ds.$$

Stochastic equations (LOBUZOV [1982]):

$$X(s,t) = x + \int_0^t b(X(x,l))\,dW(l) + \int_0^t \sigma(X(x,l))\,dl.$$

Stochastic equations (LOBUZOV [1982]):

$$Y(z,t) = x + \int_0^t a(Y(z,l),s-l)\,dW(l)$$

$$+ \int_0^t \sigma(Y(z,l),s-l)\,dl.$$

Stochastic equations (DALETSKIJ-TSVINTARNAJA [1982]):

$$x(t_1,t_2) - x(t_1^0,t_2^0) =$$

$$\int_{(t_1^0,t_2^0)}^{(t_1,t_2)} \{a_1[u_1,u_2,x(u_1,u_2)]\,du_1 + a_2[u_1,u_2,x(u_1,u_2)]\,du_2$$

$$+ b_1[u_1,u_2,x(u_1,u_2)]\,d\omega_1(u_1) + b_2[u_1,u_2,x(u_1,u_2)]\,d\omega_2(u_2)\}.$$

Parameter-dependent problems (LORD [1980]):

$$u(t, \lambda) = g(t, \lambda) + \lambda \int_0^1 k(t, y, u(y, \lambda)) \, dy.$$

Abstract Volterra equations (KIFFE [1980]):

$$u(t, x) = \int_0^t a(t - s)\{Au(s, x) + g[u(s, x)]\} \, ds = f(t, x).$$

Hammerstein equations (POVOLOTSKIJ-KALITVIN [1985, 1985a]):

$$x(t_1, t_2) = \int_{T_1} k_1(t_1, s) f_1[s, t_2, x(s, t_2)] \, ds$$

$$+ \int_{T_2} k_2(t_2, t) f_2[t_1, t, x(t_1, t)] \, dt.$$

Volterra-Sobolev equations (MURESAN [1982, 1984]):

$$u(t, x) = f(x) + \int_a^t K[s, x, u(s, x), u(x, s)] \, ds.$$

Volterra-Hammerstein equations (THIEME [1980]):

$$u(t, x) = u_0(t, x) + \int_0^t \int_D f[y, u(t - s, y)] h(x, y, s) \, ds \, dy.$$

Biomathematics (HADELER [1986]):

$$u_t(t, x) = u(t, x) \int_{-r}^0 u(t, x + s) \, d\sigma(s)$$

$$- \int_0^1 u(t, y) \int_{-r}^0 u(t, y + s) \, d\sigma(s) \, dy \, u(t, x).$$

Biomathematics (BÜRGER [1986]):

$$\frac{\partial p(x, t)}{\partial t} = [m(x) - \overline{m}(t)] p(x, t) + \mu[u * p(x, t) - p(x, t)],$$

$$u * p(x, t) = \int_{-\infty}^{\infty} u(x - y) p(y, t) \, dy,$$

$$\overline{m}(t) = \int_{-\infty}^{\infty} m(x) p(x, t) \, dx.$$

Boltzmann equations (HEROD [1983]):

$$\frac{\partial Y(t,x)}{\partial t} + Y(t,x) = \int_x^\infty \frac{1}{y} \int_0^y Y(t, y - z)Y(t,z)\, dz\, dy.$$

Random integral inclusion (PHAN [1984, 1984a]):

$$x(\omega, t) \in a(\omega, t) + \int_{T_0(\omega)}^t m(\omega, s, t)F[\omega, s, x(\omega, s)]\, ds.$$

Radiation problems (KELLEY [1980]):

$$H(x, \mu) = \int_0^1 \psi(\nu)\frac{d\nu}{\nu} \int_x^\mu H(t, \mu)H(t, \nu)e^{-(t-x)(\frac{1}{\mu} - \frac{1}{\nu})}\omega(t)\, dt.$$

Some applications of partial integral equations will be studied in § 20 in order to illustrate our abstract mathematical results. Since these equations are the result of a mathematical modelling process, it is clear that the data involved (kernel functions, multiplicator functions, inhomogeneities, numerical parameters etc.) have a precise, say, *physical* or *biological meaning*. Consequently, one should expect that also the results which are provided by the mathematical theory have then a *physical* or *biological interpretation*. However, the sample results we are going to discuss in the last sections of this book are certainly far from being exhaustive, and we strongly believe that this may be a source of an extensive and fruitful research in the future.

Chapter I

Equations of

Barbashin Type

§ 1. Differential equations in Banach spaces

In this section we show how to write an integro-differential equation of Barbashin type as a differential equation in a suitable Banach
space, and we discuss some problems occurring in this connection.
Moreover, we study the evolution operator (Cauchy operator) of the
resulting inhomogeneous differential equation which allows us to give
an explicit integral representation of the solution of an initial value problem. Many contributions contained in this and the following
sections are due to DIALLO [1988, 1988a, 1988b, 1989].

1.1. Integro-differential equations of Barbashin type

In this subsection we study the linear integro-differential equation of
Barbashin type

$$(1.1) \qquad \frac{\partial x(t,s)}{\partial t} = c(t,s)x(t,s) + \int_a^b k(t,s,\sigma)x(t,\sigma)\,d\sigma + f(t,s).$$

Here $c = c(t,s), k = k(t,s,\sigma)$, and $f = f(t,s)$ are given real functions
on $J \times [a,b], J \times [a,b] \times [a,b]$, and $J \times [a,b]$, respectively, with $-\infty < a <
b < \infty$ and J being a bounded or unbounded interval; the function
$x = x(t,s)$ is unknown. In case $f(t,s) \equiv 0$ equation (1.1) is called
homogeneous, in the other case *nonhomogeneous*.

One basic tool in the study of the Barbashin equation (1.1) is to treat
this equation as a linear *differential equation* in some appropriate
Banach space X. Formally, this may be done by writing (1.1) simply
as

$$(1.2) \qquad \frac{dx}{dt} = A(t)x + f(t),$$

where the operator $A(t) : X \to X$ is given by

$$(1.3) \qquad A(t)x(s) = c(t,s)x(s) + \int_a^b k(t,s,\sigma)x(\sigma)\,d\sigma \qquad (t \in J),$$

and the function $f(t) \in X$ by

$$(1.4) \qquad f(t) = f(t,\cdot) \qquad (t \in J).$$

In other words, we *identify* the real function $(t, s) \mapsto x(t, s)$ on $J \times [a, b]$ with the X-valued function $t \mapsto x(t, \cdot)$ on J, where X may be, at least formally, an arbitrary Banach space of real functions on $[a, b]$. However, this makes sense only under certain hypotheses on the functions c, k and f which guarantee that the vector function (1.4) and the operator function (1.3) take values in the chosen Banach space X and in the space $\mathcal{L}(X)$ of all bounded linear operators on X, respectively. Moreover, even if these hypotheses are satisfied, the equations (1.1) and (1.2) need not be equivalent. First of all, equation (1.1) may have a solution x which may not be regarded as a vector function with values in X (for instance, because the solution $x(t, \cdot)$ does not belong to X for some $t \in J$). Second, the solutions of (1.2) need not be (classical) solutions of (1.1) (for instance, because the derivatives $\partial x / \partial t$ in (1.1) and dx / dt in (1.2) are not meant in the same sense).

The simplest situation occurs if X is the *Chebyshev space* $C = C([a, b])$ of all *continuous functions* on $[a, b]$. In this case the operator $A(t)$ and the function $f(t)$ have to be continuous, and one can give a precise description (see e.g. KRASNOSEL'SKIJ [1966], HILLE-PHILLIPS [1957]) of all continuously differentiable vector functions with values in C. In fact, each such function $x = x(t)$ $(t \in J)$ corresponds precisely to one function $x = x(t, s)$ of two variables $(t \in J, s \in [a, b])$ such that both x and $\partial x / \partial t$ are continuous on $J \times [a, b]$; simply put $x(t, s) := x(t)(s)$. This observation implies the following

Lemma 1.1. *Any solution $x = x(t, s)$ of the integro-differential equation (1.1) defines a solution $x(t) := x(t, \cdot)$ of the differential equation (1.2) in the space $X = C$ if and only if both functions $(t, s) \mapsto x(t, s)$ and $(t, s) \mapsto \partial x(t, s) / \partial t$ are continuous. Conversely, any solution $x = x(t)$ of the differential equation (1.2) in the space $X = C$ defines a solution $x(t, s) := x(t)(s)$ of the integro-differential equation (1.1).*

If the space C is replaced by the *Lebesgue space* $L_p = L_p([a, b])$ of all p-summable $(1 \le p < \infty)$ or essentially bounded $(p = \infty)$ functions on $[a, b]$, the situation becomes much more complicated. The main difficulty is here due to the fact that the elements of L_p are not just functions, but *classes* of equivalent functions (i.e. functions which coincide almost everywhere). Thus, if one has to consider an L_p-valued function $x = x(t)$ on J as a real function $x = x(t, s)$ of two variables on $J \times [a, b]$, one has to choose a representation of $x(t) \in X$ in such a way that the function $(t, s) \mapsto x(t, s)$ has "nice" properties (at least measurability, say). This problem was studied in BURGESS [1954] and HILLE-PHILLIPS [1957]; in particular, it is shown there that, given any differentiable L_p-valued function $x(t)$ $(t \in J)$, one may find a measurable function $x = x(t, s)$ on $J \times [a, b]$ which is absolutely continuous in t for all $s \in [a, b]$ and admits a partial derivative $\partial x(t, s)/\partial t$ such that $x(t)$ is equivalent to $x(t, \cdot)$, and $x'(t)$ is equivalent to $\partial x(t, \cdot)/\partial t$. This result, however, does not guarantee in general that, if X is a Banach space and $x = x(t)$ is an X-valued continuously differentiable solution of (1.2) on J, then the corresponding real function $x(t, s)$ is a solution of (1.1) on $J \times [a, b]$. One may only claim that x satisfies (1.1) almost everywhere; in what follows, we shall call such functions *generalized solutions* of (1.1). Thus, a generalized solution $x = x(t, s)$ of (1.1) is characterized by three properties: (a) x is measurable on $J \times [a, b]$; (b) $x(\cdot, s)$ is absolutely continuous on J for each $s \in [a, b]$; (c) x satisfies the equation (1.1) almost everywhere on $J \times [a, b]$.

For further reference, we introduce some notation. Given a Banach space X of real functions over $[a, b]$, by $C_t(X)$ we denote the space of all functions of two variables $x : J \times [a, b] \to \mathbb{R}$ such that $t \mapsto x(t, \cdot)$ is continuous from J into X. Likewise, by $C_t^1(X)$ we denote the space of all $x \in C_t(X)$ such that $x(\cdot, s)$ is continuously differentiable on J for each $s \in [a, b]$ and $\partial x/\partial t$ belongs to $C_t(X)$ as well.

Using this terminology, we may state the following L_p analogue of Lemma 1.1:

Lemma 1.2. *Any generalized solution $x = x(t, s)$ of the integro-differential equation (1.1) defines a solution $x(t) := x(t, \cdot)$ of the differential equation (1.2) in the space $X = L_p$ ($1 \leq p \leq \infty$) if and only if $x \in C_t^1(L_p)$. Conversely, any solution $x = x(t)$ of the differential equation (1.2) in the space $X = L_p$ defines a generalized solution $x(t, s) := x(t)(s)$ of the integro-differential equation (1.1).*

Similar statements hold not only for the Lebesgue space L_p, but also for *Orlicz spaces* and their generalizations, for *Marcinkiewicz spaces*, *Lorentz spaces*, general *ideal spaces*, and other spaces of measurable functions (see § 4). A large class of spaces in which an analogue of Lemma 1.2 holds (so-called *spaces of L type*) is described in HILLE-PHILLIPS [1957]. We shall treat this problem more systematically in Subsection 4.1.

We point out that Lemma 1.1 and Lemma 1.2 are true only if we consider just classical solutions of the differential equation (1.2) in the usual sense, i.e. as continuously differentiable functions $x = x(t)$ which satisfy (1.2) for all $t \in J$. The interpretation of such solutions in the Carathéodory sense, i.e. as absolutely continuous functions $x = x(t)$ which satisfy (1.2) only almost everywhere on J will not be considered.

The Barbashin equation (1.1) is of course particularly easy to handle if either $c(t, s) \equiv 0$ or $k(t, s, \sigma) \equiv 0$. Observe that the special case $c(t, s) \equiv 0$ may always be achieved by the "variation of constants" idea. In fact, let us define $y : J \times [a, b] \to \mathbb{R}$ by the formula

$$y(t, s) := \frac{x(t, s)}{e(t, s)}$$

with

$$e(t, s) := \exp \int_0^t c(\xi, s) \, d\xi.$$

Under the weak assumption that $\partial y / \partial t$ exists (in the same sense as $\partial x / \partial t$) we get then that (1.1) is equivalent to the reduced equation

$$\frac{\partial y(t, s)}{\partial t} = \int_a^b \tilde{k}(t, s, \sigma) y(t, \sigma) \, d\sigma + \tilde{f}(t, s),$$

where

$$\tilde{k}(t, s, \sigma) := \frac{e(t, \sigma)}{e(t, s)} k(t, s, \sigma), \qquad \tilde{f}(t, s) := \frac{f(t, s)}{e(t, s)}.$$

Conversely, in some cases it might be a good idea first to transform (1.1) in this way, but then to transform the reduced equation back in the opposite direction with a "better" function $c = c(t, s)$. We point out, however, that this is a purely formal simplification inasmuch as the "structural" properties of the function c may change.

1.2. The evolution operator

Let us recall the basic properties of linear differential equations like (1.2) with a strongly continuous operator function $A = A(t)$ in a Banach space X. As usual, by $\mathfrak{L}(X, Y)$ we denote the linear space of all bounded linear operators from X into Y; in particular, $\mathfrak{L}(X, X) =: \mathfrak{L}(X)$. An operator function $A : J \to \mathfrak{L}(X)$ is called *strongly continuous* if, for each $x \in X$, the map $t \mapsto A(t)x$ is continuous from J into X. (By "abuse of language", we say then that "the operator function $A = A(t)$ is strongly continuous in X".) It is not hard to see that the integral equation

$$(1.5) \qquad Z(t) = I + \int_\tau^t A(\xi) Z(\xi) \, d\xi$$

has a solution $U = U(t, \tau)$ $(t, \tau \in J)$ which is unique in the class of all integrable operator functions over J with values in the space $\mathfrak{L}(X)$. This solution may be obtained either as limit (in $\mathfrak{L}(X)$) of the successive approximations

$$(1.6) \quad Z_{n+1}(t) = I + \int_\tau^t A(\xi) Z_n(\xi) \, d\xi \quad (n = 0, 1, \ldots; Z_0(t) = I),$$

or as limit (also in $\mathfrak{L}(X)$) of the series

$$(1.7) \qquad U(t, \tau) = I + \sum_{n=1}^\infty \int_\tau^t \cdots \int_\tau^{t_{n-1}} A(t_1) \ldots A(t_n) \, dt_n \ldots dt_1$$

which converges uniformly on J. Since

$$\left\| \int_\tau^t \cdots \int_\tau^{t_{n-1}} A(t_1) \ldots A(t_n) \, dt_n \ldots dt_1 \right\|$$

$$(1.8) \qquad \leq \int_\tau^t \cdots \int_\tau^{t_{n-1}} \|A(t_1)\| \ldots \|A(t_n)\| \, dt_n \ldots dt_1$$

$$= \frac{1}{n!} \left[\int_\tau^t \|A(\xi)\| \, d\xi \right]^n \quad (n = 1, 2, \ldots)$$

and

$$(1.9) \qquad \exp \int_\tau^t \|A(\xi)\| \, d\xi = \sum_{n=0}^\infty \frac{1}{n!} \left[\int_\tau^t \|A(\xi)\| \, d\xi \right]^n,$$

the estimate

$$(1.10) \qquad \|U(t,\tau)\| \leq \exp \int_\tau^t \|A(\xi)\| \, d\xi \quad (t,\tau \in J)$$

holds. We summarize the basic properties of the operator function $U = U(t,\tau)$ in the following two lemmas.

Lemma 1.3. *Suppose that the operator function $t \mapsto A(t)$ is strongly continuous in X. Then the operator function $(t,\tau) \mapsto U(t,\tau)$ is continuous on $J \times J$, and the vector function $(t,\tau) \mapsto U(t,\tau)x$ is differentiable in t and τ for each $x \in X$. Moreover, the operator function $\partial U(t,\tau)/\partial t$ is continuous in τ, uniformly with respect to $t \in J$, and strongly continuous in t, uniformly with respect to $\tau \in J$. Finally, the operator function $\partial U(t,\tau)/\partial \tau$ is continuous in t, uniformly with respect to $\tau \in J$, and strongly continuous in τ, uniformly with respect to $t \in J$.*

Lemma 1.4. *Suppose that the operator function $t \mapsto A(t)$ is strongly continuous in X. Then the operator function $(t,\tau) \mapsto U(t,\tau)$ has the following properties:*
(a) $U(t,t) = I \qquad (t \in J)$;
(b) $U(t,s)U(s,\tau) = U(t,\tau) \qquad (t,\tau,s \in J)$;
(c) $U(t,\tau)^{-1} = U(\tau,t) \qquad (t,\tau \in J)$.

Both lemmas are proved by standard arguments. Under weaker assumptions (Bochner-integrability of the map $t \mapsto A(t)$) Lemma 1.3 and Lemma 1.4 have been proved in DALETSKIJ-KREJN [1970], under stronger assumptions (continuity of the map $t \mapsto A(t)$ in $\mathfrak{L}(X)$) in CARTAN [1967].

The operator function $U = U(t,\tau)$ is usually called *evolution operator* or *Cauchy operator*. The first name is explained by the following fundamental

Theorem 1.1. *Suppose that the operator function $t \mapsto A(t)$ is strongly continuous in X. Then, for each fixed $\tau \in J, x_\tau \in X$, and $f \in C_t(X)$, the differential equation*

$$(1.11) \qquad \frac{dx}{dt} = A(t)x + f(t)$$

has a unique solution $x = x(t)$ satisfying the initial condition

$$(1.12) \qquad x(\tau) = x_\tau.$$

This solution is given by

$$(1.13) \qquad x(t) = U(t, \tau)x_\tau + \int_\tau^t U(t, \xi)f(\xi)\, d\xi.$$

The proof of Theorem 1.1 is the same as in the classical theory of differential equations; under somewhat different assumptions the statement may be found in DALETSKIJ-KREJN [1970] and CARTAN [1967].

In some situations it is more convenient to consider, instead of the operator function $U = U(t, \tau)$, the operator function of one variable

$$(1.14) \qquad U(t) = U(t, t_0),$$

where t_0 is some fixed element of J (usually $t_0 = 0$). By Lemma 1.4, the operator function $U(t, \tau)$ may be recovered from $U(t)$ by means of the formula

$$(1.15) \qquad U(t, \tau) = U(t)U(\tau)^{-1}.$$

In this way, the problem of analyzing the operator function (1.7) of two variables t and τ is equivalent to studying the operator function (1.14) of one variable t.

1.3. Special cases and examples

We describe now an important special case when the evolution operator $U(t, \tau)$ may be represented in a more explicit form.

Suppose that the operator (1.3) does not depend on t, i.e. $A(t) \equiv A$. Throughout the following, this case will be called *stationary* (or *constant*), while the case $A = A(t)$ will be called *non-stationary* (or *variable*). In the stationary case (see e.g. DALETSKIJ-KREJN [1970]) we have

$$(1.16) \qquad U(t,\tau) = e^{A(t-\tau)};$$

here the operator function e^{As} is defined by the usual exponential series

$$(1.17) \qquad e^{As} = \sum_{n=0}^{\infty} \frac{1}{n!} s^n A^n$$

or the Riesz integral

$$(1.18) \qquad e^{As} = \frac{1}{2\pi} \int_{\Gamma} e^{\lambda s} R(\lambda, A) d\lambda,$$

where $R(\lambda, A) = (\lambda I - A)^{-1}$ is the resolvent of the operator A, and Γ is any simple closed contour moving counterclockwise around the spectrum $\sigma(A)$ of A in the complex plane.

In the scalar case, the evolution operator is simply given by

$$(1.19) \qquad U(t,\tau) = \exp\left\{ \int_{\tau}^{t} A(\xi)\, d\xi \right\};$$

in the general case, however, (1.19) is not true. A necessary and sufficient condition for (1.19) is that the family of operators $\{A(t) : t \in J\}$ *commutes* (ERUGIN [1963], HILLE-PHILLIPS [1957]), i.e.

$$(1.20) \qquad A(t)A(\tau) = A(\tau)A(t) \quad (t, \tau \in J);$$

this is a quite restrictive condition.

As is pointed out in ERUGIN [1963], the condition

$$(1.21) \qquad A(t)\left(\int_{t_0}^{t} A(\xi)\, d\xi \right) = \left(\int_{t_0}^{t} A(\xi)\, d\xi \right) A(t) \quad (t \in J)$$

is sufficient for the representation of the operator (1.14) in the form

$$U(t) = \exp\left\{\int_{t_0}^t A(\xi)\,d\xi\right\},$$

and hence, by (1.15), of the operator (1.7) in the form

$$(1.22) \qquad U(t,\tau) = \exp\left\{\int_{t_0}^t A(\xi)\,d\xi\right\}\exp\left\{-\int_{t_0}^\tau A(\xi)\,d\xi\right\}.$$

However, the condition (1.21) is *not* necessary for the representation (1.22) to hold.

There are some other special cases when the operator function $U = U(t,\tau)$ may be expressed through the operator function $A = A(t)$ in a rather simple manner. For a rather complete survey of such cases we refer to ERUGIN [1963] and the references therein.

1.4. Some auxiliary results

Observe that the operator function $t \mapsto A(t)$ defined in (1.3) has a very special structure: it is representable as a sum

$$(1.23) \qquad\qquad A(t) = C(t) + K(t)$$

of two operator function, namely the *multiplication operator*

$$(1.24) \qquad\qquad C(t)x(s) = c(t,s)x(s)$$

induced by a fixed *coefficient* (or *multiplier*) $c = c(t,s)$, and the *integral operator*

$$(1.25) \qquad\qquad K(t)x(s) = \int_a^b k(t,s,\sigma)x(\sigma)\,d\sigma,$$

induced by a fixed *kernel* $k = k(t,s,\sigma)$. In what follows, we call an operator of the form (1.23), with $C(t)$ given by (1.24) and $K(t)$ given by (1.25), a *Barbashin operator*. As a consequence of the special form of a Barbashin operator, one could expect a richer theory for the differential equation (1.2) in case of a Barbashin operator $A(t)$ than in case of an arbitrary bounded linear operator $A(t)$. This is in fact true, and this will be illustrated over and over again in this and the following chapters.

To begin with, one could expect, for example, that the evolution operator operator $U(t, \tau)$ associated with the linear differential equation

$$(1.26) \qquad \frac{dx}{dt} = [C(t) + K(t)]x + f(t)$$

has more specific properties than the evolution operator of (1.2) for general $A(t)$. In particular, it seems interesting to compare the evolution operator $U(t, \tau)$ of (1.26) with the evolution operator $U_0(t, \tau)$ of the "reduced" equation

$$(1.27) \qquad \frac{dx}{dt} = C(t)x + f(t).$$

First of all, we point out that the evolution operator $U_0(t, \tau)$ of (1.27) may be given explicitly; it is also a multiplication operator:

Lemma 1.5. *Suppose that $c : J \times [a, b] \to \mathbb{R}$ is bounded, and the corresponding operator function (1.24) is strongly continuous in X. Then the evolution operator $U_0 = U_0(t, \tau)$ of equation (1.27) is given by*

$$(1.28) \qquad U_0(t, \tau)x(s) = e(t, \tau, s)x(s),$$

where

$$(1.29) \qquad e(t, \tau, s) := \exp\left\{ \int_\tau^t c(\xi, s)\, d\xi \right\}.$$

$\square$ Fix $x \in X$. By (1.7) we have

$$U_0(t, \tau)x = x + \sum_{n=1}^\infty \int_\tau^t \cdots \int_\tau^{t_{n-1}} C(t_1) \cdots C(t_n)x\, dt_n \cdots dt_1,$$

where the series converges in X. Since the function c is bounded we conclude that

$$x(s) + \sum_{n=1}^\infty \int_\tau^t \cdots \int_\tau^{t_{n-1}} c(t_1, s) \cdots c(t_n, s)x(s)\, dt_n \cdots dt_1$$

$$= x(s) + \sum_{n=1}^\infty \frac{1}{n!} \left[\int_\tau^t c(\xi, s)\, d\xi \right]^n x(s) = e(t, \tau, s)x(s)$$

converges for all $s \in [a, b]$, so the limits must coincide. ∎

The multiplier (1.29) is of fundamental importance in the study of equation (1.1) and will be used over and over again in what follows.

We observe that the evolution operators of (1.27) and (1.26) are related by the equality

$$(1.30) \qquad U(t, \tau) = U_0(t, \tau) + \int_\tau^t U_0(t, \xi) K(\xi) U(\xi, \tau) \, d\xi$$

(see VALEEV-ZHAUTYKOV [1974], DALETSKIJ-KREJN [1970]).

The main problem we shall be interested in below is the following. Let $\mathfrak{M}$ be some subclass of the space $\mathfrak{L}(X)$ of bounded linear operators on X, and suppose that $K(t) \in \mathfrak{M}$ for all $t \in J$. Is it true that the difference $U(t, \tau) - U_0(t, \tau)$ belongs then also to $\mathfrak{M}$ for all $t, \tau \in J$? In other words, are the properties of the difference $A(t) - C(t)$ "inherited" by the difference $U(t, \tau) - U_0(t, \tau)$ of the corresponding evolution operators?

The answer is positive if $\mathfrak{M}$ is some closed ideal in $\mathfrak{L}(X)$, for example, the ideal of compact operators. For the following, however, we need a "refinement" of this statement. In the sequel, let $\mathfrak{L}_0(X)$ be a fixed Banach algebra which is continuously imbedded in the algebra $\mathfrak{L}(X)$.

Lemma 1.6. *Suppose that the operator function $t \mapsto A(t)$ is bounded on J with values in $\mathfrak{L}_0(X)$. Then the operator function $(t, \tau) \mapsto U(t, \tau)$ is bounded on $J \times J$ and takes also values in $\mathfrak{L}_0(X)$.*

□ To see this, observe that the series (1.7) converges (in the algebra $\mathfrak{L}_0(X)$) uniformly on $J \times J$, by the estimate (1.7); therefore its limit belongs to $\mathfrak{L}_0(X)$ for all $t \in J$ and $\tau \in J$. Since $\mathfrak{L}_0(X)$ is continuously imbedded in $\mathfrak{L}(X)$, the series (1.7) converges in $\mathfrak{L}(X)$ as well; but its limit in $\mathfrak{L}(X)$ is just $U(t, \tau)$. This shows hat $U(t, \tau) \in \mathfrak{L}_0(X)$ for all $t, \tau \in J$. Furthermore, the continuity of the map $(t, \tau) \mapsto U(t, \tau)$ between $J \times J$ and $\mathfrak{L}_0(X)$ follows from the equality

$$U(t, \tau) = I + \int_\tau^t A(\xi) U(\xi, \tau) \, d\xi,$$

which obviously holds in $\mathfrak{L}_0(X)$ as well. ∎

Suppose now that $\mathfrak{M}$ is a closed ideal in $\mathfrak{L}_0(X)$, where $\mathfrak{L}_0(X)$ is continuously imbedded in $\mathfrak{L}(X)$ as before.

Theorem 1.2. *Suppose that the operator function $t \mapsto C(t)$ is bounded on J with values in $\mathfrak{L}_0(X)$, and the operator function $t \mapsto K(t)$ is bounded on J with values in $\mathfrak{M}$. Then*

$$(1.31) \qquad\qquad U(t,\tau) - U_0(t,\tau) \in \mathfrak{M}$$

for all $(t,\tau) \in J \times J$.

◻ By Lemma 1.6, applied to the operators $C(t)$ and $C(t)+K(t)$, both evolution operators $U_0(t,\tau)$ and $U(t,\tau)$ belong to $\mathfrak{L}_0(X)$ for $t,\tau \in J$. The relation $K(t) \in \mathfrak{M}\ (t \in J)$ implies then that $U_0(t,\xi)K(\xi)U(\xi,\tau) \in \mathfrak{M}$ as well $(\tau \leq \xi \leq t)$. It remains to use the equality (1.28) from Lemma 1.5 which implies that the difference $U(t,\tau) - U_0(t,\tau)$ is the limit (in $\mathfrak{L}_0(X)$) of linear combinations of operators $U_0(t,\xi)K(\xi)U(\xi,\tau)\ (\tau \leq \xi \leq t)$. ∎

Theorem 1.2 contains, of course, the well known fact that *the compactness of $K(t)$ implies the compactness of $U(t,\tau) - U_0(t,\tau)$*, as well as similar statements which may be obtained choosing $\mathfrak{L}_0(X) = \mathfrak{L}(X)$.

1.5. Degenerate kernels

As a model example we shall consider now and in the sequel the case of a *degenerate kernel k*, i.e.

$$(1.32) \qquad\qquad k(t,s,\sigma) = \sum_{j=1}^{n} \alpha_j(t,s)\beta_j(\sigma),$$

where the functions $\alpha_j : J \times [a,b] \to \mathbb{R}$ and $\beta_j : [a,b] \to \mathbb{R}$ are supposed to be bounded for the time being. The set $\mathfrak{M}_n$ of corresponding integral operators

$$K(t)x(s) = \sum_{j=1}^{n} \alpha_j(t,s) \int_a^b \beta_j(\sigma)x(\sigma)\,d\sigma$$

is obviously a closed ideal in $\mathfrak{L}(X)$: in fact, an operator $K \in \mathfrak{L}(X)$ belongs to $\mathfrak{M}_n$ if and only if its range has dimension $\leq n$. From Theorem 1.2 we conclude that the evolution operator of equation (1.26) has then the form

$$(1.33) \qquad U(t,\tau)x(s) = U_0(t,\tau)x(s) + H(t,\tau)x(s),$$

where $U_0(t,\tau)$ is given by (1.28) and $H(t,\tau)$ is an integral operator

$$(1.34) \qquad H(t,\tau)x(s) = \int_a^b h(t,\tau,s,\sigma)x(\sigma)\,d\sigma$$

with degenerate kernel

$$(1.35) \qquad h(t,\tau,s,\sigma) = \sum_{j=1}^n \gamma_j(t,\tau,s)\delta_j(t,\tau,\sigma).$$

More specifically, we have to choose γ_j and δ_j such that

$$0 = \frac{\partial}{\partial t}U(t,\tau)x(s) - A(t)U(t,\tau)x(s)$$

$$= \sum_{j=1}^n \int_a^b \delta_j(t,\tau,\sigma)x(\sigma)\,d\sigma \left[\frac{\partial}{\partial t}\gamma_j(t,\tau,s) - c(t,s)\gamma_j(t,\tau,s)\right]$$

$$+ \sum_{j=1}^n \left[\int_a^b \frac{\partial}{\partial t}\delta_j(t,\tau,\sigma)x(\sigma)\,d\sigma\right]\gamma_j(t,\tau,s)$$

$$- \sum_{j=1}^n \left[\int_a^b \beta_j(\sigma)e(t,\tau,\sigma)x(\sigma)\,d\sigma\right]\alpha_j(t,s)$$

$$- \sum_{j,k=1}^n \left[\int_a^b \beta_j(\sigma)\gamma_k(t,\tau,\sigma)\,d\sigma\right]\left[\int_a^b \delta_k(t,\tau,\sigma)x(\sigma)\,d\sigma\right]\alpha_j(t,s).$$

This is a complicated integro-differential equation for the unknown functions γ_j and δ_j ($j = 1,...,n$). We shall return to this equation later (see § 8); here we just confine ourselves to a very elementary example for $n = 1$.

Example 1.1. Let $c(t,s) := 4t$ and $k(t,s,\sigma) := -6ts\sigma$ over $[a,b] = [0,1]$, i.e. we consider the integro-differential equation

$$(1.36) \qquad \frac{\partial x(t,s)}{\partial t} = 4tx(t,s) - 6ts \int_0^1 \sigma x(t,\sigma)\, d\sigma.$$

Denoting the integral in (1.36) by $\xi(t)$ we get

$$\xi'(t) = \int_0^1 \sigma \frac{\partial x(t,\sigma)}{\partial t}\, d\sigma = \int_0^1 \sigma[4tx(t,\sigma) - 6t\sigma\xi(t)]\, d\sigma = 2t\xi(t)$$

hence $\xi(t) = e^{t^2}$. Thus we arrive at the linear first-order differential equation

$$\frac{\partial x(t,s)}{\partial t} = 4tx(t,s) - 6tse^{t^2}$$

for x, with solution $x(t,s) = 3se^{t^2}$. By what has been observed above, the evolution operator for (1.36) has the form

$$U(t,\tau)x_0(s) = e^{2(t^2-\tau^2)}x_0(s) + \int_0^1 h(t,\tau,s,\sigma)x_0(\sigma)\, d\sigma,$$

where the kernel h is of the form (1.35), i.e.

$$h(t,\tau,s,\sigma) = \gamma(t,\tau,s)\delta(t,\tau,\sigma),$$

and $x_0(s) = x(0,s)$. A straightforward, but somewhat cumbersome calculation leads to the functions

$$\gamma(t,\tau,s) = [e^{(t^2-\tau^2)} - e^{2(t^2-\tau^2)}]s, \quad \delta(t,\tau,\sigma) = 3\sigma.$$

In particular, the unique solution of the equation (1.36) with initial condition $x(0,s) = s$ is given by $x(t,s) = se^{t^2}$.

1.6. Symmetric kernels

Let us consider the equation

$$(1.37) \qquad \frac{\partial x(t,s)}{\partial t} = c(t)x(t,s) + \int_a^b k(s,\sigma)x(t,\sigma)\, d\sigma + f(s)$$

with initial condition

(1.38) $$x(0, s) = x_0(s),$$

for $0 \leq t \leq T$ and $a \leq s \leq b$, where $c : [0, T] \rightarrow \mathbb{R}$ is continuous, and $f : [a, b] \rightarrow \mathbb{R}$ and $k : [a, b] \times [a, b] \rightarrow \mathbb{R}$ are square integrable. In addition, we suppose now that the kernel k is symmetric. Under this assumption, a solution of the problem (1.37)/(1.38) may be constructed by expanding the kernel into a series of eigenfunctions. Following VASIL'EVA-TIKHONOV [1989], we shall seek the solution x in the form

(1.39) $$x(t, s) = z(t, s) + \tilde{x}(t, s),$$

where $\tilde{x}$ is a solution of the problem

$$\begin{cases} \dfrac{\partial \tilde{x}(t, s)}{\partial t} = c(t)\tilde{x}(t, s) + f(s) & \text{if } 0 \leq t \leq T, a \leq s \leq b, \\[2mm] \tilde{x}(0, s) = x_0(s) & \text{if } a \leq s \leq b. \end{cases}$$

Then

(1.40)
$$\tilde{x}(t, s) = x_0(s) \exp\left\{ \int_0^t c(\tau)\, d\tau \right\}$$
$$+ f(s) \int_0^t \exp\left\{ \int_\tau^t c(\xi)\, d\xi \right\} d\tau,$$

and z is a solution, for $a \leq s \leq b$ and $0 \leq t \leq T$, of the problem

(1.41)
$$\begin{cases} \dfrac{\partial z(t, s)}{\partial t} = c(t)z(t, s) + \displaystyle\int_a^b k(s, \sigma)x(t, \sigma)\, d\sigma + f(s), \\[2mm] z(0, s) = 0; \end{cases}$$

here we have put

$$f(t, s) := \int_a^b k(s, \sigma)\tilde{x}(t, \sigma)\, d\sigma.$$

We shall seek $z(t, s)$ in the form

(1.42) $$z(t, s) = \sum_{n=1}^{\infty} z_n(t)k_n(s),$$

where the eigenfunctions k_n of the kernel k are supposed to form an orthonormal basis in $L^2([a,b])$. Substituting (1.42) into (1.41) we be obtain

$$(1.43) \qquad \frac{dz_n(t)}{dt} = c(t)z_n(t) + \lambda_n z_n(t) + \lambda_n \tilde{x}_n(t), \quad z_n(0) = 0,$$

where

$$(1.44)$$
$$\tilde{x}_n(t) = \int_a^b \tilde{x}(t,s)k_n(s)\,ds$$
$$= x_n(s)\exp\left\{\int_0^t c(\tau)\,d\tau\right\} + f_n\int_0^t \exp\left\{\int_\tau^t c(\xi)\,d\xi\right\}d\tau,$$

with x_n and f_n ($n = 1, 2, \ldots$) being the Fourier coefficients of x_0 and f, respectively.

Obviously, a solution of the problem (1.43) is given by

$$(1.45) \qquad z_n(t) = \lambda_n \int_0^t \exp\left\{\int_\tau^t c(\xi)\,d\xi + \lambda_n(t-\tau)\right\}\tilde{x}_n(\tau)\,d\tau.$$

From (1.45) and (1.44) it follows that the series (1.42) converges in L_2, uniformly with respect to $t \in [0,T]$. Consequently, $z \in C_t(L_2)$, where $C_t(X)$ is defined as in Subsection 1.1. Analogously, the convergence of the series

$$\sum_{n=1}^\infty \frac{dz_n(t)}{dt}k_n(s)$$

in L_2, uniformly with respect to $t \in [0,T]$, follows from (1.45), (1.44) and (1.43). This implies again that $\partial z/\partial t \in C_t(L_2)$, i.e. $z \in C_t^1(L_2)$. We conclude that the unique solution of the problem (1.37)/(1.38) is given by (1.39), where $\tilde{x}$ is the function (1.40), and z is the series (1.42). The functions z_n from (1.42) in turn have the form (1.45), where λ_n is the n-th eigenvalue of the kernel k with eigenfunction k_n, and $\tilde{x}_n$ is given by (1.44).

§ 2. Barbashin equations in the space C

In this section we give a systematic account of the basic properties of Barbashin operators in the Chebyshev space C of continuous functions over $[a, b]$. In particular, we derive sufficient conditions (which sometimes are also necessary) for the strong continuity or norm-continuity of Barbashin operators in both the stationary and non-stationary case. A representation theorem for the evolution operator in the space C is also given.

2.1. Linear operators in the space C

Consider the stationary Barbashin operator

$$(2.1) \qquad Ax(s) = c(s)x(s) + \int_a^b k(s, \sigma)x(\sigma)\, d\sigma;$$

where $c = c(s)$ and $k = k(s, \sigma)$ are given measurable functions on $[a, b]$ and $[a, b] \times [a, b]$, respectively. We are interested in conditions on c and k under which the operator (2.1) is bounded in the space C of all continuous functions on $[a, b]$. The complete answer to this problem may be given rather easily by means of the classical Radon theorem (GLIVENKO [1936], DUNFORD-SCHWARTZ [1963], KANTO-ROVICH-AKILOV [1977], NATANSON [1950], RADON [1919]). Consider the functions

$$(2.2) \qquad \alpha(s) := c(s) + \int_a^b k(s, \sigma)\, d\sigma,$$

and

$$(2.3) \qquad \beta(z, s) := \int_a^z [c(s)\chi(s, \sigma) + (z - \sigma)k(s, \sigma)]\, d\sigma,$$

$(a \leq z \leq b)$, with $\chi(s, \cdot)$ denoting the characteristic function of the unbounded interval $[s, \infty)$, as well as the function

$$(2.4) \qquad \gamma(s) := |c(s)| + \int_a^b |k(s, \sigma)|\, d\sigma.$$

Lemma 2.1. *The linear operator* (2.1) *is continuous in the space* C *if and only if the functions* (2.2) *and* (2.3) *are continuous, and the* *function (2.4) is bounded. In this case, the operator (2.1) is bounded with norm*

$$(2.5) \qquad \|A\|_{\mathcal{L}(C)} = \sup_{a \leq s \leq b} \gamma(s).$$

□ First of all, we show that the operator (2.1) is bounded whenever it is defined on the space C. In fact, in this case it follows that, applying the operator (2.1) to the function $x_1(s) \equiv 1$, the function $k(s, \cdot)$ is Lebesgue integrable on $[a, b]$ for each $s \in [a, b]$, and the function $w : [a, b] \to \mathbb{R}$ defined by

$$w(s) := \int_a^b |k(s, \sigma)| \, d\sigma$$

is measurable, by Fubini's theorem. This implies that (2.1) is continuous, considered as an operator from the space C into the space S of all measurable functions on $[a, b]$; consequently, (2.1) is closed as an operator in the space C. The continuity of (2.1) follows now from the closed graph theorem.

Let us rewrite (2.1) in the form

$$(2.6) \qquad Ax(s) = \int_a^b x(\sigma) \, dg(s, \sigma),$$

where

$$(2.7) \qquad g(s, \sigma) = c(s)\chi(s, \sigma) + \int_a^\sigma k(s, \tau) \, d\tau.$$

By the Riesz representation theorem on the general form of continuous linear functionals on the space C the equality (2.6) may be regarded as the *Radon representation* (see again GLIVENKO [1936]) of the bounded linear operator (2.1). But Radon's theorem implies also that the vector function $g^* : [a, b] \to BV$ defined by

$$(2.8) \qquad g^*(s)(\sigma) := g(s, \sigma)$$

with values in the space $BV = BV([a, b])$ of all functions of bounded variation on $[a, b]$ is weakly continuous.

Various conditions for the weak continuity of BV-valued vector functions are reported in GLIVENKO [1936]. The most appropriate one is the *Riesz criterion* which states that the vector function (2.8) is weakly continuous on BV if and only if (2.8) is bounded and the scalar functions $s \mapsto g^*(s)(b) - g^*(s)(a)$ and $s \mapsto \int_a^z g^*(s)(\sigma) \, d\sigma$ are continuous. By standard formulas (see e.g. NATANSON [1950]), the total variation of $g^*(s)$ on $[a, b]$ is

$$\|g^*(s)\|_{BV} = \text{Var}_{[a,b]} g^*(s) = |c(s)| + \int_a^b |k(s,\sigma)| \, d\sigma = \gamma(s);$$

consequently, the boundedness of the vector function (2.8) in BV is equivalent to the boundedness of the function (2.4). Moreover, by the obvious formulas

$$g^*(s)(b) - g^*(s)(a) = c(s) + \int_a^b k(s,\sigma) \, d\sigma = \alpha(s),$$

and

$$\int_a^z g^*(s)(\sigma) \, d\sigma = \int_a^z [c(s)\chi(s,\sigma) + \int_a^\sigma k(s,\tau) d\tau] \, d\sigma = \beta(z,s),$$

the boundedness of the operator (2.1) on C implies the continuity of both functions (2.2) and (2.3). Thus we have shown that, whenever the operator (2.1) acts in the space C, the functions (2.2) and (2.3) are continuous, and the function (2.4) is bounded.

The converse is also true: if the functions (2.2) and (2.3) are continuous and the function (2.4) is bounded, then the vector function (2.8) is weakly continuous on the space BV, and this is in turn equivalent to the fact that the operator (2.1) acts in the space C.

It remains to prove the relation (2.5). But this is an immediate consequence of the formula

$$\|A\|_{\mathcal{L}(C)} = \sup_{a \le s \le b} \sup_{\|x\|_C \le 1} \int_a^b x(\sigma) \, dg^*(s)(\sigma)$$

$$= \sup_{a \le s \le b} \|g^*(s)\|_{BV} = \sup_{a \le s \le b} \gamma(s).$$

The proof is complete . ∎

As a matter of fact, the Radon representation of a bounded linear operator is not unique. However, Theorem 2.1 implies that the representation (2.1) for the operator A is unique. In fact, if there were two representations

$$Ax(s) = c_j(s)x(s) + \int_a^b k_j(s,\sigma)x(\sigma)\,d\sigma \quad (j = 1,2),$$

we would have

$$[c_1(s) - c_2(s)]x(s) + \int_a^b [k(s,\sigma) - k_2(s,\sigma)]x(\sigma)\,d\sigma \equiv 0.$$

Consequently, by Lemma 2.1 (applied to the zero operator) we would get

$$|c_1(s) - c_2(s)| + \int_a^b |k_1(s,\sigma) - k_2(s,\sigma)|\,d\sigma = 0 \quad (a \le s \le b),$$

and hence $c_1(s) = c_2(s)$ for all $s \in [a,b]$, as well as $k_1(s,\sigma) = k_2(s,\sigma)$ for all $s \in [a,b]$ and almost all $\sigma \in [a,b]$.

2.2. The spaces $\mathfrak{L}_b(C)$ and $\mathfrak{L}_m(C) \oplus \mathfrak{L}_i(C)$

The set of all operators of type (2.1) is a subspace of the space $\mathfrak{L}(C)$ which we shall denote throughout by $\mathfrak{L}_b(C)$ (where the subscript b stands for "Barbashin operator"). Lemma 2.1 gives a precise characterization of all operators in the class $\mathfrak{L}_b(C)$. In particular, we have the following

Lemma 2.2. *The subspace $\mathfrak{L}_b(C)$ is closed in $\mathfrak{L}(C)$.*

□ In fact, suppose that

$$A_n x(s) = c_n(s)x(s) + \int_a^b k_n(s,\sigma)x(\sigma)\,d\sigma \quad (n = 1,2,\ldots)$$

is a sequence of operators in $\mathfrak{L}_b(C)$ converging (in $\mathfrak{L}(C)$) to some $A \in \mathfrak{L}(C)$. Then $(A_n)_n$ is a Cauchy sequence in $\mathfrak{L}_b(C)$, and the relation

$$\lim_{m,n\to\infty} \sup_{a\le s\le b} \left[|c_m(s) - c_n(s)| + \int_a^b |k_m(s,\sigma) - k_n(s,\sigma)|\,d\sigma \right] = 0$$

holds, by Lemma 2.1. This relation is equivalent to both

$$(2.9) \qquad \lim_{m,n\to\infty} \sup_{a\le s\le b} |c_m(s) - c_n(s)| = 0$$

and

$$(2.10) \qquad \lim_{m,n\to\infty} \sup_{a\le s\le b} \int_a^b |k_m(s,\sigma) - k_n(s,\sigma)|\, d\sigma = 0.$$

Now, (2.9) shows that $(c_n)_n$ is a Cauchy sequence in the space of all bounded functions on $[a,b]$ (with the supremum norm). Since this space is complete, we have

$$(2.11) \qquad \lim_{n\to\infty} \sup_{a\le s\le b} |c_n(s) - c(s)| = 0$$

for some bounded (measurable) function c on $[a,b]$. Similarly, (2.10) shows that $(k_n)_n$ is a Cauchy sequence in the space $[L_1 \to L_\infty]$ with mixed norm (see Subsection 12.1 below). Since this space is also complete, we have again

$$\lim_{n\to\infty} ||k_n - \tilde{k}||_{[L_1\to L_\infty]}$$

$$= \lim_{n\to\infty} \operatorname*{ess\,sup}_{a\le s\le b} \int_a^b |k_n(s,\sigma) - \tilde{k}(s,\sigma)|\, d\sigma = 0$$

for some measurable function $\tilde{k} \in [L_1 \to L_\infty]$. This means that

$$\lim_{n\to\infty} \sup_{s\in[a,b]\setminus D_0} \int_a^b |k_n(s,\sigma) - \tilde{k}(s,\sigma)|\, d\sigma = 0,$$

for some null set $D_0 \subset [a,b]$. For $s \in D_0$, the sequence $(k_n(s,\cdot))_n$ is Cauchy in the space L_1, by (2.10); consequently, there is another function $\hat{k} = \hat{k}(s,\sigma)$ such that

$$\lim_{n\to\infty} ||k_n(s,\cdot) - \hat{k}(s,\cdot)||_{L_1} = 0,$$

uniformly with respect to $s \in D_0$, again by (2.10). The function

$$k(s,\sigma) = \begin{cases} \tilde{k}(s,\sigma) & \text{if } s \in [a,b]\setminus D_0, \\ \hat{k}(s,\sigma) & \text{if } s \in D_0 \end{cases}$$

is then measurable, by construction, and satisfies

$$(2.12) \qquad \lim_{n \to \infty} \sup_{a \le s \le b} \int_a^b |k_n(s, \sigma) - k(s, \sigma)| \, d\sigma = 0.$$

Consider now the linear operator

$$(2.13) \qquad Ax(s) = c(s)x(s) + \int_a^b k(s, \sigma)x(\sigma) \, d\sigma;$$

we claim that this operator belongs to $\mathfrak{L}_b(C)$ and is the limit (in $\mathfrak{L}_b(C)$) of the sequence $(A_n)_n$. To prove that $A \in \mathfrak{L}_b(C)$ it suffices to show that, for any fixed $x \in C$, the sequence $(A_n x)_n$ converges uniformly to Ax. But this follows immediately from (2.11), (2.12), and the obvious estimate

$$|A_n x(s) - Ax(s)| \le |c_n(s) - c(s)| \, |x(s)|$$

$$+ \int_a^b |k_n(s, \sigma) - k(s, \sigma)| \, |x(\sigma)| \, d\sigma.$$

Finally, by Lemma 2.1 we have

$$\|A_n - A\|_{\mathfrak{L}(C)} = \sup_{a \le s \le b} [|c_n(s) - c(s)|$$

$$+ \int_a^b |k_n(s, \sigma) - k(s, \sigma)| \, d\sigma],$$

and thus the convergence of $(A_n)_n$ to A in the space $\mathfrak{L}_b(C)$ is a consequence of (2.11) and (2.12). ∎

The space $\mathfrak{L}(C)$, equipped with the composition of operators as multiplication, is a Banach algebra. It is natural to ask whether or not the subclass $\mathfrak{L}_b(C)$ is also an algebra, or even an ideal in $\mathfrak{L}(C)$.

To see that $\mathfrak{L}_b(C)$ is *not* an ideal in $\mathfrak{L}(C)$ is easy: just consider the composition of some operator from $\mathfrak{L}_b(C)$ with the evaluation operator $Ex(s) = x(s_0)$ at some point $s_0 \in [a, b]$. Nevertheless, the following holds.

Lemma 2.3. *The subspace* $\mathfrak{L}_b(C)$ *is a subalgebra of* $\mathfrak{L}(C)$. *More precisely, if* $A, B \in \mathfrak{L}_b(C)$ *are given by*

$$Ax(s) = c(s)x(s) + \int_a^b k(s, \sigma)x(\sigma)\, d\sigma$$

and

$$Bx(s) = d(s)x(s) + \int_a^b l(s, \sigma)x(\sigma)\, d\sigma,$$

then $AB \in \mathfrak{L}_b(C)$ *is given by*

(2.14) $$ABx(s) = e(s)x(s) + \int_a^b m(s, \sigma)x(\sigma)d\sigma,$$

where

$$e(s) := c(s)d(s)$$

and

$$m(s, \sigma) := c(s)l(s, \sigma) + k(s, \sigma)d(\sigma) + \int_a^b k(s, \xi)l(\xi, \sigma)\, d\xi.$$

☐ The formula (2.14) follows from

$$ABx(s) = c(s)d(s)x(s) + \int_a^b c(s)l(s, \sigma)x(\sigma)\, d\sigma$$

$$+ \int_a^b k(s, \sigma)d(\sigma)x(\sigma)\, d\sigma + \int_a^b k(s, \xi)\left[\int_a^b l(\xi, \sigma)x(\sigma)d\sigma\right]\, d\xi,$$

where the order of integration may be changed by Fubini's theorem; in fact, the equality

$$\int_a^b \int_a^b |k(s, \xi)|\, |l(\xi, \sigma)|\, |x(\sigma)|\, d\sigma\, d\xi$$

$$= \int_a^b |k(s, \xi)| \int_a^b |l(\xi, \sigma)|\, |x(\sigma)|\, d\sigma\, d\xi$$

implies that the function $(\sigma, \xi) \mapsto k(s, \xi)l(\xi, \sigma)x(\sigma)$ is integrable on $[a, b] \times [a, b]$ for each $s \in [a, b]$. By (2.14), we see that $AB \in \mathfrak{L}_b(C)$. ■

By definition, every operator $A \in \mathfrak{L}_b(C)$ is the sum of a multiplication operator C generated by some multiplier $c = c(s)$, and an integral operator K generated by some kernel $k = k(s, \sigma)$. One should expect that, whenever A is a bounded linear operator in the space C, each of the corresponding parts C and K are also bounded linear operators in the space C. Surprisingly enough, this is not true!

Example 2.1. Consider the linear operator A defined by

$$Ax(s) := \begin{cases} x(s) & \text{if } 0 \leq s \leq \dfrac{1}{2}, \\ \dfrac{2}{2s-1} \displaystyle\int_{1/2}^{s} x(\sigma)\, d\sigma & \text{if } \dfrac{1}{2} < s \leq 1. \end{cases}$$

By Lemma 2.1, this operator acts in the space $C = C([0,1])$ and has the form (2.1), where

$$c(s) = \begin{cases} 1 & \text{if } 0 \leq s \leq \frac{1}{2}, \\ 0 & \text{if } \frac{1}{2} < s \leq 1, \end{cases}$$

and

$$k(s,\sigma) = \begin{cases} \frac{2}{2s-1} & \text{if } \frac{1}{2} < \sigma \leq s \leq 1, \\ 0 & \text{otherwise,} \end{cases}$$

and satisfies $\|A\|_{\mathfrak{L}(C)} = 1$. Nevertheless, since the function c is discontinuous at $s = \frac{1}{2}$, the corresponding multiplication operator $Cx(s) = c(s)x(s)$ (and hence also the integral operator $K = A - C$ with kernel k) does *not* act in the space C.

Thus the problem arises to study those operators $A \in \mathfrak{L}_b(C)$ more in detail for which both terms in (2.1) are bounded linear operators in the space C. To this end, denote by $\mathfrak{L}_m(C)$ the class of all multiplication operators

$$Cx(s) = c(s)x(s),$$

and by $\mathfrak{L}_i(C)$ the class of all integral operators

$$Kx(s) = \int_a^b k(s,\sigma)x(\sigma)\, d\sigma$$

in $\mathfrak{L}(C)$. The operators we are interested in belong to the proper subclass $\mathfrak{L}_m(C) \oplus \mathfrak{L}_i(C)$ of $\mathfrak{L}_b(C)$. From Lemma 2.1 and Lemma 2.2 we get the following

Lemma 2.4. *The subspace $\mathfrak{L}_m(C) \oplus \mathfrak{L}_i(C)$ is closed in $\mathfrak{L}_b(C)$.*

□ Observe, first of all, that an operator $A \in \mathfrak{L}_b(C)$ belongs to $\mathfrak{L}_m(C) \oplus \mathfrak{L}_i(C)$ if and only if the multiplicator function $c = c(s)$ is continuous on $[a, b]$. Now, if the successive approximations

$$A_n x(s) = c_n(s)x(s) + \int_a^b k_n(s, \sigma)x(\sigma)\, d\sigma \quad (n = 0, 1, \ldots)$$

belong to $\mathfrak{L}_m(C) \oplus \mathfrak{L}_i(C)$ and converge (in $\mathfrak{L}(C)$) to some operator

$$A x(s) = c(s)x(s) + \int_a^b k(s, \sigma)x(\sigma)\, d\sigma$$

in $\mathfrak{L}_b(C)$, then, by formula (2.5) of Lemma 2.1,

$$\lim_{n \to \infty} \sup_{a \le s \le b} |c_n(s) - c(s)| = 0,$$

and thus the function c is continuous as well. This shows that actually $A \in \mathfrak{L}_m(C) \oplus \mathfrak{L}_i(C)$. ∎

An additional statement about the algebraic properties of the class $\mathfrak{L}_m(C) \oplus \mathfrak{L}_i(C)$ is given in the following

Lemma 2.5. *The subspace $\mathfrak{L}_i(C)$ is an ideal in both the algebra $\mathfrak{L}_m(C) \oplus \mathfrak{L}_i(C)$ and the algebra $\mathfrak{L}_b(C)$. The subspace $\mathfrak{L}_m(C) \oplus \mathfrak{L}_i(C)$ is a subalgebra of the algebra $\mathfrak{L}_b(C)$.*

□ We show that $\mathfrak{L}_i(C)$ is an ideal in $\mathfrak{L}_m(C) \oplus \mathfrak{L}_i(C)$. To this end, let K be an integral operator with kernel $k = k(s, \sigma)$, and B an arbitrary operator from the class $\mathfrak{L}_m(C) \oplus \mathfrak{L}_i(C)$, i.e.

$$B x(s) = d(s)x(s) + \int_a^b l(s, \sigma)x(\sigma)\, d\sigma.$$

By (2.14), the operators KB and BK have the form

$$K B x(s) = \int_a^b [k(s, \sigma)d(\sigma) + \int_a^b k(s, \xi)l(\xi, \sigma)\, d\xi]x(\sigma)\, d\sigma,$$

and

$$BKx(s) = \int_a^b [d(s)k(s,\sigma) + \int_a^b l(s,\xi)k(\xi,\sigma)\,d\xi]x(\sigma)\,d\sigma,$$

and hence both belong to $\mathfrak{L}_i(C)$. The fact that $\mathfrak{L}_i(C)$ is an ideal in $\mathfrak{L}_b(C)$ is proved in the same way. Finally, if A and B are two operators in $\mathfrak{L}_m(C) \oplus \mathfrak{L}_i(C)$, their "multiplicative parts" $Cx(s) = c(s)x(s)$ and $Dx(s) = d(s)x(s)$, say, are generated by continuous functions c and d; consequently, by the continuity of the product $c(s)d(s)$ and formula (2.4) we have $AB \in \mathfrak{L}_m(C) \oplus \mathfrak{L}_i(C)$ and $BA \in \mathfrak{L}_m(C) \oplus \mathfrak{L}_i(C)$ as well. ∎

We remark that the set $\mathfrak{L}_i(C)$ of all integral operators is not an ideal in the algebra $\mathfrak{L}(C)$ of all bounded linear operators on C.

2.3. Compact and weakly compact operators

Let us recall some further facts from the theory of bounded linear operators in the space C of continuous functions. An operator A in C is called *compact* if A maps bounded sets into precompact sets, and *weakly compact* if A maps bounded sets into weakly precompact sets. By the well known Grothendieck theorem (see e.g. DUNFORD-SCHWARTZ [1963]), a bounded linear operator A in C is weakly compact if and only if A maps weakly convergent sequences into strongly convergent sequences. In particular, this implies the classical Dunford-Pettis theorem which states that the composition of two weakly compact operators in C is a compact operator.

It is well known that the classes $\mathfrak{L}_c(C)$ and $\mathfrak{L}_w(C)$ of all compact respectively weakly compact operators in the space C is an ideal in the algebra $\mathfrak{L}(C)$.

We are mainly interested in compact or weakly compact operators of the form (2.1), of course, i.e. in the classes $\mathfrak{L}_b(C) \cap \mathfrak{L}_c(C)$ and $\mathfrak{L}_b(C) \cap \mathfrak{L}_w(C)$. We are going to formulate two criteria for the (weak) compactness of an operator $A \in \mathfrak{L}_b(C)$.

Lemma 2.6. *A linear operator $A \in \mathfrak{L}_b(C)$ is compact if and only if $c(s) \equiv 0$ and the L_1-valued vector function $s \mapsto k(s,\cdot)$ is continuous,* *i.e.*

(2.15) $$\lim_{\xi \to s} \int_a^b |k(\xi, \sigma) - k(s, \sigma)|\, d\sigma = 0 \quad (a \le s \le b).$$

□ The assertion follows from the following general compactness criterion for bounded linear operators on the space C (see e.g. DUNFORD-SCHWARTZ [1963]): the operator (2.1) is compact in C if and only if the BV-valued vector function (2.8) is continuous. Since the function (2.8) is also C-weakly continuous, this is in turn equivalent to the norm-continuity of (2.8). But in the norm of the space BV we have

$$\|g^*(\xi) - g^*(s)\|_{BV} = \mathrm{Var}_{[a,b]}[g(\xi, \cdot) - g(s, \cdot)]$$

$$= |c(\xi)| + |c(s)| + \int_a^b |k(\xi, \sigma) - k(s, \sigma)|\, d\sigma,$$

(see e.g. NATANSON [1950]), and thus the continuity of (2.8) at each $s \in [a, b]$ is obviously equivalent to the requirement that $c(s) \equiv 0$. ∎

Lemma 2.7. *A linear operator $A \in \mathfrak{L}_b(C)$ is weakly compact if and only if $c(s) \equiv 0$ and the L_1-valued vector function $s \mapsto k(s, \cdot)$ has a weakly compact range, i.e.*

(2.16) $$\lim_{\mu(D) \to 0} \sup_{a \le s \le b} \int_D |k(s, \sigma)|\, d\sigma = 0,$$

where $\mu(D)$ denotes the Lebesgue measure of $D \subseteq [a, b]$.

□ Here we apply the following general criterion for the weak compactness of bounded linear operators on C (see again DUNFORD-SCHWARTZ [1963]): the operator (2.1) is weakly compact in C if and only if the BV-valued vector function (2.8) is C^{**}-weakly continuous. By the Lebesgue theorem (see e.g. NATANSON [1950]), the space BV is a direct sum of the subspace AC of all absolutely continuous functions, the subspace SC of all singularly continuous functions, and the subspace PC of all piecewise constant functions. The formula (2.7) implies, in turn, that the vector function (2.8) is a direct sum of the two functions

(2.17) $$g_{ac}(s, \sigma) = \int_a^\sigma k(s, \tau)\, d\tau, \quad g_{pc}(s, \sigma) = c(s)\chi(s, \sigma),$$

where g_{ac} is absolutely continuous and g_{pc} is the "piecewise constant" part of g on $[a, b]$. Consequently, from Lebesgue's theorem it follows that the vector function (2.8) is C^{**}-weakly continuous if and only if both components in (2.17) are C^{**}-weakly continuous. Now, the C^{**}-weak continuity of g_{ac} in BV is equivalent to the weak continuity of the L_1-valued vector function $s \mapsto k(s, \cdot)$, and this is in turn equivalent, by the De la Vallée-Poussin theorem (NATANSON [1950]) to the weak compactness of the range of this function, i.e. to (2.16). On the other hand, the C^{**}-weak continuity of g_{pc} in BV is equivalent to the weak compactness (in C) of the multiplication operator $Cx(s) = c(s)x(s)$. But this operator maps weakly convergent (i.e. pointwise convergent) sequences in C into strongly convergent (i.e. uniformly convergent) sequences in C if and only if $c(s) \equiv 0$. ∎

Lemma 2.6 and Lemma 2.7 imply, in particular, that any (weakly) compact operator $A \in \mathfrak{L}_b(C)$ is necessarily an integral operator; the converse, of course, is not true. In other words, we have

$$\mathfrak{L}_b(C) \cap \mathfrak{L}_c(C) = \mathfrak{L}_i(C) \cap \mathfrak{L}_c(C) \subset \mathfrak{L}_i(C) \cap \mathfrak{L}_w(C)$$

$$= \mathfrak{L}_b(C) \cap \mathfrak{L}_w(C) \subset \mathfrak{L}_i(C),$$

where both inclusions are strict. From Lemma 2.5 we may deduce, in addition, the following

Lemma 2.8. *The classes $\mathfrak{L}_i(C) \cap \mathfrak{L}_c(C)$ and $\mathfrak{L}_i(C) \cap \mathfrak{L}_w(C)$ are closed ideals in the spaces $\mathfrak{L}_b(C)$ and $\mathfrak{L}_m(C) \oplus \mathfrak{L}_i(C)$.*

2.4. Strongly continuous operator functions

Consider now the non-stationary operator function

$$(2.18) \qquad A(t)x(s) = c(t, s)x(s) + \int_a^b k(t, s, \sigma)x(\sigma) \, d\sigma$$

defined for $t \in J$ and taking values in $\mathfrak{L}_b(C)$. In this subsection we shall be concerned with conditions on the functions $c = c(t, s)$ and $k = k(t, s, \sigma)$ under which the operator function $A = A(t)$ is strongly continuous in the space $\mathfrak{L}(C)$ of bounded linear operators in C. In analogy to the functions (2.2), (2.3), and (2.4), we now introduce the functions

$$(2.19) \qquad \alpha(t,s) := c(t,s) + \int_a^b k(t,s,\sigma) \, d\sigma,$$

$$(2.20) \qquad \beta(z,t,s) := \int_a^z [c(t,s)\chi(s,\sigma) + (z-\sigma)k(t,s,\sigma)] \, d\sigma$$

$(a \le z \le b)$, and

$$(2.21) \qquad \gamma(t,s) := |c(t,s)| + \int_a^b |k(t,s,\sigma)| \, d\sigma.$$

Theorem 2.1. *The operator function* (2.18) *is strongly continuous in the space* C *if and only if the functions* (2.19) *and* (2.20) *are continuous, and the function* (2.21) *is bounded on each bounded subset of* $J \times [a,b]$.

☐   In the proof of Lemma 2.1 we established, for fixed $t \in J$, the relations

$$\alpha(t,s) = A(t)x_1(s),$$

$$\beta(z,t,s) = A(t)\xi_z(s),$$

and

$$\sup_{a \le s \le b} \gamma(t,s) = \|A(t)\|_{\mathfrak{L}(C)};$$

here $x_1(s) \equiv 1$ and $\xi_z(s) := \max\{0, z-s\}$. By the classical Banach-Steinhaus theorem, these relations imply the "only if" part of the assertion.

To prove the "if" part it suffices to show, by (2.5) and again by the Banach-Steinhaus theorem, that, given some dense subset $M \subset C$, the functions $t \mapsto A(t)x$ are continuous on J for each $x \in M$. For simplicity, as dense subset M we take the system of all piecewise

linear functions on $[a, b]$. Given $x \in M$, we may write x in the form
(see e.g. GLIVENKO [1936]),

$$x(s) = \lambda + \sum_{j=1}^{n} \lambda_j \max\{0, z_j - s\},$$

i.e. as a linear combination of the functions x_1 and $\xi_{z_1}, \ldots, \xi_{z_n}$ (see
above). But the functions $t \mapsto A(t)x_1 = \alpha(t, \cdot)$ and $t \mapsto A(t)\xi_{z_j} = \beta(z_j, t, \cdot)$ $(j = 1, \ldots, n)$ are continuous on J, by hypothesis, and hence
$t \mapsto A(t)x$ is continuous for any $x \in M$, since $A(t)$ is linear. This
completes the proof. ■

2.5. Norm-continuous operator functions

Lemma 2.1 was the main tool in the proof of Theorem 2.1. This
proposition allows us also to formulate rather easily a criterion for
the continuity of the operator function (2.18) in norm which means
that

$$\lim_{t \to t_0} \|A(t) - A(t_0)\| = 0 \quad (t_0 \in J).$$

Of course, norm-continuity is stronger than strong continuity; therefore the hypotheses in the following theorem are more restrictive
than those given in Theorem 2.1.

Theorem 2.2. *Suppose that the operator function* (2.18) *belongs to*
$\mathfrak{L}_b(C)$ *for each* $t \in J$. *Then this operator function is continuous in the
norm of the space* $\mathfrak{L}(C)$ *if and only if the following three conditions
are satisfied:*

(a) *the function* $t \mapsto c(t, s)$ *is continuous on* J, *uniformly with respect
to* $s \in [a, b]$;

(b) *the function* $t \mapsto k(t, s, \sigma)$ *is continuous on* J, *as well as continuous in measure as a function of* σ, *uniformly with respect to*
$s \in [a, b]$, *i.e.*

$$\lim_{t \to t_0} \sup_{a \le s \le b} \mu(\{\sigma : |k(t, s, \sigma) - k(t_0, s, \sigma)| > h\}) = 0 \quad (h > 0);$$

(c) *the function* $t \mapsto k(t, s, \sigma)$ *is continuous on* J, *as well as absolutely
continuous in mean as a function of* σ, *uniformly with respect to*

$s \in [a, b]$, *i.e.*

$$\lim_{t \to t_0} \lim_{\mu(D) \to 0} \sup_{a \leq s \leq b} \int_D |k(t, s, \sigma) - k(t_0, s, \sigma)| \, d\sigma = 0.$$

□ We prove first the necessity of the conditions (a), (b), and (c). If the operator function (2.18) is norm-continuous in $\mathfrak{L}_b(C)$, the equality

$$\lim_{t \to t_0} \sup_{a \leq s \leq b} \left[|c(t, s) - c(t_0, s)| + \int_a^b |k(t, s, \sigma) - k(t_0, s, \sigma)| \, d\sigma \right] = 0$$

holds for each $t_0 \in J$, by Lemma 2.1. This implies that (a) holds and, moreover,

$$(2.22) \qquad \lim_{t \to t_0} \sup_{a \leq s \leq b} \int_a^b |k(t, s, \sigma) - k(t_0, s, \sigma)| \, d\sigma = 0.$$

By the obvious estimates

$$h\mu(\{\sigma : |k(t, s, \sigma) - k(t_0, s, \sigma)| \geq h\})$$

$$\leq \int_a^b |k(t, s, \sigma) - k(t_0, s, \sigma)| \, d\sigma$$

and

$$\int_D |k(t, s, \sigma) - k(t_0, s, \sigma)| \, d\sigma \leq \int_a^b |k(t, s, \sigma) - k(t_0, s, \sigma)| \, d\sigma,$$

(2.22) implies (b) and (c).

Conversely, suppose that the conditions (a), (b), and (c) hold. Again by Lemma 2.1, it suffices to show (2.22). To this end, fix $\varepsilon > 0$, let $h = \varepsilon/2(b - a)$, and consider the estimate

$$\int_a^b |k(t, s, \sigma) - k(t_0, s, \sigma)| \, d\sigma$$

$$(2.23)$$

$$\leq h(b - a) + \int_{D(t, t_0, s, \sigma)} |k(t, s, \sigma) - k(t_0, s, \sigma)| d\sigma,$$

where $D(t, t_0, s, h) = \{\sigma : |k(t, s, \sigma) - k(t_0, s, \sigma)| > h\}$. By condition (b), we have

$$\lim_{t \to t_0} \sup_{a \leq s \leq b} \mu(D(t, t_0, s, h)) = 0,$$

hence, by (c), also

$$\lim_{t \to t_0} \sup_{a \leq s \leq b} \int_{D(t, t_0, s, h)} |k(t, s, \sigma) - k(t_0, s, \sigma)| \, d\sigma = 0.$$

Consequently, from (2.23) we conclude that

$$\sup_{a \leq s \leq b} \int_a^b |k(t, s, \sigma) - k(t_0, s, \sigma)| d\sigma \leq \varepsilon$$

for t sufficiently close to t_0. Since $\varepsilon > 0$ was arbitrary, we are done. ∎

In applications it often happens that the kernel $\sigma \mapsto k(t, s, \sigma)$ has an absolutely continuous L_1-norm for each $t \in J$, uniformly with respect to $s \in [a, b]$, i.e.

$$(2.24) \qquad \lim_{\mu(D) \to 0} \sup_{a \leq s \leq b} \int_D |k(t, s, \sigma)| \, d\sigma = 0.$$

In this case, condition (c) of Theorem 2.2 may be replaced by the simpler condition

$$(2.25) \qquad \lim_{\substack{\mu(D) \to 0}} \sup_{\substack{a \leq s \leq b \\ |\tau - t| \leq 1}} \int_D |k(\tau, s, \sigma)| \, d\sigma = 0,$$

which usually may be verified quite easily by means of majorant techniques (see e.g. KRASNOSEL'SKIJ-ZABREJKO-PUSTYL'NIK-SOBOLEV-SKIJ [1966]).

2.6. Representation of the evolution operator

Let us return to the differential equation (1.2) with operator function (1.3) in the space C of continuous functions. In this subsection we shall suppose throughout that the hypotheses of Theorem 2.1 are satisfied, and hence the operator function (1.3) is strongly continuous in C.

Since the operator function (1.3) is representable as a sum of the multiplication operator (1.23) and the integral operator (1.24), it is natural to expect that also the evolution operator $U = U(t, \tau)$ for the equation (1.2) admits a representation as a sum of the evolution operator $U_0 = U_0(t, \tau)$ of the "reduced" equation

$$(2.26) \qquad \frac{dx}{dt} = C(t)x$$

and some integral operator

$$(2.27) \qquad H(t, \tau)x(s) = \int_a^b h(t, \tau, s, \sigma)x(\sigma)\, d\sigma.$$

This is in fact a consequence of Theorem 1.2 and Lemma 2.8. Indeed, the following stronger assertion is true:

Theorem 2.3. *Suppose that the operator function* (2.18) *is strongly continuous in the space* C. *Then the evolution operator* $U = U(t, \tau)$ *for the differential equation* (1.2) *belongs to* $\mathfrak{L}_b(C)$ *for all* $t, \tau \in J$, *i.e. admits a representation*

$$(2.28) \qquad U(t, \tau)x(s) = e(t, \tau, s)x(s) + H(t, \tau)x(s),$$

where $e(t, \tau, s)$ *is given by* (1.29), *and* H *is defined by* (2.27) *with* $h = h(t, \tau, s, \sigma)$ *being a measurable function on* $J \times J \times [a, b] \times [a, b]$. *Moreover,* $U(t, \tau) \in \mathfrak{L}_m(C) \oplus \mathfrak{L}_i(C)$ *if* $A(t) \in \mathfrak{L}_m(C) \oplus \mathfrak{L}_i(C)$, $H(t, \tau) \in \mathfrak{L}_i(C) \cap \mathfrak{L}_w(C)$ *if* $K(t) \in \mathfrak{L}_i(C) \cap \mathfrak{L}_w(C)$, *and* $H(t, \tau) \in \mathfrak{L}_i(C) \cap \mathfrak{L}_c(C)$ *if* $K(t) \in \mathfrak{L}_i(C) \cap \mathfrak{L}_c(C)$.

□   As was shown in Subsection 1.2, the evolution operator $U(t, \tau)$ for equation (1.2) is representable as a series (1.7) which converges in the norm of the space $\mathfrak{L}(C)$, uniformly for t in a bounded interval. By Lemma 2.3, each of the operators $A(t_1) \cdots A(t_n)$ $(n = 1, 2, \ldots)$ occuring in (1.7) is strongly continuous as a function of $t_1, \ldots, t_n \in J$ and belongs to $\mathfrak{L}_b(C)$, i.e. admits a representation

$$A(t_1) \cdots A(t_n)x(s) = c(t_1, s) \ldots c(t_n, s)x(s)$$

$$(2.29) \qquad\qquad + \int_a^b r_n(t_1, \ldots, t_n, s, \sigma)x(\sigma)\, d\sigma;$$

here $r_n = r_n(t_1, \ldots, t_n, s, \sigma)$ $(n = 1, 2, \ldots)$ are measurable functions which satisfy the estimate

(2.30)
$$\sup_{a \leq s \leq b} \int_a^b |r_n(t_1, \ldots t_n, s, \sigma)| \, d\sigma$$
$$\leq \|A(t_1)\|_{\mathfrak{L}(C)} \cdots \|A(t_n)\|_{\mathfrak{L}(C)},$$

by Lemma 2.1. Integrating (2.29) over $t_1, \ldots, t_n$ yields

(2.31)
$$\left[\int_\tau^t \int_\tau^{t_1} \cdots \int_\tau^{t_{n-1}} A(t_1) \ldots A(t_n) \, dt_n \ldots dt_1 \right] x(s)$$
$$= \frac{1}{n!} \left[\int_\tau^t c(\xi, s) \, d\xi \right]^n x(s) +$$
$$\int_a^b \left[\int_\tau^t \cdots \int_\tau^{t_{n-1}} r_n(t_1, \ldots, t_n, s, \sigma) \, dt_n \ldots dt_1 \right] x(\sigma) \, d\sigma,$$

where the change of order of integration is justified by (2.30) and Fubini's theorem. Again by (2.30), we may change the integrals (2.31) with the series (1.7) and obtain (2.28), where H is given by (2.27) with

$$h(t, \tau, s, \sigma)$$

(2.32)
$$:= \sum_{n=1}^\infty \int_\tau^t \int_\tau^{t_1} \cdots \int_\tau^{t_{n-1}} r_n(t_1, \ldots, t_n, s, \sigma) \, dt_n \ldots dt_1.$$

The last part follows from Theorem 1.2, applied to $\mathfrak{L}_0(C) = \mathfrak{L}_b(C)$ and $\mathfrak{M} = \mathfrak{L}_i(C)$, $\mathfrak{M} = \mathfrak{L}_i(C) \cap \mathfrak{L}_w(C)$, or $\mathfrak{M} = \mathfrak{L}_i(C) \cap \mathfrak{L}_c(C)$. ■

2.7. Stamping and infra-stamping operators

The main result in the preceding subsections was Lemma 2.1 and, in particular, formula (2.5) for the norm of the operator (2.1) which we used several times. In the special case $c(s) \equiv 1$, this formula reduces to

(2.33)
$$\|I + K\| = 1 + \|K\|,$$

since

$$\|K\|_{\mathfrak{L}(C)} = \sup_{a \le s \le b} \int_a^b |k(s,\sigma)| \, d\sigma.$$

The somewhat surprising equality (2.33) was noticed first, in case of general completely continuous operators K, by DAUGAVET [1963]. Afterwards, (2.33) was proved by LOZANOVSKIJ [1966] for integral operators K, and by KRASNOSEL'SKIJ [1967] for a fairly general class of operators called *stamping operators*; for some more related results, see also KHOLUB [1987], KAMOWITZ [1984], and SYNNATZSCHKE [1984]. It is natural to expect that the formula (2.5) carries over also to operators of the form

$$(2.34) \qquad Ax(s) = c(s)x(s) + Kx(s),$$

where K is not necessarily a linear integral operator. In this case, (2.5) should hold true if the definition (2.4) of the function γ is replaced with the more general definition

$$(2.35) \qquad \gamma(s) := |c(s)| + \|\kappa(s)\|,$$

where $\kappa(s)$ is the linear functional defined by $k(s)x := Kx(s)$ on the space C.

Recall that, by Radon's theorem, the linear operator K admits a representation

$$(2.36) \qquad Kx(s) = \int_a^b x(\sigma) \, dg(s,\sigma),$$

where $g(\cdot, \sigma)$ is weakly continuous, $g(s, \cdot)$ is of bounded variation on $[a,b]$, and the functional

$$(2.37) \qquad \kappa(s) = \mathrm{Var}_{[a,b]} g(s, \cdot)$$

is upper semi-continuous. From (2.36) we conclude that the operator admits a similar representation

$$(2.38) \qquad Ax(s) = \int_a^b x(\sigma) \, d[c(s)\chi(s,\sigma) + g(s,\sigma)];$$

with $\chi = \chi(s,\sigma)$ denoting the characteristic function of the triangular domain $\{(s,\sigma) : a \leq s \leq \sigma \leq b\}$. Again by Radon's theorem, the bounded linear functional $\alpha(s)$ defined by $\alpha(s) := Ax(s)$ on C admits a representation

$$(2.39) \qquad \alpha(s) = \mathrm{Var}_{[a,b]}[c(s)\chi(s,\cdot) + g(s,\cdot)].$$

Finally, we associate with the operator (2.36) a function $\delta_K : [a,b] \to \mathbb{R}$ defined by

$$(2.40) \qquad \delta_K(s) := g(s,s+0) - g(s,s-0).$$

Lemma 2.9. *The equality*

$$(2.41) \qquad \alpha(s) - \kappa(s) = |c(s) + \delta_K(s)| - |\delta_K(s)|$$

holds for $a \leq s \leq b$.

□ Let $a = s_0 < s_1 < \ldots < s_{m-1} < s_m = b$ be an arbitrary fixed partition of the interval $[a,b]$. We have

$$\sum_{j=1}^{m}[c(s)\chi(s,s_j) - c(s)\chi(s,s_{j-1}) + g(s,s_j) - g(s_{j-1})]$$

$$\leq |c(s)\chi(s,s_k) - c(s)\chi(s,s_{k-1}) + g(s,s_k) - g(s,s_{k-1})|$$

$$+ \sum_{\substack{j=1 \\ j \neq k}}^{m} |c(s)\chi(s,s_j) - c(s)\chi(s,s_{j-1}) + g(s,s_j) - g(s,s_{j-1})|$$

$$= |c(s) + g(s,s_k) - g(s,s_{k-1})| - |g(s,s_k) - g(s,s_{k-1})|$$

$$+ \sum_{j=1}^{m} |g(s,s_j) - g(s,s_{j-1})|$$

$$\leq |c(s) + g(s,s_k) - g(s,s_{k-1})|$$

$$- |g(s,s_k) - g(s,s_{k-1})| + \mathrm{Var}_{[a,b]}g(s,\cdot);$$

here the index k is chosen in such a way that $s_{k-1} < s \le s_k$. Since the partition $\{s_0, \ldots, s_m\}$ was arbitrary, and since

$$\lim_{\sigma \uparrow s} g(s, \sigma) = g(s, s - 0), \quad \lim_{\sigma \downarrow s} g(s, \sigma) = g(s, s + 0)$$

we conclude that

$$\alpha(s) = \mathrm{Var}_{[a,b]} |c(s)\chi(s, \cdot) + g(s, \cdot)|$$

$$\le |c(s) + g(s, s + 0) - g(s, s - 0)|$$

$$-|g(s, s + 0) - g(s, s - 0)| + \mathrm{Var}_{[a,b]} g(s, \cdot)$$

$$= |c(s) + \delta_K(s)| - |\delta_K(s)| + \kappa(s).$$

Similarly, the estimate

$$\sum_{j=1}^{m} |g(s, s_j) - g(s, s_{j-1})|$$

$$\le -|c(s) + g(s, s_k) - g(s, s_{k-1})| + |g(s, s_k) - g(s, s_{k-1})|$$

$$+ \sum_{j=1}^{m} |c(s)\chi(s, s_j) - c(s)\chi(s, s_{j-1}) + g(s, s_j) - g(s, s_{j-1})|$$

for $s_{k-1} < s \le s_k$ implies that

$$\kappa(s) = \mathrm{Var}_{[a,b]} g(s, \cdot) \le -|c(s) + g(s, s + 0) - g(s, s - 0)|$$

$$+ |g(s, s + 0) - g(s, s - 0)| + \mathrm{Var}_{[a,b]} [c(s)\chi(s, \cdot) + g(s, \cdot)]$$

$$= -|c(s) + \delta_K(s)| + |\delta_K(s)| + \alpha(s).$$

Consequently, (2.41) holds. ∎

Let us call an operator K as in (2.36) *infra-stamping* if the function (2.40) is zero almost everywhere on $[a, b]$.

Theorem 2.4. *Suppose that the function $c = c(s)$ is continuous, and the operator K is infra-stamping. Then formula (2.5) holds with γ given by (2.35).*

□ Obviously, it suffices to show that

(2.42) $|c(s)| + \mathrm{Var}_{[a,b]} g(s, \cdot) \leq \sup_{a \leq s \leq b} \alpha(s).$

As already mentioned, the function $\gamma(s)$ defined by (2.35) is upper semi-continuous; consequently, for any $\varepsilon > 0$, the set D_ε of all $s \in [a, b]$ such that $\gamma(s) > \sup\{\gamma(\sigma) - \varepsilon : a \leq \sigma \leq b\}$ is non-empty and open. Choose a point $s_\varepsilon \in D_\varepsilon$ where the function (2.40) vanishes, i.e. $\delta_K(s_\varepsilon) = 0$. For any $s \in [a, b]$, we have then

$$|c(s)| + \mathrm{Var}_{[a,b]} g(s, \cdot) \leq \sup_{a \leq s \leq b} \gamma(s) \leq \gamma(s_\varepsilon) + \varepsilon$$

$$= |c(s_\varepsilon)| + \mathrm{Var}_{[a,b]} g(s_\varepsilon, \cdot) + \varepsilon$$

$$= |c(s_\varepsilon) + \delta_K(s_\varepsilon)| - |\delta_K(s_\varepsilon)| + \mathrm{Var}_{[a,b]} g(s_\varepsilon, \cdot) + \varepsilon,$$

hence, by Lemma 2.1,

$$|c(s)| + \mathrm{Var}_{[a,b]} g(s, \cdot) \leq \mathrm{Var}_{[a,b]}[c(s)\chi(s_\varepsilon, \cdot) + g(s_\varepsilon, \cdot)]$$

$$+ \varepsilon \leq \alpha(s_\varepsilon) + \varepsilon \leq \sup_{a \leq s \leq b} \alpha(s) + \varepsilon.$$

Since $\varepsilon > 0$ was arbitrary, (2.42) is proved. ∎

As was shown by DIALLO-ZABREJKO [1987], every integral operator or stamping operator (in Krasnosel'skij's sense) is infra-stamping. In this way, Theorem 2.4 generalizes and extends the results by DAUGAVET [1963], KRASNOSEL'SKIJ [1967], and LOZANOVSKIJ [1966]. On the other hand, there exist infra-stamping operators which are neither integral operators nor stamping; just consider the operator $Kx(s) = x(s/2)$ on the space $C = C([0, 1])$. Theorem 2.4 is mentioned in DIALLO-ZABREJKO [1987] even for the space $C = C(Q)$ of continuous functions on an arbitrary compact metric space Q. In contrast to Lemma 2.1, however, the continuity of the function $c = c(s)$ is required in Theorem 2.4. As will be shown in the next chapter, this requirement is superfluous if K is considered as a bounded linear operator from C into L_∞, and the function c, rather than being continuous, has merely the property that the operator (2.34) acts in the space C. This situation is typical in many applications of Lemma 2.1.

§ 3. Barbashin equations in Lebesgue spaces

In this section we give a systematic account of the basic properties of Barbashin operators in the Lebesgue space L_p of p-summable $(1 \leq p < \infty)$ or essentially bounded $(p = \infty)$ functions over $[a, b]$. In particular, we derive sufficient conditions (which sometimes are also necessary) for the strong continuity or norm-continuity of Barbashin operators in both the stationary and non-stationary case. To this end, we study some special kernel classes and make use of interpolation theorems for integral operators. A representation theorem for the evolution operator in the space L_p is also given.

3.1. Linear operators in the space L_p

Let us recall some basic properties of bounded linear operators of the form

$$(3.1) \qquad Ax(s) = c(s)x(s) + \int_a^b k(s, \sigma)x(\sigma) \, d\sigma$$

in the Lebesgue spaces L_p $(1 \leq p \leq \infty)$; as in § 2, we assume that $c = c(s)$ and $k = k(s, \sigma)$ are given measurable functions on $[a, b]$ and $[a, b] \times [a, b]$, respectively.

We are especially interested in conditions on the functions c and k which ensure the boundedness of the operator (3.1) in some Lebesgue space L_p of p-integrable $(1 \leq p < \infty)$ or essentially bounded $(p = \infty)$ functions on $[a, b]$. In contrast to the situation in the space C, there are no simple theorems on the general form of bounded linear operators in L_p (see e.g. DUNFORD-SCHWARTZ [1963], KRASNO-SEL'SKIJ-ZABREJKO-PUSTYL'NIK-SOBOLEVSKIJ [1966]). This is the reason why the theory of operators of the form (3.1) to be developed below is not as simple and complete as the corresponding theory in the space C developed in the preceding paragraph. No systematic account on the operator (3.1) in Lebesgue spaces is known in the literature. Nevertheless, there is a vast literature on linear *integral operators* (i.e. operators (3.1) with $c(s) \equiv 0$); we refer, in particular, to the monographs KRASNOSEL'SKIJ-ZABREJKO-PUSTYL'NIK-SOBOLEVSKIJ [1966], HALMOS-SUNDER [1978], KOROTKOV [1983].

We shall show that the basic facts from KRASNOSEL'SKIJ-ZABREJKO-PUSTYL'NIK-SOBOLEVSKIJ [1966] on linear integral operators carry over to the more general operators (3.1) as well.

One of the most important facts on linear integral operators in Lebesgue spaces is the classical Banach theorem on the "automatic" continuity of such operators (see e.g. BANACH [1932], KRASNOSEL'SKIJ-ZABREJKO-PUSTYL'NIK-SOBOLEVSKIJ [1966]). This theorem holds also for operators of the form (3.1):

Lemma 3.1. *Suppose that the linear operator* (3.1) *acts in the space* L_p $(1 \leq p \leq \infty)$. *Then the operator* (3.1) *is continuous.*

□ Fix $x \in L_p$; by hypothesis, we have then $Ax \in L_p$, with A given by (3.1). In particular, the function $k(s, \cdot)x(\cdot)$ is integrable on [a,b] for almost all $s \in [a, b]$, and hence also the function $|k(s, \cdot)| \, |x(\cdot)|$. By Fubini's theorem, the function $v : [a, b] \to \mathbb{R}$ defined by

$$v(s) := \int_a^b |k(s, \sigma)| \, |x(\sigma)| \, d\sigma$$

is measurable on [a,b].

To prove the continuity of the operator (3.1), it suffices to prove its closedness, by the closed graph theorem. Thus, let $(x_n)_n$ be a sequence in L_p converging to zero, say, and assume that the sequence $(Ax_n)_n$ converges to some function $z \in L_p$. Without loss of generality, we may suppose (see e.g. KRASNOSEL'SKIJ-ZABREJKO-PUSTYL'NIK-SOBOLEVSKIJ [1966]) that the functions x_n are uniformly bounded on $[a, b]$ by some function $u \in L_p$, i.e. $|x_n(s)| \leq |u(s)|$ almost everywhere on $[a, b]$. The convergence of the sequence $(x_n)_n$ to zero in the norm of L_p implies its convergence in measure. Consequently, the sequences $c(\cdot)x_n(\cdot)$ and $k(s, \cdot)x_n(\cdot)$ converge in measure, too. Since

$$|k(s, \sigma)x_n(\sigma)| \leq |k(s, \sigma)| \, |u(\sigma)| \qquad (n = 1, 2, \ldots),$$

the sequence $(v_n)_n$ of functions

$$v_n(s) := \int_a^b k(s, \sigma)x_n(\sigma) \, d\sigma$$

converges, for almost all $s \in [a, b]$, to zero, by Lebesgue's theorem on dominated convergence. Altogether, this shows that the sequence

$$Ax_n(s) = c(s)x_n(s) + \int_a^b k(s, \sigma)x_n(\sigma)\, d\sigma$$

converges in measure to zero; consequently, $z(s) = 0$ almost everywhere on $[a, b]$, and so we are done. ■

Recall that a linear operator A in the space L_p is called *regular* if A may be written as a difference of two positive linear operators in L_p; a regular operator is always bounded. Given a regular linear operator A in L_p, one may define its *module* $|A|$ as the minimal of all positive bounded linear operators $\tilde{A}$ in L_p satisfying

(3.2) $|Ax| \leq \tilde{A}(|x|) \qquad (x \in L_p)$.

It was shown in KANTOROVICH-VULIKH-PINSKER [1950] and KRAS-NOSEL'SKIJ-ZABREJKO-PUSTYL'NIK-SOBOLEVSKIJ [1966] that a linear integral operator K generated by a kernel $k = k(s, \sigma)$ is regular in L_p if and only if the linear integral operator $]K[$ generated by the kernel $|k| = |k(s, \sigma)|$ also acts in L_p; moreover, $|K| =]K[$ in this case. The same is true for the operator (3.1):

Lemma 3.2. *Suppose that the linear operator* (3.1) *acts in the space* L_p $(1 \leq p \leq \infty)$. *Then the operator* (3.1) *is regular if and only if the linear operator* $]A[$ *defined by*

(3.3) $]A[x(s) = |c(s)||x(s)| + \int_a^b |k(s, \sigma)||x(\sigma)|\, d\sigma;$

acts in L_p *as well; moreover, the equality*

(3.4) $|A| =]A[$

holds in this case.

□ The inequality $|Ax| \leq]A[|x|$ is obvious. Therefore, if the operator (3.3) acts in the space L_p, then the operator (3.1) is regular and satisfies the estimate $|A| \leq]A[$; here the inequality is meant in the sense of the ordering on $\mathfrak{L}(L_p)$ induced by the cone of positive linear operators in L_p.

Conversely, suppose that (3.1) is regular, and let $x \in L_p$ be a non-negative function Since L_p is a *K-space of countable type* (for the terminology see e.g. DUNFORD-SCHWARTZ [1963] or KANTOROVICH-VULIKH-PINSKER [1950]), there exists a countable subset M of the conic interval $[-x, x] = \{z : z \in L_p, |z| \leq x\}$ such that

$$w = \sup\{|Az| : z \in [-x, x]\} = \sup\{|Az| : z \in M\}.$$

Without loss of generality we may suppose that M is dense (with respect to convergence in measure) in $[-x, x]$.

Denote by $\mathfrak{D}$ the system of all subintervals $D = [\alpha, \beta] \subseteq [a, b]$ with rational endpoints α, β, and let

$$M^* = \{x \operatorname{sign}(c\chi_D) + z\chi_{[a,b]\backslash D} : z \in M, D \in \mathfrak{D}\};$$

here χ_D is, of course, the characteristic function of $D \in \mathfrak{D}$. The set M^* is then also countable and dense (in measure) in the conic intervall $[-x, x]$. Observe that, given $s \in [a, b]$, we may find a sequence of sets $D_n \in \mathfrak{D}$ such that $s \in D_n$ and $\mu(D_n) \to 0$ as $n \to \infty$.

The least upper bound of a countable subset of L_p may be "calculated" for almost all $s \in [a, b]$. Since $M \subseteq M^* \subseteq [-x, x]$, we have

$$w(s) = \sup\{|Az(s)| : z \in M^*\}.$$

Now fix $s \in [a, b]$, let $(z_n)_n$ be a sequence in M converging (in measure) to the function $\operatorname{sign} k(s, \cdot)x(\cdot)$, and let $(D_n)_n$ be a sequence in $\mathfrak{D}$ such that $s \in D_n$ ($n = 1, 2, \ldots$) and $\mu(D_n) \to 0$ as $n \to \infty$. Then the functions $z_n^* = x \operatorname{sign}(c\chi_{D_n}) + z_n\chi_{[a,b]\backslash D_n}$ also converge (in measure) to the function $\operatorname{sign} k(s, \cdot)x(\cdot)$. Since

$$\left| c(s)z_n^*(s) + \int_a^b k(s, \sigma)z_n^*(\sigma)\, d\sigma \right|$$

$$= \left| |c(s)|z_n^*(s) + \int_a^b k(s, \sigma)z_n^*(\sigma)\, d\sigma \right|$$

and

$$\lim_{n\to\infty} \int_a^b k(s,\sigma)z_n^*(\sigma)\,d\sigma = \int_a^b |k(s,\sigma)|x(\sigma)\,d\sigma,$$

by Lebesgue's theorem, we obtain

$$\lim_{n\to\infty} |c(s)z_n^*(s) + \int_a^b k(s,\sigma)z_n^*(\sigma)\,d\sigma|$$

$$= |c(s)|x(s) + \int_a^b |k(s,\sigma)|x(\sigma)\,d\sigma.$$

From this it follows that

$$w(s) = \sup\{|Az(s)| : z \in M^*\} \geq \lim_{n\to\infty} |Az_n^*(s)|$$

$$= |c(s)|x(s) + \int_a^b |k(s,\sigma)|x(\sigma)\,d\sigma =]A[x(s).$$

Consequently, for almost all $s \in [a,b]$ we have that $]A[\,x(s) \leq w(s)$, and hence $]A[\,x \in L_p$. In other words, the linear operator (3.3) maps any nonnegative function $x \in L_p$ again into a function $]A[\,x \in L_p$. Since every L_p function is representable as a difference of two nonnegative L_p functions, the operator (3.3) acts in the space L_p.

It remains to observe that any positive linear operator $\tilde{A}$ satisfying (3.2) in L_p has the property that $w(s) \leq \tilde{A}x(s)$, and so

$$]A[x(s) \leq \tilde{A}x(s)$$

for all $x \in L_p$. This shows that $]A[$ is the minimal of all operators satisfying (3.2), and thus $]A[\leq |A|$ as claimed. ∎

In what follows, we identify the *dual space* L_p^* in case $1 \leq p < \infty$, as usual, with the space L_q with $q = p/(p-1)$; in fact, the duality

$$\langle x, y \rangle = \int_a^b x(s)y(s)\,ds$$

gives a 1-1 representation of any bounded linear functional $l \in L_p^*$ in the form

(3.5) $l(x) = \langle x, l_0 \rangle,$

where $l_0 \in L_q$ with $||l||_{L_p^*} = ||l_0||_{L_q}$. Thus, if the operator A in (3.1) acts in the space L_p $(1 \leq p < \infty)$, its *adjoint* A^* acts in the space L_q $(q = p/(p-1))$; the question arises whether or not A^* admits also a representation in the form (3.1).

Recall that the dual space L_∞^* of L_∞ may *not* be identified with the space L_1. However, L_1 is a closed subspace of L_∞^*, inasmuch as every element $l_0 \in L_1$ defines a bounded linear functional l on L_∞ by (3.5), where $||l||_{L_\infty^*} = ||l_0||_1$. In this connection, L_1 is sometimes called the *Köthe dual* (or *associate space*) of the space L_∞ and denoted by L_∞'; thus, we have $L_p^* = L_p' = L_q$ $(q = p/(p-1))$ in case $1 \leq p < \infty$, and $L_\infty^* \supset L_\infty' = L_1$ in case $p = \infty$. In the latter case, it is more natural to consider, rather than the adjoint A^* of the operator (3.1) on L_∞^*, the restriction A' of A^* on the subspace $L_\infty' = L_1$; A' is sometimes called the *Köthe adjoint* (or *associate operator*) of A. Thus, if A is defined by (3.1), we are interested in the existence of its Köthe adjoint A', and also in conditions which allow us to represent A' again in the form (3.1).

In ZABREJKO [1968] it is shown that, if K is a linear integral operator with kernel $k = k(s,\sigma)$ in L_p $(1 \leq p < \infty)$, then the Köthe adjoint K' always exists, but is not necessarily an integral operator. However, for every $y \in L_q$ $(q = p/(p-1))$ which belongs to the domain of definition of the linear integral operator

$$K^{\#}y(s) = \int_a^b k^{\#}(s,\sigma)y(\sigma)\,d\sigma$$

generated by the kernel $k^{\#} = k^{\#}(s,\sigma) = k(\sigma,s)$, one has $K'y = K^{\#}y$. In particular, if the operator K is regular then the operator $K^{\#}$ is defined for each $y \in L_q$, and thus $K' = K^{\#}$ on the whole space L_q. A similar result holds for the operator (3.1), as was shown in KALITVIN-ZABREJKO [1991]:

Lemma 3.3. *Suppose that the linear operator* (3.1) *acts in the space* L_p $(1 \leq p \leq \infty)$. *Then the operator* (3.1) *admits the Köthe adjoint*

$$(3.6) \qquad\qquad A'y(s) = A^{\#}y(s),$$

where

$$(3.7) \qquad A^\# y(s) = c(s)y(s) + \int_a^b k^\#(s,\sigma)y(\sigma)\,d\sigma$$

for any function $y \in L_q$ for which the right-hand side of (3.6) makes sense. In particular, if the operator (3.1) is regular, then (3.6) holds for all $y \in L_q$.

In connection with the results given above, the problem arises how to find simple conditions (at least sufficient) on the functions $c = c(s)$ and $k = k(s,\sigma)$ under which the operator (3.1) is continuous or regular in some L_p space ($1 \le p \le \infty$). Such conditions may be found by combining analogous conditions for each of the two components of (3.1), i.e. the multiplication operator

$$(3.8) \qquad\qquad Cx(s) = c(s)x(s)$$

and the integral operator

$$(3.9) \qquad\qquad Kx(s) = \int_a^b k(s,\sigma)x(\sigma)\,d\sigma.$$

By means of Kantorovich's regularity theorems for bounded linear operators in L_∞, one may conclude from Lemma 3.2 and Lemma 3.3 that, in case $p = 1$ or $p = \infty$, the operator (3.1) acts in L_p if and only if both components (3.8) and (3.9) act in L_p. In case $1 < p < \infty$, however, an analogous result is not known. On the one hand, a necessary and sufficient condition for the linear multiplication operator (3.8) to act in L_p is, of course, that $c \in L_\infty$. On the other hand, a necessary and sufficient condition on the kernel $k = k(s,\sigma)$ for the linear integral operator (3.9) to act in L_p is known only in case $p = 1$ or $p = \infty$. Some *sufficient* conditions on k, however, which guarantee that the operator (3.9) acts in L_p for $1 \le p \le \infty$ will be given in the next subsections of this chapter.

3.2. The space $\mathcal{L}_b^r(L_p)$

In analogy to what we have done in Subsection 2.2 by introducing the class $\mathfrak{L}_b(C)$, we could try to develop a parallel theory on linear operators of type (3.1) in the Lebesgue spaces L_p.

However, this would be extremely difficult and unsatisfactory, mainly because the class of *all* linear integral operators in Lebesgue spaces has not been very well studied up to the present. Nevertheless, since most integral operators which occur in significant applications are *a priori* known to be regular, one could be less ambitious and try at least to develop a theory of *regular* linear operators of type (3.1). It turns out, in fact, that this may be done quite successfully.

In the sequel, we denote by $\mathfrak{L}_b^r(L_p)$ the class of all regular Barbashin operators (i.e. regular operators of the form (3.1)) which act in the space L_p ($1 \leq p \leq \infty$); of course, this is a subspace of the space $\mathfrak{L}(L_p)$ of all bounded linear operators in L_p. Observe that the space $\mathfrak{L}_b^r(L_p)$ is *not* closed in $\mathfrak{L}(L_p)$ (see e.g. KRASNOSEL'SKIJ-ZABREJKO-PUSTYL'-NIK-SOBOLEVSKIJ [1966], HALMOS-SUNDER [1978]), and thus a direct analogue to Lemma 2.2. for Lebesgue spaces is not true.

A somewhat weaker statement, however, is possible. To this end, let us introduce the norm

$$(3.10) \qquad ||A||_{\mathfrak{L}_b^r(L_p)} = ||\,|A|\,||_{\mathfrak{L}(L_p)}$$

on the space $\mathfrak{L}_b^r(L_p)$, where $|A|$ denotes the module of A (see Lemma 3.2).

Lemma 3.4. *The subspace $\mathfrak{L}_b^r(L_p)$ with norm (3.10) is a Banach space which is continuously imbedded in $\mathfrak{L}(L_p)$.*

□ First of all, observe that the continuous imbedding

$$(3.11) \qquad \mathfrak{L}_b^r(L_p) \subseteq \mathfrak{L}(L_p), \qquad ||A||_{\mathfrak{L}(L_p)} \leq ||A||_{\mathfrak{L}_b^r(L_p)}$$

follows from the definition of $|A|$ (see (3.3) and (3.4)). Further, for any $A \in \mathfrak{L}_b^r(L_p)$ we have, by Lemma 3.2,

$$(3.12) \qquad ||C||_{\mathfrak{L}_b^r(L_p)} \leq ||A||_{\mathfrak{L}_b^r(L_p)}, \qquad ||K||_{\mathfrak{L}_b^r(L_p)} \leq ||A||_{\mathfrak{L}_b^r(L_p)}$$

with C and K being the components (3.8) and (3.9) of the operator A. Suppose now that

$$A_n x(s) = c_n(s)x(s) + \int_a^b k_n(s,\sigma)x(\sigma)\,d\sigma,$$

is a Cauchy sequence of operators $A_n \in \mathfrak{L}_b^r(L_p)$. By (3.11), the sequence $(A_n)_n$ is also Cauchy in $\mathfrak{L}(L_p)$, and hence converges (in $\mathfrak{L}(L_p)$) to some operator $A \in \mathfrak{L}(L_p)$. On the other hand, by (3.12), both components C_n and K_n corresponding to A_n are Cauchy sequences in $\mathfrak{L}_b^r(L_p)$. Now, the obvious equality

$$||C_i - C_j||_{\mathfrak{L}_b^r(L_p)} = ||c_i - c_j||_{L_\infty}$$

shows that the sequence of functions $(c_n)_n$ is also Cauchy in L_∞; by the completeness of the space L_∞, this sequence converges (in L_∞) to some $c \in L_\infty$. Again by the equality

$$||C - C_n||_{\mathfrak{L}_b^r(L_p)} = ||c - c_n||_{L_\infty}$$

it follows that the sequence of operators $(C_n)_n$ converges (in $\mathfrak{L}_b^r(L_p)$) to the operator $Cx(s) = c(s)x(s)$.

To analyze the convergence behaviour of the sequence $(K_n)_n$, observe first that

$$(3.13) \qquad \begin{aligned} &\left|\left| \int_a^b |k_i(\cdot, \sigma) - k_j(\cdot, \sigma)| x(\sigma) \, d\sigma \right|\right|_{L_p} \\ &\qquad \leq ||K_i - K_j||_{\mathfrak{L}_b^r(L_p)} \quad (||x||_{L_p} \leq 1) \end{aligned}$$

and hence, by the particular choice $x(s) \equiv 1$,

$$(3.14) \qquad \int_a^b \int_a^b |k_i(s, \sigma) - k_j(s, \sigma)| \, ds \, d\sigma \leq ||K_i - K_j||_{\mathfrak{L}_b^r(L_p)}.$$

The last estimate implies that the sequence of functions $(k_n)_n$ is Cauchy in the space $L_1 = L_1([a, b] \times [a, b])$. Since this space is complete, too, this sequence converges (in L_1) to some $k \in L_1$. By passing, if necessary, to a subsequence we may assume, without loss of generality, that $(k_n)_n$ converges to k almost everywhere on $[a, b] \times [a, b]$. In particular, the sequence $(k_n(s, \cdot))_n$ converges, for almost all $s \in [a, b]$, almost everywhere on $[a, b]$ to the function $k(s, \cdot)$.

Now let x be an arbitrary nonnegative function in the unit ball of L_p. Setting $i = n$ in (3.13) and letting $j \to \infty$ yields, by Fatou's lemma,

$$\left|\left| \int_a^b |k_n(\cdot, \sigma) - k(\cdot, \sigma)| x(\sigma) \, d\sigma \right|\right|_{L_p} \leq \varlimsup_{j \to \infty} ||K_n - K_j||_{\mathfrak{L}_b^r(L_p)}.$$

The right-hand side of this inequality tends to zero as $n \to \infty$, since the sequence $(k_n)_n$ is Cauchy in $\mathfrak{L}_b^r(L_p)$. Moreover, the right-hand side does not depend on x; therefore,

$$\lim_{n \to \infty} \sup_{\substack{||x||_{L_p} \leq 1 \\ x \geq 0}} || \int_a^b |k_n(\cdot, \sigma) - k(\cdot, \sigma)| x(\sigma) \, d\sigma ||_{L_p} = 0,$$

and hence, by the obvious inequality

$$| \int_a^b |k_n(s, \sigma) - k(s, \sigma)| x(\sigma) \, d\sigma | \leq \int_a^b |k_n(s, \sigma) - k(s, \sigma)| |x(\sigma)| \, d\sigma,$$

it follows that

$$\lim_{n \to \infty} \sup_{||x||_{L_p} \leq 1} || \int_a^b |k_n(\cdot, \sigma) - k(\cdot, \sigma)| x(\sigma) \, d\sigma ||_{L_p} = 0.$$

This means exactly that the sequence $(K_n)_n$ converges (in $\mathfrak{L}_b^r(L_p)$) to the linear integral operator K generated by the kernel $k = k(s, \sigma)$, i.e. to the operator (3.9). We have shown that the operators C_n and K_n converge in $\mathfrak{L}_b^r(L_p)$ to the operators C and K, respectively, and thus the operators $A_n = C_n + K_n$ converge to the operator $A = C + K$. ∎

The space $\mathfrak{L}(L_p)$ of all bounded linear operators in L_p is a Banach algebra with respect to the usual composition of operators. It would be useful to know if the same is true for the space $\mathfrak{L}_b^r(L_p)$. Moreover, the same question as in § 2 arises, whether or not the subspace of integral operators (or regular integral operators) in L_p is an ideal in $\mathfrak{L}(L_p)$ or $\mathfrak{L}_b^r(L_p)$. It turns out that the situation here is not as simple as in the space C treated in § 2. For instance (KRASNOSEL'-SKIJ-ZABREJKO-PUSTYL'NIK-SOBOLEVSKIJ [1966]), the composition of two integral operators is not necessarily again an integral operator. Nevertheless, the following result holds which is parallel to Lemma 2.3 and Lemma 2.5:

Lemma 3.5. *The subspace $\mathfrak{L}_b^r(L_p)$ is a subalgebra of $\mathfrak{L}(L_p)$, and the subspace $\mathfrak{L}_i(L_p)$ is an ideal in the algebra $\mathfrak{L}_b^r(L_p)$. More precisely, if*

$A, B \in \mathfrak{L}_b^r(L_p)$ *are given by*

$$Ax(s) = c(s)x(s) + \int_a^b k(s,\sigma)x(\sigma)\,d\sigma$$

and

$$Bx(s) = d(s)x(s) + \int_a^b l(s,\sigma)x(\sigma)\,d\sigma$$

then $AB \in \mathfrak{L}_b^r(L_p)$ *is given by*

(3.15) $$ABx(s) = e(s)x(s) + \int_a^b m(s,\sigma)x(\sigma)\,d\sigma,$$

where

$$e(s) := c(s)d(s)$$

and

$$m(s,\sigma) := c(s)l(s,\sigma) + k(s,\sigma)d(\sigma) + \int_a^b k(s,\xi)l(\xi,\sigma)\,d\xi.$$

☐ The formula (3.15) follows from

$$ABx(s) = c(s)d(s)x(s) + \int_a^b c(s)l(s,\sigma)x(\sigma)d\sigma$$
$$+ \int_a^b k(s,\sigma)d(\sigma)\,d\sigma + \int_a^b k(s,\xi)\left[\int_a^b l(\xi,\sigma)x(\sigma)\,d\sigma\right]d\xi;$$

here one may change the order of integration by Fubini's theorem; in fact, the regularity of the operators A and B implies the regularity of the integral operators K (with kernel k) and L (with kernel l), and the equalitiy

$$\int_a^b \int_a^b |k(s,\xi)l(\xi,\sigma)x(\sigma)|\,d\sigma\,d\xi$$

$$= \int_a^b |k(s,\xi)|\left[\int_a^b |l(\xi,\sigma)||x(\sigma)|\,d\sigma\right]d\xi = |K|\,|L|\,|x(s)|$$

implies that the function $(\sigma,\xi) \mapsto k(s,\xi)l(\xi,\sigma)x(\sigma)$ is integrable on $[a,b] \times [a,b]$ for almost all $s \in [a,b]$. By (3.15), we have that

$AB \in \mathfrak{L}_b^r(L_p)$. Moreover, the evident inequality $|AB| \leq |A| \, |B|$ and the definition (3.10) of the norm in $\mathfrak{L}_b^r(L_p)$ imply that

$$\|AB\|_{\mathfrak{L}_b^r(L_p)} = \| \, |AB| \, \|_{\mathfrak{L}(L_p)} \leq \| \, |A||B| \, \|_{\mathfrak{L}(L_p)}$$

and

$$\| \, |A| \, \|_{\mathfrak{L}(L_p)} \| \, |B| \, \|_{\mathfrak{L}(L_p)} = \|A\|_{\mathfrak{L}_b^r(L_p)} \|B\|_{\mathfrak{L}_b^r(L_p)},$$

and thus the first assertion is proved. The second assertion follows immediately from (3.15): in fact, if A, say, is a "pure" integral operator, then $c(s) \equiv 0$, hence $c(s)d(s) \equiv 0$, and thus AB is also an integral operator. ■

3.3. Sufficient conditions for regularity

In this subsection we shall formulate some important sufficient conditions for the operator (3.1) to belong to the class $\mathfrak{L}_b^r(L_p)$. Such conditions may be obtained, loosely speaking, from sufficient conditions for the integral operator (3.9) to be regular in the space L_p (KRASNOSEL'SKIJ-ZABREJKO-PUSTYL'NIK-SOBOLEVSKIJ [1966], ZA-BREJKO-KOSHELEV-KRASNOSEL'SKIJ-MIKHLIN-RAKOVSHCHIK-STETSENKO [1968]). For our purpose, however, it is more convenient to formulate such conditions through imbedding theorems for classes of kernels on $[a,b] \times [a,b]$. Following KRASNOSEL'SKIJ-ZABREJKO-PUSTYL'NIK-SOBOLEVSKIJ [1966], we denote by $\mathfrak{Z}_p$ ($1 \leq p \leq \infty$) the (Zaanen) class of all measurable functions $z = z(s, \sigma)$ on $[a, b] \times [a, b]$ for which the norm

$$(3.16) \qquad \|z\|_{\mathfrak{Z}_p} = \sup_{\substack{\|x\|_{L_p} \leq 1 \\ \|y\|_{L_q} \leq 1}} \int_a^b \int_a^b |z(s, \sigma)x(\sigma)y(s)| \, d\sigma \, ds$$

($q = p/(p-1)$) is finite. The importance of this class is seen from the fact that *every function in $\mathfrak{Z}_p$ is the kernel of some regular linear integral operator in L_p*, and, conversely, *every kernel of a regular linear integral operator in L_p is a function in $\mathfrak{Z}_p$*.

Combining this fact with Lemma 3.2 we arrive at the following important

Lemma 3.6. *The linear operator*

$$(3.17) \qquad Ax(s) = c(s)x(s) + \int_a^b k(s, \sigma)x(\sigma)\, d\sigma$$

belongs to the class $\mathfrak{L}_b^r(L_p)$ if and only if $c \in L_\infty$ and $k \in \mathfrak{Z}_p$. Moreover, in this case the two-sided estimate

$$(3.18) \qquad \frac{1}{2}(\|c\|_{L_\infty} + \|k\|_{\mathfrak{Z}_p}) \leq \|A\|_{\mathfrak{L}_b^r(L_p)} \leq \|c\|_{L_\infty} + \|k\|_{\mathfrak{Z}_p}$$

holds.

We can reformulate this result as follows. If we denote, in analogy to Subsection 2.3, by $\mathfrak{L}_m(L_p)$ the class of all multiplication operators (3.8) in L_p (note that a multiplication operator in L_p is always regular), and by $\mathfrak{L}_i^r(L_p)$ the class of all regular integral operators (3.9) in L_p, then the decomposition

$$\mathfrak{L}_b^r(L_p) = \mathfrak{L}_m(L_p) \oplus \mathfrak{L}_i^r(L_p)$$

holds. In particular, we cannot find a parallel counterexample to Example 2.1 for regular Barbashin operators in Lebesgue spaces.

At first glance it seems that the problem of describing the class $\mathfrak{L}_b^r(L_p)$ is completely solved by Lemma 3.6. However, there is an essential flaw: it is very hard, in general, to find explicit formulas or estimates for the norm of a function in $\mathfrak{Z}_p$, and just the problem of characterizing the elements in $\mathfrak{Z}_p$ is not solved even for very simple functions arising frequently in applications (e.g. functions with polynomial or logarithmic singularities on the diagonal $s = \sigma$). Conditions which are merely *sufficient*, of course, are well known (see KRASNOSEL'-SKIJ-ZABREJKO-PUSTYL'NIK-SOBOLEVSKIJ [1966] and below).

Some more auxiliary classes of kernels on $[a, b] \times [a, b]$ are in order. Given a measurable set $D \subseteq [a, b]$, let

$$(3.19) \qquad P_D x(s) := \begin{cases} x(s) & \text{if} \quad s \in D, \\ 0 & \text{if} \quad s \notin D \end{cases}$$

denote the *characteristic operator* of $D \subseteq [a, b]$, i.e. the multiplication operator generated by the *characteristic function* χ_D of D. By 3_p^- we denote the space of all $z \in 3_p$ such that

$$(3.20) \qquad \lim_{\mu(D)\to 0} \sup_{\substack{\|x\|_{L_p} \leq 1 \\ \|y\|_{L_q} \leq 1}} \int_a^b \int_a^b |z(s,\sigma) P_D x(\sigma) y(s)|\, d\sigma\, ds = 0.$$

Similarly, by 3_p^+ we denote the space of all $z \in 3_p$ such that

$$(3.21) \qquad \lim_{\mu(D)\to 0} \sup_{\substack{\|x\|_{L_p} \leq 1 \\ \|y\|_{L_q} \leq 1}} \int_a^b \int_a^b |z(s,\sigma) x(\sigma) P_D y(s)|\, d\sigma\, ds = 0.$$

Finally, let

$$3_p^0 := 3_p^- \cap 3_p^+.$$

It is well known that in case $1 < p < \infty$ all three classes $3_p^-, 3_p^+$ and 3_p^0 coincide and consist of all kernels which generate *compact* regular integral operators (3.9) in L_p (more precisely, they generate all regular operators (3.9) such that both K and $|K|$ are compact). In case $p = 1$ we have $3_1^- = \{0\}$, and 3_1^+ consists of all kernel functions which generate *weakly compact* integral operators in L_1. Similarly, in case $p = \infty$ we have $3_\infty^+ = \{0\}$, and 3_∞^- consists of all kernels which generate weakly compact integral operators in L_∞.

As already mentioned, the kernels $k \in 3_p$ are precisely those which generate integral operators $K \in \mathcal{L}_i^r(L_p)$. Let us denote by $\mathcal{L}_i^{r,-}(L_p)$, $\mathcal{L}_i^{r,+}(L_p)$, and $\mathcal{L}_i^{r,0}(L_p)$ the class of all integral operators $K \in \mathcal{L}_i^r(L_p)$ which are generated by kernels $k \in 3_p^-, 3_p^+$, and 3_p^0, respectively. From well known properties of compact and weakly compact operators we get the following useful result (which, by the way, may also be obtained directly):

Lemma 3.7. *The subspaces $\mathcal{L}_i^{r,-}(L_p)$, $\mathcal{L}_i^{r,+}(L_p)$, and $\mathcal{L}_i^{r,0}(L_p)$ are ideals in the algebra $\mathcal{L}_b^r(L_p)$.*

Now we introduce some important subclasses of the kernel class 3_p. For $1 \leq p, q \leq \infty$, denote by $U(p, q)$ the linear space of all measurable functions $z = z(s, \sigma)$ on $[a, b] \times [a, b]$ for which the so-called *mixed norm*

$$(3.22) \qquad ||z||_{U(p,q)} := ||s \mapsto ||z(s, \cdot)||_{L_p}||_{L_q}$$

is finite; similarly, denote by $V(p, q)$ the linear space of all measurable functions $z = z(s, \sigma)$ on $[a, b] \times [a, b]$ for which the mixed norm

$$(3.23) \qquad ||z||_{V(p,q)} := ||\sigma \mapsto ||z(\cdot, \sigma)||_{L_p}||_{L_q}$$

is finite. The functionals (3.22) and (3.23) are norms which turn $U(p, q)$ and $V(p, q)$ into Banach spaces.

The following two propositions provide some relations between the kernel classes introduced so far; they may be regarded as reformulation of well known sufficient acting conditions for integral operators in L_p, expressed in terms of the corresponding kernels (HILLE-TAMARKIN [1930], see also KANTOROVICH-AKILOV [1977] or KRASNOSEL'SKIJ-ZABREJKO-PUSTYL'NIK-SOBOLEVSKIJ [1966]):

Lemma 3.8. *Let* $1 \leq p \leq \infty$ *and* $q = p/(p-1)$. *Then* $U(q, p)$ *is continuously imbedded in* $\mathfrak{Z}_p$, *i.e.*

$$(3.24) \qquad ||z||_{\mathfrak{Z}_p} \leq ||z||_{U(q,p)} \qquad (z \in U(q, p)).$$

Moreover, $U(1, \infty) = \mathfrak{Z}_\infty$, $U(q, p) \subseteq \mathfrak{Z}_p^0$ *for* $1 < p < \infty$, *and* $U(\infty, 1) \subseteq \mathfrak{Z}_1^+$.

Lemma 3.9. *Let* $1 \leq p \leq \infty$ *and* $q = p/(p-1)$. *Then* $V(p, q)$ *is continuously imbedded in* $\mathfrak{Z}_p$, *i.e.*

$$(3.25) \qquad ||z||_{\mathfrak{Z}_p} \leq ||z||_{V(p,q)} \qquad (z \in V(p, q)).$$

Moreover, $V(1, \infty) = \mathfrak{Z}_1$, $V(p, q) \subseteq \mathfrak{Z}_p^0$ *for* $1 < p < \infty$, *and* $V(\infty, 1) \subseteq \mathfrak{Z}_\infty^-$.

The following result is much more precise and difficult and is essentially due to Kantorovich (see KANTOROVICH [1956], KANTORO-VICH-AKILOV [1977], KRASNOSEL'SKIJ-ZABREJKO-PUSTYL'NIK-SOBOLEVSKIJ [1966]):

Lemma 3.10. *Let $p, p_0, p_1, q_0, q_1 \in [1, \infty]$ and $\lambda \in [0, 1]$ be real numbers satisfying*

$$1 - \frac{1-\lambda}{p_0} - \frac{\lambda}{q_1} \leq \frac{1}{p} \leq \frac{1-\lambda}{q_0} + \frac{\lambda}{p_1}.$$

Then $U(p_0, q_0) \cap V(p_1, q_1)$ is imbedded in $\mathfrak{Z}_p$ and

(3.26) $$\|z\|_{\mathfrak{Z}_p} \leq c \|z\|_{U(p_0, q_0)}^{1-\lambda} \|z\|_{V(p_1, q_1)}^{\lambda}.$$

Moreover, $U(p_0, q_0) \cap V(p_1, q_1) \subseteq \mathfrak{Z}_p^0$ if either $q_0 < \infty$ or $q_1 < \infty$.

Observe that Lemma 3.10 is a kind of "interpolation" between Lemma 3.8 and Lemma 3.9; in fact, for $\lambda = 0, p_0 = q$, and $q_0 = p$, (3.26) coincides with (3.24), while for $\lambda = 1, p_1 = p$, and $q_1 = q$, (3.26) coincides with (3.25).

Lemma 3.10 makes it possible to treat kernels $z = z(s, \sigma)$ with polynomial singularities on the diagonal $s = \sigma$ as elements of $\mathfrak{Z}_p$. Nevertheless, Lemma 3.10 does not allow us to recover *all* kernels from $\mathfrak{Z}_p$, in particular, not those which generate *noncompact* integral operators in L_p (observe that the inclusion $\mathfrak{Z}_p^0 \subseteq \mathfrak{Z}_p$ is always strict for $1 < p < \infty$).

Other descriptions of the elements of $\mathfrak{Z}_p$ are possible by means of *interpolation theorems* in Lebesgue spaces (see e.g. BERGH-LÖFSTRÖM [1976], KREJN-PETUNIN-SEMENOV [1978]). The first theorems of this type have been established by Riesz and Marcinkiewicz, afterwards by Stein and Weiss, Dikarev and Matsaev; a certain completion of the theory was reached with the "sharp" interpolation theorem by OVCHINNIKOV [1983]. In what follows, we shall give a formulation of this theorem in the spirit of the book ZABREJKO-KOSHELEV-KRAS-NOSEL'SKIJ-MIKHLIN-RAKOVSHCHIK-STETSENKO [1968].

Denote by $W(\alpha, \beta)$ $(0 \leq \alpha, \beta \leq 1)$ the linear space of all measurable functions $z = z(s, \sigma)$ on $[a, b] \times [a, b]$ for which the norm

$$\|z\|_{W(\alpha, \beta)}$$

(3.27)
$$= \sup_{D_1, D_2 \subseteq [a,b]} \mu(D_1)^{-\alpha} \mu(D_2)^{-\beta} \int_{D_1} \int_{D_2} |z(s, \sigma)| \, d\sigma \, ds$$

is finite; similarly, by $W^0(\alpha, \beta)$ we mean the subspace of all $z \in W(\alpha, \beta)$ satisfying

$$(3.28) \qquad \lim_{\substack{\mu(D_1) \to 0 \\ \mu(D_2) \to 0}} \mu(D_1)^{-\alpha} \mu(D_2)^{-\beta} \int_{D_1} \int_{D_2} |z(s, \sigma)| \, d\sigma ds = 0.$$

Lemma 3.11. *Let* $\alpha_0, \alpha_1, \beta_0, \beta_1 \in [0, 1], p \in [1, \infty],$ *and* $\lambda \in [0, 1]$ *be real numbers satisfying*

$$1 - (1 - \lambda)\alpha_0 - \lambda\alpha_1 \leq \frac{1}{p} \leq (1 - \lambda)\beta_0 + \lambda\beta_1.$$

Then $W(\alpha_0, \beta_0) \cap W(\alpha_1, \beta_1)$ *is imbedded in* $\mathfrak{Z}_p$ *and*

$$(3.29) \qquad ||z||_{\mathfrak{Z}_p} \leq c \, ||z||_{W(\alpha_0, \beta_0)}^{1 - \lambda} ||z||_{W(\alpha_1, \beta_1)}^{\lambda}$$

for some $c > 0$. *Moreover,* $W^0(\alpha_0, \beta_0) \cap W(\alpha_1, \beta_1) \subseteq \mathfrak{Z}_p^0$ *if* $\beta_0 > 0,$ *and* $W(\alpha_0, \beta_0) \cap W^0(\alpha_1, \beta_1) \subseteq \mathfrak{Z}_p^0$ *if* $\beta_1 > 0.$

All results concerning the kernel classes $U(p, q), V(p, q),$ or $W(\alpha, \beta)$ given in this subsection provide conditions on the parameters p, q, α, β under which these classes are subspaces of $\mathfrak{Z}_p$ (or even $\mathfrak{Z}_p^0$). Each of these kernel classes induces, of course, a corresponding space of linear integral operators (3.9). The question arises which of these space are *ideals* (or at least *subalgebras*) of the algebra $\mathfrak{L}_b^r(L_p)$. For some of these classes (e.g. for $U(p, q)$ and $V(p, q)$), the answer is positive; this follows from basic properties of so-called *u-bounded* and *v-cobounded* operators in Lebesgue spaces (see e.g. KRASNOSEL'SKIJ-ZABREJKO-PUSTYL'NIK-SOBOLEVSKIJ [1966]). For some other classes, the answer is unknown.

Let us discuss still another aspect of the operator (3.1). In Lemma 2.1 we established the explicit formula (2.5) for the norm of the operator (3.1) in the space C which may be expressed rather simply in terms of the corresponding functions $c = c(s)$ and $k = k(s, \sigma)$. A parallel formula may be obtained in the space L_p, at least in the "extreme" cases $p = 1$ and $p = \infty$.

Lemma 3.12. *Suppose that the linear operator (3.17) acts in the space L_1 or in the space L_∞. Then the operator (3.17) is bounded with norm*

$$\|A\|_{\mathfrak{L}(L_1)} = \|\gamma_1\|_{L_\infty}$$

respectively

$$\|A\|_{\mathfrak{L}(L_\infty)} = \|\gamma_\infty\|_{L_\infty},$$

where

$$(3.30) \qquad \gamma_1(s) = |c(s)| + \int_a^b |k(\sigma, s)|\, d\sigma$$

and

$$(3.31) \qquad \gamma_\infty(s) = |c(s)| + \int_a^b |k(s, \sigma)|\, d\sigma.$$

☐ The boundedness of the operator (3.17) follows from Lemma 3.1; moreover, by Lemma 3.7, the estimates

$$(3.32) \qquad \|A\|_{\mathfrak{L}(L_1)} \le \|\gamma_1\|_{L_\infty}$$

and

$$(3.33) \qquad \|A\|_{\mathfrak{L}(L_\infty)} \le \|\gamma_\infty\|_{L_\infty}$$

hold. Let first $p = \infty$. Since the unit ball in L_∞ consists of all functions $|x| \le 1$, we have, by Lemma 3.2,

$$\sup_{\|x\|_{L_\infty} \le 1} |Ax(s)| = \gamma_\infty(s).$$

Further, since for any subset M of L_∞ we have (see e.g. KANTORO-VICH-VULIKH-PINSKER [1950] or KANTOROVICH-AKILOV [1977])

$$\sup\{\|x\| : x \in M\} = \|\sup\{|x| : x \in M\}\|,$$

we conclude that

$$\|A\|_{\mathfrak{L}(L_\infty)} = \sup_{\|x\|_{L_\infty} \le 1} \|Ax\|_{L_\infty} = \|\,|A|x_1\|_{L_\infty} = \|\gamma_\infty\|_{L_\infty},$$

where $x_1(s) \equiv 1$. This proves the assertion in case $p = \infty$. In case $p = 1$, the statement follows from Lemma 3.3 by passing to the Köthe dual $L_1 = L'_\infty$ of L_∞ and the Köthe adjoint A' of A. ■

Observe that, in the special case $c(s) \equiv 1$, the formulas (3.30) and (3.31) become

$$\|I + K\| = 1 + \|K\|$$

with K given by (3.9); this is of course parallel to the situation in the space C (see (2.33) in Subsection 2.7). Such formulas have been obtained by LOZANOVSKIJ [1966] by slightly different methods; these methods, however, apply also to more general operators of the form (3.17).

3.4. Strongly continuous operator functions

Consider again the operator function

$$(3.34) \qquad A(t)x(s) = c(t,s)x(s) + \int_a^b k(t,s,\sigma)x(\sigma)\,d\sigma.$$

on a bounded or unbounded interval J. By Lemma 3.6, formula (3.34) defines an operator function A with values in $\mathfrak{L}(L_p)$ if $c(t,\cdot) \in L_\infty$ and $k(t,\cdot,\cdot) \in \mathfrak{Z}_p$ for $t \in J$. The purpose of this subsection is to find conditions on the functions c and k under which this operator function is strongly continuous. If the operator function (3.34) takes values not only in $\mathfrak{L}(L_p)$, but in the subclass $\mathfrak{L}_b^r(L_p)$, then (3.34) may be represented, by Lemma 3.6, as sum of the two simpler operator functions

$$(3.35) \qquad\qquad C(t)x(s) = c(t,s)x(s),$$

and

$$(3.36) \qquad\qquad K(t)x(s) = \int_a^b k(t,s,\sigma)x(\sigma)\,d\sigma$$

therefore the problem reduces to studying the strong continuity of (3.35) and (3.36) separately.

Lemma 3.13. *Let $c_n \in L_\infty$ $(n = 1, 2, \ldots)$ and $c \in L_\infty$, and let C_n and C, respectively, be the corresponding multiplication operators (3.8). Then the sequence $(C_n)_n$ converges strongly in L_p $(p < \infty)$ to C if and only if the sequence $(c_n)_n$ is bounded in L_∞ and converges to c either in measure or in L_1; likewise, the sequence $(C_n)_n$ converges strongly in L_∞ to C if and only if $(c_n)_n$ converges to c in L_∞.*

□ Let first $p < \infty$. The necessity of the boundedness of $(c_n)_n$ follows then from the Banach-Steinhaus theorem on the uniform boundedness of strongly convergent operator sequences and the trivial equality $\|C_n\|_{\mathcal{L}(L_p)} = \|c_n\|_{L_\infty}$. The necessity of the convergence of $(c_n)_n$ either in measure or in L_1 is a consequence of the obvious equalities $C_n x_1 = c_n$ and $C x_1 = c$ ($x_1(s) \equiv 1$). Conversely, the sufficiency of these two conditions follows from the Lebesgue theorem on dominated convergence.

In case $p = \infty$ the proof is even simpler: indeed, the statement follows directly from the equality $\|(C_n - C)x_1\|_{L_\infty} = \|c_n - c\|_{L_\infty}$. ■

Lemma 3.14. *Let $k_n \in \mathfrak{Z}_p$ $(n = 1, 2, \ldots)$ and $k \in \mathfrak{Z}_p$, and let K_n and K, respectively, be the corresponding integral operators (3.9). Then the sequence $(K_n)_n$ converges strongly in L_p to K if and only if the sequence $(k_n)_n$ is bounded in $\mathfrak{Z}_p$ and if, in case $p < \infty$, the sequence*

$$\kappa_n(s, z) = \int_a^z k_n(s, \sigma)\, d\sigma \qquad (a \le z \le b)$$

converges in L_p to the function

$$\kappa(s, z) = \int_a^z k(s, \sigma)\, d\sigma \qquad (a \le z \le b).$$

□ The proof is a simple consequence of the Banach - Steinhaus theorem and the fact that the linear hull of the characteristic functions $\chi_{[a,z]}$ ($a \le z \le b$) is dense in L_p for $p < \infty$, while the linear hull of all characteristic functions χ_D ($D \subseteq [a, b]$) is dense in L_∞. ■

Combining Lemma 3.13 and Lemma 3.14 we arrive at a (sufficient) condition for the strong continuity of the operator function (3.34) in L_p which we state as

Theorem 3.1. *Suppose that $c = c(t, s)$ and $k = k(t, s, \sigma)$ are measurable on $J \times [a, b]$ and $J \times [a, b] \times [a, b]$, respectively, and satisfy the following conditions:*

(a) for each $t \in J$, the function $c(t, \cdot)$ belongs to L_∞, and in case $p < \infty$ the function $t \mapsto c(t, \cdot)$ is bounded from J into L_∞ and continuous from J into L_1, while in case $p = \infty$ the function $t \mapsto c(t, \cdot)$ is continuous from J into L_∞;

(b) for each $t \in J$, the function $k(t, \cdot, \cdot)$ belongs to $\mathfrak{Z}_p$, and the function $t \mapsto k(t, \cdot, \cdot)$ is bounded from J into $\mathfrak{Z}_p$;

(c) in case $p < \infty$, the function

$$t \mapsto \int_a^c k(t, \cdot, \sigma)\, d\sigma$$

is continuous from J into L_p for each $c \in [a, b]$, while in case $p = \infty$ the function

$$t \mapsto \int_D k(t, \cdot, \sigma)\, d\sigma$$

is continuous from J into L_∞ for each $D \subseteq [a, b]$.

Then the operator function $A = A(t)$ defined by (3.34) is strongly continuous in L_p.

The hypotheses of Theorem 3.1 are rather easy to verify; the only difficult part is the verification of (b), since the norm of the class $\mathfrak{Z}_p$ is somewhat complicated. In most applications, however, one may use the Lemmas 3.8 - 3.11 proved in the preceding subsection.

We point out that the conditions (a) - (c) of Theorem 3.1 are not only sufficient in case $p = 1$ and $p = \infty$, but also necessary; this follows from the fact that $\mathfrak{L}_b(L_p)$ and $\mathfrak{L}_b^r(L_p)$ coincide in this case, and from the Banach-Steinhaus theorem.

3.5. Norm-continuous operator functions

Generally speaking, necessary and sufficient conditions neither for the strong continuity of the operator function (3.34), nor for its continuity in the norm of $\mathfrak{L}(L_p)$, are easy to obtain. It is easy, however, to find conditions which are merely sufficient. Moreover, if we consider (3.34) as an operator function with values in $\mathfrak{L}_b^r(L_p)$, rather than $\mathfrak{L}(L_p)$, then it is in fact possible to give continuity conditions which are both necessary and sufficient. More precisely, such conditions may be given for all operator functions whose integral part (3.36) is, for fixed $t \in J$ an integral operator in $\mathfrak{L}_i^{r,0}(L_p)$; this situation occurs very often in applications.

Lemma 3.15. *Let $c_n \in L_\infty$ $(n = 1, 2, \ldots)$ and $c \in L_\infty$, and let C_n and C, respectively, be the corresponding multiplication operators (3.8). Then the sequence $(C_n)_n$ converges in the norm of $\mathfrak{L}(L_p)$ to C if and only if the sequence $(c_n)_n$ converges to c in L_∞.*

□ The statement follows trivially from the fact that the norm of the operator (3.8) in $\mathfrak{L}(L_p)$ is exactly $||c||_{L_\infty}$. ■

Recall (KRASNOSEL'SKIJ-ZABREJKO-PUSTYL'NIK-SOBOLEVSKIJ [1966]) that a set $M \subset L_p$ is called *absolutely bounded* if

$$\lim_{\mu(D) \to 0} \sup_{x \in M} \int_D |x(s)|^p \, ds = 0,$$

i.e. the elements of M have *uniformly absolutely continuous norms.* Likewise, a set $M \subset \mathfrak{Z}_p$ is called *n-absolutely bounded* if

$$(3.37) \qquad \sup_{z \in M} \sup_{\substack{||x||_{L_p} \leq 1 \\ ||y||_{L_q} \leq 1}} \int_a^b \int_a^b |z(s, \sigma) P_{D_1} x(\sigma) P_{D_2} y(s)| \, d\sigma \, ds \to 0$$

as $\mu(D_1) \to 0$ and $\mu(D_2) \to 0$. It is not hard to see that every n-absolutely bounded set in $\mathfrak{Z}_p$ is bounded in norm; the converse is, of course, false.

Lemma 3.16. *For $1 < p < \infty$, let $k_n \in \mathfrak{Z}_p^0$ $(n = 1, 2, \ldots)$ and $k \in \mathfrak{Z}_p^0$, and let K_n and K, respectively, be the corresponding integral operators (3.9). Then the sequence $(K_n)_n$ converges in the norm of $\mathfrak{L}_i^{r,0}(L_p)$ to K if and only if the sequences $(u_n)_n$ and $(v_n)_n$ defined by*

$$(3.38) \qquad u_n(s) = \int_a^b |k_n(s, \sigma)| \, d\sigma, \qquad v_n(\sigma) = \int_a^b |k_n(s, \sigma)| \, ds$$

are absolutely bounded in L_p and L_q, respectively, and the sequence
$(k_n)_n$ converges in L_1 to k and is n-absolutely bounded in 3_p^0.

☐ The necessity of the given conditions follows from the continuity of the imbedding $3_p \subseteq L_1$ and from the absolute boundedness of a convergent sequence in L_p (see e.g. KRASNOSEL'SKIJ-ZABREJKO-PUSTYL'NIK-SOBOLEVSKIJ [1966]). For instance, the first condition follows from the elementary estimates

$$\|P_D u_n\|_{L_p} \leq \|P_D(\int_a^b |k_n(\cdot,\sigma) - k(\cdot,\sigma)|d\sigma)\|_{L_p}$$

$$+\|P_D(\int_a^b |k(s,\sigma)| \, d\sigma)\|_{L_p}$$

$$\leq \sup_{\|y\|_{L_q}\leq 1} \int_a^b \int_a^b |[k_n(s,\sigma) - k(s,\sigma)]y(s)| \, d\sigma \, ds$$

$$+ \sup_{\|y\|_{L_q}\leq 1} \int_a^b \int_a^b |k(s,\sigma)P_D y(s)| d\sigma \, ds$$

$$\leq (b-a)^{1/p}\|k_n - k\|_{3_p} + \|P_D|K|x_1\|_{L_p}$$

with $x_1(s) \equiv 1$.

The proof of the sufficiency is harder. Observe, first of all, that, by the convergence of $(k_n)_n$ to k in L_1, we have

$$(3.39) \qquad \lim_{n\to\infty} \int_a^b \int_a^b |k_n(s,\sigma) - k(s,\sigma)| \, d\sigma \, ds = 0.$$

Further, the absolute boundedness of the sequence $(u_n)_n$ in L_p (see (3.38)) implies that of the sequence $(u_n^*)_n$ defined by

$$u_n^*(s) = \int_a^b |k_n(s,\sigma) - k(s,\sigma)| \, d\sigma.$$

On the other hand, this sequence converges in L_1 to zero, by (3.39), hence also in measure; consequently (see e.g. KRASNOSEL'SKIJ-ZA-BREJKO-PUSTYL'NIK-SOBOLEVSKIJ [1966]), this sequence converges in L_p to zero as well. Since

$$\|u_n^*\|_{L_p} = \sup_{\|y\|_{L_q}\leq 1} \int_a^b \int_a^b |[k_n(s,\sigma) - k(s,\sigma)]y(s)| \, d\sigma \, ds,$$

we conclude that

$$(3.40) \qquad \lim_{n \to \infty} \sup_{\|y\|_{L_q} \leq 1} \int_a^b \int_a^b |[k_n(s,\sigma) - k(s,\sigma)]y(s)| \, d\sigma \, ds = 0.$$

Similarly, the absolute boundedness of the sequence $(v_n)_n$ in L_q (see (3.38)) implies that

$$(3.41) \qquad \lim_{n \to \infty} \sup_{\|x\|_{L_p} \leq 1} \int_a^b \int_a^b |[k_n(s,\sigma) - k(s,\sigma)]x(\sigma)| \, d\sigma \, ds = 0.$$

Finally, the n-absolute boundedness of the sequence $(k_n)_n$ in 3_p^0 means that, given $\varepsilon > 0$, we may find a $\delta > 0$ such that for $\mu(D_1) \leq \delta$ and $\mu(D_2) \leq \delta$ we have

$$\sup_{\substack{\|x\|_{L_p} \leq 1 \\ \|y\|_{L_q} \leq 1}} \int_a^b \int_a^b |[k_n(s,\sigma) - k(s,\sigma)]P_{D_1}x(\sigma)P_{D_2}y(s)| \, d\sigma \, ds \leq \varepsilon.$$

For this δ we may find in turn a number $h > 0$ such that, for any $x \in L_p$ with $\|x\|_{L_p} \leq 1$ and $y \in L_1$ with $\|y\|_{L_q} \leq 1$, we have $\mu(D(x)) \leq \delta$ and $\mu(D(y)) \leq \delta$, where

$$D(z) = \{t : a \leq t \leq b, |z(t)| > h\}.$$

Consequently,

$$(3.42) \qquad \int_a^b \int_a^b |[k_n(s,\sigma) - k(s,\sigma)]| P_{D(x)}x(\sigma)P_{D(y)}y(s) \, d\sigma \, ds < \varepsilon$$

for $||x||_{L_p} \leq 1$ and $||y||_{L_q} \leq 1$. The elementary inequality

$$\int_a^b \int_a^b |[k_n(s,\sigma) - k(s,\sigma)]x(\sigma)y(s)|d\sigma \, ds$$

$$\leq h^2 \int_a^b \int_a^b |k_n(s,\sigma) - k(s,\sigma)| \, d\sigma \, ds$$

$$+ h \int_a^b \int_a^b |[k_n(s,\sigma) - k(s,\sigma)]y(s)| \, d\sigma \, ds$$

$$+ h \int_a^b \int_a^b |[k_n(s,\sigma) - k(s,\sigma)]x(\sigma)|d\sigma \, ds$$

$$+ \int_a^b \int_a^b |[k_n(s,\sigma) - k(s,\sigma)]||P_{D(x)}x(\sigma)P_{D(y)}y(s)| \, d\sigma \, ds$$

implies, together with (3.39) - (3.42) that

$$\varlimsup_{n \to \infty} \sup_{\substack{||x||_{L_p} \leq 1 \\ ||y||_{L_q} \leq 1}} \int_a^b \int_a^b |[k_n(s,\sigma) - k(s,\sigma)]x(\sigma)y(s)| \, d\sigma \, ds \leq \varepsilon.$$

Since $\varepsilon > 0$ was arbitrary, we conclude that

$$\lim_{n \to \infty} \sup_{\substack{||x||_{L_p} \leq 1 \\ ||y||_{L_q} \leq 1}} \int_a^b \int_a^b |[k_n(s,\sigma) - k(s,\sigma)]x(\sigma)y(s)| \, d\sigma \, ds = 0$$

and thus the proof is complete. ∎

Combining Lemma 3.3 and Lemma 3.4 we arrive at a (necessary and sufficient) condition for the continuity of the operator function (3.34) in the norm of $\mathcal{L}_b^r(L_p)$ which is parallel to Theorem 2.2.

Theorem 3.2. *Let $1 < p < \infty$, and suppose that $c = c(t,s)$ and $k = k(t,s,\sigma)$ are measurable on $J \times [a,b]$ and $J \times [a,b] \times [a,b]$, respectively, and satisfy the following conditions:*

(a) for each $t \in J$, the function $c(t,\cdot)$ belongs to L_∞, and the function $t \mapsto c(t,\cdot)$ is continuous from J into L_∞;

(b) *for each $t \in J$, the function $k(t, \cdot, \cdot)$ belongs to 3_p^0, and the function $t \mapsto k(t, \cdot, \cdot)$ is continuous from J into L_1;*

(c) *the functions*

$$t \mapsto \int_a^b |k(t, \cdot, \sigma)| \, d\sigma$$

and

$$t \mapsto \int_a^b |k(t, s, \cdot)| \, ds$$

map every bounded subinterval of J into an absolutely bounded subset of L_p and L_q, respectively;

(d) *the function $t \mapsto k(t, \cdot, \cdot)$ maps every bounded subset of J into an n-absolutely bounded subset of 3_p^0.*

Then the operator function $A = A(t)$ defined by (3.34) is continuous in the norm of $\mathfrak{L}_b^r(L_p)$.

Conversely, if the operator function (3.34) is continuous in the norm of $\mathfrak{L}_b^r(L_p)$, then the corresponding functions $c = c(t, s)$ and $k = k(t, s, \sigma)$ satisfy the conditions (a) - (d).

The hypotheses of Theorem 3.2 are rather easy to verify; as in Theorem 3.1, one may use the Lemmas 3.8 - 3.11 to verify (b) and (d).

The case when the integral part (3.36) of (3.34) belongs, for each $t \in J$, only to $\mathfrak{L}_i^r(L_p)$ instead of $\mathfrak{L}_i^{r,0}(L_p)$, is not covered by Theorem 3.2. As a matter of fact, the continuity of (3.34) is usually proved by showing that the function $t \mapsto c(t, \cdot)$ is continuous from J into L_∞, and the function $t \mapsto k(t, \cdot, \cdot)$ is continuous from J into 3_p. But just the verification of the second of these properties is extremely difficult, because there is no analogue to Lemma 3.4 for integral operators $\mathfrak{L}_i^r(L_p)$ instead of $\mathfrak{L}_i^{r,0}(L_p)$. By the way, in some cases one may also use the Lemmas 3.8 - 3.11 to prove the continuity of the function $t \mapsto k(t, \cdot, \cdot)$ from J into the kernel classes $U(q, p), V(p, q)$, or $W(\alpha, \beta)$ which are imbedded in the class 3_p (see Subsection 3.3).

The case $p = 1$ or $p = \infty$ (which is not covered by Theorem 3.2) is rather easy to deal with. In fact, the equalities $\mathfrak{L}_b(L_1) = \mathfrak{L}_b^r(L_1)$ and

$\mathfrak{L}_b(L_\infty) = \mathfrak{L}_b^r(L_\infty)$ hold, as well as the formulas

$$(3.43) \qquad \|A\|_{\mathfrak{L}(L_1)} = \|\, |c(\cdot)| + \int_a^b |k(\sigma, \cdot)|\, d\sigma \|_{L_\infty}$$

and

$$(3.44) \qquad \|A\|_{\mathfrak{L}(L_\infty)} = \|\, |c(\cdot)| + \int_a^b |k(\cdot, \sigma)|\, d\sigma \|_{L_\infty}$$

for the norm of the operator (3.1). This fact allows us to get rather easily conditions for the functions $c = c(t,s)$ and $k = k(t,s,\sigma)$, both necessary and sufficient, under which the operator function (3.34) is norm continuous. We state this as

Theorem 3.3. *Let $p = 1$ or $p = \infty$, and suppose that $c = c(t,s)$ and $k = k(t,s,\sigma)$ are measurable on $J \times [a,b]$ and $J \times [a,b] \times [a,b]$, respectively, and satisfy the following conditions:*

(a) the function $t \mapsto c(t,\cdot)$ is continuous from J into L_∞;

(b) the function $t \mapsto k(t,\cdot,\cdot)$ is continuous from J into $V(1,\infty)$ [respectively into $U(1,\infty)$].

Then the operator function $A = A(t)$ defined in (3.34) is continuous in the norm of $\mathfrak{L}(L_1)$ [respectively of $\mathfrak{L}(L_\infty)$].

Conversely, if the operator function (3.34) is continuous in the norm of $\mathfrak{L}(L_1)$ [respectively of $\mathfrak{L}(L_\infty)$], then the corresponding functions $c = c(t,s)$ and $k = k(t,s,\sigma)$ satisfy the conditions (a) and (b).

☐ The proof is a straightforward consequence of (3.43), (3.44), and Lemma 3.12. ∎

3.6. Representation of the evolution operator

Let us return to the study of the linear differential equation (1.2) with operator function (1.3) in the Lebesgue space L_p ($1 \le p \le \infty$). In what follows, we shall suppose that the hypotheses of Theorem 3.1 are satisfied, and thus (3.34) is a strongly continuous operator function with values in the space $\mathfrak{L}_b^r(L_p)$. As in the case of the space C (see Subsection 2.6) it is natural to expect that the evolution operator

$U = U(t, \tau)$ for equation (1.2) is again representable as sum of the evolution operator $U_0 = U_0(t, \tau)$ of the "reduced" equation

$$(3.45) \qquad \frac{dx}{dt} = C(t)x$$

and some integral operator

$$(3.46) \qquad H(t, \tau)x(s) = \int_a^b h(t, \tau, s, \sigma)x(\sigma)\, d\sigma.$$

This is in fact true also in the space L_p and follows from Theorem 1.2, Lemma 3.5, and Lemma 3.7; one can say even more:

Theorem 3.4. *Suppose that the operator function (3.34) is strongly continuous in $\mathfrak{L}_b^r(L_p)$. Then the evolution operator $U = U(t, \tau)$ for the differential equation (1.2) belongs to $\mathfrak{L}_b^r(L_p)$ for each $t, \tau \in J$, i.e. admits a representation*

$$(3.47) \qquad U(t, \tau)x(s) = e(t, \tau, s)x(s) + H(t, \tau)x(s),$$

where $e(t, \tau, s)$ is given by (1.29), and H is defined by (3.46) with $h = h(t, \tau, s, \sigma)$ being a measurable function on $J \times J \times [a, b] \times [a, b]$. Moreover, $H(t, \tau) \in \mathfrak{L}_i^{r,-}(L_p)$ if $K(t) \in \mathfrak{L}_i^{r,-}(L_p)$, $H(t, \tau) \in \mathfrak{L}_i^{r,+}(L_p)$ if $K(t) \in \mathfrak{L}_i^{r,+}(L_p)$, and $H(t, \tau) \in \mathfrak{L}_i^{r,0}(L_p)$ if $K(t) \in \mathfrak{L}_i^{r,0}(L_p)$.

□ Recall (see Subsection 1.2) that the evolution operator $U(t, \tau)$ for equation (1.2) is representable as a series (1.7) which converges, uniformly on every bounded subinterval of J, in the norm of the space $\mathfrak{L}(L_p)$. By Lemma 3.5, every operator $A(t_1) \cdots A(t_n)$ $(n = 1, 2, \ldots)$ in (1.7) belongs to $\mathfrak{L}_b^r(L_p)$, i.e. admits a representation

$$A(t_1) \cdots A(t_n)x(s) = c(t_1, s) \ldots c(t_n, s)x(s)$$

$$(3.48)$$

$$+ \int_a^b r_n(t_1, \ldots, t_n, s, \sigma)x(\sigma)\, d\sigma,$$

where the function $r_n = r_n(t_1, \ldots, t_n, s, \sigma)$ $(n = 1, 2, \ldots)$ is measurable, and the function $r_n(t_1, \ldots, t_n, \cdot, \cdot)$ belongs to $\mathfrak{Z}_p$ with

$$(3.49) \qquad \|r_n(t_1, \ldots, t_n, \cdot, \cdot)\|_{\mathfrak{Z}_p} \leq \|A(t_1)\|_{\mathfrak{L}(L_p)} \cdots \|A(t_n)\|_{\mathfrak{L}(L_p)}.$$

Integrating (3.48) over $t_1, \ldots, t_n$ yields

$$\left[\int_\tau^t \int_\tau^{t_1} \cdots \int_\tau^{t_{n-1}} A(t_1) \cdots A(t_n) \, dt_n \cdots dt_1 \right] x(s)$$

(3.50)
$$= \frac{1}{n!} \left[\int_\tau^t c(\xi, s) \, d\xi \right]^n x(s) +$$

$$\int_a^b \left[\int_\tau^t \cdots \int_\tau^{t_{n-1}} r_n(t_1, \ldots, t_n, s, \sigma) \, dt_n \cdots dt_1 \right] x(\sigma) \, d\sigma;$$

here the order of integration may be changed by Fubini's theorem and the inequality (3.49). Interchanging the integration (3.50) with the series (1.7) we get (3.47) with

$$h(t, \tau, s, \sigma)$$

(3.51)
$$:= \sum_{n=1}^\infty \int_\tau^t \int_\tau^{t_1} \cdots \int_\tau^{t_{n-1}} r_n(t_1, \ldots, t_n, s, \sigma) \, dt_n \ldots dt_1.$$

The proof is complete. ■

A comparison with Theorem 2.3 shows that the proof of Theorem 3.4 is essentially the same as in the case of the space C. The only difference consists in the fact that the convergence of the operator series for $U(t, \tau)$, uniformly on every bounded subinterval of J, is proved for the series (2.32) in the norm of $U(1, \infty)$, and for the series (3.51) in the norm of $\mathfrak{Z}_p$. Since $U(1, \infty) = \mathfrak{Z}_\infty$, Theorem 3.4 is more general than Theorem 2.3, at least formally. It is worth mentioning, however, that Theorem 2.3 covers all strongly continuous operator functions with values in $\mathfrak{L}(C)$, while Theorem 3.4 applies only (except for the "degenerate" cases $p = 1$ and $p = \infty$) to restricted classes of strongly continuous operator functions in L_p, namely those which are actually representable as regular integral operators in L_p.

§ 4. Barbashin equations in ideal spaces

In this section we try to generalize some results of § 3 in the framework of so-called ideal spaces (Banach lattices) of measurable functions. Examples of ideal spaces are the Lebesgue spaces considered in the last paragraph, Orlicz spaces (which are useful in problems with strong nonlinearities), or Lorentz and Marcinkiewicz spaces (which arise in interpolation theory for linear operators).

4.1. Ideal spaces

Many of the results presented in § 3 carry over from Lebesgue spaces to larger classes of spaces which frequently occur in applications. In this subsection we develop the notions which will be needed to formulate the corresponding results.

Let Ω be a nonempty set with a σ-algebra $\mathfrak{A}$ of subsets of Ω (called "measurable sets") and a σ-finite σ-additive measure μ on $\mathfrak{A}$. Since in all applications which follow Ω will be a bounded domain in Euclidean space, we shall assume that $\mu(\Omega) < \infty$. We also suppose throughout that the measure μ is atom-free, although many of the reults which follow also hold for measures with atoms.

By $\mathfrak{S} = \mathfrak{S}(\Omega)$ we denote the set of all μ-measurable functions $x :$ $\Omega \to \mathbb{R}$ equipped with the usual algebraic operations and the metric

$$(4.1) \qquad \rho(x, y) := \inf_{0 < h < \infty} \{h + \mu(\{s : s \in \Omega, |x(s) - y(s)| > h\})\}$$

or the matric

$$\hat{\rho}(x, y) := \int_\Omega \frac{|x(s) - y(s)|}{1 + |x(s) - y(s)|} \, d\mu(s).$$

To be precise, the elements of $\mathfrak{S}$ are *classes* of functions, since we have to identify two functions which differ only on a nullset. Obviously, convergence in the metric (4.1) is just *convergence in measure* (see e.g. DUNFORD-SCHWARTZ [1963]). There is also a natural *ordering* on the metric space $\mathfrak{S}$: we write $x \leq y$ if $x(s) \leq y(s)$ almost everywhere on Ω.

An *ideal space* (or *Banach lattice*) over Ω is a Banach space $X \subset \mathfrak{S}(\Omega)$ with the property that $x \in X$, $z \in \mathfrak{S}$ and $|z| \leq |x|$ implies that $z \in X$

and $||z||_X \leq ||x||_X$. The most prominent example of an ideal space is of course the *Lebesgue space* $L_p = L_p(\Omega)$ $(1 \leq p \leq \infty)$ with norm

$$
(4.2) \qquad ||x||_{L_p} = \begin{cases} \left\{ \int_\Omega |x(s)|^p \, d\mu(s) \right\}^{1/p} & \text{if } 1 \leq p < \infty, \\[2mm] \text{ess sup } \{|x(s)| : s \in \Omega\} & \text{if } p = \infty, \end{cases}
$$

which we used throughout in § 3. An important generalization of the space L_p is the *Orlicz space* L_M which in its simplest variant is defined as follows (see e.g. KRASNOSEL'SKIJ-RUTITSKIJ [1958]): Let $M : \mathbb{R} \to [0, \infty)$ be a *Young function*, i.e. a convex even function which is increasing on $[0, \infty)$ and satisfies

$$
\lim_{u \to 0} \frac{M(u)}{u} = 0, \qquad \lim_{u \to \infty} \frac{M(u)}{u} = \infty.
$$

A function $x \in \mathfrak{S}(\Omega)$ belongs to $L_M(\Omega)$ if the norm

$$
(4.3) \qquad ||x||_{L_M} := \inf \left\{ k : k > 0, \int_\Omega M[x(s)/k] \, d\mu(s) \leq 1 \right\}
$$

or, equivalently, the norm

$$
(4.4) \qquad |||x|||_{L_M} := \inf_{0 < k < \infty} \frac{1}{k} \left[1 + \int_\Omega M[kx(s)] \, d\mu(s) \right]
$$

is finite. (The equivalence of (4.3) and (4.4) follows from the two-sided estimate $||x||_{L_M} \leq |||x|||_{L_M} \leq 2||x||_{L_M}$.) Of course, the choice $M(u) := |u|^p$ $(1 < p < \infty)$ just gives the Lebesgue space L_p. From the viewpoint of applications, however, the Orlicz space L_M is more interesting if the generating function M is of non-polynomial growth (e.g. $M(u) = e^{|u|} - |u| - 1$).

There are two other types of ideal spaces which are particularly important in *interpolation theory* (see e.g. KREJN-PETUNIN-SEMENOV [1978]). Given $x \in \mathfrak{S}(\Omega)$, denote by $x^* : [0, \mu(\Omega)] \to [0, \infty)$ the *decreasing rearrangement* of x, i.e. the unique decreasing function which is equi-measurable with x. (Recall that two functions x and y are called *equi-measurable* if $\mu(\{s : |x(s)| > h\}) = \mu(\{s : |y(s)| > h\})$ for all

$h > 0$.) Given $\alpha \in (0,1)$, the *Lorentz space* $\Lambda_\alpha = \Lambda_\alpha(\Omega)$ consists of all functions $x \in \mathfrak{S}(\Omega)$ for which the norm

$$(4.5) \qquad \|x\|_{\Lambda_\alpha} := \alpha \int_0^\infty \frac{x^*(t)}{t^{1-\alpha}}\, dt$$

is finite, while the *Marcinkiewicz space* $M_\alpha = M_\alpha(\Omega)$ is defined by the norm

$$(4.6) \qquad \|x\|_{M_\alpha} := \sup_{0 < t < \infty} \frac{1}{t^{1-\alpha}} \int_0^t x^*(\tau)\, d\tau.$$

These spaces are closely related to the Lebesgue space L_p inasmuch as

$$(4.7) \qquad \Lambda_\alpha \subseteq L_{1/\alpha} \subseteq M_\alpha \subseteq \Lambda_{\alpha+\varepsilon} \qquad (\varepsilon > 0)$$

(continuous imbeddings). Observe that in all ideal spaces mentioned so far any two equi-measurable functions x and y have the same norm; ideal spaces with this property are called *symmetric spaces*. In particular, in a symmetric space X the norm $\|\chi_D\|_X$ of the characteristic function χ_D of a set $D \in \mathfrak{A}$ depends only on the measure (but not on the "position") of D. Therefore we can define the so-called *fundamental function* φ_X of a symmetric space X by

$$(4.8) \qquad \varphi_X(\delta) := \|\chi_D\|_X \qquad (D \in \mathfrak{A},\ \mu(D) = \delta).$$

For example, an easy computation shows that

$$\varphi_{L_p}(\delta) = \delta^{1/p}, \quad \varphi_{L_M}(\delta) = \frac{1}{M^{-1}(1/\delta)}, \quad \varphi_{\Lambda_\alpha}(\delta) = \varphi_{M_\alpha}(\delta) = \delta^\alpha;$$

here by φ_{L_M} we mean the fundamental function of L_M with respect to the norm (4.3). A typical example of an ideal space which is *not* symmetric may be obtained as follows. Let X be a fixed ideal space and $w : \Omega \to (0,\infty)$ a weight function. The *weighted ideal space* $X(w)$ is defined by the norm

$$(4.9) \qquad \|x\|_{X(w)} := \|wx\|_X.$$

Such spaces will be considered, for example, in Subsection 9.5 below.

The following Lemma 4.1 gives some kind of "substitute" for the fundamental function in an ideal space which is not symmetric:

Lemma 4.1. *In every ideal space X one can define a monotonically increasing function $\varphi_X : [0,\infty) \to [0,1]$ such that $\varphi_X(0) = 0, \varphi_X(\delta) > 0$ for $\delta > 0$, and*

$$||\chi_D||_X \geq \varphi_X(\mu(D)) \qquad (D \in \mathfrak{A}).$$

□ For $\delta \geq 0$ we define

$$\varphi_X(\delta) := \min\{1, \inf_{\mu(D)\geq\delta} ||\chi_D||_X\}.$$

It is clear that this function is increasing with $\varphi_X(0) = 0$. Assume that $\varphi_X(\delta) = 0$ for some $\delta > 0$. Then there exists a sequence of sets $E_1, E_2, \ldots \in \mathfrak{A}$ with $||\chi_{E_n}||_X \leq 2^{-n}$ and $\mu(E_n) \geq \delta$ $(n = 1, 2, \ldots)$. For $D_n := E_{n+1} \cup E_{n+2} \cup \ldots$ we have then $D_1 \supseteq D_2 \supseteq D_3 \supseteq \ldots \supseteq D_n$, $\mu(D_n) < \infty$, and $||\chi_{D_n}||_X < \infty$, i.e. $\chi_{D_n} \in X$. Denote the intersection of all sets D_n by D. We get then, on the one hand,

$$\mu(D) = \lim_{n\to\infty} \mu(D_n) \geq \delta,$$

and, on the other,

$$||\chi_D||_X \leq ||\chi_{D_n}||_X \leq \sum_{k=n+1}^{\infty} 2^{-k} = 2^{-n},$$

i.e. $\mu(D) = 0$. This contradiction shows that $\varphi_X(\delta) > 0$ for $\delta > 0$, and thus φ_X has all the required properties. ■

Lemma 4.1 implies, in particular, the important fact that *every ideal space X is continuously imbedded in the space $\mathfrak{S}$ with metric* (4.1). Indeed, if $(x_n)_n$ is a sequence converging in the norm of X to zero, for the set $M_n(h) := \{s : s \in \Omega, |x_n(s)| > h\}$ we get the estimate

$$||x_n||_X \geq ||h\chi_{M_n(h)}||_X \geq h\varphi_X(\mu(M_n(h))) \quad (h > 0),$$

and hence $\mu(M_n(h)) \to 0$, as $n \to \infty$, for any $h > 0$.

We need some more notions and results from the theory of ideal spaces (see e.g. LUXEMBURG-ZAANEN [1971], ZAANEN [1983], ZABREJKO [1974], or VÄTH [1997]). Recall that the *support* supp x of a measurable function $x : \Omega \to \mathbb{R}$ is the set of all $s \in \Omega$ such that $x(s) \neq 0$; the support is of course defined only up to nullsets. If N is any set of measurable functions we define the support of N by

$$\text{supp } N := \text{supp} [\sup \{\chi_{\text{supp} x} : x \in N\}].$$

Thus, every function $x \in N$ vanishes outside supp N, and for any set $D \subseteq \text{supp } N$ of positive measure one can find a subset $D_0 \subseteq D$, also of positive measure, and a function $x_0 \in N$ for which $D_0 \subseteq \text{supp } x_0$.

Given an ideal space X over Ω, a *unit in* X is, by definition, a non-negative function $u \in X$ such that supp $u = $ supp X. One can show (ZABREJKO [1974]) that *units exist in every ideal space*.

An element $x \in X$ is said to have an *absolutely continuous norm* if

$$\lim_{\mu(D)\to 0} ||P_D x||_X = 0,$$

where P_D is the multiplication operator (3.19). The set X^0 of all functions with an absolutely continuous norm in X is a closed (ideal) subspace of X; we call X^0 the *regular part* of X. The subspace X^0 has many remarkable properties. In particular, convergent sequences in X^0 admit an easy characterization: a sequence $(x_n)_n$ in X^0 converges to $x \in X$ (actually, $x \in X^0$) if and only if $(x_n)_n$ converges to x in measure and

$$\lim_{\mu(D)\to 0} \sup_n ||P_D x_n||_X = 0,$$

i.e. the elements x_n have *uniformly absolutely continuous norms*. Further, the space X^0 is separable if and only if the underlying space $\mathfrak{S}$ with metric (4.1) is separable.

Unfortunately, the subspace X^0 can be much smaller than the whole space X. An ideal space X is called *regular* if $X = X^0$, and *quasi-regular* if supp $X = $ supp X^0; the latter condition means that X^0 is dense in X with respect to convergence in measure.

Let us give some examples in the case of a bounded domain Ω with μ being the Lebesgue measure. The Lebesgue space L_p is regular for $1 \leq p < \infty$, but not for $p = \infty$; in fact, $L_\infty^0 = \{0\}$. More generally, the Orlicz space L_M with either the norm (4.3) or the norm (4.4) is regular if and only if the Young function M satisfies a Δ_2-condition (see KRASNOSEL'SKIJ-RUTITSKIJ [1958] or RAO-REN [1991]). The Lorentz space Λ_α with norm (4.5) is regular, the Marcinkiewicz space M_α with norm (4.6) is not; more precisely, the functions $x \in M_\alpha^0$ are characterized by the property

$$\lim_{t \to 0} \frac{1}{t^{1-\alpha}} \int_0^t x^*(\tau) \, d\tau = 0.$$

We still need some other notions. An ideal space is called *almost perfect* if, given a sequence $(x_n)_n$ in X which converges in measure to $x \in X$, one has

$$\|x\|_X \leq \varliminf_{n \to \infty} \|x_n\|_X,$$

i.e. the norm in X has the *Fatou property*. Moreover, X is called *perfect* if, given a sequence $(x_n)_n$ in X which converges in measure to $x \in \mathfrak{S}$ and satisfies

$$\varliminf_{n \to \infty} \|x_n\|_X < \infty$$

one has $x \in X$ and the above inequality holds. It is easy to see that a space is almost perfect if and only if its unit ball is a closed set in X, and perfect if and only if its unit ball is a closed set in $\mathfrak{S}$ (with respect to convergence in measure). Every regular space is almost perfect; the converse is false.

Let X be an ideal space over some set Ω with measure μ. By X' we denote the set of all functions $y \in \mathfrak{S}(\Omega)$, vanishing outside supp X, for which

$$\langle x, y \rangle := \int_\Omega x(s)y(s) \, d\mu(s) < \infty.$$

Equipped with the natural norm

$$\|y\|_{X'} := \sup \{\langle x, y \rangle : \|x\|_X \leq 1\},$$

the set X' becomes an ideal space which is always perfect; we call X' the *Köthe dual* (or *associate space*) to X. The associate space X' is a closed (possibly strict) subspace of the usual dual space X^*, and coincides with X^* if and only if X is regular.

For example, with Ω as above for $1 \leq p < \infty$ we have $L_p' = L_p^* = L_q$ with $p^{-1} + q^{-1} = 1$, while $L_\infty' = L_1 \neq L_\infty^*$ (see Subsection 3.1). More generally, the Käthe dual of the Orlicz space L_M with norm (4.3) (respectively with norm (4.4)) is just the Orlicz space L_N with norm (4.4) (respectively with norm (4.3)), generated by the *associate Young function*

$$N(v) := \sup \{|uv| - M(u) : 0 < u < \infty\}.$$

Finally, the Lorentz space Λ_α and the Marcinkiewicz space $M_{1-\alpha}$ are associate to each other for $0 < \alpha < 1$ (see e.g. KREJN-PETUNIN-SEMENOV [1978]).

4.2. Functions of two variables and vector functions

Let X be an ideal space over $\Omega = [a, b]$ with the Lebesgue measure μ, and J a bounded or unbounded interval. As in Subsection 1.1., by $C_t(X)$ we denote the space of all product-measurable functions of two variables $x : J \times [a, b] \to \mathbb{R}$ such that $t \mapsto x(t, \cdot)$ is continuous from J into X. Likewise, by $C_t^1(X)$ we denote the space of all $x \in C_t(X)$ such that $x(\cdot, s)$ is absolutely continuous on J for each $s \in [a, b]$ and $\partial x / \partial t$ belongs to $C_t(X)$ as well. For example, in case $X = L_p$ we get again the spaces $C_t(L_p)$ and $C_t^1(L_p)$ considered in Lemma 1.2. In particular, we have seen in Lemma 1.2 that the solutions $x \in C_t^1(L_p)$ of the Barbashin equation

$$(4.10) \qquad \frac{\partial x(t, s)}{\partial t} = c(t, s)x(t, s) + \int_a^b k(t, s, \sigma)x(t, \sigma)\, d\sigma + f(t, s)$$

give rise to C^1-solutions x of the differential equation

$$(4.11) \qquad \frac{dx}{dt} = [C(t) + K(t)]x + f(t)$$

in the Banach space $X = L_p$, and vice versa. We show now that an analogous fact is true in any ideal space X.

Let us denote, as usual, by $C(J, X)$ [respectively, $C^1(J, X)$] the set of all continuous [respectively, continuously differentiable] functions on J with values in the Banach space X.

Lemma 4.2. *For any ideal space X over $[a, b]$, the spaces $C(J, X)$ and $C_t(X)$ are isomorphic under the canonical map $\Phi : C(J, X) \to C_t(X)$ defined by*

$$(4.12) \qquad \Phi x(t, s) := x(t)(s).$$

$\square$ Fix $x \in C(J, X)$. We have to prove that there exists a product-measurable function $y : J \times [a, b] \to \mathbb{R}$ such that $\|x(t) - y(t, \cdot)\|_X = 0$ for all $t \in J$. Without loss of generality we may assume that $J = [0, T]$. Let $\{t_0, t_1, ..., t_n\}$ be an equidistant partition of $[0, T]$, and define

$$x_n(t, s) := \begin{cases} x(t_{k-1}, s) & \text{if } t_{k-1} \leq t < t_k \ (k = 1, ..., n), \\ x(T, s) & \text{if } t = T. \end{cases}$$

Then x_n is product-measurable and $\|x_n(t, \cdot) - x(t, \cdot)\|_X \to 0$, uniformly in $t \in [0, T]$, since $x \in C(J, X)$. In particular, we know that

$$(4.13) \qquad \|x_n(t, \cdot) - x_m(t, \cdot)\|_X \to 0 \quad (m, n \to \infty).$$

We claim that (4.13) implies the convergence of $x_n - x_m$ to zero in the product measure on $J \times [a, b]$. To see this, for $h > 0$ let

$$M_{n,m}(h) := \{(t, s) : t \in J, a \leq s \leq b, |x_n(t, s) - x_m(t, s)| > h\}.$$

Lemma 4.1 implies that, for any $h > 0$,

$$\|x_n(t, \cdot) - x_m(t, \cdot)\|_X \geq \|h\chi_{M_{n,m}(h)}\|_X$$

$$\geq h\varphi_X(\mu(\{s : a \leq s \leq b, (t, s) \in M_{n,m}(h)\}))$$

$$= h\varphi_X\left(\int_a^b \chi_{M_{n,m}(h)}(t, s)\, ds\right).$$

We conclude that, for any $\delta > 0$,

$$\mu\left(\left\{t : \varphi_X\left(\int_a^b \chi_{M_{n,m}(h)}(t, s)\, ds\right) \geq \varphi_X(\delta)\right\}\right) \to 0 \quad (m, n \to \infty),$$

hence also

$$\mu\left(\left\{t : \int_a^b \chi_{M_{n,m}(h)}(t, s)\, ds \geq \delta\right\}\right) \to 0 \quad (m, n \to \infty),$$

by the monotonicity of φ_X. From Lebesgue's dominated convergence theorem it follows that

$$\int_0^T \int_a^b \chi_{M_{n,m}(h)}(t, s)\, ds\, dt \to 0 \quad (m, n \to \infty),$$

and the Fubini-Tonelli theorem in turn implies that $\mu(M_{n,m}(h)) \to 0$ as $m, n \to \infty$.

Now, since $x_n - x_m$ tends to zero in measure, we find a subsequence $(x_{n_k})_k$ and a product-measurable function z such that $x_{n_k}(t, s) \to z(t, s)$ almost everywhere on $J \times [a, b]$. Moreover, we may choose a nullset $N \subset J$ with the property that, for each $t \in J \setminus N$, the sequence $(x_{n_k}(t, \cdot))_k$ converges almost everywhere on $[a, b]$ to $z(t, \cdot)$. Since $\|x_n(t, \cdot) - x(t, \cdot)\|_X \to 0$, from the remark after Lemma 4.1 we conclude that $x_n(t, \cdot) \to x(t, \cdot)$ in measure, and hence $z(t, \cdot) = x(t, \cdot)$ almost everywhere on $[a, b]$ for $t \in J \setminus N$. Thus, the function y defined by

$$y(t, s) := \begin{cases} z(t, s) & \text{if } t \in J \setminus N, \\ x(t, s) & \text{if } t \in N \end{cases}$$

has the required properties.

We have shown that (4.12) is in fact well-defined as a map from $C(J, X)$ into the space of (classes of) product-measurable functions on $J \times [a, b]$ with values in X. The continuous dependence of $\Phi x(t, s)$ on t is an obvious consequence of the definition of $C(J, X)$. Furthermore, Φ is clearly a bijection with inverse map

$$\Phi^{-1} x(t)(s) := x(t, s),$$

and so we are done. ■

We point out that the nontrivial part in the proof of Lemma 4.2 is the fact that the element Φx is, up to equivalence, *product-measurable* on $J \times [a, b]$. As a matter of fact, it is *not* true that each $x : J \times [a, b] \to \mathbb{R}$ such that $t \mapsto x(t, \cdot)$ is continuous from J into L_1, say, is product-measurable: for instance, following Sierpinski's classical counterexample (see e.g. RUDIN [1973]) one may construct a non-measurable subset $Q \subset [0, 1] \times [0, 1]$ such that each section $Q_s = \{t : (t, s) \in Q\}$ is at most countable, and then take $x = \chi_Q$.

We state now a parallel result for the spaces $C^1(J, X)$ and $C_t^1(X)$. Following HILLE-PHILLIPS [1957] we call a Banach space $X \subset \mathfrak{S}(\Omega)$ a *space of type L* if the map (4.12) is an isomorphism between $C(J, X)$ and $C_t(X)$ and, in addition, if

$$(4.14) \qquad \left(\int_J x(t) \, dt \right)(s) = \int_J \Phi x(t, s) \, dt \qquad (a \le s \le b)$$

for any $x \in C(J, X)$; here the integral at the left-hand side of (4.14) is the Bochner integral of the X-valued function x, while the integral at the right-hand side is the (parameter dependent) Lebesgue integral of the real function Φx. In the same way as was shown on pp. 69/70 of HILLE-PHILLIPS [1957] for the space $X = L_p$, one can prove that *every ideal space is of type L*. From this we may deduce, in particular, the following useful

Lemma 4.3. *For any ideal space X, the spaces $C^1(J, X)$ and $C_t^1(X)$ are isomorphic under the canonical map* (4.12).

□ The fact that $\Phi(C^1(J, X)) \subseteq C_t^1(X)$ is essentially proved in Theorem 3.4.2 of HILLE-PHILLIPS [1957]. It remains to show that, given $x \in C_t^1(X)$, the function $\Phi^{-1} x : J \to X$ (which exists and belongs to $C(J, X)$, by Lemma 4.2) is actually continuously differentiable with derivative $\partial x / \partial t$. But for fixed $t \in J$ we have, for some τ between t

and $t + h$,

$$\left\| \Phi^{-1}x(t+h) - \Phi^{-1}x(t) - h\frac{\partial x(t,\cdot)}{\partial t} \right\|_X$$

$$= \left\| \int_t^{t+h} \left[\frac{\partial x(\tau,\cdot)}{\partial t} - \frac{\partial x(t,\cdot)}{\partial t} \right] d\tau \right\|_X$$

$$\leq | \int_t^{t+h} \left\| \frac{\partial x(\tau,\cdot)}{\partial t} - \frac{\partial x(t,\cdot)}{\partial t} \right\|_X d\tau | = o(h) \qquad (h \to 0),$$

since $\partial x / \partial t \in C_t(X)$. ∎

4.3. Barbashin operators in ideal spaces

Now we discuss some basic properties of bounded linear operators of the form

$$(4.15) \qquad Ax(s) = c(s)x(s) + \int_a^b k(s,\sigma)x(\sigma)\, d\sigma$$

or, more generally, operator functions of the form

$$(4.16) \qquad A(t)x(s) = c(t,s)x(s) + \int_a^b k(t,s,\sigma)x(\sigma)\, d\sigma$$

in an ideal space X over $\Omega = [a,b]$. In the special case $X = L_p([a,b])$ we obtain of course the main results of § 3.

The operator A in (4.15) and the operator function $A(t)$ in (4.16) may be represented again as sums $A = C + K$ and $A(t) = C(t) + K(t)$, respectively, where

$$(4.17) \qquad Cx(s) = c(s)x(s),$$

$$(4.18) \qquad Kx(s) = \int_a^b k(s,\sigma)x(\sigma)\, d\sigma,$$

$$(4.19) \qquad C(t)x(s) = c(t,s)x(s),$$

and

$$(4.20) \qquad K(t)x(s) = \int_a^b k(t, s, \sigma)x(\sigma)\, d\sigma.$$

Precisely as in Lemma 3.1 one can show that, *whenever the linear operator* (4.15) *acts in an ideal space, this operator is continuous.*

As in § 3 we denote by $\mathfrak{L}_b(X)$ the class of all bounded linear operators of type (4.15) in X, and by $\mathfrak{L}_b^r(X)$ the subclass of all *regular* operators $A \in \mathfrak{L}_b(X)$ (which means that A may be written as difference of two positive linear operators in X). The *module* $|A|$ of a regular operator A is defined as in Subsection 3.1.

Lemma 4.4. *Suppose that the linear operator* (4.15) *acts in an ideal space X. Then this operator is regular if and only if the linear operator*

$$(4.21) \qquad]A[\, x(s) = |c(s)|x(s) + \int_a^b |k(s, \sigma)|x(\sigma)\, d\sigma$$

acts in X as well; moreover, the equality $|A| =]A[$ holds in this case.

□ The proof of this lemma repeats literally the proof of Lemma 3.2. ∎

Let us now pass to the non-stationary Barbashin operator (4.16). It is clear that the family of multiplication operators (4.19) acts in any ideal space if $c \in L_\infty([0, T] \times [a, b])$. The converse is also true:

Lemma 4.5. *Let X be an ideal space over $[a, b]$ and $c \in C_t(X)$. Then* (4.19) *defines a strongly continuous operator family in X if and only if the function c is essentially bounded on $[0, T] \times [a, b]$.*

□ Suppose that (4.19) defines a strongly continuous operator family in X. By the uniform boundedness principle we have $\|C(t)\|_{\mathfrak{L}(X)} \le M < \infty$ for all $t \in [0, T]$. We claim that then also $\|c\|_{L_\infty} \le M$. In fact, if this were false, the set

$$D_\varepsilon(t) := \{s : a \le s \le b, |c(t, s)| \ge M + \varepsilon\}$$

would have positive measure for some $t \in [0, T]$ and some $\varepsilon > 0$. Given any unit u in X, for the function $x := P_{D_\varepsilon(t)} u$ we get then

$$\|C(t)x\|_X \ge \|(M + \varepsilon)x\|_X > M\|x\|_X \ge \|C(t)\|_{\mathfrak{L}(X)}\|x\|_X,$$

a contradiction. ∎

Together with the family of operators (4.20), consider the operator

$$(4.22) \qquad \hat{K}x(t, s) = \int_a^b k(t, s, \sigma)x(t, \sigma)\, d\sigma$$

which acts in *spaces of functions of two variables*. The following lemma shows that there is a close relation between the operators (4.20) and (4.22); we shall return to this relation in § 10 below.

Lemma 4.6. *Let X be an ideal space over $[a, b]$. Then (4.20) defines a strongly continuous operator function in X if and only if the operator (4.22) maps the space $C_t(X)$ into itself.*

□ Suppose first that $K(t)$ is strongly continuous in X. For any $x \in C_t(X)$ we have then

$$\|\hat{K}x(t, \cdot) - \hat{K}x(t_0, \cdot)\|_X$$

$$\leq \left\|\int_a^b k(t, \cdot, \sigma)[x(t, \sigma) - x(t_0, \sigma)]\, d\sigma\right\|_X$$

$$+ \left\|\int_a^b [k(t, \cdot, \sigma) - k(t_0, \cdot, \sigma)]x(t_0, \sigma)\, d\sigma\right\|_X$$

$$\leq \|K(t)\|_{\mathcal{L}(X)}\|x(t, \cdot) - x(t_0, \cdot)\|_X$$

$$+ \|[K(t) - K(t_0)]x(t_0, \cdot)\|_X \to 0$$

as $t \to t_0$, by the uniform boundedness principle. Conversely, from $\hat{K}(C_t(X)) \subseteq C_t(X)$ it follows that $\|\hat{K}x\|_{C_t(X)} \leq \|\hat{K}\|\, \|x\|_{C_t(X)}$ and

$$\|\hat{K}x(t, \cdot) - \hat{K}x(t_0, \cdot)\|_X \to 0 \qquad (t \to t_0).$$

For any fixed $z \in X$ we have then for the function $x(t, s) := z(s)$

$$\|K(t)z\|_X = \|\hat{K}x(t, \cdot)\|_X \leq \|\hat{K}\|\, \|z\|_X$$

and

$$\|K(t)z - K(t_0)z\|_X = \|\hat{K}x(t, \cdot) - \hat{K}x(t_0, \cdot)\|_X \to 0$$

as $t \to t_0$. This shows that $K(t)$ is strongly continuous in X. ∎

4.4. The space $\mathcal{L}_b^r(X)$

In analogy what we have done in Subsection 3.2 by introducing the class $\mathcal{L}_b^r(X)$, we could try to develop a parallel theory on linear operators of type (4.15) or (4.16) in arbitrary ideal spaces X. So we denote by $\mathcal{L}_b^r(X)$ the class of all stationary regular Barbashin operators (i.e. operators of the form (4.15)) which act in the space X; of course, this is a subspace of the space $\mathcal{L}(X)$ of all bounded linear operators in X.

Since the results which follow (and their proofs) are completely analogous to those for the case $X = L_p$, we will state the corresponding lemmas without proofs. We will point out only the differences (if there are any).

As before, we equip $\mathcal{L}_b^r(X)$ with the norm

$$(4.23) \qquad \|A\|_{\mathcal{L}_b^r(X)} = \| \, |A| \, \|_{\mathcal{L}(X)},$$

where $|A|$ denotes the module of A (see Lemma 3.2).

Lemma 4.7. *The subspace $\mathcal{L}_b^r(X)$ with norm (3.10) is a Banach space which is continuously imbedded in $\mathcal{L}(X)$.*

Lemma 4.8. *The subspace $\mathcal{L}_b^r(X)$ is a subalgebra of $\mathcal{L}(X)$, and the subspace $\mathcal{L}_i(X)$ of all integral operators in X is an ideal in the algebra $\mathcal{L}_b^r(X)$. More precisely, if $A, B \in \mathcal{L}_b^r(X)$ are given by*

$$Ax(s) = c(s)x(s) + \int_a^b k(s, \sigma)x(\sigma)\, d\sigma$$

and

$$Bx(s) = d(s)x(s) + \int_a^b l(s, \sigma)x(\sigma)\, d\sigma$$

then $AB \in \mathcal{L}_b^r(X)$ is given by

$$(4.24) \qquad ABx(s) = e(s)x(s) + \int_a^b m(s, \sigma)x(\sigma)\, d\sigma,$$

where

$$e(s) := c(s)d(s)$$

and

$$m(s, \sigma) := c(s)l(s, \sigma) + k(s, \sigma)\,d(\sigma) + \int_a^b k(s, \xi)l(\xi, \sigma)\,d\xi.$$

4.5. Sufficient conditions for regularity

In this subsection we repeat the sufficient conditions for the operator (4.15) to belong to the class $\mathfrak{L}_b^r(X)$. Such conditions may be obtained, loosely speaking, from sufficient conditions for the integral operator (4.18) to be regular in the space X (ZABREJKO [1966, 1968]). We formulate such conditions again through imbedding theorems for classes of kernels on $[a, b] \times [a, b]$. Let us denote by $\mathfrak{Z}(X)$ the class of all measurable functions $z = z(s, \sigma)$ on $[a, b] \times [a, b]$ for which the norm

$$(4.25) \qquad ||z||_{\mathfrak{Z}(X)} = \sup_{\substack{||x||_X \leq 1 \\ ||y||_{X'} \leq 1}} \int_a^b \int_a^b |z(s, \sigma)x(\sigma)y(s)|\,d\sigma\,ds$$

is finite, where X' is the Köthe dual (see Subsection 4.1) of X. Of course, in case $X = L_p$ this gives the kernel class $\mathfrak{Z}_p$ introduced in Subsection 3.3.

Lemma 4.9. *The linear operator*

$$(4.26) \qquad Ax(s) = c(s)x(s) + \int_a^b k(s, \sigma)x(\sigma)\,d\sigma$$

belongs to the class $\mathfrak{L}_b^r(X)$ if and only if $c \in L_\infty$ and $k \in \mathfrak{Z}(X)$. Moreover, in this case the two-sided estimate

$$(4.27) \qquad \frac{1}{2}(||c||_{L_\infty} + ||k||_{\mathfrak{Z}(X)}) \leq ||A||_{\mathfrak{L}_b^r(X)} \leq ||c||_{L_\infty} + ||k||_{\mathfrak{Z}(X)}$$

holds.

We can reformulate this result as follows. If we denote, in analogy to Subsection 3.3, by $\mathfrak{L}_m(X)$ the class of all multiplication operators

(4.17) in X, and by $\mathfrak{L}_i^r(X)$ the class of all regular integral operators (4.18) in X, then the decomposition

$$\mathfrak{L}_b^r(X) = \mathfrak{L}_m(X) \oplus \mathfrak{L}_i^r(X)$$

holds.

Given a measurable set $D \subseteq [a, b]$, let P_D denote the multiplication operator (3.19). By $3^-(X)$ we denote the space of all $z \in 3(X)$ such that

$$(4.28) \qquad \lim_{\mu(D) \to 0} \sup_{\substack{\|x\|_X \leq 1 \\ \|y\|_{X'} \leq 1}} \int_a^b \int_a^b |z(s, \sigma) P_D x(\sigma) y(s)| \, d\sigma \, ds = 0;$$

similarly, by $3^+(X)$ we denote the space of all $z \in 3(X)$ such that

$$(4.29) \qquad \lim_{\mu(D) \to 0} \sup_{\substack{\|x\|_X \leq 1 \\ \|y\|_{X'} \leq 1}} \int_a^b \int_a^b |z(s, \sigma) x(\sigma) P_D y(s)| \, d\sigma \, ds = 0;$$

finally, let

$$3^0(X) := 3^-(X) \cap 3^+(X).$$

As already mentioned, the kernels $k \in 3(X)$ are precisely those which generate integral operators $K \in \mathfrak{L}_i^r(X)$. Let us denote by $\mathfrak{L}_i^{r,-}(X), \mathfrak{L}_i^{r,+}(X)$, and $\mathfrak{L}_i^{r,0}(X)$ the class of all integral operators $K \in \mathfrak{L}_i^r(X)$ which are generated by kernels $k \in 3^-(X), 3^+(X)$, and 3_p^0, respectively. The following is then parallel to Lemma 3.7:

Lemma 4.10. *The subspaces $\mathfrak{L}_i^{r,-}(X), \mathfrak{L}_i^{r,+}(X)$, and $\mathfrak{L}_i^{r,0}(X)$ are ideals in the algebra $\mathfrak{L}_b^r(X)$.*

One may also give a refined description of the kernel classes $3(X)$, $3^-(X)$, and $3_+(X)$ by means of ideal spaces of kernels with mixed norm. Since such spaces will be systematically studied later (Subsection 12.1), we shall not consider them here.

4.6. Representation of the evolution operator

Let us now briefly study the non-stationary Barbashin equation (4.10) in an ideal space X. As in the case of the spaces C (see Subsection 2.6) and L_p (see Subsection 3.6) it is natural to expect that the evolution operator $U = U(t, \tau)$ for the equation

$$(4.30) \qquad \frac{dx}{dt} = A(t)x,$$

with $A(t)$ given by (4.16), is again representable as sum of the evolution operator $U_0 = U_0(t, \tau)$ of the "reduced" equation

$$(4.31) \qquad \frac{dx}{dt} = C(t)x,$$

with $C(t)$ given by (4.19), and some integral operator

$$(4.32) \qquad H(t, \tau)x(s) = \int_a^b h(t, \tau, s, \sigma)x(\sigma) \, d\sigma.$$

This is in fact true if the operator function (4.16) is strongly continuous in X, i.e. the map $t \mapsto A(t)x$ is continuous from J into X:

Theorem 4.1. *Suppose that the operator function (4.16) is strongly continuous in $\mathfrak{L}_b^r(L_p)$. Then the evolution operator $U = U(t, \tau)$ for the differential equation (4.30) belongs to $\mathfrak{L}_b^r(X)$ for each $t, \tau \in J$, i.e. admits a representation*

$$(4.33) \qquad U(t, \tau)x(s) = e(t, \tau, s)x(s) + H(t, \tau)x(s),$$

where $e(t, \tau, s)$ is given by (1.29), and H is defined by (3.46) with $h = h(t, \tau, s, \sigma)$ being a measurable function on $J \times J \times [a, b] \times [a, b]$. Moreover, $H(t, \tau) \in \mathfrak{L}_i^{r,-}(X)$ if $K(t) \in \mathfrak{L}_i^{r,-}(X)$, $H(t, \tau) \in \mathfrak{L}_i^{r,+}(X)$ if $K(t) \in \mathfrak{L}_i^{r,+}(X)$, and $H(t, \tau) \in \mathfrak{L}_i^{r,0}(X)$ if $K(t) \in \mathfrak{L}_i^{r,0}(X)$.

4.7. Some spectral theory for Barbashin operators

Spectral properties constitute an extremely important part of the theory of bounded linear operators in Banach spaces. Since we have dealt throughout with operators of the form $A = C + K$ in ideal spaces, where C is the multiplication operator (4.17), and K is a compact operator (typically, the integral operator (4.18)), we shall concentrate now on spectral properties of such operators A. All results of this subsection are taken from BIBERDORF-VÄTH [1999].

We will assume throughout that X is an ideal space over $[a, b]$ with full support. As in Subsection 1.1, we identify the stationary Barbashin equation

$$(4.34) \qquad \frac{\partial x(t, s)}{\partial t} = c(s)x(t, s) + \int_a^b k(s, \sigma)x(t, \sigma)\, d\sigma$$

with the differential equation

$$(4.35) \qquad \frac{dx}{dt} = Ax$$

in X; here A is of course the sum of the two operators (4.17) and (4.18).

We will assume that $c \in L_\infty([a, b])$, and that $K : X \to X$ is compact; for our results we do not need that K is an integral operator. Let us call

$$(4.36) \qquad \operatorname{ess} c(a, b) := \{u : \operatorname*{ess\,inf}_{s \in [a,b]} |c(s) - u| = 0\}$$

the *essential range* of the function c.

Recall that a point λ in the complex plane is a *Fredholm point* of an operator $A \in \mathcal{L}(X)$, if $A - \lambda I$ has closed range, the codimension m of the range of $A - \lambda I$ is finite, and the dimension n of the null space of $A - \lambda I$ is finite. If, additionally, $m = n$, we call λ a *Fredholm point of index* 0 (see also Subsection 14.1 below).

Let $\sigma_{ew}(A)$ be the *essential spectrum* of A in the sense of WOLF [1959], i.e. the complement of the set of all Fredholm points, and $\sigma_{es}(A)$ be the essential spectrum of A in the sense of SCHECHTER [1965], i.e. the complement of all Fredholm points of index 0. It is known (SCHECHTER [1965], see also KATO [1966], KREJN [1971], or AKHMEROV-KAMENSKIJ-POTAPOV-RODKINA-SADOVSKIJ [1986]) that both sets are invariant under compact perturbations.

Theorem 4.2. *The equality*

$$(4.37) \qquad \operatorname{ess} c(a, b) = \sigma_{ew}(A) = \sigma_{es}(A)$$

holds.

☐ Since $\sigma_{ew}(A) = \sigma_{ew}(A - K)$ and $\sigma_{es}(A) = \sigma_{es}(A - K)$, by the compactness of K, it is sufficient to prove the statement for $K = 0$. Let first $\lambda \notin \text{ess } c(a, b)$. Then $\text{ess inf } |c(s) - \lambda| = \delta > 0$ implies that

$$\underset{s \in [a,b]}{\text{ess sup}} |c(s) - \lambda|^{-1} = \frac{1}{\delta} < \infty.$$

Thus,

$$(4.38) \qquad (C - \lambda)^{-1}x(s) = \frac{x(s)}{c(s) - \lambda}$$

defines a bounded operator, i.e. $C - \lambda$ is an isomorphism, and hence Fredholm of index 0. Conversely, assume that $\lambda \in \text{ess } c(a, b)$ is a Fredholm point of C. The nullspace of $C - \lambda$ contains all functions of X vanishing outside $E = \{s : c(s) = \lambda\}$. Thus, if E has positive measure, the nullspace has infinite dimension, a contradiction. But if E is a nullset, $C - \lambda$ is 1-1. Since, by assumption, the range Y of $C - \lambda$ is closed, the inverse operator (4.38) is bounded as a mapping between Y and X, by the open mapping theorem. By assumption, the set

$$D := \{s : (c(s) - \lambda)^{-1} > \|(C - \lambda)^{-1}\| + 1\}$$

has positive measure. Choose a function $x \in X$, $x \neq 0$, vanishing outside D. Then $y = (C - \lambda)x \in Y$ and $y \neq 0$; but since y vanishes outside D, we have

$$\|(C - \lambda)^{-1}y\|_X \geq (\|C - \lambda\|^{-1} + 1)\|y\|_X,$$

a contradiction.

We have shown that $\sigma_{es}(A) \subseteq \text{ess } c(a, b) \subseteq \sigma_{ew}(A)$. The statement now follows from the inclusion $\sigma_{ew}(A) \subseteq \sigma_{es}(A)$ which is always true. ■

We remark that Theorem 4.2 also holds true for the space $X = C([a, b])$ (if A maps X into itself, of course); to see this, just replace in the proof "E has positive measure" throughout by "E has interior points", and observe that the function c is continuous, so D is open.

Theorem 4.2 has important consequences for the exponential stability of the Barbashin equation (4.36); we shall return to this later (Subsection 5.3).

Let us now turn to the full spectrum of the Barbashin operator $A = C + K$. If K is an integral operator with a degenerate kernel, we may reduce the problem to a finite-dimensional system:

Example 4.1. Suppose that the kernel k in (4.18) has the form

$$(4.39) \qquad k(s, \sigma) = \sum_{i=1}^{n} a_i(s) b_i(\sigma),$$

where the functions a_i belong to X and the functions b_i to the associate space X' (see Subsection 4.1). To determine $\sigma(A)$ it suffices to calculate the eigenvalues $\lambda \notin \operatorname{ess} c(a, b)$. For such λ there exists $x \neq 0$ with

$$c(s)x(s) + \sum_{i=1}^{n} a_i(s) \int_a^b b_i(\sigma)x(\sigma)\, d\sigma = \lambda x(s).$$

Putting

$$(4.40) \qquad \beta_i = \int_a^b b_i(\sigma)x(\sigma)\, d\sigma \qquad (i = 1, \dots, n)$$

we find by $\operatorname*{ess\,inf}_{s \in [a,b]} |\lambda - c(s)| > 0$ that

$$(4.41) \qquad x(s) = \sum_{j=1}^{n} \frac{a_j(s)}{\lambda - c(s)} \beta_j.$$

The equations (4.40) now become

$$(4.42) \qquad \Gamma(\lambda)\beta = \beta,$$

where β is the column vector consisting of β_i $(i = 1, \dots, n)$, and $\Gamma(\lambda)$ is the matrix consisting of $\gamma_{ij}(\lambda)$ $(i, j = 1, \dots, n)$ with

$$\gamma_{ij}(\lambda) := \int_a^b \frac{b_i(\sigma)a_j(\sigma)}{\lambda - c(\sigma)}\, d\sigma.$$

Equation (4.42) implies $\beta \neq 0$, since $x \neq 0$. Thus, a necessary condition for $\lambda \notin \mathrm{ess}\, c(a, b)$ to be an eigenvalue is

$$(4.43) \qquad\qquad \det(\Gamma(\lambda) - I) = 0.$$

But (4.43) is also sufficient, since in this case there exists a nontrivial solution $\beta = (\beta_1, \ldots, \beta_n)$ of (4.42), and a straightforward calculation shows that (4.41) is an eigenfunction of A for λ. In other words, we have proved that

$$(4.44) \qquad \sigma(A) = \mathrm{ess}\, c(a, b) \cup \{\lambda : \det(\Gamma(\lambda) - I) = 0\}.$$

In the special case $n = 1$, i.e. $k(s, \sigma) = a(s)b(\sigma)$, the equality (4.44) simply reduces to

$$(4.45) \qquad \sigma(A) = \mathrm{ess}\, c(a, b) \cup \{\lambda : \int_a^b \frac{a(\sigma)b(\sigma)}{\lambda - c(\sigma)}\, d\sigma = 1\}.$$

Example 4.1 may be used to estimate numerically the borders of $\sigma(A)$: any compact set R of resolvent points of A still belongs to the resolvent set under small (depending on R) perturbations of the operator (see e.g. Theorem 3.1 in KATO [1966]). In particular, for estimates it suffices to approximate $A = C + K$ by simpler operators (in the operator norm).

If K may be approximated by finite rank operators, and if the ideal space X is regular (see Subsection 4.1), the kernels of the approximating operators have the form (4.39). Indeed, a finite rank operator K_0 may be written as

$$K_0 x = \sum_{i=1}^{n} a_i l_i(x)$$

with $a_i \in X$, and continuous linear functionals l_i ($i = 1, \ldots, n$). By the regularity of X, any continuous linear functional on X has the form

$$l_i(x) = \langle b_i, x \rangle = \int_a^b b_i(\sigma)x(\sigma)\, d\sigma$$

with $b_i \in X'$ (see e.g. ZAANEN [1967] or ZABREJKO [1974]). Thus, for the spectrum of the approximating operators one may use Example 4.1.

In the next theorem we give an estimate for the spectrum of a Barbashin operator.

Theorem 4.3. *The inclusion*

$$(4.46) \qquad \sigma(A) \subseteq \text{ess } c(a, b) + \{\lambda : |\lambda| \le ||K||\}$$

holds.

□ Assume that the statement is false. Then there is some $\lambda \in \sigma(A) \setminus$ ess $c(a, b)$ and some $\delta > 0$ with

$$|\lambda - c(s)| \ge ||K|| + \delta$$

for almost all s. Since λ belongs to the point spectrum of A, there is some $x \in X$, $x \ne 0$, with

$$c(s)x(s) + Kx(s) = \lambda x(s)$$

for almost all s. In particular, we have

$$|Kx(s)| = |[\lambda - c(s)]x(s)| \ge (||K|| + \delta)|x(s)|.$$

But this implies that

$$||K|| \, ||x||_X \ge ||Kx||_X \ge (||K|| + \delta)||x||_X,$$

contradicting $x \ne 0$. ∎

Again, Theorem 4.3 also holds true for the space $X = C([a, b])$. Theorem 4.3 shows that, if K is "small", the spectrum of A is mainly given by c. For some important cases we may also give sharper estimates:

Example 4.2. If X is a Hilbert space and C and K are self-adjoint, then

$$\sigma(A) \subseteq [\text{ess inf } c(s) + \min \sigma(K), \text{ess sup } c(s) + \max \sigma(K)].$$

This follows from the equalities $\langle Ax, x \rangle = \langle Cx, x \rangle + \langle Kx, x \rangle$ and $\sigma(C) = \text{ess } c(a, b)$.

Example 4.3. If K is a Volterra integral operator, i.e.

$$Kx(s) = \int_a^s k(s,\sigma)x(\sigma)\,d\sigma,$$

and X is regular (see Subsection 4.1), or if $X = L_\infty([a,b])$ or $X = C([a,b])$, then $\sigma(A) = \text{ess}\,c(a,b)$. In fact, assume that there exists some $\lambda \in \sigma(A) \setminus \text{ess}\,c(a,b)$. Then λ is an eigenvalue of A, i.e. there is some $x_\lambda \neq 0$ satisfying

(4.47) $$0 = (A - \lambda)x_\lambda(s) = [c(s) - \lambda]x_\lambda(s) + Kx_\lambda(s)$$

almost everywhere on $[a,b]$. Since $\text{ess}\inf_{s\in[a,b]} |c(s) - \lambda| > 0$ we see that the linear operator (4.38) is bounded. Hence, $(C - \lambda)^{-1}K$ again is a compact Volterra operator. Thus it has spectral radius 0 (see ZA-BREJKO [1967, 1967a]). In particular, -1 can not be an eigenvalue of $(C - \lambda)^{-1}K$, i.e. x_λ can not satisfy (4.47).

By the last examples one might suspect that it is always possible to replace $\|K\|$ by the *spectral radius* $r_\sigma(K)$ of K in (4.46), i.e.

(4.48) $$\sigma(A) \subseteq \text{ess}\,c(a,b) + \{\lambda : |\lambda| \le r_\sigma(K)\}.$$

However, this is not true in general. It is not even true that the maximal real part of $\sigma(A)$ is bounded by $\|c\|_{L_\infty} + r_\sigma(K)$ (as in the Hilbert space case for selfadjoint operators). We give a counterexample for a class of integral operators with degenerate kernels:

Example 4.4. Consider the degenerate integral the operator

$$Kx(s) = \int_{-1}^1 a(s)b(\sigma)x(\sigma)\,d\sigma.$$

Example 4.1 shows, for $c \equiv 0$, that the spectrum of this operator consists of the points 0 and the point

$$\mu := \int_{-1}^1 a(\sigma)b(\sigma)\,d\sigma.$$

We now consider the special L_∞-function

$$c(s) := \begin{cases} -1 & \text{if } s < 0, \\ 0 & \text{if } s \geq 0, \end{cases}$$

i.e. ess $c(a,b) = \{-1, 0\}$. Example 4.1 shows that $\lambda \neq -1, 0$ belongs to $\sigma(A)$ if and only if

$$(4.49) \qquad \frac{\alpha}{\lambda+1} + \frac{\beta}{\lambda} = 1,$$

where we have put

$$\alpha := \int_{-1}^{0} a(\sigma)b(\sigma)\, d\sigma, \qquad \beta := \int_{0}^{1} a(\sigma)b(\sigma)\, d\sigma.$$

Given any number $\lambda \geq \delta > 0$, choose functions a and b, such that

$$\alpha = -\delta(\lambda+1), \qquad \beta = (\delta+1)\lambda.$$

Then (4.49) holds, i.e. λ is an eigenvalue of the corresponding operator A. However, since $\mu = \alpha + \beta$, the spectrum of the integral operator is just $\{0, \alpha+\beta\} = \{0, \lambda-\delta\}$. In particular, if $r_\sigma(K)$ denotes the spectral radius of the integral operator K, the number $\lambda = ||c||_{L_\infty} + r_\sigma(K) + \delta$ belongs to the spectrum of A. This means that the difference of the maximal real value in $\sigma(A)$ and $||c||_{L_\infty} + r_\sigma(K)$ is positive and may even be arbitrarily large.

There is an interesting relation of the previous results with the theory of measures of noncompactness (see e.g. SADOVSKIJ [1972] or AKHMEROV-KAMENSKIJ-POTAPOV-RODKINA-SADOVSKIJ [1986]). Given a bounded set M in a Banach space X, the *measure of noncompactness* of M is defined by

$$(4.50) \qquad \gamma(M) := \inf\{\varepsilon : \varepsilon > 0, M \text{ has a finite } \varepsilon\text{-net}\},$$

where by a finite ε-net we mean, as usual, a covering by finitely many balls of radius $\varepsilon > 0$. Similarly, the measure of noncompactness of an operator $A \in \mathfrak{L}(X)$ is defined by

$$(4.51) \qquad \begin{aligned} \gamma(A) &:= \inf\{k : k > 0, \gamma(A(M)) \leq k\gamma(M) \\ &\quad \text{for all bounded } M \subseteq X\}. \end{aligned}$$

The name "measure of noncompactness" is of course motivated by the fact that $\gamma(M) = 0$ if and only if M has compact closure, and $\gamma(A) = 0$ if and only if A is a compact operator. Obviously, the measure of noncompactness of a bounded linear operator may be estimated from above by its norm. Here is an example of an operator, where this estimate is sharp:

Lemma 4.11. *The multiplication operator* (4.17) *satisfies* $\gamma(C) = ||C|| = ||c||_{L_\infty}$.

$\square$ Since the equality $||C|| = ||c||_{L_\infty}$ and the estimate $\gamma(C) \leq ||C||$ are clear, it suffices to show that $\gamma(C) \geq ||c||_{L_\infty}$. Let $0 < \delta < ||c||_{L_\infty}$ and put $D = \{s : |c(s)| \geq ||c||_{L_\infty} - \delta\}$. Denote by Y the restriction of X to functions vanishing outside D. Since Y still has infinite dimension, its unit ball M satisfies $\gamma(M) > 0$. Moreover, if $\{z_1, \ldots, z_m\}$ is a finite ε-net for $C(M)$, then the set $\{P_D(z_1/c), \ldots, P_D(z_m/c)\}$ (P_D as in (3.19)) gives a finite η-net of M with $\eta = \varepsilon/(||c||_{L_\infty} - \delta)$. This shows that

$$\gamma(M) \leq \frac{\gamma(C(M))}{||c||_{L_\infty} - \delta}.$$

We conclude that $\gamma(C) \geq ||c||_{L_\infty}$ as claimed. ∎

Lemma 4.11 shows that, loosely speaking, multiplication operators are as "noncompact" as possible. In fact, there is no compact operator K such that the distance $||C - K||$ of C to K is strictly smaller than the distance $||C||$ of C to the zero operator.

Given a bounded linear operator A, the number

$$r_{ew}(A) = \sup\{|\lambda| : \lambda \in \sigma_{ew}(A)\}$$

is called the *radius of the essential spectrum* of A ; this number satisfies the important relation

(4.52) $$r_{ew}(A) = \lim_{n \to \infty} \sqrt[n]{\gamma(A^n)}$$

(see e.g. Theorem 2.6.11 of AKHMEROV-KAMENSKIJ-POTAPOV-ROD-KINA-SADOVSKIJ [1986]) which is of course analogous to the well-known Gel'fand formula for the radius of the whole spectrum. In fact, we have $\gamma(A^n) = ||c||_{L_\infty}^n$, hence $r_{ew}(A) = ||c||_{L_\infty}$ as in Theorem 4.2. To see this, observe that $A^n = C^n + K_n$ with some compact operator K_n, and thus $\gamma(A^n) = \gamma(C^n)$. The assertion follows now from Lemma 4.11.

Chapter II

Theory of Linear

Barbashin Equations

§ 5. Stability of solutions

In this section we study various stability properties of the differential equation

$$(5.1) \qquad \frac{dx}{dt} = A(t)x$$

containing a Barbashin operator $A(t) = C(t)+K(t)$ in a Banach space X. The most complete information may be obtained in two cases, namely either if the Barbashin operator is stationary (i.e. $A(t) \equiv A$), or if $K(t) \equiv 0$. These two cases, however, are rather trivial. The general case may be treated by considering $K(t)$ as a "perturbation" of $C(t)$. Finally, we make some remarks on the existence and properties of Ljapunov functions associated with equation (5.1); details may be found in DIALLO [1989], DIALLO-ZABREJKO [1990], and ERMOLOVA [1995], see also KHALILOV [1956, 1961].

5.1. Ljapunov and Ljapunov-Bohl exponents

In order to study stability properties of the solutions of the Barbashin equation

$$(5.2) \qquad \frac{\partial x(t,s)}{\partial t} = c(t,s)x(t,s) + \int_a^b k(t,s,\sigma)x(t,\sigma)\,d\sigma$$

we first recall some general notions and results from general stability theory for the differential equation (5.1) which is equivalent to (5.2) in many function spaces. To this end, we suppose that the operator function $A = A(t)$ in (5.1) is strongly continuous in some Banach space X. As model cases, one may always think of $X = C([a,b])$ or $X = L_p([a,b])$ $(1 \leq p \leq \infty)$.

In spite of its simple form, the investigation of the differential equation (5.1) from the viewpoint of stability encounters many specific difficulties, which are mentioned in the books BARBASHIN [1967, 1970], and which may be the reason for the fact that only very rudimentary results on the stability of (5.1) are known. Also today, there exist only two effective methods from the vast theory of stability (see e.g. DALETSKIJ-KREJN [1970]) which apply to equation (5.1), namely Ljapunov-Bohl exponents and Ljapunov functions.

Let $U = U(t, \tau)$ be the evolution operator for equation (5.1) (see Subsection 1.2). Recall that the *Ljapunov exponent* of (5.1) is defined by

$$(5.3) \qquad \omega = \varlimsup_{t \to \infty} \frac{\log \|U(t, 0)\|}{t},$$

while the *Ljapunov-Bohl* exponent of (5.1) is defined by

$$(5.4) \qquad \omega^* = \varlimsup_{\substack{\tau \to \infty \\ t - \tau \to \infty}} \frac{\log \|U(t, \tau)\|}{t - \tau}.$$

The importance of the exponent (5.3) in stability theory was first observed by Ljapunov. For instance, the inequality $\omega \le 0$ is necessary for the stability of the trivial (and hence any) solution of equation (5.1), and the inequality $\omega < 0$ is sufficient. To study stability properties of perturbed equations (in particular, inhomoneneous linear or quasilinear differential equations), one has to consider the more general exponent (5.4). For instance, the inequality $\omega^* < 0$ implies the asymptotic (and even exponential) stability of the trivial solution of equation (5.1), or of small perturbations of (5.1).

One may study classes of operator functions $A(t)$ for which the exponents (5.3) and (5.4) are finite and may be found rather explicitly. A simple sufficient condition for the finiteness of both ω and ω^* is, by (1.10),

$$(5.5) \qquad \sup_{0 \le t < \infty} \int_t^{t+1} \|A(\xi)\| \, d\xi < \infty.$$

In particular, if the operator function in (5.1) is *stationary* (i.e. $A(t) \equiv A$), both ω and ω^* are finite and coincide; in fact, (1.16) implies then that

$$(5.6) \qquad \omega = \omega^* = \varlimsup_{t \to \infty} \frac{\log \|e^{At}\|}{t} = \sup \{\operatorname{Re} \lambda : \lambda \in \sigma(A)\}.$$

The formula (5.6) is extremely important and natural, since it gives a link between the *spectral properties* of the operator A and the *stability properties* of the (stationary) equation

$$(5.7) \qquad \frac{dx}{dt} = Ax.$$

One should remark that the corresponding problem for linear integro-differential equations is much harder. In fact, for the integro-differential equation (5.1) the operator function $A = A(t)$ has the form

$$(5.8) \qquad A(t) = C(t) + K(t),$$

where

$$(5.9) \qquad C(t)x(s) = c(t,s)x(s)$$

and

$$(5.10) \qquad K(t)x(s) = \int_a^b k(t,s,\sigma)x(\sigma)\,d\sigma.$$

Even in the stationary case (i.e. $C(t) \equiv C$ and $K(t) \equiv K$), the spectrum $\sigma(A)$ of the operator sum $A = C + K$ may be very complicated. Of course, if the multiplier $c(s) \equiv c$ is constant, and the kernel $k = k(s,\sigma)$ generates a (stationary) compact linear integral operator K, one may use the obvious fact that

$$\sigma(A) = c + \sigma(K).$$

In this case, one may apply the well-known classical spectral theory of compact linear operators. Some more information on the spectral properties of the operator (5.8) in the stationary case may be found in Subsection 4.7.

5.2. Perturbation of the Ljapunov-Bohl exponent

The result discussed in the last subsection are not encouraging, since the explicit calculation of the Ljapunov exponent and Ljapunov-Bohl exponent is one of the most difficult problems in the theory of differential equations in Banach spaces. Nevertheless, one may obtain some stability theorems for equation (5.2) observing that, by (5.8), *this integro-differential equation may be considered as an "integral perturbation" of the simpler linear differential equation*

$$(5.11) \qquad \frac{dx}{dt} = C(t)x.$$

As we have seen before, the evolution operator for (5.11) may be given explicitly by

$$(5.12) \qquad U_0(t, \tau)x(s) = e(t, \tau, s)x(s),$$

where

$$e(t, \tau, s) = \exp\left\{\int_\tau^t c(\xi, s)\, d\xi\right\}$$

(see Lemma 1.5). Consequently, the Ljapunov exponent ω_0 of equation (5.11) in $X = L_p$, say, may be "calculated" by the formula

$$\omega_0 = \varlimsup_{t\to\infty} \frac{1}{t} \log \|e(t, 0, \cdot)\|_{L_\infty} = \varlimsup_{t\to\infty} \operatorname*{ess\,sup}_{a\le s\le b} \left|\frac{1}{t}\int_0^t c(\xi, s)\, d\xi\right|.$$

One could expext that there is some chance to obtain also explicit formulas for the Ljapunov-Bohl exponent of (5.11), and that small perturbations of (5.11) by integral terms like (5.10) should lead to small changes of the Ljapunov-Bohl exponent of (5.11).

It is known (DALETSKIJ-KREJN [1970]) that the Ljapunov exponent (5.3) is not stable under small perturbations, but the Ljapunov-Bohl exponent (5.4) is. Let us recall a basic result whose proof may be found in DALETSKIJ-KREJN [1970] (but our proof which builds on some "Gronwall inequality argument" is much simpler).

Lemma 5.1. *Suppose that the evolution operator $U_0(t, \tau)$ of the differential equation (5.11) satisfies an estimate*

$$(5.13) \qquad \|U_0(t, \tau)\| \le M e^{\delta(t-\tau)} \qquad (0 \le \tau \le t < \infty).$$

Then the evolution operator $U(t, \tau)$ of the differential equation (5.1) satisfies

$$(5.14) \qquad \|U(t, \tau)\| \le M \exp\left\{\delta(t - \tau) + M \int_\tau^t \|K(\xi)\|\, d\xi\right\}.$$

$\square$ From the equality

$$U(t, \tau) = U_0(t, \tau) + \int_\tau^t U_0(t, \xi)K(\xi)U(\xi, \tau)d\xi$$

(see (1.29)) we get, by (5.13), the estimate

$$\|U(t,\tau)\| \le M e^{\delta(t-\tau)} + \int_\tau^t M e^{\delta(t-\xi)} \|K(\xi)\| \, \|U(\xi,\tau)\| \, d\xi.$$

Putting

$$W(t,\tau) = e^{-\delta(t-\tau)} \|U(t,\tau)\|,$$

we may rewrite this in the form

$$(5.15) \qquad W(t,\tau) \le M + M \int_\tau^t \|K(\xi)\| W(\xi,\tau) \, d\xi.$$

Since the scalar function

$$r(t) = M + M \int_\tau^t \|K(\xi)\| W(\xi,\tau) \, d\xi$$

satisfies $r'(t) \le M\|K(t)\| r(t)$ and $r(\tau) = M$, we conclude that

$$r(t) \le M \exp M \int_\tau^t \|K(\xi)\| \, d\xi,$$

and hence, again by (5.15), that

$$W(t,\tau) \le M \exp M \int_\tau^t \|K(\xi)\| \, d\xi \qquad (0 \le \tau \le t < \infty).$$

This proves the assertion. ∎

Observe that in case $c \in L_\infty([a,b] \times [0,\infty))$ the estimate (5.13) is always true (with $M = 1$ and $\delta = \|c\|_{L_\infty}$).

5.3. Application to Barbashin operators

Lemma 5.1 implies that the Ljapunov-Bohl exponent ω^* of the linear integro-differential equation (5.1) satisfies

$$(5.16) \qquad \omega^* \le \delta + M \varlimsup_{\substack{\tau \to \infty \\ t-\tau \to \infty}} \frac{1}{t-\tau} \int_\tau^t \|K(\xi)\| \, d\xi$$

with δ as in (5.13). Since the number δ may be taken arbitrarily close to the Ljapunov-Bohl exponent ω^* of the linear differential equation

(5.11), we may deduce from Lemma 5.1 that the Ljapunov-Bohl exponent is in fact stable under small perturbations with respect to the quasinorm

$$(5.17) \qquad ||K||_* = \varlimsup_{\substack{\tau \to \infty \\ t-\tau \to \infty}} \frac{1}{t-\tau} \int_\tau^t ||K(\xi)|| \, d\xi.$$

For the integro-differential equation (5.1), this can be made more precise. First, we mention the following simple

Lemma 5.2. *Let X be either the space C or the space L_p ($1 \leq p \leq \infty$), and let $U_0(t,\tau)$ be defined by (5.12). Then the Ljapunov-Bohl exponent of equation (5.11) is given by*

$$(5.18) \qquad \omega_0^* = \varlimsup_{\substack{\tau \to \infty \\ t-\tau \to \infty}} \operatorname{ess\,sup}_{a \leq s \leq b} \frac{1}{t-\tau} \int_\tau^t c(\xi,s) \, d\xi.$$

Moreover, the estimate

$$(5.19) \qquad ||U_0(t,\tau)|| \leq e^{\omega_0^{**}(t-\tau)} \qquad (0 \leq \tau \leq t < \infty)$$

holds, where

$$(5.20) \qquad \omega_0^{**} = \sup_{0 \leq \tau \leq t < \infty} \operatorname{ess\,sup}_{a \leq s \leq b} \frac{1}{t-\tau} \int_\tau^t c(\xi,s) \, d\xi.$$

□ For X as above, the norm (in $\mathcal{L}(X)$) of the multiplication operator by a fixed function coincides with the L_∞-norm of this function. Consequently, from (5.12) it follows that

$$||U_0(t,\tau)|| = \operatorname{ess\,sup}_{a \leq s \leq b} \left(\exp \int_\tau^t c(\xi,s) \, d\xi \right) = \exp \left(\operatorname{ess\,sup}_{a \leq s \leq b} \int_\tau^t c(\xi,s) \, d\xi \right).$$

Thus, (5.18) follows from the definition (5.4) of the Ljapunov-Bohl exponent. The estimate (5.19) is obvious. ■

We point out that the two numbers (5.18) and (5.20) may be different.

Theorem 5.1. *Let X be either the space C or the space L_p $(1 \leq p \leq \infty)$. Suppose that the multiplier $c = c(t,s)$ and the kernel $k = k(t,s,\sigma)$ are such that*

$$(5.21) \qquad \omega_0^{**} + ||K||_* < 0,$$

where ω_0^{**} is given by (5.20) and $||K||_*$ by (5.17). Then the Ljapunov-Bohl exponent of the integro-differential equation (5.1) is negative; consequently, the trivial solution of (5.1) is exponentially stable.

☐ The statement follows immediately from the estimates (5.16) and (5.19). ∎

As the definition (5.20) shows, the number ω_0^{**} does not depend on the space X, but only on the multiplier $c = c(t,s)$. The number (5.17), however, depends on both the kernel $k = k(t,s,\sigma)$ and the space X. For example, for $X = C$ or $X = L_\infty$ we have

$$(5.22) \qquad ||K||_* = \varlimsup_{\substack{\tau \to \infty \\ t-\tau \to \infty}} \frac{1}{t-\tau} \int_\tau^t \operatorname*{ess\,sup}_{a \leq s \leq b} \int_a^b |k(\xi,s,\sigma)| \, d\sigma \, d\xi,$$

while for $X = L_1$ we have

$$(5.23) \qquad ||K||_* = \varlimsup_{\substack{\tau \to \infty \\ t-\tau \to \infty}} \frac{1}{t-\tau} \int_\tau^t \operatorname*{ess\,sup}_{a \leq \sigma \leq b} \int_a^b |k(\xi,s,\sigma)| \, ds \, d\xi.$$

In case $X = L_p$ $(1 < p < \infty)$ the situation is more complicated, since norms of integral operators are then difficult to compute. However, in many situations it suffices to use the upper estimate

$$(5.24) \qquad ||K||_* \leq \varlimsup_{\substack{\tau \to \infty \\ t-\tau \to \infty}} \frac{1}{t-\tau} \int_\tau^t \hat{k}(\xi) \, d\xi$$

with

$$\hat{k}(\xi) := \sup_{\substack{||x||_{L_p} \leq 1 \\ ||y||_{L_q} \leq 1}} \int_a^b \int_a^b |k(\xi,s,\sigma)x(\sigma)y(s)| \, d\sigma \, ds,$$

or to use the norms of the classes $U(p,q), V(p,q), U(p_0,q_0) \cap V(p_1,q_1)$, or $W(\alpha_0,\beta_0) \cap W(\alpha_1,\beta_1)$ discussed in Subsection 3.3.

Theorem 5.2. *Let X be either the space C or the space L_p $(1 \le p \le \infty)$. Suppose that the multiplier $c = c(t, s)$ and the kernel $k = k(t, s, \sigma)$ are such that*

$$(5.25) \qquad\qquad \omega_0^* < \infty$$

and

$$(5.26) \qquad\qquad \|K\|_* = 0.$$

Then the Ljapunov-Bohl exponent of the integro-differential equation (5.1) coincides with ω_0^*; consequently, the trivial solution of (5.1) is exponentially stable if $\omega_0^* < 0$.

□ The statement follows again from the estimate (5.16), since (5.13) holds for arbitrary $\delta > \omega_0^*$ (and appropriate $M > 0$ depending on δ). ∎

As before, (5.25) is independent of the space X, but (5.26) is not. For instance, (5.26) reduces in $X = C$ or $X = L_\infty$ to

$$(5.27) \qquad \varlimsup_{\substack{\tau \to \infty \\ t-\tau \to \infty}} \frac{1}{t - \tau} \int_\tau^t \operatorname*{ess\,sup}_{a \le s \le b} \int_a^b |k(\xi, s, \sigma)| \, d\sigma \, d\xi = 0,$$

and in $X = L_1$ to

$$(5.28) \qquad \varlimsup_{\substack{\tau \to \infty \\ t-\tau \to \infty}} \frac{1}{t - \tau} \int_\tau^t \operatorname*{ess\,sup}_{a \le \sigma \le b} \int_a^b |k(\xi, s, \sigma)| \, ds \, d\xi = 0.$$

Moreover, the condition

$$(5.29) \qquad \varlimsup_{\substack{\tau \to \infty \\ t-\tau \to \infty}} \frac{1}{t - \tau} \int_\tau^t \hat{k}(\xi) \, d\xi = 0$$

with

$$\hat{k}(\xi) = \sup_{\substack{\|x\|_{L_p} \le 1 \\ \|y\|_{L_q} \le 1}} \int_a^b \int_a^b |k(\xi, s, \sigma) x(\sigma) y(s)| \, d\sigma \, ds$$

as above is sufficient for (5.26) in $X = L_p$ $(1 < p < \infty)$. Observe that (5.26) is automatically satisfied if either

$$(5.30) \qquad\qquad \int_0^\infty \|K(t)\| \, dt < \infty$$

or, more generally,

$$(5.31) \qquad \lim_{t \to \infty} \|K(t)\| = 0.$$

These conditions are not very restrictive, are much simpler than (5.26), and may be verified immediately in terms of the kernel $k = k(t, s, \sigma)$.

5.4. Ljapunov functions

A nonnegative differentiable function $H = H(x)$ on a Banach space X is called *Ljapunov function* for the differential equation (5.1) if

(a) for any sequence $(x_n)_n$ in X, $\lim_{n \to \infty} H(x_n) = 0$ implies that $\lim_{n \to \infty} x_n = 0$,

and there exists a non-negative function $N = N(x)$ on X such that

(b) $H'(x)(A(t)x) \leq -N(x) \quad (x \in X)$,

and

(c) $\lim_{n \to \infty} N(x_n) = 0$ implies that $\lim_{n \to \infty} x_n = 0$.

As was already observed by Ljapunov, the existence of such a function H for (5.1) implies the asymptotic stability of the trivial solution of (5.1) under small perbutations of $A(t)$ and of initial values. Moreover, he proved that in the stationary case $A(t) \equiv A$ the existence of such a function H is also necessary for asymptotic stability.

Ljapunov functions are an extremely useful tool for studying the stability of several typical systems of differential equations (see e.g. BARBASHIN [1967, 1970]). However, when passing from finite systems to infinite systems of equations (in particular, to integro-differential equations), the theory of Ljapunov functions seems to fail. One exception is provided by equations in *Hilbert space*, where all basic results carry in fact over from finite to infinite systems. Some information in this direction, related with quadratic Ljapunov functions, may be found in the monographs AKHIEZER-GLAZMAN [1977] and DALETSKIJ-KREJN [1970]. In general Banach spaces, and even in very simple spaces like C or L_p $(1 \leq p \leq \infty, p \neq 2)$, Ljapunov functions have not been investigated for several reasons. It is clear, for instance, that the main difficulties are due to the degeneracy of smooth functionals H satisfying (a), (b), and (c) as above on these spaces. To see this, consider the quadratic functional

$$(5.32) \qquad \Phi(x) = \int_a^b d(s)x^2(s)\,ds + \int_a^b \int_a^b l(s,\sigma)x(\sigma)x(s)\,d\sigma\,ds,$$

which is a natural generalization of quadratic forms on Euclidean space. Obviously, this functional cannot be defined on the whole space L_p for $1 \le p < 2$ (except for the trivial case $d(s) \equiv 0$) : in fact, the function d in the first term of (5.32) may be viewed as a bounded linear functional on $L_{p/2}$, and $L_p^* = \{0\}$ for $0 < p < 1$ (see e.g. DUNFORD-SCHWARTZ [1962]). On the other hand, if $l = l(s,\sigma)$ is, say, a continuous function on $[a,b] \times [a,b]$, the second term of (5.32) cannot satisfy condition (a) of a Ljapunov function on $X = C$ or $X = L_p$ for $2 < p \le \infty$. In fact, this condition is equivalent to an estimate

$$(5.33) \qquad\qquad\qquad \Phi(x) \ge \alpha \|x\|_X^2$$

for some $\alpha > 0$, since (5.32) is quadratic in x. But (5.33) means that the norm $\|\cdot\|_X$ is equivalent to the norm

$$\||x\||_X := \left\{ \int_a^b d(s)x^2(s)ds + \int_a^b \int_a^b l(s,\sigma)x(\sigma)x(s)d\sigma\,ds \right\}^{1/2},$$

which turns X into a Hilbert space, a contradiction. This reasoning shows that the functional (5.32) may be a Ljapunov function only in case $X = L_2$. Moreover, in case $X = L_2$ the function $d = d(s)$ in (5.32) is necessarily bounded (since $x^2 \in L_1$ and hence $d \in L_1^* = L_\infty$), and the function $l = l(s,\sigma)$ belongs to the kernel class $\mathfrak{Z}_2$ (by the Banach-Steinhaus theorem). Unfortunately, to show that the functional Φ satisfies the estimate (5.33) remains a difficult problem even in case $X = L_2$.

Lemma 5.3. *Suppose that the function Φ given by (5.32) is defined on $X = L_2$. Then the operator*

$$(5.34) \qquad\qquad Tx(s) = d(s)x(s) + \int_a^b l(s,\sigma)x(\sigma)\,d\sigma$$

is self-adjoint in L_2, and the relation

$$(5.35) \qquad \inf \{\Phi(x) : ||x||_{L_2} = 1\} = \inf \{|\lambda| : \lambda \in \sigma(T)\}$$

holds.

Since Lemma 5.3 is a special case of a well-known result on the spectrum of self-adjoint operators in Hilbert spaces, we drop the proof. Observe, however, that the problem of calculating the spectrum of the operator (5.34) is not at all trivial. Even worse, the following lemma shows that only in the case when d is bounded away from zero, i.e.

$$(5.36) \qquad \operatorname*{ess\,inf}_{a \leq s \leq b} d(s) > 0$$

one has a chance to verify (5.33).

Lemma 5.4. *Suppose that $d \in L_\infty$ and $l \in 3_2^0$. Then the estimate (5.33) implies necessarily that*

$$(5.37) \qquad \alpha \leq \operatorname*{ess\,inf}_{a \leq s \leq b} d(s)$$

□   Let D_n be a sequence of subsets of positive measure with the property that

$$d(\sigma) < \operatorname*{ess\,inf}_{a \leq s \leq b} d(s) + \frac{1}{n}$$

for $\sigma \in D_n$ $(n = 1, 2, \ldots)$. Consider the sequence of functions $x_n = \mu(D_n)^{-1/2} \chi_{D_n}$. Obviously $||x_n||_{L_2} = 1$ and

$$\operatorname*{ess\,inf}_{a \leq s \leq b} d(s) \leq \int_a^b d(s) x_n^2(s) \, ds$$

$$= \frac{1}{\mu(D_n)} \int_{D_n} d(s) \, ds \leq \operatorname*{ess\,inf}_{a \leq s \leq b} d(s) + \frac{1}{n}.$$

Consequently,

$$(5.38) \qquad \lim_{n \to \infty} \int_a^b d(s) x_n^2(s) \, ds = \operatorname*{ess\,inf}_{a \leq s \leq b} d(s).$$

Since the kernel l belongs to $\mathfrak{Z}_2^0$, the corresponding integral operator L is compact in L_2. Moreover, since the sequence $(x_n)_n$ converges weakly in L_2 to 0, $(Lx_n)_n$ converges in the norm of L_2 to 0, and hence $\langle Lx_n, x_n \rangle \to 0$ as $n \to \infty$. But

$$\langle Lx_n, x_n \rangle = \int_a^b \int_a^b l(s, \sigma) x_n(\sigma) x_n(s) \, d\sigma \, ds,$$

hence

(5.39) $$\lim_{n \to \infty} \int_a^b \int_a^b l(s, \sigma) x_n(\sigma) x_n(s) \, d\sigma \, ds = 0.$$

Combining (5.38) and (5.39), we conclude that

$$\lim_{n \to \infty} \Phi(x_n) = \operatorname*{ess\,inf}_{a \le s \le b} d(s),$$

which implies (5.37). ∎

We remark that Lemma 5.4 may also be proved by means of general perturbation theorems for spectra of linear operators (see e.g. DUNFORD-SCHWARTZ [1962] or KATO [1966]).

As already observed, the application of Lemma 5.3 is difficult, since it is hard to compute the spectrum of the operator (5.34). It is in general much easier to study the integral operator L generated by the kernel $l = l(s, \sigma)$. For instance, sometimes one can determine the greatest lower bound $m_-(L)$ of L which is, at least for compact operators L, a non-negative number. In particular, $m_-(L) = 0$ if L in non-negative definite. The estimate

(5.40) $$\operatorname*{ess\,inf}_{a \le s \le b} d(s) + m_-(L) > 0$$

implies then obviously the estimate (5.33).

So far we have analyzed condition (a) in the definition of a Ljapunov function Φ as given in (5.32); let us now discuss condition (b). If the operator T is given by (5.34), a straightforward calculation shows that

$$\Phi'(x)(A(t)x) = \langle [T, A(t)]x, x \rangle,$$

where we have put

(5.41) $[T, A(t)] := TA(t) + A^*(t)T.$

Using Lemma 3.5 we may write the functional $\hat{\Phi}(x) = \Phi'(x)(A(t)x)$ in the form

$$\hat{\Phi}(x) = \int_a^b 2c(t,s)d(s)x^2(s)\,ds$$

$$+ \int_a^b \int_a^b l(s,\sigma)c(t,\sigma)x(\sigma)x(s)\,d\sigma\,ds$$

$$+ \int_a^b \int_a^b [c(t,s)l(s,\sigma) + d(s)k(t,s,\sigma) + k(t,\sigma,s)d(\sigma)]x(\sigma)x(s)\,d\sigma\,ds$$

$$+ \int_a^b \int_a^b \left\{ \int_a^b [l(s,\xi)k(t,s,\xi) + k(t,\xi,s)l(\xi,\sigma)]\,d\xi \right\} x(\sigma)x(s)\,d\sigma\,ds.$$

This shows that $\hat{\Phi}$ is also a quadratic functional like Φ, i.e.

$$\hat{\Phi}(x) = \int_a^b \hat{d}(t,s)x^2(s)\,ds + \int_a^b \int_a^b \hat{l}(t,s,\sigma)x(\sigma)x(s)\,d\sigma\,ds,$$

where

$$\hat{d}(t,s) = 2c(t,s)d(s)$$

and

$$\hat{l}(t,s,\sigma) = l(s,\sigma)c(t,\sigma) + c(t,s)l(s,\sigma) + d(s)k(t,s,\sigma)$$

$$+ k(t,\sigma,s)d(\sigma) + \int_a^b [l(s,\xi)k(t,s,\xi) + k(t,\xi,\sigma)l(\xi,\sigma)]\,d\xi.$$

Analogously to (5.34), we may associate with the functional $\hat{\Phi}$ the operator function

(5.42) $\hat{T}(t)x(s) = \hat{d}(t,s)x(s) + \int_a^b \hat{l}(t,s,\sigma)x(\sigma)\,d\sigma;$

by the definition of $\hat{\Phi}$ we have then

(5.43) $\hat{T}(t) = [T, A(t)].$

If this is a Ljapunov function, we should have

$$(5.44) \qquad \langle \hat{T}(t)x, x \rangle \leq -\beta \|x\|_2^2 \quad (0 \leq t < \infty)$$

with some $\beta > 0$. For proving (5.44), we may follow the same reasoning as above for proving (5.33). We summarize with the following

Theorem 5.3. *Suppose that $c(t, \cdot) \in L_\infty$ and $k(t, \cdot, \cdot) \in \mathfrak{Z}_2$ for any $t \in [0, \infty)$, and that the corresponding operator function (5.42) is strongly continuous in the space L_2. Assume that, moreover, $d \in L_\infty$ and $l \in \mathfrak{Z}_2$. Then the quadratic functional Φ defined by (5.32) is a Ljapunov function for the differential equation (5.1) in L_2 if and only if the estimates*

$$(5.45) \qquad \inf \sigma(T) \geq \alpha$$

and

$$(5.46) \qquad \sup \sigma(\hat{T}(t)) \leq -\beta \quad (0 \leq t < \infty)$$

hold for some $\alpha, \beta > 0$. For (5.45) and (5.46) in turn the estimates

$$\operatorname*{ess\,inf}_{a \leq s \leq b} d(s) > 0$$

and

$$2 \sup_{0 \leq t < \infty} \operatorname*{ess\,sup}_{a \leq s \leq b} c(t, s) d(s) \leq -\beta$$

are necessary, while the estimates

$$\operatorname*{ess\,inf}_{a \leq s \leq b} d(s) + m_-(L) > 0$$

and

$$2 \sup_{0 \leq t < \infty} \operatorname*{ess\,sup}_{a \leq s \leq b} c(t, s) d(s) + m_+([L, C(t)]$$

$$+ [D, K(t)] + [L, K(t)]) \leq -\beta$$

are sufficient; here D is defined by $Dx(s) = d(s)x(s)$, and $K(t)$ by (5.10).

§ 6. Continuous dependence on parameters

In this section we consider the linear differential equation

(6.1) $$\frac{dx}{dt} = A(t,\varepsilon)x$$

with an operator function $A = A(t,\varepsilon)$ depending continuously on a scalar parameter ε. We are interested in the continuous dependence of the evolution operator $U = U(t,\tau,\varepsilon)$ and of the solution $x = x(t,\varepsilon)$ of (6.1), considered as functions of ε. More regular (e.g. smooth) dependence is also studied.

6.1. Continuous dependence of the evolution operator

One of the most important topics for applying differential equations in Banach spaces (in particular, linear differential equations) is the continuous dependence of the evolution operator on parameters, as a consequence of the continuous dependence of the right-hand side of such equations on these parameters. A classical result in this spirit is the following (see e.g. DALETSKIJ-KREJN [1970]): the uniformly continuous dependence of the right-hand side of a differential equation on a parameter implies the uniformly continuous dependence of the corresponding evolution operator on this parameter. For many applications, however, this statement is not sufficient. Beginning with the pioneering work of BOGOLJUBOV [1945] (see also BOGOLJUBOV-MITROPOL'SKIJ [1963] and MITROPOL'SKIJ [1971]), many results on the continuous dependence of solutions to differential equations have been obtained in situations when the corresponding right-hand side depends on a parameter only in some weak "averaged" sense. Some rather sophisticated theorems, especially for the evolution operator of linear differential equations, have been proved by KURZWEIL [1957] and afterwards by LEVIN [1967]. Our Lemma 4.1 below reduces in fact in the scalar case to Levin's result: although in LEVIN [1967] only the case of finite linear systems is considered, the proof carries over to general systems without any change.

Let $J = [0,T]$, and consider in a Banach space X the linear differential equation (6.1) with an operator function $A = A(t,\varepsilon)$ depending continuously on a scalar parameter $\varepsilon \in [-\varepsilon_0, \varepsilon_0]$. We are interested in conditions for the continuous dependence (in the norm of $\mathfrak{L}(X)$) of the corresponding evolution operator $U(t, \tau, \varepsilon)$ at $\varepsilon = 0$, say. To this end, we put

$$(6.2) \qquad B(t, \varepsilon) := A(t, \varepsilon) - A(t, 0).$$

Lemma 6.1. *Suppose that one of the following four conditions holds:*

$$(6.3) \qquad \sup_{|\varepsilon| \le \varepsilon_0} \int_0^T ||A(t, \varepsilon)|| \, dt < \infty;$$

$$(6.4) \qquad \lim_{\varepsilon \to 0} \int_0^T \int_0^t ||B(\xi, \varepsilon)B(t, \varepsilon)|| \, d\xi \, dt = 0;$$

$$(6.5) \qquad \lim_{\varepsilon \to 0} \int_0^T \int_0^t ||B(t, \varepsilon)B(\xi, \varepsilon)|| \, d\xi \, dt = 0;$$

$$(6.6) \qquad \lim_{\varepsilon \to 0} \int_0^T \left\{ \int_0^t ||B(\xi, \varepsilon)|| \, d\xi \, ||B(t, \varepsilon)|| \right.$$
$$\left. - ||B(t, \varepsilon)|| \int_0^t ||B(\xi, \varepsilon)|| \, d\xi \right\} dt = 0.$$

Then the relations

$$(6.7) \qquad \lim_{\varepsilon \to 0} \sup_{0 \le t \le T} \left|\left| \int_0^t [A(\xi, \varepsilon) - A(\xi, 0)] \, d\xi \right|\right| = 0$$

and

$$(6.8) \qquad \lim_{\varepsilon \to 0} \sup_{0 \le t, \tau \le T} ||U(t, \tau, \varepsilon) - U(t, \tau, 0)|| = 0$$

are equivalent.

The most convenient of all conditions in Lemma 6.1 to verify is (6.3); this condition is usually fulfilled in many applications. The most interesting condition, however, is (6.6). First of all, this condition is fulfilled if the evolution operator is defined by formula (1.19), and thus in this case the statement of Lemma 6.1 holds true. Moreover, (6.6) is always satisfied in the case of a finite system which is obtained in the usual way from the higher order equation

$$\frac{d^n x}{dt^n} + a_{n-1}(t)\frac{d^{n-1}x}{dt^{n-1}} + \ldots + a_1(t)\frac{dx}{dt} + a_0(t)x = 0$$

with commuting coefficients, i.e.

$$a_j(t)a_k(\xi) = a_k(\xi)a_j(t) \quad (j, k = 1, 2, \ldots, n)$$

which is of course a trivial requirement in the scalar case.

6.2. Application to Barbashin operators

The results developed in § 2 and § 3 allow us to apply Lemma 6.1 to operator functions of the form

(6.9) $$A(t, \varepsilon) = C(t, \varepsilon) + K(t, \varepsilon),$$

where

$$C(t, \varepsilon)x(s) := c(t, s, \varepsilon)x(s),$$

$$K(t, \varepsilon)x(s) := \int_a^b k(t, s, \sigma, \varepsilon)x(\sigma)\,d\sigma,$$

and $c : J \times [a, b] \times [-\varepsilon_0, \varepsilon_0] \to \mathbb{R}$ and $k : J \times [a, b] \times [a, b] \times [-\varepsilon_0, \varepsilon_0] \to \mathbb{R}$ are measurable functions. In order to not overburden the notation, we confine ourselves to the case when the first condition (6.3) of Lemma 6.1 holds.

It is clear that (6.3) depends on the underlying space X, since it contains the norm in $\mathfrak{L}(X)$. One may unify the notation, however, by using the norm (3.16) for the class $\mathfrak{Z}_p$ $(1 \leq p \leq \infty)$ and choosing $p = \infty$ in case $X = C$.

Lemma 6.2. *Suppose that the functions* $c = c(t, s, \varepsilon)$ *and* $k = k(t, s, \sigma, \varepsilon)$ *satisfy the estimates*

$$(6.10) \qquad \sup_{|\varepsilon| \leq \varepsilon_0} \int_0^T \operatorname{ess\,sup}_{a \leq s \leq b} |c(\xi, s, \varepsilon)| \, d\xi < \infty$$

and

$$(6.11) \qquad \sup_{|\varepsilon| \leq \varepsilon_0} \int_0^T \|k(\xi, \cdot, \cdot, \varepsilon)\|_{3_p} \, d\xi < \infty.$$

Then the operator function (6.9) *fulfills* (6.3) *in case* $X = L_p$ *for* $1 \leq p < \infty$, *as well as in case* $X = L_\infty$ *or* $X = C$.

☐ The proof of Lemma 6.2 is an easy consequence of Lemma 2.1 and Lemma 3.6. ∎

To verify (6.11) one has to calculate again norms in the class 3_p which is not easy at all. Nevertheless, in case $p = 1$ condition (6.11) is equivalent to the simpler condition

$$(6.12) \qquad \sup_{|\varepsilon| \leq \varepsilon_0} \int_0^T \left\{ \operatorname{ess\,sup}_{a \leq \sigma \leq b} \int_a^b |k(\xi, s, \sigma, \varepsilon)| \, ds \right\} d\xi < \infty,$$

and in case $p = \infty$ to the condition

$$(6.13) \qquad \sup_{|\varepsilon| \leq \varepsilon_0} \int_0^T \left\{ \operatorname{ess\,sup}_{a \leq s \leq b} \int_a^b |k(\xi, s, \sigma, \varepsilon)| \, d\sigma \right\} d\xi < \infty,$$

by Lemma 3.12. To verify (6.11) in case $1 < p < \infty$ one may again use Lemmas 3.8 - 3.11.

Theorem 6.1. *Suppose that the functions* $c = c(t, s, \varepsilon)$ *and* $k = k(t, s, \sigma, \varepsilon)$ *satisfy the hypotheses of Theorem 2.1 or Theorem 3.1, as well as the estimates* (6.10) *and* (6.11). *Then the evolution operator* $U(t, \tau, \varepsilon)$ *depends continuously on* ε *at* $\varepsilon = 0$, *uniformly with respect to* $(t, \tau) \in J \times J$, *if*

$$(6.14) \qquad \lim_{\varepsilon \to 0} \sup_{0 \leq t \leq T} \operatorname{ess\,sup}_{a \leq s \leq b} \int_0^t [c(\xi, s, \varepsilon) - c(\xi, s, 0)] \, d\xi = 0$$

and

(6.15) $$\lim_{\varepsilon \to 0} \sup_{0 \le t \le T} \left\| \int_0^t [k(\xi, \cdot, \cdot, \varepsilon) - k(\xi, \cdot, \cdot, 0)] \, d\xi \right\|_{3p} = 0.$$

□ We claim that (6.14) and (6.15) are equivalent to (6.7). For this we have to prove that

(6.16)
$$\left\{ \int_0^t A(\xi, \varepsilon) \, d\xi \right\} x(s) = \left\{ \int_0^t c(\xi, s, \varepsilon) \, d\xi \right\} x(s)$$
$$+ \int_a^b \left\{ \int_0^t k(\xi, s, \sigma, \varepsilon) \, d\xi \right\} x(\sigma) \, d\sigma,$$

by Lemma 2.1 and Lemma 3.6. But by (6.10), (6.11) and Fubini's theorem we may interchange the order of integration over $\xi \in [0, t]$ and $\sigma \in [a, b]$ in the last term of (6.16). The assertion follows now from Lemma 6.1. ■

By Lemma 3.8 and Lemma 3.9, condition (6.15) is for $p = 1$ equivalent to

(6.17) $$\operatorname{ess\,sup}_{a \le \sigma \le b} \int_a^b \left| \int_0^t [k(\xi, s, \sigma, \varepsilon) - k(\xi, s, \sigma, 0)] \, d\xi \right| ds \to 0,$$

as $\varepsilon \to 0$, and for $p = \infty$ to

(6.18) $$\operatorname{ess\,sup}_{a \le s \le b} \int_a^b \left| \int_0^t [k(\xi, s, \sigma, \varepsilon) - k(\xi, s, \sigma, 0)] \, d\xi \right| d\sigma \to 0,$$

as $\varepsilon \to 0$, uniformly in $t \in [0, T]$. In case $1 < p < \infty$ it is again useful to employ Lemmas 3.8 - 3.11 for verifying (6.15).

6.3. The first Bogoljubov theorem

Now we apply the above results on the continuous dependence of the resolvent operator on a parameter to a special, but important case, namely

(6.19) $$A(t, \varepsilon) = A_0(t) + B(\tfrac{t}{\varepsilon}).$$

Of course, the right hand side of (6.19) is defined only for $\varepsilon \neq 0$; in addition we put

(6.20)
$$A(t,0) = A_0(t).$$

For further use, we need a technical lemma.

Lemma 6.3. *Suppose that the operator function $B : [0,\infty) \to \mathcal{L}(X)$ is integrable on each bounded interval. Then the two relations*

(6.21)
$$\lim_{\varepsilon \to 0} \sup_{0 \le t \le T} \left|\left| \int_0^t B(\tfrac{\xi}{\varepsilon})\, d\xi \right|\right| = 0$$

and

(6.22)
$$\lim_{\tau \to \infty} \frac{1}{\tau} \left|\left| \int_0^\tau B(t)\, dt \right|\right| = 0$$

are equivalent.

□ The fact that (6.21) implies (6.22) is obvious; we show that (6.22) implies (6.21). Given $\eta > 0$, choose $\tau_\eta > 0$ such that

$$\frac{1}{\tau} \left|\left| \int_0^\tau B(t)\, dt \right|\right| \le \frac{\eta}{T}$$

for $\tau \ge \tau_\eta$. For $\varepsilon > 0$ and $t \le \varepsilon \tau_\eta$ we have then

$$\left|\left| \int_0^t B(\tfrac{\xi}{\varepsilon})\, d\xi \right|\right| \le \int_0^t \left|\left| B(\tfrac{\xi}{\varepsilon}) \right|\right| d\xi$$

$$= \varepsilon \int_0^{t/\varepsilon} \left|\left| B(s) \right|\right| ds \le \varepsilon \int_0^{\tau_\eta} \left|\left| B(s) \right|\right| ds \le \eta,$$

provided that we choose

(6.23)
$$\varepsilon \le \eta \int_0^{\tau_\eta} \left|\left| B(s) \right|\right| ds.$$

On the other hand, for $\varepsilon \tau_\eta < t \le T$ we have, by (6.22),

$$\left|\left| \int_0^t B(\tfrac{\xi}{\varepsilon})\, d\xi \right|\right| = \varepsilon \le \int_0^{t/\varepsilon} \left|\left| B(s) \right|\right| ds \le \varepsilon \frac{t}{\varepsilon} \frac{\eta}{T} \le \eta.$$

Consequently, (6.22) implies that

$$\left\| \int_0^t B(\tfrac{\xi}{\varepsilon})\, d\xi \right\| \leq \eta$$

for all $t \in [0, T]$, and thus (6.21) holds. ∎

Lemma 6.3 shows that the additional convention (6.20) is natural if and only if the operator function B has zero integral mean on the half-axis $[0, \infty)$.

From Lemma 6.1 we get the following

Lemma 6.4. *Suppose that the operator functions $A : [0, \infty) \to \mathcal{L}(X)$ and $B : [0, \infty) \to \mathcal{L}(X)$ are strongly continuous and integrable on each bounded interval. Assume, moreover, that the limit*

$$(6.24) \qquad\qquad B := \lim_{\tau \to \infty} \frac{1}{\tau} \int_0^\tau B(t)\, dt$$

exists. Denote by $U(t, \tau; \varepsilon)$ the resolvent operator of the linear differential equation

$$(6.25) \qquad\qquad \frac{dx}{dt} = [A(t) + B(\tfrac{t}{\varepsilon})]x$$

for $\varepsilon \neq 0$, and of the equation

$$(6.26) \qquad\qquad \frac{dx}{dt} = [A(t) + B]x$$

for $\varepsilon = 0$. Then the operator function $\varepsilon \mapsto U(t, \tau; \varepsilon)$ is continuous at $\varepsilon = 0$.

The assertion of Lemma 6.4 is the so-called "first Bogoljubov theorem" in the averaging theory for linear differential equations. A characteristic feature of this theory is the appearance of the "fast time scale" t/ε in the right hand side of a differential equation. The first Bogoljubov theorem asserts, loosely speaking, that all terms containing such a fast time scale may be replaced by their integral averages without essentially changing the corresponding solutions.

To apply the first Bogoljubov theorem to integro-differential equations of Barbashin type, we have to calculate the integral average of
the non-stationary Barbashin operator

$$(6.27) \qquad B(t)u(s) = d(t,s)u(s) + \int_a^b l(t,s,\sigma)u(\sigma)\,d\sigma$$

in a suitable function space (usually, in C or L_p for $1 \leq p \leq \infty$). This
is possible, in particular, if the limits

$$(6.28) \qquad d_0(s) = \lim_{\tau \to \infty} \frac{1}{\tau} \int_0^\tau d(\xi,s)\,d\xi$$

and

$$(6.29) \qquad l_0(s,\sigma) = \lim_{\tau \to \infty} \frac{1}{\tau} \int_0^\tau l(\xi,s,\sigma)\,d\xi$$

exist. In this case it is natural to expect that the average of the
operator function (6.27) is simply

$$(6.30) \qquad B_0 u(s) = d_0(s)u(s) + \int_a^b l_0(s,\sigma)u(\sigma)\,d\sigma.$$

This is in fact true if the limit (6.28) exists in the norm of L_∞, and the
limit (6.29) exists in the norm of the Zaanen class $\mathfrak{Z}(X)$ (see (4.25)).

As an application, consider the equation

$$(6.31) \qquad \frac{dx}{dt} = \varepsilon[A(t)x + f(t)]$$

which is one of the simplest equations of nonlinear mechanics (see e.g.
BOGOLJUBOV-MITROPOL'SKIJ [1963]). Under natural hypotheses on
the functions A and f in (6.31), the solutions of (6.31) converge on
every bounded interval, as $\varepsilon \to 0$, to constant functions. For equation
(6.31), however, it is more interesting to find solutions "on bounded,
but arbitrarily large time intervals".

This name is explained as follows. After the substitution

$$(6.32) \qquad \varepsilon^\theta t = \tau$$

with some fixed $\theta \in (0, \infty)$, equation (6.31) becomes

(6.33)
$$\frac{dx}{d\tau} = \varepsilon^{1-\theta}[A(\tau\varepsilon^{-\theta})x + f(\tau\varepsilon^{-\theta})].$$

If a solution of (6.33) is considered on a bounded interval $0 \leq \tau \leq T$, the corresponding solution of the original equation (6.31) is considered on the interval $0 \leq t \leq T\varepsilon^{-\theta}$ which "blows up" for $\varepsilon \to 0$. This effect is called solvability "on bounded, but arbitrarily large time intervals" in nonlinear mechanics.

For $0 < \theta < 1$, the substitution (6.32) is not interesting, since then all solutions of (6.33) on bounded time intervals converge uniformly to zero.

In case $\theta > 1$, however, (6.33) is essentially an equation containing a small parameter in the highest derivative, and thus we end up with a singular perturbation problem. Finally, for $\theta = 1$ the equation (6.33) is easier to deal with; let us discuss this case more in detail. equation (6.33) has then the form

(6.34)
$$\frac{dx}{d\tau} = A(\tfrac{\tau}{\varepsilon})x + f(\tfrac{\tau}{\varepsilon}),$$

and we are interested in the bahaviour of possible solutions of this equation as $\varepsilon \to 0$. From our previous discussion one should expect that the solutions of (6.34) tend, as $\varepsilon \to 0$, to the solutions of the stationary equation

(6.35)
$$\frac{dx}{d\tau} = A_0 x + f_0,$$

where

(6.36)
$$A_0 = \lim_{\tau \to \infty} \frac{1}{\tau} \int_0^\tau A(\xi)\, d\xi$$

and

(6.37)
$$f_0 = \lim_{\tau \to \infty} \frac{1}{\tau} \int_0^\tau f(\xi)\, d\xi.$$

In fact, the solutions of (6.34) and (6.35) which satisfy the same initial condition $x(0) = x_0$ may be represented by means of the corresponding resolvent operators in the form

$$(6.38) \qquad x_\varepsilon(\tau) = U(\varepsilon; \tau, 0)x_0 + \int_0^\tau U(\varepsilon; \tau, \sigma)f(\tfrac{\sigma}{\varepsilon})\, d\sigma$$

and

$$(6.39) \qquad x_0(\tau) = U(0; \tau, 0)x_0 + \int_0^\tau U(0; \tau, \sigma)f_0(\sigma)\, d\sigma,$$

respectively. Let us study these solutions on $[0, T]$ for some $T > 0$. From Lemma 6.4 it follows that

$$(6.40) \qquad \lim_{\varepsilon \to 0} \sup_{0 \le \sigma, \tau \le T} \|U(\varepsilon; \tau, \sigma) - U(0; \tau, \sigma)\| = 0.$$

Consequently,

$$(6.41) \qquad \begin{aligned} &\sup_{0 \le \sigma, \tau \le T} \left\| \int_0^\tau [U(\varepsilon; \tau, \sigma) - U(0; \tau, \sigma)]f(\tfrac{\sigma}{\varepsilon})\, d\sigma \right\| \le \\ &\sup_{0 \le \sigma, \tau \le T} \|U(\varepsilon; \tau, \sigma) - U(0; \tau, \sigma)\| \int_0^\tau \|f(\tfrac{\sigma}{\varepsilon})\|\, d\sigma \to 0 \end{aligned}$$

as $\varepsilon \to 0$, provided that the function f satisfies

$$(6.42) \qquad \varlimsup_{\tau \to \infty} \int_0^\tau \|f(\sigma)\|\, d\sigma < \infty.$$

Observing that $U(0; t, s) = e^{A_0(t-s)}$ we thus arrive at the equality

$$(6.43) \qquad x_\varepsilon(\tau) - x(\tau) = w_\varepsilon(\tau) + \int_0^\tau e^{A_0(\tau - \sigma)}[f(\tfrac{\sigma}{\varepsilon}) - f_0(\sigma)]\, d\sigma,$$

where w_ε satisfies

$$\lim_{\varepsilon \to 0} \sup_{0 \le \tau \le T} \|w_\varepsilon(\tau)\| = 0.$$

Integrating now the right hand side of

$$(6.44) \qquad \begin{aligned} &\int_0^\tau e^{A_0(\tau - \sigma)}[f(\tfrac{\sigma}{\varepsilon}) - f_0(\sigma)]\, d\sigma \\ &= \int_0^\tau e^{A_0(\tau - \sigma)} \frac{d}{d\sigma} \left\{ \int_0^\sigma [f(\tfrac{\xi}{\varepsilon}) - f_0(\xi)]\, d\xi \right\} d\sigma \end{aligned}$$

by parts, we end up with

$$\int_0^\tau e^{A_0(\tau-\sigma)}[f(\tfrac{\sigma}{\varepsilon}) - f_0(\sigma)]\,d\sigma$$

$$= \int_0^\tau [f(\tfrac{\xi}{\varepsilon}) - f_0(\xi)]\,d\xi + \int_0^\tau e^{A_0(\tau-\sigma)} A_0 \int_0^\sigma [f(\tfrac{\xi}{\varepsilon}) - f_0(\xi)]\,d\xi\,d\sigma.$$

By (6.37), (6.42), and Lemma 6.3 we have

$$\lim_{\varepsilon\to 0}\ \sup_{0\le\sigma\le T}\ \left\|\int_0^\sigma [f(\tfrac{\xi}{\varepsilon}) - f_0(\xi)]\,d\xi\right\| = 0,$$

hence

$$\lim_{\varepsilon\to 0}\ \sup_{0\le\tau\le T}\ \left\|\int_0^\tau e^{A_0(\tau-\sigma)}[f(\tfrac{\sigma}{\varepsilon}) - f_0(\sigma)]\,d\sigma\right\| = 0.$$

But this means precisely that

(6.45)
$$\lim_{\varepsilon\to 0} x_\varepsilon(t) = x_0(t)$$

in the space $C([0,T],X)$, where x_ε denotes the solution of (6.34) with initial condition $x(0) = x_0$, and x_0 denotes the solution of (6.35) with the same initial condition. Observe that the latter solution may be written explicitly as

$$x_0(\tau) = e^{A_0\tau}x_0 + \left(\int_0^\tau e^{A_0(\tau-\sigma)}\,d\sigma\right) f_0(\tau)$$

or, in case of invertible A_0 even simpler as

$$x_0(\tau) = e^{A_0\tau}[x_0 + A_0^{-1}f_0(\tau)] - A_0^{-1}f_0(\tau).$$

We summarize our discussion with the following

Theorem 6.2. *Suppose that the operator function $A : [0,\infty) \to \mathfrak{L}(X)$ and the vector function $f : [0,\infty) \to X$ are strongly continuous and integrable on each bounded interval. Assume, moreover, that the limits (6.36) and (6.37) exist, and that condition (6.43) is satisfied. Then the solution x_ε of the initial value problem*

$$\frac{dx}{dt} = \varepsilon[A(t) + f(t)], \qquad x(0) = x_0$$

satisfies, for any fixed $T > 0$, the relation

$$\lim_{\varepsilon \to 0} \sup_{0 \le t \le T/\varepsilon} \|x_\varepsilon(t) - e^{\varepsilon A_0 t}[x_0 + \left(\int_0^{\varepsilon t} e^{-\varepsilon A_0 s}\, ds\right) f_0]\| = 0.$$

Theorem 6.2 is an improvement of Bogoljubov's first theorem on the averaging of solutions "on bounded, but arbitrarily large time intervals". Another application of averaging principles to Barbashin equations may be found in IVANITSKIJ [1971, 1973] and LEDOVSKA-JA [1987].

6.4. Smooth dependence

Consider now the inhomogeneous differential equation

$$(6.46) \qquad \frac{dx}{dt} = A(t, \varepsilon) + f(t, \varepsilon) \qquad (|\varepsilon| \le \varepsilon_0).$$

By $x = x(t, \varepsilon)$ we denote the solution of (6.46) which satisfies the initial condition

$$(6.47) \qquad x(\tau, \varepsilon) = x_{\tau,\varepsilon} \in X.$$

In this subsection we are interested in conditions under which the smooth dependence of the operator function $A(t, \varepsilon)$, the function $f(t, \varepsilon)$, and the initial value $x_{\tau,\varepsilon}$ on ε implies the smooth dependence of the solution $x(t, \varepsilon)$ on ε.

Let

$$B(t, \varepsilon) := A(t, \varepsilon) - A(t, 0),$$

$$g(t, \varepsilon) := f(t, \varepsilon) - f(t, 0),$$

and

$$z_{\tau,\varepsilon} := x_{\tau,\varepsilon} - x_{\tau,0}.$$

An easy calculation shows that the difference $z(t, \varepsilon) := x(t, \varepsilon) - x(t, 0)$ solves the initial value problem

$$\begin{cases} \dfrac{dz}{dt} = A(t, 0)z(t, \varepsilon) + B(t, \varepsilon)x(t, \varepsilon) + g(t, \varepsilon), \\[2mm] z(\tau, \varepsilon) = z_{\tau,\varepsilon}. \end{cases}$$

By Theorem 1.1 this solution is representable as

(6.48)
$$z(t,\varepsilon) = U(t,\tau)z_{\tau,\varepsilon}+$$
$$\int_\tau^t U(t,\xi)B(\xi,\varepsilon)x(\xi,\varepsilon)\,d\xi + \int_\tau^t U(t,\xi)g(\xi,\varepsilon)\,d\xi,$$

where $U(t,\tau) = U(t,\tau,0)$ is the evolution operator of the "limit equation"

(6.49)
$$\frac{dx}{dt} = A(t,0)x + f(t,0).$$

But the right-hand side of (6.48) converges to zero, uniformly in $(t,\tau) \in [0,T] \times [0,T]$, since the evolution operator is bounded. Consequently, we have proved the following

Theorem 6.3. *Suppose that the functions* $c = c(t,s,\varepsilon)$ *and* $k = k(t,s,\sigma,\varepsilon)$ *satisfy the hypotheses of Theorem 2.1 or Theorem 3.1, that*

$$\lim_{\varepsilon\to0} \int_0^T \|B(t,\varepsilon)\|_{\mathfrak{L}(X)}\,dt = \lim_{\varepsilon\to0} \int_0^T \|g(t,\varepsilon)\|_X\,dt = 0,$$

and $x_{\tau,\varepsilon} \to x_{\tau,0}$ *as* $\varepsilon \to 0$. *Then* $x(t,\varepsilon) \to x(t,0)$, *uniformly on* $[0,T]$, *as* $\varepsilon \to 0$.

Theorem 6.3 may be sharpened in various directions. For stating this more precisely, let us introduce the "data-operators"

(6.50) $\mathbf{A} : [-\varepsilon_0,\varepsilon] \to L_1([0,T],\mathfrak{L}(X)), \quad \mathbf{A}(\varepsilon) := A(\cdot,\varepsilon),$

(6.51) $\mathbf{f} : [-\varepsilon_0,\varepsilon] \to L_1([0,T],X), \quad \mathbf{f}(\varepsilon) := f(\cdot,\varepsilon),$

(6.52) $\mathbf{x}_\tau : [-\varepsilon_0,\varepsilon] \to X, \quad \mathbf{x}_\tau(\varepsilon) := x_{\tau,\varepsilon},$

and the "solution-operator"

(6.53) $\mathbf{x} : [-\varepsilon_0,\varepsilon] \to C([0,T],X), \quad \mathbf{x}(\varepsilon) := x(\cdot,\varepsilon).$

Theorem 6.3 states then that *the continuity of the operators* (6.50), (6.51), *and* (6.52) *implies the continuity of the operator* (6.53). A similar result holds for Hölder continuity:

Theorem 6.4. *Suppose that the operators* (6.50), (6.51), *and* (6.52) *are Hölder continuous with respect to* ε. *Then the operator* (6.53) *is also Hölder continuous with respect to* ε.

□ Recall that a continuous function $\Phi : [a, b] \to Z$ (Z Banach space) is called Hölder continuous (with Hölder exponent $\alpha > 0$) if

$$\|\Phi(\varepsilon) - \Phi(\delta)\|_Z \leq c|\varepsilon - \delta|^\alpha \quad (a \leq \varepsilon, \delta \leq b)$$

for some $c > 0$ independent of ε and δ. Now, from the estimate (6.41) and the boundedness of the evolution operator $U(t, \tau)$ it follows that

$$\|x(\cdot, \varepsilon) - x(\cdot, \delta)\|_{C([0,T],X)} \leq M\|x_{\tau,\varepsilon} - x_{\tau,\delta}\|_X$$

$$+M \int_0^T \|f(\xi, \varepsilon) - f(\xi, \delta)\|_X \, d\xi$$

$$+M\|x(\cdot, \varepsilon)\|_{C([0,T],X)} \int_0^T \|A(\xi, \varepsilon) - A(\xi, \delta)\|_{\mathcal{L}(X)} \, d\xi,$$

where M is some positive constant independent of $\varepsilon, \delta \in [-\varepsilon_0, \varepsilon]$. This proves the assertion. ■

§ 7. Bounded and periodic solutions

In this section we consider the Barbashin equation

$$(7.1) \qquad \frac{\partial x(t,s)}{\partial t} = c(t,s)x(t,s) + \int_a^b k(t,s,\sigma)x(t,\sigma)\,d\sigma + f(t,s)$$

for $(t,s) \in J \times [a,b]$, where J is an unbounded interval; typical examples are $J = \mathbb{R}$ or $J = [t_0, \infty)$ for some fixed t_0. We derive sufficient conditions for the boundedness of solutions of (7.1) on J. The main tools are (nonclassical) fixed point principles and majorization techniques.

We also briefly sketch conditions for the existence of periodic (in t) solutions of the Barbashin equation (7.1), provided that the multiplier c, the kernel k, and the inhomogeneity f are periodic with the same period. This problem is reduced, as usual, to the study of fixed points of the corresponding Poincaré operator.

7.1. A fixed point principle

To study the existence of bounded and periodic solutions of the Barbashin equation (7.1), where the variable t is assumed to run over an unbounded interval J, we need a fixed point principle which we recall in this subsection. By $BC = BC(J \times [a,b])$ we denote the Banach space of all bounded continuous functions $x : J \times [a,b] \to \mathbb{R}$ with the usual norm

$$(7.2) \qquad \|x\|_{BC} = \sup_{t \in J} \max_{a \leq s \leq b} |x(t,s)|,$$

and by $C = C(J \times [a,b])$ the locally convex space of all continuous functions $x : J \times [a,b] \to \mathbb{R}$ with the system of seminorms

$$(7.3) \qquad p_n(x) = \sup_{t \in J_n} \max_{a \leq s \leq b} |x(t,s)| \quad (n = 1,2,...),$$

where we take $J_n = [-n,n]$ in case $J = \mathbb{R}$ and $J_n = [t_0, t_0 + n]$ in case $J = [t_0, \infty)$. Since the system (7.3) is countable, we may define a metric on this space by putting

$$(7.4) \qquad d(x,y) = \sum_{n=1}^{\infty} \frac{1}{2^n} \frac{p_n(x-y)}{1 + p_n(x-y)}.$$

In a straightforward way one can prove the following

Lemma 7.1. *The space $C(J \times [a,b])$ equipped with the system (7.3) of seminorms is a locally convex complete metric space. A sequence $x_k \in C(J \times [a,b])$ converges to $x \in C(J \times [a,b])$ if and only if the restriction $x_k|_{J_n \times [a,b]}$ converges to $x|_{J_n \times [a,b]}$ in $C(J_n \times [a,b])$ for each n. Moreover, a subset $M \subseteq C(J \times [a,b])$ is bounded (precompact) if and only if each restricted set $\{x|_{J_n \times [a,b]} : x \in M\}$ is a bounded (respectively precompact) subset of $C(J_n \times [a,b])$.*

Using this lemma, one easily obtains a result for operators Φ, not necessarily linear, which map functions of $C(J \times [a,b])$ into real functions on $J \times [a,b]$, such that $\Phi x|_{J_n \times [a,b]}$ depends only on $x|_{J_n \times [a,b]}$. This means that Φ defines, for each $n \in \mathbb{N}$, an operator Φ_n which maps functions of $C(J_n \times [a,b])$ into real functions on $J_n \times [a,b]$. We will call such operators Φ *Volterra-like* in the sequel.

Lemma 7.2. *Let Φ be a Volterra-like operator. Then Φ maps $C(J \times [a,b])$ into itself if and only if each Φ_n maps $C(J_n \times [a,b])$ into itself. Moreover, $\Phi : C(J \times [a,b]) \to C(J \times [a,b])$ is continuous, bounded or compact iff each $\Phi_n : C(J_n \times [a,b]) \to C(J_n \times [a,b])$ is.*

To state the fixed point theorem we need in what follows we have to recall some definitions. Recall that, given a bounded set M in a Banach space X, the measure of noncompactness of M is defined by (4.50). Obviously, we have $\gamma(M) = 0$ if and only if the set M is precompact.

Now suppose that $\Phi : X \to X$ is a (not necessarily linear) operator with the property that

$$(7.5) \qquad \qquad \gamma(\Phi(M)) \leq q\gamma(M)$$

for any bounded set $M \subset X$, where $q < 1$ is independent of M; such operators are usually called *condensing* (SADOVSKIJ [1972]). Intuitively speaking, the condition (7.5) means that the operator Φ makes every bounded set M "slightly more compact". For example, every *compact operator* satisfies (7.5) (with $q = 0$), as well as every *contraction*, i.e.

$$(7.6) \qquad ||\Phi(x) - \Phi(y)|| \leq q||x - y|| \quad (x, y \in X),$$

where $q < 1$ is independent of x and y. Similarly, if Φ is a sum $\Phi = \Phi_1 + \Phi_2$ of a contraction Φ_1 and a compact operator Φ_2, then Φ is condensing. We will consider such operators later (e.g. in Theorem 8.1 below).

The measure of noncompactness (7.4) may also be defined in locally convex spaces. If $\{p_n : n = 1, 2, \ldots\}$ is a system of seminorms generating the locally convex topology in a space X, the measure of noncompactness of a set $M \subset X$ is defined by

$$(7.7) \qquad \begin{aligned} \gamma_n(M) = \inf \{&\varepsilon : \varepsilon > 0, M \text{ admits a finite } \varepsilon - \text{net} \\ &\text{with respect to } p_n \text{ in } X\} \end{aligned}$$

$(n = 1, 2, \ldots)$. Likewise, an operator Φ in such a space X is called condensing if

$$(7.8) \qquad \gamma_n(\Phi(M)) \leq q_n \gamma_n(M) \quad (n = 1, 2, \ldots)$$

with $q_n < 1$ for each n. Important examples are again *compact operators* and *generalized contractions*, i.e.

$$(7.9) \qquad p_n(\Phi(x) - \Phi(y)) \leq q_n p_n(x - y) \quad (x, y \in X)$$

with $q_n < 1$. The following fixed point principle goes essentially back to DARBO [1955]:

Theorem 7.1. *Let X be a Banach space, $M \subset X$ a closed convex bounded set, and $\Phi : M \to M$ a continuous condensing operator. Then Φ has a fixed point in M.*

$\square$ We briefly sketch the idea of the proof. Define a sequence $(M_k)_k$ of subsets of X inductively by

$$M_0 = M, \; M_1 = \overline{co}\,\Phi(M), \ldots, M_{k+1} = \overline{co}\,\Phi(M_k),$$

where $\overline{co}\, M$ denotes the closed convex hull of M. Every set M_k is nonempty, convex, closed, bounded and invariant under Φ, and thus the same is true for the set

$$M_\infty = \bigcap_{k=1}^{\infty} M_k.$$

Moreover, the relation $\gamma(M_k) \leq q^k \gamma(M)$ implies that $\gamma(M_\infty) = 0$, i.e. M_∞ is compact. The assertion follows now from Schauder's classical fixed point principle applied to $\Phi : M_\infty \to M_\infty$. ∎

There is also a variant of Theorem 7.1 for locally convex spaces (see SADOVSKIJ [1972]):

Theorem 7.2. *Let X be a separable locally convex metric linear space, $M \subset X$ a closed convex bounded set, and $\Phi : M \to M$ a continuous condensing operator. Then Φ has a fixed point in M.*

7.2. Application to Barbashin operators

Consider now the Barbashin equation (7.1) on $J \times [a, b]$, subject to the initial condition

(7.10) $\qquad x(t_0, s) = x_0(s) \qquad (a \leq s \leq b),$

where t_0 is fixed and $x_0 : [a, b] \to \mathbb{R}$ is a given continuous function. Integrating (7.1) over $[t_0, t]$ yields

(7.11) $\qquad x(t, s) = \hat{C}x(t, s) + \hat{K}x(t, s) + \hat{f}(t, s),$

where we have put

$$\hat{C}x(t, s) = \int_{t_0}^{t} c(\tau, s)x(\tau, s)\, d\tau,$$

$$\hat{K}x(t, s) = \int_{t_0}^{t} \int_{a}^{b} k(\tau, s, \sigma)x(\tau, \sigma)\, d\sigma\, d\tau,$$

and

$$\hat{f}(t, s) = x_0(s) + \int_{t_0}^{t} f(\tau, s)\, d\tau.$$

The following lemma describes a connection between the operator equation (7.11) and the initial value problem (7.1)/(7.10).

Lemma 7.3. *Suppose that for each $s \in [a, b]$ the operator*

$$K_s x(t) = \int_a^b k(t, s, \sigma) x(\sigma) d\sigma$$

is bounded from $C([a, b])$ into $C(J_n)$ for each n, and that each of the functions $c(\cdot, s)$ and $f(\cdot, s)$ is continuous on J. Then every continuous solution of (7.11) is a classical solution of (7.1)/(7.10).

□ Let x be continuous on $J \times [a, b]$ and $s \in [a, b]$ be fixed. We write $x_t := x(t, \cdot)$. For $t \to \bar{t} \in J_n$ we have

$$\left| \int_a^b k(t, s, \sigma) x(t, \sigma)\, d\sigma - \int_a^b k(\bar{t}, s, \sigma) x(\bar{t}, \sigma)\, d\sigma \right|$$

$$= |K_s x_t(t) - K_s x_{\bar{t}}(\bar{t})| \leq |K_s(x_t - x_{\bar{t}})(t)| + |K_s x_{\bar{t}}(t) - K_s x_{\bar{t}}(\bar{t})| \to 0.$$

Thus the right hand side of (7.1) is continuous in t. ∎

We set $X := C(J \times [a, b])$ and are interested in conditions under which $\hat{C} : X \to X$ is bounded and $\hat{K} : X \to X$ is compact. Since $\hat{C}$ and $\hat{K}$ are Volterra-like, this can be verified by means of Lemma 7.2.

Lemma 7.4. *Suppose that k is measurable,*

$$S_n := \sup_{s \in [a, b]} \int_{J_n} \int_a^b |k(t, s, \sigma)|\, d\sigma\, dt < \infty \quad (n = 1, 2, ...),$$

and

$$\lim_{s \to \bar{s}} \int_{J_n} \int_a^b |k(t, s, \sigma) - k(t, \bar{s}, \sigma)|\, d\sigma\, dt = 0$$

for each $\bar{s} \in [a, b]$. Then $\hat{K}$ maps X into X and is compact.

□ Let $n \in \mathbb{N}$, $M > 0, \varepsilon > 0$, and $(\bar{t}, \bar{s}) \in J_n \times [a, b]$ be arbitrary. Choose $\delta > 0$ so small that for $|s - \bar{s}| \leq \delta$ we have

$$\int_{J_n} \int_a^b |k(\tau, s, \sigma) - k(\tau, \bar{s}, \sigma)|\, d\sigma\, d\tau \leq \varepsilon.$$

Consequently, for all $D \subseteq J_n$ with $\mathrm{mes}\, D \leq \delta$ (without loss of generality) and $|s - \bar{s}| \leq \delta$, we have

$$\int_D \int_a^b |k(\tau, s, \sigma)|\, d\sigma\, d\tau \leq \varepsilon + \int_D \int_a^b |k(\tau, \bar{s}, \sigma)|\, d\sigma\, d\tau \leq 2\varepsilon.$$

For each $x \in C(J_n \times [a, b])$ with $\|x\|_{C(J_n \times [a,b])} \leq M$ and each $(t, s) \in J_n \times [a, b]$ with $|t - \bar{t}| \leq \delta$, $|s - \bar{s}| \leq \delta$ this implies:

$$|\hat{K}x(t, s) - \hat{K}x(\bar{t}, \bar{s})| \leq M \int_{J_n} \int_a^b |k(\tau, s, \sigma) - k(\tau, \bar{s}, \sigma)|\, d\sigma\, d\tau + 2M\varepsilon \leq 3M\varepsilon.$$

This means, that the set $\{\hat{K}x : \|x\|_{C(J_n \times [a,b])} \leq M\}$ is equicontinuous in $J_n \times [a, b]$. Since this set is also bounded (by $S_n M$), the statement follows from Lemma 7.2 and the Arzelà-Ascoli compactness criterion. ∎

Lemma 7.5. *Suppose that the following three conditions are satisfied:*

(a) *the function* $\alpha : J \times [a, b] \to \mathbb{R}$ *defined by*

$$\alpha(t, s) = \int_{t_0}^t c(\tau, s)\, d\tau$$

is continuous;

(b) *the function* $\beta : J \times J \times [a, b] \to \mathbb{R}$ *defined by*

$$\beta(t, z, s) = \begin{cases} \displaystyle\int_{t_0}^{\min\{t,z\}} (z - \tau) c(\tau, s)\, d\tau & \text{if } z \geq t_0, \\[2mm] \displaystyle\int_z^t (z - \tau) c(\tau, s)\, d\tau & \text{if } t \leq z < t_0, \\[2mm] 0 & \text{if } z < \min\{t, t_0\} \end{cases}$$

is continuous;

(c) *the number*

(7.12)
$$q_n = \sup_{a \leq s \leq b} \int_{J_n} |c(\tau, s)|\, d\tau$$

is finite for each n.

Then $\hat{C}$ maps X into X and is bounded.

□ It suffices to prove that the mapping $B : [a, b] \to \mathfrak{L}(C(J_n))$ defined by

$$B(s)x(t) := \int_0^t c(\tau, s)x(\tau)\, d\tau$$

is strongly continuous and bounded for fixed n. Indeed, if we write now $x_s := x(\cdot, s)$ we have for $(t, s) \to (\bar{t}, \bar{s}) \in J_n \times [a, b]$ that

$$|\hat{C}x(t, s) - \hat{C}x(\bar{t}, \bar{s})| = |B(s)x_s(t) - B(\bar{s})x_{\bar{s}}(\bar{t})|$$

$$\leq |B(s)(x_s - x_{\bar{s}})(t)| + |(B(s) - B(\bar{s}))x_{\bar{s}}(t)|$$

$$+ |B(\bar{s})x_{\bar{s}}(t) - B(\bar{s})x_{\bar{s}}(\bar{t})| \to 0.$$

The strong continuity of B is, by Theorem 2.1, equivalent to the conditions (a) - (c) above on J_n (without loss of generality we may assume that $t_0 \in J_n$). These conditions imply also the boundedness of B, and so we are done. ■

Theorem 7.3. *Let the hypotheses of Lemmas 7.3, 7.4 and 7.5 be satisfied, and assume that $q_n < 1$ for each n, with q_n given by (7.12). Suppose that there exists a nonempty, closed, bounded and convex subset $M \subseteq C(J \times [a, b])$ such that $\hat{C}x + \hat{K}x + \hat{f} \in M$ for all $x \in M$. Then (7.1)/(7.10) has a (classical) solution in M.*

□ Define $\Phi : X \to X$ by

$$(7.13) \qquad\qquad \Phi(x) = \hat{C}x + \hat{K}x + \hat{f}$$

and fix $B \subseteq M$. By Lemma 7.4, the set $\hat{K}(B)$ is precompact in X, and hence $\gamma_n(\hat{K}(B)) = 0$ for all n. On the other hand, Lemma 7.5 implies that

$$p_n(\hat{C}x - \hat{C}y) = \sup_{t \in J_n} \sup_{a \leq s \leq b} \left| \int_{t_0}^t c(\tau, s)[x(\tau, s) - y(\tau, s)]\, d\tau \right|$$

$$\leq q_n p_n(x - y),$$

hence $\gamma_n(\hat{C}(B)) \leq q_n \gamma_n(B)$. Altogether we have

$$\gamma_n(\Phi(B)) \leq \gamma_n(\hat{C}(B)) + \gamma_n(\hat{K}(B)) + \gamma_n(\hat{f}) \leq q_n \gamma_n(B),$$

i.e. (7.8) is true. The assertion follows now from Theorem 7.2. ■

In particular, if M contains only bounded functions, we get the existence of a bounded solution. The choice of M depends on k, c and f. For example, assume that $J = [t_0, \infty)$, and the measurable functions k, c, and f are nonnegative. Assume, furthermore, that there exists a measurable function $y : J \times [a, b] \to [0, \infty)$ which is bounded on compact subsets and satisfies

$$\int_{t_0}^{t} c(\tau, s) y(\tau, s) \, d\tau + \int_{t_0}^{t} \int_{a}^{b} k(\tau, s, \sigma) y(t, \sigma) \, d\sigma \, d\tau$$

$$+ \hat{f}(t, s) \leq y(t, s).$$

Then, under the assumptions of the previous theorem, the set

$$M := \{x : x \in C(J \times [a, b]), 0 \leq x(t, s) \leq y(t, s)\}$$

satisfies $AM \subseteq M$, i.e. (7.1) has a nonnegative solution, which is bounded by y (since the local boundedness of y implies the boundedness of M).

Brute-force-estimates on norms lead to a stronger result, since then we can directly apply Banach's fixed point theorem:

Theorem 7.4. *Let k and c be measurable and such that*

$$q := \operatorname*{ess\,sup}_{a \leq s \leq b} \left(\int_{J} |c(\tau, s)| \, d\tau + \int_{J} \int_{a}^{b} |k(\tau, s, \sigma)| \, d\sigma \, d\tau \right) < 1.$$

Suppose that

$$\hat{f}(t, s) := \int_{t_0}^{t} f(\tau, s) \, d\tau + x_0(s)$$

is essentially bounded on $J \times [a, b]$. Then the initial value problem (7.1)/(7.10) has exactly one generalized measurable essentially bounded solution x, namely the unique measurable essentially bounded solution of (7.11). A bound is given by

$$\beta := \frac{1}{1 - q} \|\hat{f}\|_{L_\infty}.$$

Furthermore, if Φ given by (7.13) maps $C(J \times [a,b])$ into itself, x is continuous. Consequently, under the conditions of Lemma 7.3, x is a classical solution of (7.1)/(7.10).

☐ The generalized measurable essentially bounded solutions of (7.1)/(7.10) are exactly those of (7.11). So we only consider (7.11). Now apply Banach's fixed point theorem to $\Phi : X \to X$, where $X = L_\infty(J \times [a,b])$, $X = \{x : x \in L_\infty(J \times [a,b]), ||x||_{L_\infty} \le \beta\}$ respectively $X = L_\infty(J \times [a,b]) \cap C(J \times [a,b])$. In the second case A maps X into itself, because $||x||_{L_\infty} \le \beta$ implies

$$||Ax||_{L_\infty} \le q||x||_{L_\infty} + ||\hat{f}||_{L_\infty} \le q\beta + ||\hat{f}||_{L_\infty} = \beta.$$

This proves the assertion. ■

7.3. The use of majorant functions

We return to the initial value problem

$$(7.14) \qquad \frac{dx}{dt} = A(t)x + f(t), \quad x(t_0) = x_0,$$

where as before

$$A(t)x(t,s) = c(t,s)x(t,s) + \int_a^b k(t,s,\sigma)x(t,\sigma)\,d\sigma + f(t,s)$$

for $t \in J = [t_0, \infty)$, $a \le s \le b$. The following reasoning is motivated by the fact that the *scalar* differential equation

$$\frac{dx}{dt} = a(t)x(t) + f(t) \quad (t_0 \le t < \infty)$$

has, for each $f \in BC(J)$, only bounded solutions if $a(t) \le \alpha < 0$ for $t_0 \le t < \infty$. One might expect that a similar result for (7.14) is true if $c(t,s) \le \gamma < 0$ and $|k(t,s,\sigma)|$ is small in some sense. To get a general result let us call a function $M : J \times [a,b] \to [0,\infty)$ a *majorant function* for (7.14) if

$$(7.15) \qquad \frac{\partial M(t,s)}{\partial t} > c(t,s)M(t,s) + k_M(t,s)$$

for all $(t, s) \in J \times [a, b]$, where we have used the abbreviation

$$k_M(t, s) := \int_a^b |k(t, s, \sigma)| \, M(t, \sigma) \, d\sigma.$$

Theorem 7.5. *Suppose that there exists a majorant function M for (7.14) such that the number*

$$(7.16) \qquad \Sigma := \sup_{\substack{t \geq t_0 \\ a \leq s \leq b}} \left| \frac{f(t, s)}{c(t, s)M(t, s) + k_M(t, s) - \partial M(t, s)/\partial t} \right|$$

is finite. Let $C > \Sigma$ be arbitrary. Then any solution x of (7.14) whose initial value x_0 is M-majorized, i.e.

$$|x_0(s)| \leq CM(t_0, s) \quad (a \leq s \leq b),$$

satisfies

$$|x(t, s)| \leq CM(t, s) \quad (t \in J, \ a \leq s \leq b).$$

☐ Let x be a solution of (7.14), and denote by t the infimum of all τ such that $|x(\tau, s)| > CM(\tau, s)$ for some s. From $x(t, s) = CM(t, s)$ it follows that

$$\frac{\partial x(t, s)}{\partial t} = c(t, s)x(t, s) + \int_a^b k(t, s, \sigma)x(t, \sigma) \, d\sigma + f(t, s)$$

$$\leq Cc(t, s)M(t, s) + Ck_M(t, s) + f(t, s) < C\frac{\partial M(t, s)}{\partial t}.$$

Similarly, for $x(t, s) = -CM(t, s)$ we get

$$\frac{\partial x(t, s)}{\partial t} = c(t, s)x(t, s) + \int_a^b k(t, s, \sigma)x(t, \sigma) \, d\sigma + f(t, s)$$

$$\geq Cc(t, s)M(t, s) - Ck_M(t, s) - f(t, s) > -C\frac{\partial M(t, s)}{\partial t}.$$

This contradiction proves the assertion. ∎

For the left-open interval $J = (-\infty, t_0]$ one gets an analogous result if one defines a *minorant function* $m : J \times [a, b] \to [0, \infty)$ by the property

(7.17) $$\frac{\partial m(t, s)}{\partial t} < c(t, s)m(t, s) - k_m(t, s)$$

with

$$k_m(t, s) := \int_a^b |k(t, s, \sigma)| \, m(t, \sigma) \, d\sigma,$$

and replaces (7.16) by

(7.18) $$\Sigma := \sup_{\substack{t \le t_0 \\ a \le s \le b}} \frac{|f(t, s)|}{c(t, s)m(t, s) - k_m(t, s) - \partial m(t, s)/\partial t}.$$

We remark that one may replace the derivative of M in (7.15) by the right hand side derivative, and the derivative of m in (7.17) by the left hand side derivative; this is useful in some applications.

Theorem 7.5 typically applies to Barbashin operators in the space $C([a, b])$. We give now a parallel result for the space $X = L_p([a, b])$. Consider in X the differential equation

(7.19) $$\frac{dx}{dt} = A(t)x + f(t)$$

$(t \in J)$, where each $A(t)$ maps X into X and $f(t) \in X$. We look for conditions which assure that the L_p-norm $||x(t)||_{L_p}$ is bounded or dominated by a fixed function $y(t)$. For this we can use the previous ideas together with the following

Lemma 7.6. Let $x : [\alpha, \beta] \to X = L_p([a, b])$ be differentiable (in α and β we consider the one-sided derivative). Define

$$\frac{d^+||x(t)||_{L_p}}{dt} := \lim_{h \to 0+} \frac{||x(t + h)||_{L_p} - ||x(t)||_{L_p}}{h} \qquad (t < \beta)$$

and

$$\frac{d^-||x(t)||_{L_p}}{dt} := \lim_{h \to 0+} \frac{||x(t)||_{L_p} - ||x(t - h)||_{L_p}}{h} \qquad (t > \alpha).$$

Then for all t with $x(t) \neq 0$ we have

(7.20)
$$\frac{d^{\pm}||x(t)||_{L_p}}{dt} =$$

$$||x(t)||_{L_p}^{1-p} \int_a^b \frac{dx(t)}{dt}(s)|x(t,s)|^{p-1}\operatorname{sign}x(t,s)\,ds$$

if $1 < p < \infty$, and

(7.21)
$$\frac{d^{\pm}||x(t)||_{L_1}}{dt} = \int_a^b \frac{dx(t)}{dt}(s)y_{\pm}(t,s)\,ds,$$

where

$$y_{\pm}(t,s) := \begin{cases} \operatorname{sign} x(t,s) & \text{if } x(t,s) \neq 0, \\ \pm\operatorname{sign} \frac{dx(t)}{dt}(s) & \text{if } x(t,s) = 0. \end{cases}$$

□ Evidently,

$$||x(t)||_{L_p}\frac{d^{\pm}||x(t)||_{L_p}}{dt} = \langle x'(t), x(t)\rangle_{\pm},$$

with $\langle y, x\rangle_+ := \max x^*(y)$ and $\langle y, x\rangle_- := \min x^*(y)$, where the maximum respectively minimum is taken over all values x^* of the duality map of x (see e.g. MARTIN [1976]). Writing out the explicit form of the duality map in L_p ($1 \leq p < \infty$) gives (7.20) and (7.21). ■

Theorem 7.6. *Let $J = [t_0, \infty)$ and $X = L_p([a,b])$ with $1 \leq p < \infty$. Suppose that there exists a positive continuous function $y : J \to (0, \infty)$ such that the right hand side derivative d^+y/dt exists everywhere, and such that $x \in X$, $||x||_{L_p}$ implies*

(7.22)
$$||x||_{L_p}^{1-p} \int_a^b [A(t)x(s) + f(t,s)]|x(s)|^{p-1}\operatorname{sign} x(s)\,ds$$

$$< \frac{d^+y(t)}{dt}$$

if $1 < p < \infty$, *and*

(7.23) $$\int_a^b [A(t)x(s) + f(t,s)]z_+(t,s)\,ds < \frac{d^+y(t)}{dt}$$

if $p = 1$, *where*

$$z_+(t,s) := \begin{cases} \operatorname{sign} x(s) & \text{if } x(s) \neq 0, \\ \operatorname{sign} [A(t)x(s) + f(t,s)] & \text{if } x(s) = 0. \end{cases}$$

Then each pointwise solution x *of* (7.19) *with* $\|x(t_0)\|_{L_p} \leq y(t_0)$ *satisfies* $\|x(t)\|_{L_p} \leq y(t)$ *for all* $t \in J$.

□ By Lemma 7.6 and our choice of y we have

$$\frac{d^+ \|x(t)\|_{L_p}}{dt} < \frac{d^+ y(t)}{dt},$$

from which the assertion follows. ∎

For $J = (-\infty, t_0]$ we get an analogous result if we replace (7.22) and (7.23) by

(7.24) $$\|x\|_{L_p}^{1-p} \int_a^b [A(t)x(s) + f(t,s)]|x(s)|^{p-1}\operatorname{sign} x(s)\,ds$$
$$> \frac{d^- y(t)}{dt}$$

if $1 < p < \infty$, *and*

(7.25) $$\int_a^b [A(t)x(s) + f(t,s)]z_-(t,s)\,ds > \frac{d^- y(t)}{dt}$$

if $p = 1$, *where*

$$z_-(t,s) := \begin{cases} \operatorname{sign} x(s) & \text{if } x(s) \neq 0, \\ -\operatorname{sign} [A(t)x(s) + f(t,s)] & \text{if } x(s) = 0. \end{cases}$$

Theorem 7.6 is particularly useful in case $y(t) \equiv const.$ (which often occurs in applications). For example, in the Hilbert space case $p = 2$ condition (7.22) is satisfied for $y(t) \equiv R$ if $\langle A(t)x, x \rangle < -\|f(t)\|_{L_2}\|x\|_{L_2}$ for $\|x\|_{L_2} = R$.

7.4. The Green function

In this subsection we consider the operator function

$$(7.26) \qquad A(t)x(s) = c(t,s)x(s) + \int_a^b k(t,s,\sigma)x(\sigma)\,d\sigma$$

for t on the whole real line $J = \mathbb{R}$. Assume that (7.26) defines a strongly continuous operator function with values in $\mathfrak{L}(X)$, where either $X = C$ or $X = L_p$ ($1 \leq p \leq \infty$). We are interested in conditions on the multiplier $c = c(t,s)$ and the kernel $k = k(t,s,\sigma)$ under which there exists a *Green function* for the bounded solutions of the differential equation (7.19) in the space X (and hence for the bounded solutions of the corresponding integro-differential equation of Barbashin type).

The basic results on the existence and properties of Green's function for bounded solutions may be found in DALETSKIJ-KREJN [1970]. For example, it is well-known that the Green function $G = G(t, \tau)$ exists if and only if (7.19) has, for each $f \in BC(\mathbb{R})$, a unique bounded solution x on $\mathbb{R}$; in this case, the integral representation

$$(7.27) \qquad x(t) = \int_{-\infty}^{\infty} G(t,\tau)f(\tau)\,d\tau$$

holds. Unfortunately, a simple procedure for finding the Green function G is not known. Even in the stationary case $A(t) \equiv A$, i.e.

$$Ax(s) = c(s)x(s) + \int_a^b k(s,\sigma)x(\sigma)\,d\sigma$$

the situation is rather complicated, but the existence of G may be guaranteed by the following (intrinsic) characterization: *The Green function exists if and only if the spectrum $\sigma(A)$ of A does meet the imaginary axis.* On the other hand, as already pointed out, the determination of $\sigma(A)$ is far from being trivial. In the special case of a constant multiplier $c(s) \equiv c$ one may of course use the relation $\sigma(A) = c + \sigma(K)$. In another special case when the integral operator K is compact, the boundedness of the function $s \mapsto 1/c(s)$ is necessary for the spectrum $\sigma(A)$ not to meet the imaginary axis.

The situation is particularly pleasant if we know not only that $\sigma(A)$ does not contain purely imaginary points, but also know the projections P_- and P_+ corresponding to the parts $\sigma_-(A) = \{\lambda : \lambda \in \sigma(A), \operatorname{Re}\lambda < 0\}$ and $\sigma_+(A) = \{\lambda : \lambda \in \sigma(A), \operatorname{Re}\lambda > 0\}$, respectively, of the spectrum. In this case the Green function may be given explicitly by

$$(7.28) \qquad G(t, \tau) = \begin{cases} -e^{A(t-\tau)} P_+ & \text{if } -\infty < \tau \leq t < \infty, \\ e^{A(t-\tau)} P_- & \text{if } -\infty < t \leq \tau < \infty. \end{cases}$$

The case of a non-stationary operator function (7.26) is much more difficult. In this case the existence of Green's function is equivalent to the so-called *exponential dichotomy* of the homogeneous differential equation

$$(7.29) \qquad \frac{dx}{dt} = A(t)x.$$

To verify the exponential dichotomy condition, however, one has to know again the evolution operator $U = U(t, \tau)$ for (7.29) (see Subsection 1.2), and so this criterion is only of theoretical interest.

It turns out that the problem of finding Green's function may be attacked successfully by the same idea that we employed for asymptotic stability in Subsection 5.3: the operator function $A = A(t)$ given in (7.26) may be viewed as an "integral perturbation" of the simpler operator function

$$(7.30) \qquad C(t)x(s) = c(t, s)x(s).$$

In case of the multiplication operator (7.30) it is easy to write down the Green function $G_0 = G_0(t, \tau)$ for the corresponding differential equation

$$(7.31) \qquad \frac{dx}{dt} = C(t)x + f(t).$$

For instance, the following holds (APPELL-DE PASCALE-DIALLO [1994]):

Lemma 7.7. *Suppose that the limit*

$$(7.32) \qquad c_0(s) := \lim_{T \to \infty} \frac{1}{T} \int_t^{t+T} c(\xi, s)\, d\xi$$

exists uniformly with respect to $t \in \mathbb{R}$, and the function $s \mapsto 1/c_0(s)$ is essentially bounded. Then the Green function $G_0 = G_0(t, \tau)$ for the differential equation (7.31) *exists and has the form*

$$G_0(t, \tau) f(s) = g_0(t, \tau, s) f(s),$$

where
(7.33)

$$g_0(t, \tau, s) = \begin{cases} -e(t, \tau, s) & \text{if } -\infty < \tau \le t < \infty \text{ and } c_0(s) > 0, \\ +e(t, \tau, s) & \text{if } -\infty < t \le \tau < \infty \text{ and } c_0(s) < 0, \\ 0 & \text{otherwise}, \end{cases}$$

with $e = e(t, \tau, s)$ given by (1.29).

The statement of Lemma 7.7 is true in any of the spaces C and L_p ($1 \le p \le \infty$). However, if the function c_0 in (7.32) is continuous, the function $s \mapsto 1/c_0(s)$ may be bounded, of course, only if c_0 does not change its sign. In this case the function g_0 in (7.33) becomes somewhat simpler.

7.5. Comparison of Green functions

In order to compare the Green function G_0 of the "reduced" equation (7.31) and the Green function G of the "full" equation (7.19) it is useful to work with special norms. As in Subsection 3.3, by $\mathfrak{Z}_p$ ($1 \le p \le \infty$) we denote the set of all measurable kernel functions $z : [a, b] \times [a, b] \to \mathbb{R}$ for which the norm (3.16) is finite.

Let $c : \mathbb{R} \times [a, b] \to \mathbb{R}$ and $k : \mathbb{R} \times [a, b] \times [a, b] \to \mathbb{R}$ be measurable, and put

$$(7.34) \qquad \gamma(s) = \begin{cases} \displaystyle \sup_{t \in \mathbb{R}} \int_{-\infty}^{t} g_0(t, \tau, s) \, d\tau & \text{if} \quad c_0(s) < 0, \\[4mm] \displaystyle \sup_{t \in \mathbb{R}} \int_{t}^{\infty} g_0(t, \tau, s) \, d\tau & \text{if} \quad c_0(s) > 0, \end{cases}$$

$$(7.35) \qquad \delta = \sup_{t \in \mathbb{R}} \| c(t, \cdot) \|_\infty,$$

$$(7.36) \qquad \mu_0 = \sup_{t \in \mathbb{R}} \| k(t, \cdot, \cdot) \|_{3p},$$

and

$$(7.37) \qquad \mu = \sup_{-\infty < \alpha < \beta < \infty} \frac{1}{1 - \alpha + \beta} \left\| \int_{\alpha}^{\beta} k(\xi, \cdot, \cdot) \, d\xi \right\|_{3p}.$$

Under the hypotheses of Lemma 7.7, the differential operator

$$(7.38) \qquad L_0 x = \frac{dx}{dt} - C(t)x$$

is then continuously invertible with

$$(7.39) \qquad \| L_0^{-1} \|_{\mathcal{L}(BC)} \leq \| \gamma \|_{L_\infty}.$$

The following theorem gives condition on the numbers (7.35), (7.36) and (7.37) under which the differential operator

$$(7.40) \qquad Lx = \frac{dx}{dt} - A(t)x$$

is continuously invertible as well:

Theorem 7.7. *Suppose that the functions* $c = c(t, s)$ *and* $k = k(t, s, \sigma)$ *satisfy the hypotheses of Theorem 2.1 (of Theorem 3.1, respectively). Assume, morover, that*

$$(7.41) \qquad (1 + \delta) \, \delta \, \| \gamma \|_{L_\infty}^2 \, \mu_0 \, \mu < 1.$$

Then the differential equation (7.19) admits a Green function $G = G(t, \tau)$ in the space C (in the space L_p, respectively).

□ We consider only the case $X = C$, the case $X = L_p$ is similar. We claim that, under the hypothesis (7.41), the norm of $L_0^{-1} K$ in the space $BC = BC(\mathbb{R})$ is small, where

$$K(t)x(s) = \int_a^b k(t, s, \sigma)x(\sigma)\, d\sigma.$$

To prove this we derive some norm estimates. First, let $BC^1 = BC^1(\mathbb{R})$ be the space of all continuously differentiable functions x with $x \in BC$ and $x' \in BC$, equipped with the norm

$$(7.42) \qquad \|x\|_{BC^1} = \max\{\|x\|_{BC}, \|x'\|_{BC}\}.$$

Moreover, by $B = B(\mathbb{R})$ we denote the space BC with the special norm

$$(7.43) \qquad \|x\|_B = \sup_{-\infty < \alpha < \beta < \infty} \frac{1}{1 + \beta - \alpha} \left| \int_\alpha^\beta x(\tau)\, d\tau \right|.$$

The space B which is sometimes called *Bogoljubov space*, is of fundamental importance in the theory of differential equations: in fact, if the Green function G_0 for the operator (7.38) exists, the formula

$$(7.44) \qquad L_0^{-1} f(t) = \int_{-\infty}^\infty G_0(t, \tau)f(\tau)\, d\tau$$

defines an isomorphism from $B(\mathbb{R})$ into $BC^1(\mathbb{R})$ (see ZABREJKO-MAZEL'-TRETJAKOVA [1984], ZABREJKO-PETROVA [1980]). Now, the obvious identity

$$\frac{d}{dt} L_0^{-1} f(t) = f(t) + C(t)L_0^{-1} f(t) \quad (f \in BC)$$

gives

$$\|L_0^{-1}\|_{\mathfrak{L}(BC, BC^1)} \leq (1 + \delta)\|L_0^{-1}\|_{\mathfrak{L}(BC)} \leq (1 + \delta)\|\gamma\|_{L_\infty},$$

by (7.39). Moreover, from the definitions (7.36) and (7.37) we see that

$$(7.45) \qquad \|K\|_{\mathfrak{L}(BC)} \leq \mu_0, \qquad \|K\|_{\mathfrak{L}(BC^1, B)} \leq \mu.$$

Finally, integrating by parts one obtains

$$\|L_0^{-1} f\|_{BC} \le \delta \|\gamma\|_{L_\infty} \|f\|_B \quad (f \in BC),$$

hence

(7.46) $$\|L_0^{-1}\|_{\mathfrak{L}(B,BC)} \le \delta \|\gamma\|_{L_\infty}.$$

Combining now (7.44), (7.45) and (7.46) we conclude that

$$\|(L_0^{-1} K)^2\|_{\mathfrak{L}(BC)} = \|L_0^{-1} K L_0^{-1} K\|_{\mathfrak{L}(BC)}$$

$$\le \|L_0^{-1}\|_{\mathfrak{L}(B,BC)} \|K\|_{\mathfrak{L}(BC^1,B)} \|L_0^{-1}\|_{\mathfrak{L}(BC,BC^1)} \|K\|_{\mathfrak{L}(BC)} < 1,$$

by (7.41). Consequently, the Neumann series

$$(I - L_0^{-1} K)^{-1} = \sum_{k=0}^{\infty} (L_0^{-1} K)^k$$

converges in $\mathfrak{L}(BC)$. But

$$(I - L_0^{-1} K)^{-1} = (I - L_0^{-1} [A(t) - C(t)])^{-1} = L^{-1} L_0,$$

and so we are done. ∎

The crucial point in the hypotheses of Theorem 7.7 is the calculation of the constants (7.36) and (7.37); this may be done by means of Lemma 2.1 in the space C, and Lemmas 3.8-3.12 in the space L_p.

7.6. The second Bogoljubov theorem

Now we study the "stability" of the Green function for equations depending on a parameter ε, i.e.

(7.47) $$\frac{dx}{dt} = A(t, \varepsilon) x + f(t),$$

where

(7.48) $$A(t, \varepsilon) x(s) = c(t, s, \varepsilon) x(s) + \int_a^b k(t, s, \sigma, \varepsilon) x(\sigma) \, d\sigma.$$

As before, we suppose that the operator function $t \mapsto A(t, \varepsilon)$ is strongly continuous in $X = C$ or $X = L_p$ for each $\varepsilon \in [-\varepsilon_0, \varepsilon_0]$.

There are many results on the continuous dependence of the Green function $G = G(t, \tau, \varepsilon)$ of (7.47), the most interesting being probably those given by BOGOLJUBOV [1945]. We remark that for the continuous dependence of $G(t, \tau, \varepsilon)$ with respect to ε it suffices that $A(t, \varepsilon)$ is just integrable with respect to ε (see e.g. BOGOLJUBOV-MITROPOL'SKIJ [1963], MITROPOL'SKIJ [1971]).

Let us first recall a result on the continuous dependence of Green's function in the norm (7.43) (ZABREJKO-KOLESOV-KRASNOSEL'SKIJ [1969]):

Lemma 7.8. *Suppose that the operator function (7.48) is strongly continuous in $X = C$ or $X = L_p$ and bounded in measure. Assume that*

$$\lim_{\varepsilon \to 0} ||A(t, \varepsilon) - A(t, 0)||_B = 0$$

and the Green function $G = G(t, \tau, 0)$ for the problem

$$(7.49) \qquad \frac{dx}{dt} = A(t, 0)x + f(t)$$

exists. Then the Green function $G = G(t, \tau, \varepsilon)$ for (7.47) exists as well for $|\varepsilon| \leq \varepsilon_0$, and

$$(7.50) \qquad \lim_{\varepsilon \to 0} \sup_{-\infty < t, \tau < \infty} e^{\delta(t-\tau)} ||G(t, \tau, \varepsilon) - G(t, \tau, 0)||_{\mathcal{L}(B)} = 0$$

for any $\delta > 0$.

The application of Lemma 7.8 to the operator (7.48) immediately gives the following

Theorem 7.8. *Suppose that the operator function (7.48) is strongly continuous in $X = C$ or $X = L_p$ and bounded in measure. Assume that*

$$(7.51) \qquad \sup_{\alpha < \beta} \frac{1}{1 + \beta - \alpha} \left\| \int_\alpha^\beta [c(\xi, \cdot, \varepsilon) - c(\xi, \cdot, 0)] \, d\xi \right\|_{L_\infty} \to 0$$

and

$$(7.52) \qquad \sup_{\alpha < \beta} \frac{1}{1 + \beta - \alpha} \left\| \int_\alpha^\beta [k(\xi, \cdot, \cdot, \varepsilon) - k(\xi, \cdot, \cdot, 0)] \, d\xi \right\|_{3p} \to 0$$

as $\varepsilon \to 0$. Suppose that the Green function $G = G(t, \tau, 0)$ for the problem (7.49) exists. Then the Green function $G = G(t, \tau, \varepsilon)$ for (7.47) exists as well for $|\varepsilon| \leq \varepsilon_0$, and (7.50) holds.

Theorem 7.8 extends the famous Bogoljubov lemma to linear integro-differential equations of Barbashin type. Building on this theorem one may also establish a result which is analogous to Bogoljubov's so-called "second averaging theorem" (see e.g. BOGOLJUBOV [1945], BOGOLJUBOV-MITROPOL'SKIJ [1963], DIALLO-ZABREJKO [1987a]) for the equation

$$\frac{dx}{dt} = A\left(\frac{t}{\varepsilon}\right) x$$

(see also KRASNOSEL'SKIJ-KREJN [1956], MITROPOL'SKIJ [1971], ZABREJKO-KOLESOV-KRASNOSEL'SKIJ [1969]). Since the operator $A(t/\varepsilon)$ is not defined for $\varepsilon = 0$, we have to redefine it for $\varepsilon = 0$ by

$$(7.53) \qquad A(t, 0)x(s) = c_0(s)x(s) + \int_a^b k_0(s, \sigma)x(\sigma) \, d\sigma,$$

where $c_0 : [a, b] \to \mathbb{R}$ and $k_0 : [a, b] \times [a, b] \to \mathbb{R}$ are two functions satisfying

$$(7.54) \qquad \lim_{T \to \infty} \sup_{-\infty < t < \infty} \left\| \int_t^{t+T} [c(\xi, \cdot) - c_0(\cdot)] \, d\xi \right\|_{L_\infty} = 0$$

and

$$(7.55) \qquad \lim_{T \to \infty} \sup_{-\infty < t < \infty} \left\| \int_t^{t+T} [k(\xi, \cdot, \cdot) - k_0(\cdot, \cdot)] \, d\xi \right\|_{3p} = 0.$$

In this connection it is interesting to note that, if we replace the operator (7.53) by the non-stationary operator

$$A(t, 0)x(s) = c_0(t, s)x(s) + \int_a^b k_0(t, s, \sigma)x(\sigma) \, d\sigma,$$

the conditions (7.54) and (7.55) do not imply, in general, the conditions (7.51) and (7.52).

7.7. Application of Darbo's fixed point principle

Let us now pass to the study of *periodic solutions* of integro-differential equations of Barbashin type. Consider the equation

$$(7.56) \qquad \frac{dx}{dt} = A(t)x + f(t) \qquad (t \in \mathbb{R}),$$

where $A(t) = C(t) + K(t)$ with

$$C(t)x(s) = c(t, s)x(s)$$

and

$$K(t)x(s) = \int_a^b k(t, s, \sigma)x(\sigma)\, d\sigma$$

in some ideal space X (see Subsection 4.1). We assume throughout this paragraph that c, k and f are *periodic* with respect to t with the same period $T > 0$, i.e.

$$(7.57) \qquad \begin{aligned} c(t + T, s) &= c(t, s), \quad k(t + T, s, \sigma) = k(t, s, \sigma), \\ f(t + T, s) &= f(t, s). \end{aligned}$$

Furthermore, we suppose that the operator function $t \mapsto A(t)$ is strongly continuous and locally integrable in X, and hence the problem (7.56) with initial condition $x(0) = x_0 \in X$ has the unique solution

$$x(t) = U(t, 0)x_0 + \int_0^t U(t, \xi)f(\xi)\, d\xi;$$

we will denote this solution by $x(t; x_0)$ in what follows to emphasize the dependence on x_0.

Now consider the operator $\Phi : X \to X$ which assigns to each $x_0 \in X$ the value of the solution starting from x_0 after one period T, i.e.

$$(7.58) \qquad \Phi x_0 = x(T; x_0).$$

This operator is called *Poincaré operator* or *shift operator* (along the trajectories of the problem (7.56)). Of course, there is a 1-1 correspondence between the T-periodic solutions of (7.56) and the fixed points of this operator: x is a T-periodic solution of (7.56) if and only if x starts from a fixed point x_0 of the operator (7.58). This allows us to reduce the existence problem for T-periodic solutions to the fixed point problem for Φ. For example, the following holds.

Theorem 7.9. *Suppose that (7.57) holds for some $T > 0$, X is an ideal space or $X = C$, and $t \mapsto A(t)$ is a strongly continuous and locally integrable operator function in X. Define $H(t, \tau)$ by the equality*

$$(7.59) \qquad U(t, \tau)x(s) = e(t, \tau, s)x(s) + H(t, \tau)x(s),$$

where $e(t, \tau, s)$ is given by (1.29), and assume that one of the following conditions is satisfied:

(a) $\displaystyle \sup_{a \le s \le b} \int_0^T c(t, s)\, dt < 0, \quad H(T, 0) \in \mathcal{L}_c(X),$

(b) $\displaystyle \inf_{a \le s \le b} \int_0^T c(t, s)\, dt > 0, \quad H(0, T) \in \mathcal{L}_c(X).$

Then (7.56) has a T-periodic solution if and only if there exists a nonempty, closed, bounded and convex subset $M \subseteq X$ which is invariant under Φ (in case (a)) respectively Φ^{-1} (in case (b)), where Φ is the mapping (7.58). Moreover, if $H(T, 0) = 0$ respectively $H(0, T) = 0$ there exists at most one T-periodic solution.

☐ Suppose that (a) holds. We may write Φ as sum $\Phi = \Phi_1 + \Phi_2$ of the operators

$$\Phi_1 x_0(s) = e(T, 0, s)x_0(s)$$

and

$$\Phi_2 x_0(s) = H(T, 0)x_0(s).$$

By assumption (a), Φ_1 is a contraction with constant

$$(7.60) \qquad q := \sup_{a \le s \le b} e(T, 0, s) < 1,$$

and Φ_2 is a compact operator in X. Thus the assertion follows from Theorem 7.1.

If (b) holds we replace T by $-T$ and use the periodicity assumptions
(7.57). Finally, the equality $H(T,0) = 0$ (or $H(0,T) = 0$) means that
Φ_2 is the zero operator, and the uniqueness of the fixed point x_0 of
$\Phi = \Phi_1$ follows from Banach's contraction mapping principle. ∎

In the space $X = C$ we can verify the conditions of the previous
theorem by means of Theorem 2.1 and Theorem 7.5. As an example
we state a result for the special case, when M respectively m in
Theorem 7.5 can be chosen independent of t:

Theorem 7.10. *Let $t \mapsto A(t)$ be a strongly continuous operator function in C, $K \in \mathfrak{L}_c(C)$, and (7.57) be satisfied. Suppose that c has a
constant sign and satisfies*

$$\inf_{a \leq s \leq b} \int_0^T |c(t,s)|\, dt > 0.$$

*Assume, furthermore, for some measurable function $y : [a,b] \to (0,\infty)$,
that*

$$\int_a^b |k(t,s,\sigma)| y(\sigma)\, d\sigma < |c(t,s)| y(s)$$

for all $(t,s) \in [0,T] \times [a,b]$, and

$$\Sigma := \sup_{t,s} \frac{|f(t,s)|}{|c(t,s)| y(s) - \int_a^b |k(t,s,\sigma)| y(\sigma)\, d\sigma} < \infty.$$

Then (7.56) has a continuous T-periodic solution x which satisfies

$$|x(t,s)| \leq \Sigma y(s) \quad ((t,s) \in [0,T] \times [a,b]).$$

*Moreover, if $k(t,s,\sigma) \equiv 0$ the continuous T-periodic solution is uni-
que.*

☐   By Theorem 2.3 we can assume without loss of generality that
$H(T,0)$ and $H(0,T)$ are compact. Define

$$M := \{x : x \in C([a,b]), |x(s)| \leq \Sigma y(s) \ (a \leq s \leq b)\}.$$

If we write $M(t,\cdot)$ and $m(t,\cdot)$ instead of y, the conditions (7.15)/(7.16)
respectively (7.17)/(7.18) are satisfied, if $c < 0$ respectively $c >$

0. Thus Φ respectively Φ^{-1} leaves the set M invariant. Obviously, $k(t, s, \sigma) \equiv 0$ implies that H is the zero operator. $\blacksquare$

Example 7.1. The one-dimensional equation

$$(7.61) \qquad \frac{dx(t)}{dt} = a(t)x(t) + b(t)$$

with T-periodic and continuous a and b, a without zeros, has exactly one T-periodic solution x. This solution is bounded by $\max\limits_{0 \le t \le T} |b(t)/a(t)|$.

To see this, choose $c(t, s) \equiv a(t)$, $k(t, s, \sigma) \equiv 0$, $f(t, s) \equiv b(t)$, and $y(s) \equiv 1$.

A simple calculation shows that the (unique) fixed point of the Poincaré operator (7.58) in this example is

$$x_0 = \frac{\displaystyle\int_0^T \exp\left\{ -\int_0^t a(\tau)d\tau \right\} b(t)\, dt}{\displaystyle\exp\left\{ -\int_0^T a(t)dt \right\} - 1}.$$

In the space $X = L_p$ the conditions of Theorem 7.9 may be verified by means of Theorem 1.2 and Theorem 7.6. We will again restrict ourselves to the case, when y in Theorem 7.6 is constant.

Theorem 7.11. *Suppose that (7.57) is satisfied for some $T > 0$, $X = L_p([a, b])$ $(1 < p < \infty)$, $f : \mathbb{R} \to X$ is continuous, $c : \mathbb{R} \times [a, b] \to \mathbb{R}$ is measurable, $K(t) \in \mathcal{L}_c(X)$ for each t, and $t \mapsto A(t)$ and $t \mapsto C(t)$ are strongly continuous operator functions in X. Assume, moreover, that there exists an $R > 0$ such that one of the following two conditions is satisfied:*

(a) $\text{ess sup}\limits_{a \le s \le b} \displaystyle\int_0^T c(t, s)\, dt < 0$, *and* $x \in X$, $\|x\|_{L_p} = R$ *implies that*

$$(7.62) \qquad \int_a^b [A(t)x(s) + f(t, s)]|x(s)|^{p-1}\, \text{sign}\, x(s)\, ds < 0;$$

(b) $\operatorname*{ess\,inf}_{a\leq s\leq b} \int_0^T c(t,s)dt > 0$, and $x \in X$, $||x||_{L_p} = R$ implies that

(7.63) $\int_a^b [A(t)x(s) + f(t,s)]|x(s)|^{p-1} \operatorname{sign} x(s)\, ds > 0.$

Then (7.56) has a T-periodic solution x, satisfying $||x(t)||_{L_p} \leq R$ for all t.

◻ By Theorem 1.2 we may assume without loss of generality that $H(T,0)$ and $H(0,T)$ are in $\mathcal{L}_c(X)$. Since (7.22)/(7.23) respectively (7.24)/(7.25) are satisfied for $y \equiv R$, the set $M := \{x : x \in X, ||x||_p \leq R\}$ is invariant under Φ respectively Φ^{-1}. ∎

We remark that a similar theorem may be proved also in case $p = 1$. Here (7.62) and (7.63) have to be replaced by the conditions

$$\int_a^b [A(t)x(s) + f(t,s)]z_+(t,s)\, ds < 0$$

with

$$z_+(t,s) = \begin{cases} \operatorname{sign} x(s) & \text{if } x(s) \neq 0, \\ \operatorname{sign} f(t,s) & \text{if } x(s) = 0, \end{cases}$$

and

$$\int_a^b [A(t)x(s) + f(t,s)]z_-(t,s)\, ds > 0$$

with

$$z_-(t,s) = \begin{cases} \operatorname{sign} x(s) & \text{if } x(s) \neq 0, \\ -\operatorname{sign} f(t,s) & \text{if } x(s) = 0, \end{cases}$$

respectively.

7.8. Application of the Fredholm alternative

We consider again equation (7.56) in a Banach space X and suppose that (7.57) holds. Of course, if (7.56) has precisely one bounded solution $x : \mathbb{R} \to X$, this solution must be T-periodic. In fact, $\tilde{x} : \mathbb{R} \to X$ defined by $\tilde{x}(t) = x(t + T)$ is also bounded and hence $\tilde{x} = x$.

The condition mentioned above implies that the trivial solution is the only bounded solution of the homogeneous equation

$$(7.64) \qquad \frac{dx}{dt} = A(t)x.$$

The Fredholm theory allows us to reduce this to T-periodic solutions:

Theorem 7.12. *Let all conditions of Theorem 7.9 be satisfied. Assume, furthermore, that the only T-periodic solution of (7.64) is the trivial solution $x(t) \equiv 0$. Then (7.56) has a (unique) T-periodic solution in X.*

□ We consider only the case (a) (otherwise replace T by $-T$). The multiplication operator

$$Ex(s) = e(T, 0, s)x(s) = [U(T, 0) - H(T, 0)]x(s)$$

(see (8.4)) is a linear mapping of X into X with norm < 1, by (7.60). Thus $I - E$ is an isomorphism on X. This means that $I - U(T, 0) = (I - E) - H(T, 0)$ is Fredholm of index 0. By assumption its nullspace is trivial, which implies that the equation

$$(7.65) \qquad [I - U(T, 0)]x = \int_0^T U(T, \xi)f(\xi)\, d\xi$$

has a solution $x_0 \in X$. But this means precisely that

$$\Phi x_0 = U(T, 0)x_0 + \int_0^T U(T, \xi)f(\xi)\, d\xi = x_0$$

which proves the assertion. ■

Since $I - U(T, 0)$ is Fredholm of index 0, (7.64) can only have finitely many linear independent initial values which generate T-periodic solutions. Furthermore its range has this (finite) number as codimension, which means that in general (7.65) still remains solvable for "most" f. Remember that this is equivalent to the fact that (7.56) has a T-periodic solution for those f.

§ 8. Degenerate kernels

In this section we show how certain degeneration assumptions in Barbashin equations allow us to reduce the initial value or boundary value problem to much simpler problems or even lead to explicit solutions.

8.1. A general approach

In this section we will develop a numerical method for solving the initial value problem and the boundary value problem for the linear Barbashin integral equation. This will also reveal a deeper reason, why the initial value problem usually is uniquely solvable, and why this must not be true for the boundary value problem.

At first, we consider the equation

$$(8.1) \qquad \frac{\partial x(t,s)}{\partial t} = c(t,s)x(t,s) + \int_a^b k(t,s,\sigma)x(t,\sigma) + f(t,s),$$

subject to the initial condition

$$x(t_0, s) = \varphi(s).$$

It turns out that a straightforward degeneration assumption on the kernel k does not lead to a much simpler problem. However, a degeneration assumption on a related kernel (depending on k and c) will allow us to reduce the problem to a finite system of usual differential equations. Moreover, if c and k are "well-behaved" (e.g. continuous), it is always possible to approximate them by such degenerated kernels.

To start with, we rewrite the initial value problem for (8.1) as a fixed point problem in an ideal space X over $[a, b]$. By a solution we understand a function $x \in C_t^1(X)$ which satisfies (8.1) almost everywhere. Define the functions

$$e(t, \tau, s) = \exp\left\{ \int_\tau^t c(\xi, s)\, d\xi \right\}$$

and

$$g(t, s) = \int_{t_0}^t e(t, \tau, s) f(\tau, s)\, d\tau + e(t, t_0, s)\varphi(s),$$

as well as the operators

$$\hat{K}x(t,s) = \int_a^b k(t,s,\sigma)x(t,\sigma)d\sigma$$

and

$$Lx(t,s) = \int_{t_0}^t e(t,\tau,s)\hat{K}x(\tau,s)d\tau.$$

Observing that in case $k = 0$ the initial value problem for (8.1) has the explicit solution $x(t,s) = e(t,\tau,s)\varphi(s)$, we may apply the variations-of-constants formula and prove the following

Theorem 8.1. *Let* $\varphi \in X$, *c and k be measurable, and the correspon-ding operators C and K be locally integrable and strongly continuous in* $\mathfrak{L}(X)$. *If* $f \in C_t(X)$, *each solution of the initial value problem for* (8.1) *is also a solution of the fixed point equation*

$$(8.2) \qquad\qquad x(t,s) = Lx(t,s) + g(t,s).$$

Conversely, every solution $x \in C_t(X)$ *of* (8.2) *actually belongs to* $C_t^1(X)$ *and solves the initial value problem for* (8.1).

To solve the fixed point equation (8.2), we assume that c and k are degenerated in that sense that L degenerates to

$$Lx(t,s) = \int_{t_0}^t \int_a^b \exp\left(\int_\tau^t c(\xi,s)\,d\xi\right) k(\tau,s,\sigma)x(\tau,\sigma)\,d\sigma\,d\tau$$

$$= \sum_{j=1}^n \int_{t_0}^t \int_a^b a_j(t,\tau,\sigma)b_j(t,s)x(\tau,\sigma)\,d\sigma\,d\tau.$$

This is the case, for example, if c is independent of s, and k is dege-nerated in the sense that

$$k(t,s,\sigma) = \sum_{j=1}^n \alpha_j(t,\sigma)\beta_j(s).$$

If c and k are continuous, one might choose polynomials approxima-ting $e(t,\tau,s)k(\tau,s,\sigma)$ uniformly in (t,τ,s,σ) for bounded t,τ. In this case we can even choose $a_j(t, \tau, \sigma) = a_j(\tau, \sigma)$ independent of t; this will be important later on.

If now x solves the fixed point equation (8.2) and L is degenerated as above then x must have the form

$$x(t, s) = \sum_{k=1}^{n} \xi_k(t) b_k(t, s) + g(t, s),$$

where the functions ξ_k are unknown, but must solve

$$\xi_j(t) = \int_{t_0}^{t} \int_{a}^{b} a_j(t, \tau, \sigma) x(\tau, \sigma) \, d\sigma \, d\tau \qquad (j = 1, \ldots, n).$$

Substituting these formulas into each other we get

$$\xi_j(t) = \sum_{k=1}^{n} \int_{t_0}^{t} \gamma_{jk}(t, \tau) \xi_k(\tau) \, d\tau + \delta_j(\tau) \qquad (j = 1, \ldots, n),$$

where

$$\gamma_{jk}(t, \tau) = \int_{a}^{b} a_j(t, \tau, \sigma) b_k(\tau, \sigma) \, d\sigma$$

and

$$\delta_j(\tau) = \int_{t_0}^{t} \int_{a}^{b} a_j(t, \tau, \sigma) g(\tau, \sigma) \, d\sigma \, d\tau.$$

Writing this equation in vector form, $\xi = (\xi_1, \ldots, \xi_n)$, $\delta = (\delta_1, \ldots, \delta_n)$, and $\gamma = (\gamma_{jk})$, we thus have reduced the degenerated initial value problem to an n-dimensional Volterra integral equation of the form

$$\xi(t) = \int_{t_0}^{t} \gamma(t, \tau) \xi(\tau) \, d\tau + \delta(t).$$

Observe that, if γ and δ are smooth, this equation is uniquely solvable.

Moreover, in the special case when we can choose $a_j(t, \tau) = a_j(\tau)$ independent of t (recall that approximating smooth functions this can always be done), the function $\gamma(t, \tau) = \gamma(\tau)$ is also independent of t, and thus in view of $\delta(t_0) = 0$ the above Volterra equation reduces to an n-dimensional linear initial value problem of the form

$$\begin{cases} \xi'(\tau) = \gamma(\tau)\xi(\tau) + a(\tau), \\ \xi(t_0) = 0, \end{cases}$$

where a is the vector function defined by

$$a(\tau) = \left(\int_a^b a_1(\tau, \sigma)g(\tau, \sigma)\, d\sigma, \ldots, \int_a^b a_n(\tau, \sigma)g(\tau, \sigma)\, d\sigma \right).$$

Example 8.1. Consider the initial value problem

$$\frac{\partial x(t, s)}{\partial t} = cx(t, s) + \int_a^b \alpha(t, \sigma)\beta(s)x(t, \sigma)\, d\sigma, \quad x(t_0, s) = \varphi(s).$$

Here one may choose $n = 1$, $a_1(t, \tau, \sigma) = e^{-c\tau}\alpha(\tau, \sigma)$, and $b_1(t, s) = e^{ct}\beta(s)$. Then the above calculation shows that a solution is given by

$$x(t, s) = e^{(t-t_0)c}\varphi(s) +$$

$$e^{(t-t_0)c}\beta(s) \int_{t_0}^t \exp \int_\xi^t \int_a^b \alpha(\tau, \sigma)\beta(\sigma)\, d\sigma\, d\tau \int_a^b \alpha(\xi, \sigma)\varphi(\sigma)\, d\sigma\, d\xi.$$

This example is a special case of a class of examples discussed in Section 8.2.

Now we consider the boundary value problem analogously. Our degeneration assumption now reads

$$Lx(t, s) = \begin{cases} \displaystyle\sum_{j=1}^n \int_a^t \int_{-1}^1 a_j(t, \tau, \sigma)b_j(t, s)x(\tau, \sigma)\, d\sigma\, d\tau & \text{if } s > 0, \\[3mm] \displaystyle\sum_{j=1}^n \int_b^t \int_{-1}^1 a_j(t, \tau, \sigma)b_j(t, s)x(\tau, \sigma)\, d\sigma\, d\tau & \text{if } s < 0. \end{cases}$$

Now, if x solves the fixed point equation $x = Lx + g$, it must have the form

$$
x(t,s) = \begin{cases} \displaystyle\sum_{k=1}^{n} \xi_k(t) b_k(t,s) + g(t,s) & \text{if } s > 0, \\[4mm] \displaystyle\sum_{k=1}^{n} \eta_k(t) b_k(t,s) + g(t,s) & \text{if } s < 0, \end{cases}
$$

where again the scalar functions ξ_k and η_k are unknown, but must solve the equations

$$
\xi_j(t) = \int_a^t \int_{-1}^1 a_j(t,\tau,\sigma) x(\tau,\sigma)\, d\sigma\, d\tau \qquad (j = 1,\ldots,n)
$$

and

$$
\eta_j(t) = \int_b^t \int_{-1}^1 a_j(t,\tau,\sigma) x(\tau,\sigma)\, d\sigma\, d\tau \qquad (j = 1,\ldots,n).
$$

Substituting these formulas into each other, we get for $j = 1,\ldots,n$ the relations

$$
\xi_j(t) = \sum_{k=1}^{n} \int_a^t \gamma_{jk}^+(t,\tau) \xi_k(\tau)\, d\tau + \sum_{k=1}^{n} \int_a^t \gamma_{jk}^-(t,\tau) \eta_k(\tau)\, d\tau + \delta_j^+(t)
$$

and

$$
\eta_j(t) = \sum_{k=1}^{n} \int_b^t \gamma_{jk}^+(t,\tau) \xi_k(\tau)\, d\tau + \sum_{k=1}^{n} \int_b^t \gamma_{jk}^-(t,\tau) \eta_k(\tau)\, d\tau + \delta_j^-(t),
$$

where

$$
\gamma_{jk}^+(t,\tau) = \int_0^1 a_j(t,\tau,\sigma) b_k(\tau,\sigma)\, d\sigma,
$$

$$
\gamma_{jk}^-(t,\tau) = \int_{-1}^0 a_j(t,\tau,\sigma) b_k(\tau,\sigma)\, d\sigma,
$$

$$
\delta_j^+(t) = \int_a^t \int_{-1}^1 a_j(t,\tau,\sigma) g(\tau,\sigma)\, d\sigma\, d\tau,
$$

and

$$\delta_j^-(t) = \int_b^t \int_{-1}^1 a_j(t,\tau,\sigma)g(\tau,\sigma)\,d\sigma\,d\tau = \delta_j^+(t) - \delta_j^+(b).$$

If we write this system again in vector form, $\xi = (\xi_1,\ldots,\xi_n)$, $\eta = (\eta_1,\ldots,\eta_n)$, $\delta^\pm = (\delta_1^\pm,\ldots,\delta_n^\pm)$, and $\gamma^\pm = (\gamma_{jk}^\pm)$, we get the $2n$-dimensional Fredholm integral equation

$$\begin{pmatrix} \xi(t) \\ \eta(t) \end{pmatrix} = \begin{pmatrix} \int_a^t (\gamma^+(t,\tau)\xi(\tau) + \gamma^-(t,\tau)\eta(\tau))\,d\tau \\ \int_b^t (\gamma^+(t,\tau)\xi(\tau) + \gamma^-(t,\tau)\eta(\tau))\,d\tau \end{pmatrix} + \begin{pmatrix} \delta^+(t) \\ \delta^-(t) \end{pmatrix}.$$

Recall that a Fredholm integral equation need not be uniquely solvable, even if all parameter functions are smooth. This of course coincides with the fact that the boundary value problem need not have a unique solution. Observe again that if $a_k(t,\tau,\sigma) = a_k(\tau,\sigma)$ is independent of t (recall that if we approximate smooth functions this can always be done), then $\gamma^\pm(t,\tau) = \gamma^\pm(\tau)$ is independent of t, too.

8.2. Explicit solution for a particular class

We now give a class of Barbashin equations which can be explicitly solved. As a consequence we get many useful examples. We consider the linear Barbashin equation in the case $c(t,s) = c(t)$, $k(t,s,\sigma) = \alpha(t,\sigma)\beta(s)$, $f \equiv 0$, i.e. we consider the equation

$$(8.3) \qquad \frac{\partial x(t,s)}{\partial t} = c(t)x(t,s) + \int_a^b \alpha(t,\sigma)\beta(s)x(t,\sigma)\,d\sigma.$$

First, we will show in a formal calculation how we can explicitly solve any initial value problems concerning (8.3), i.e. how we can calculate the evolution operator $U(t,\tau)$ of this equation. Later, we will discuss the corresponding boundary value problems.

Assume that x is a (still unknown) solution of (8.3). Fix s. Then $y(t) = x(t,s)$ solves the ordinary differential equation

$$(8.4) \qquad y'(t) = c(t)y(t) + \beta(s)\tilde{f}(t),$$

where $\tilde{f}$ is not known yet. However, the variation-of-constant formula shows that the solution of (8.4) must have the form

$$y(t) = g(s) \exp\left(\int_\tau^t c(\xi)\,d\xi\right) + \beta(s)f(t),$$

where f and g are not known yet. Thus we know that x has the form

(8.5) $\qquad\qquad x(t,s) = g(s)e^{\hat{c}(t)} + \beta(s)f(t),$

where f and g have to be determined, and where we have put

$$\hat{c}(t) = \int_\tau^t c(\xi)\,d\xi.$$

Assume now that g is given. Substituting (8.5) in (8.3) yields, by dividing by $\beta(s)$ (which is allowed, unless $\beta \equiv 0$, since β is independent of t),

$$f'(t) = \left(c(t) + \int_a^b \alpha(t,\sigma)\beta(\sigma)\,d\sigma\right)f(t)$$

(8.6)

$$+e^{\hat{c}(t)}\int_a^b \alpha(t,\sigma)g(\sigma)\,d\sigma.$$

Using the abbreviation

$$\hat{a}(t) = \int_\tau^t \int_a^b \alpha(\xi,\sigma)\beta(\sigma)\,d\sigma\,d\xi,$$

we can give the general solution f of the ordinary differential equation (8.6), namely

$$f(t) = \left(\int_\tau^t \int_a^b \alpha(\xi,\sigma)g(\sigma)\,d\sigma\;e^{-\hat{a}(\xi)}d\xi + const\right)e^{\hat{c}(t)+\hat{a}(t)}.$$

Looking at (8.5) we have

$$x(t,s) = g(s)e^{\hat{c}(t)}+$$

$$\beta(s)\left(\int_\tau^t \int_a^b \alpha(\xi,\sigma)g(\sigma)d\sigma\;e^{-\hat{a}(\xi)}\,d\xi + const\right)e^{\hat{c}(t)+\hat{a}(t)}.$$

Assuming, that we want to solve the initial value problem $x(\tau, s) = \varphi(s)$ (which will give us by $U(t, \tau)\varphi(s) = x(t, s)$ the evolution operator), we may choose $g = \varphi$ and $const = 0$, since $\hat{c}(\tau) = \hat{a}(\tau) = 0$. If we assume furthermore that we may apply Fubini's theorem, we have the solution

$$U(t, \tau)\varphi(s) = e^{\hat{c}(t)}\varphi(s)+$$

$$\beta(s) \int_a^b \left(\int_\tau^t \alpha(\xi, \sigma)e^{-\hat{a}(\xi)}\, d\xi \right) \varphi(\sigma)\, d\sigma\, e^{\hat{a}(t)}e^{\hat{c}(t)}.$$

We remark that Example 8.1 might be considered as a special case of the above calculation. Observe also that the variable s in the kernel of the above integral operator is separated. We will see in the next section that this is a special case of a more general result.

Now we consider the corresponding degenerated boundary value problem

$$\begin{cases} \dfrac{\partial x(t, s)}{\partial t} = c(t)x(t, s) + \displaystyle\int_{-1}^1 \alpha(t, \sigma)\beta(s)x(t, \sigma)\, d\sigma \\[2mm] x(a, s) = \varphi(s) \text{ if } 0 < s \le 1, \\[2mm] x(b, s) = \psi(s) \text{ if } -1 \le s < 0. \end{cases}$$

The problem consists in extending φ to an initial value function on the whole interval $[-1, 1]$, such that

$$U(b, a)\varphi(s) = \psi(s) \qquad (-1 \le s < 0).$$

Since we know U, we are led to the degenerated Fredholm integral equation

$$\varphi(s) + \beta(s) \int_{-1}^0 \gamma(\sigma)\varphi(\sigma)d\sigma = C\psi(s) - D\beta(s) \qquad (-1 \le s < 0),$$

where we have put

$$\gamma(s) = \int_a^b \alpha(\xi, s) \exp\left(\int_\xi^b \int_{-1}^1 \alpha(\rho, \sigma)\beta(\sigma)\, d\sigma\, d\rho \right) d\xi,$$

$$C = \exp\left(-\int_a^b c(\xi)\,d\xi\right),$$

and

$$D = \int_0^1 \gamma(\sigma)\varphi(\sigma)\,d\sigma.$$

This is equivalent to the system

$$\begin{cases} \varphi(s) = C\psi(s) - \xi\beta(s) & \text{if } -1 \le s < 0 \\[2mm] \xi = \int_{-1}^0 \gamma(s)\varphi(s)\,ds + D, \end{cases}$$

which is satisfied if and only if φ has the form $\varphi(s) = C\psi(s) - \xi\beta(s)$ for $-1 \le s < 0$, and

$$\xi = C\int_{-1}^0 \gamma(s)\psi(s)\,ds - \xi\int_{-1}^0 \gamma(s)\beta(s)\,ds + D.$$

Thus, if

$$B = \int_{-1}^0 \gamma(s)\beta(s)\,ds \ne -1,$$

we have the unique solution

$$\varphi(s) = C\psi(s) - (1+B)^{-1}\left(C\int_{-1}^0 \gamma(\sigma)\psi(\sigma)\,d\sigma + D\right)\beta(s)$$

for $-1 \le s < 0$. But if $B = -1$, we have solutions if and only if

$$\int_{-1}^0 \gamma(s)\psi(s)\,ds = -\int_0^1 \gamma(s)\varphi(s)\,ds\,\exp\left(\int_a^b c(\xi)\,d\xi\right).$$

In the latter case, all functions of the form $\varphi(s) = C\psi(s) - \xi\beta(s)$ for $-1 \le s < 0$, with arbitrary numbers ξ, are solutions.

8.3. An abstract degeneration result

As usual, we consider the Barbashin equation

$$(8.7) \qquad\qquad \frac{dx}{dt} = A(t)x,$$

where $A(t) = C(t) + K(t)$ with

$$C(t)x(s) = c(t, s)x(s)$$

and

$$K(t)x(s) = \int_a^b k(t, s, \sigma)x(\sigma)d\sigma$$

in some space X. We will prove that a straightforward degenerati-on assumption on the kernel k already implies that the evolution operator of the Barbashin equation is degenerated in the same way. However, as already remarked in Section 8.1., this result is more of theoretical interest, since it apparently does not provide a way to simplify the calculation of the evolution operator.

Lemma 8.1. *Let X be an ideal space over $[a, b]$. Assume that c is measurable, and C is strongly continuous in $\mathcal{L}(X)$. Then the evolution operator U_0 of*

$$\frac{dx}{dt} = C(t)x$$

in X may be written, for almost all t, τ, as

$$U_0(t, \tau)x(s) = \exp\left(\int_\tau^t c(\xi, s)\, d\xi\right)x(s).$$

☐   For fixed $x \in X$ we have

$$U_0(t, \tau)x = x + \sum_{n=1}^{\infty} \int_\tau^t \cdots \int_\tau^{t_{n-1}} C(t_1)\cdots C(t_n)\dot{x}\, dt_n \ldots dt_1,$$

where the series converges in X. The function c is essentially locally bounded. By Lemma 4.2 we may write the partial sums of this series for almost all t, τ, s in the form

$$x(s) + \sum_{n=1}^{k} \int_\tau^t \cdots \int_\tau^{t_{n-1}} c(t_1, s)\cdots c(t_n, s)x(s)\, dt_n \ldots dt_1$$

$$= x(s) + \sum_{n=1}^{k} \frac{1}{n!}\left(\int_\tau^t c(\xi, s)d\xi\right)^n x(s).$$

Since these partial sums converge almost everywhere, their pointwise limit must be $U_0(t, \tau)x(s)$ almost everywhere. ∎

Now let X be a regular ideal space over $[a, b]$ (see Subsection 4.1). Then the dual space X^* may be identified with the associate space X' in the canonical way, see ZAANEN [1953] or ZABREJKO [1974].

It is a natural degeneration assumption that there exists a number n such that each $K(t)$ belongs to the set $\mathfrak{M}_n$ of all integral operators of the form

$$K x(s) = \sum_{j=1}^{n} \int_a^b \alpha_j(s)\beta_j(\sigma)x(\sigma)\, d\sigma, \qquad (\alpha_j \in X, \beta_j \in X')$$

(see Subsection 1.5). This set $\mathfrak{M}_n$ is very nice as is shown by the following

Lemma 8.2. *Each set $\mathfrak{M}_n$ is a closed ideal in $\mathfrak{L}(X)$.*

□ We first prove that

$$\mathfrak{M}_n = \mathfrak{N}_n := \{K : K \in \mathfrak{L}(X), \dim K(X) \leq n\}.$$

Indeed, given $K \in \mathfrak{N}_n$, let $\{\alpha_1, \ldots, \alpha_m\}$ be a basis of $K(X)$. Thus, $K x = l_1(x)\alpha_1 + \ldots + l_m(x)\alpha_m$, where $l_1, \ldots, l_m \in X^*$. This implies that there exist $\beta_1, \ldots, \beta_m \in X'$ such that

$$K x(t) = \sum_{j=1}^{m} \alpha_j(t) \int_a^b \beta_j(s)x(s)\, ds.$$

Assume now that $\mathfrak{N}_n$ is not closed, i.e. for some $K_j \in \mathfrak{N}_n$ we have $K_j \to K$, but $K \notin \mathfrak{N}_n$. But this means that $K(X)$ has $n + 1$ linear independent normed vectors which of course cannot be approximated simultaneously by only n linear independent normed vectors of $K_j(X)$, a contradiction. ∎

Now we apply Theorem 1.2: If $K(t)$ belongs to the closed ideal $\mathfrak{M}_n$ then also $U - U_0$ must belong to this closed ideal. Using the above formula for U_0, we get the following degeneration result:

Theorem 8.2. *Suppose that X is a regular ideal space over $[a, b]$, c is measurable, A and C are strongly continuous in $\mathfrak{L}(X)$, and*

$$k(t, s, \sigma) = \sum_{j=1}^{n} \alpha_j(t, s)\beta_j(t, \sigma),$$

where $\alpha_j(t, \cdot) \in X$ and $\beta_j(t, \cdot) \in X'$. Then the evolution operator of (8.7) has the form

$$U(t, \tau)x(s) = \exp\left(\int_{\tau}^{t} c(\xi, s) \, d\xi\right) x(s) + \sum_{j=1}^{n} \int_{a}^{b} \gamma_j(t, \tau, s)\delta_j(t, \tau, \sigma)x(\sigma) \, d\sigma,$$

where $\gamma_j(t, \tau, \cdot) \in X$ and $\delta_j(t, \tau, \cdot) \in X'$.

The result is illustrated for $n = 1$ by the class of examples given in the previous section.

§ 9. Stationary boundary value problems

In this section we study the Barbashin equation

$$(9.1) \qquad \frac{\partial x(t,s)}{\partial t} = c(s)x(t,s) + \int_{-1}^{1} k(s,\sigma)x(t,\sigma)\,d\sigma + f(t,s)$$

for $(t,s) \in Q := [0,T] \times [-1,1]$, subject to the boundary conditions

$$(9.2) \qquad \begin{cases} x(0,s) = \phi(s) & \text{if } 0 < s \leq 1, \\ x(T,s) = \psi(s) & \text{if } -1 \leq s < 0. \end{cases}$$

where $\phi : (0,1] \to \mathbb{R}$, and $\psi : [-1,0) \to \mathbb{R}$ are given functions. As in the theory of ordinary differential equations, such a boundary value problem does not always admit a solution, in contrast to the initial value problem (7.1)/(7.10). First, we reduce the problem (9.1)/(9.2) to an integral equation in the space $C_t(X)$ of functions of two variables in order to apply the classical Fredholm theory. In the second part we apply more sophisticated methods, such as fixed point principles in so-called K-normed spaces. The results of this section may be found in APPELL-KALITVIN-ZABREJKO [1994, 1996] and KALITVIN [1992].

9.1. The abstract problem

For the time being we restrict ourselves to the case of *stationary* multipliers and kernels (i.e. independent of t); the non-stationary case will be treated in § 10.

As in § 1, we rewrite equation (9.1) as a linear differential equation

$$(9.3) \qquad \frac{dx}{dt} = Ax + f(t)$$

in some Banach space $X = X([-1,1])$ of functions over $[-1,1]$, i.e. we identify the scalar functions $(t,s) \mapsto x(t,s)$ and $(t,s) \mapsto f(t,s)$ with the Banach space valued functions $t \mapsto x(t) = x(t,\cdot)$ and $t \mapsto f(t) = f(t,\cdot)$, and define the operator A by

$$(9.4) \qquad Ax(s) = c(s)x(s) + \int_{-1}^{1} k(s,\sigma)x(\sigma)\,d\sigma.$$

Similarly, the boundary conditions (9.2) may then be written in the form

$$(9.5) \qquad P_+x(0) = \phi, \quad P_-x(T) = \psi,$$

where P_+ (respectively P_-) denotes the restriction operator from $X = X([-1,1])$ onto $X_+ = X([0,1])$ (respectively onto $X_- = X([-1,0])$).

As pointed out in Subsection 1.1, a crucial point when passing from the problem (9.1)/(9.2) to the problem (9.3)/(9.5) is the appropriate choice of the function space X. It is clear that taking the space $X = C([-1,1])$ of continuous functions is too restrictive. In fact, even if all functions c, k, and f in (9.1) are zero, and the boundary functions ϕ and ψ in (9.2) are continuous, there is certainly no solution of (9.1)/(9.2) if the "glueing condition"

$$\lim_{s \to 0+} \phi(s) = \lim_{s \to 0-} \psi(s)$$

fails; a specific example for this arising in applications may be found in VLADIMIROV [1961]. Therefore we shall not study the existence problem in the space $C([-1,1])$, but in spaces of measurable functions. As in Subsection 1.1, a (generalized) solution of (9.1)/(9.2) will be a measurable function $x : Q \to \mathbb{R}$ which has the property that $x(\cdot, s)$ is absolutely continuous on $[0, T]$ for almost all $s \in [-1, 1]$, and satisfies (9.1) almost everywhere on Q and (9.2) almost everywhere on $[-1, 1]$. As a model case for the underlying Banach space X one may always think of the Lebesgue space $X = L_p([-1, 1])$ $(1 \leq p \leq \infty)$.

9.2. Equivalent operator equations

Recall that the (unique) solution of the differential equation (9.3) with initial condition $x(\tau) = x_\tau$ is given by

$$(9.6) \qquad x(t) = U(t, \tau)x_\tau + \int_\tau^t U(t, \xi)f(\xi) \, d\xi,$$

where $U = U(t, \tau)$ denotes the evolution operator (see Subsection 1.2) of (9.3). We are interested, in particular, in the *two-point boundary value problem*

$$(9.7) \qquad x(0) = x_0, \quad x(T) = x_T,$$

with $x_0 = \phi$ on $(0,1]$, $x_0 = z$ on $[-1,0)$, $x_T = y$ on $(0,1]$, and $x_T = \psi$ on $[-1,0)$; here the functions $z \in X_-$ and $y \in X_+$ will be specified below. In Subsection 19.5 we will discuss a physical problem where this kind of boundary value problem occurs.

As already observed, the unique solution of (9.4) with initial condition $x(0) = x_0$ is given by (9.6) with $\tau = 0$. Taking into account the definition (1.7) of $U(t,\tau)$, we may write the solution for $\tau = 0$ equivalently in the form

$$(9.8) \quad x(t,s) = \begin{cases} m(t,s) + h_z(t,s) & (0 < s \leq 1), \\ n(t,s) + e^{tc(s)} z(s) + h_z(t,s) & (-1 \leq s < 0), \end{cases}$$

where we have put

$$m(t,s) = e^{tc(s)} \phi(s) + \int_0^1 h(t,s,\sigma)\phi(\sigma)\,d\sigma$$

$$+ \int_0^t [e^{(t-\tau)c(s)} f(\tau,s) + \int_{-1}^1 h(t-\tau,s,\sigma) f(\tau,\sigma)\,d\sigma]\,d\tau,$$

$$h_z(t,s) = \int_{-1}^0 h(t,s,\sigma) z(\sigma)\,d\sigma,$$

and

$$n(t,s) = \int_0^1 h(t,s,\sigma)\phi(\sigma)\,d\sigma$$

$$+ \int_0^t [e^{(t-\tau)c(s)} f(\tau,s) + \int_{-1}^1 h(t-\tau,s,\sigma) f(\tau,\sigma)\,d\sigma]\,d\tau.$$

Using now also the second condition in (9.7), we arrive at the system of two linear integral equations

$$\begin{cases} y(s) = m(T,s) + h_z(T,s) & \text{if } 0 < s \leq 1, \\ e^{Tc(s)} z(s) + h_z(T,s) = \psi(s) - n(T,s) & \text{if } -1 \leq s < 0 \end{cases}$$

for the couple $(y,z) \in X_+ \times X_-$. Moreover, if we put

$$(9.9) \quad p(s,\sigma) = \frac{h(T,s,\sigma)}{e^{Tc(s)}}, \qquad r(s) = \frac{\psi(s) - n(T,s)}{e^{Tc(s)}},$$

where $-1 \leq s < 0$, the system (9.8) may in turn be written as a single equation

$$(9.10) \qquad z(s) + \int_{-1}^{0} p(s, \sigma) z(\sigma) \, d\sigma = r(s).$$

This shows that the solvability of the original problem (9.1)/(9.2) is closely related to the solvability of the integral equation of second kind (9.10). Thus, we may apply the whole arsenal of results of classical Fredholm theory. To this end, we introduce some notation. Given an ideal space X over $[-1, 1]$, we write (as in Subsection 1.1) $x \in C_t(X)$ if the function of two variables $x : Q \to \mathbb{R}$ has the property that the map $t \mapsto x(t, \cdot)$ is continuous from $J = [0, T]$ into X. If, in addition, $x(\cdot, s)$ is absolutely continuous for almost all $s \in [-1, 1]$ and $\partial x / \partial t$ also belongs to $C_t(X)$, we write $x \in C_t^1(X)$. The spaces $C_t(X)$ and $C_t^1(X)$ will be endowed with the natural norms

$$\|x\|_{C_t(X)} = \max_{0 \leq t \leq T} \|x(t, \cdot)\|_X$$

and

$$\|x\|_{C_t^1(X)} = \max_{0 \leq t \leq T} \left[\|x(t, \cdot)\|_X + \left\| \frac{\partial x(t, \cdot)}{\partial t} \right\|_X \right],$$

respectively. In the following, we shall restrict ourselves again to the model case $X = L_p([-1, 1])$ $(1 \leq p \leq \infty)$.

Theorem 9.1. *Suppose that the functions* $c : [-1, 1] \to \mathbb{R}$ *and* $k : Q \to \mathbb{R}$ *are measurable, and the operator* (9.4) *is regular in the space* $X = L_p([-1, 1])$. *Then the following three statements are equivalent:*

(a) *the integro-differential equation* (9.1) *with boundary condition* (9.2) *is (uniquely) solvable for any* $\phi \in X_+ = L_p([0, 1])$, $\psi \in X_- = L_p([-1, 0])$, *and* $f \in C_t(X)$;

(b) *the differential equation* (9.3) *with boundary condition* (9.7) *is (uniquely) solvable in* $C_t^1(X)$;

(c) *the Fredholm integral equation* (9.10) *is (uniquely) solvable in* $X_- = L_p([-1, 0])$.

□   If the problem (9.1)/(9.2) has a solution x in the sense defined at the end of the first subsection, from $f \in C_t(X)$ it follows that x

solves (9.3)/(9.7) and belongs to $C_t^1(X)$. This shows that (a) implies (b). The fact that (b) implies (c) and (c) implies (a) has already been proved in the preceding discussion. ■

Theorem 9.1 makes it possible to reduce the existence and uniqueness problem for the original Barbashin equation to that for the integral equation (9.10). In this way, we get much information on the equation (9.1) from standard results of classical functional analysis, applied to the operator

$$(9.11) \qquad Pz(s) = \int_{-1}^{0} p(s,\sigma)z(\sigma)\,d\sigma.$$

For example, the following holds.

Lemma 9.1. *Suppose that, under the hypotheses of Theorem 9.1, we have $-1 \notin \sigma(P)$. Then the problem (9.1)/(9.2) has a unique solution $x \in C_t^1(X)$ for any $\phi \in X_+, \psi \in X_-$, and $f \in C_t(X)$. This solution may be determined by formula (9.8), where z is the unique solution of (9.10).*

If the integral operator

$$(9.12) \qquad Kx(s) = \int_{-1}^{1} k(s,\sigma)x(\sigma)\,d\sigma$$

is compact in X, the integral operator (9.11) is compact in X as well (see APPELL-DIALLO-ZABREJKO [1988]). Therefore, by the classical Fredholm theory, the following holds in case $-1 \in \sigma(P)$.

Lemma 9.2. *Suppose that, under the hypotheses of Theorem 9.1, we have $-1 \in \sigma(P)$, and the operator (9.12) is compact in X. Then the problem (9.1)/(9.2) has a (non-unique) solution $x \in C_t^1(X)$ only for those $\phi \in X_+, \psi \in X_-$, and $f \in C_t(X)$, for which the function r in (9.9) is orthogonal to all solutions v of the homogeneous adjoint equation*

$$v(s) + \int_{-1}^{0} p(\sigma,s)v(\sigma)\,d\sigma = 0.$$

The results of this subsection show that, if the integral operator (9.12) is compact in X, the integro-differential equation (9.1) with boundary condition (9.2) may have exactly one solution, no solution, or a finite number of linearly independent solutions in X. The last case is covered by Lemma 9.2, but the verification of the orthogonality relation $\langle r, v \rangle = 0$ may be rather difficult. This may be avoided, however, by reducing the problem (9.1)/(9.2) not to a system of two one-dimensional integral equations, but to a single "two-dimensional" integral equation. One is lead to such an equation rather naturally when inverting directly the operator $\frac{\partial}{\partial t} - c(s)$ between suitable function spaces; we will do this in the following subsection.

9.3. Reduction to a two-dimensional integral equation

A useful method of studying the problem (9.1)/(9.2) consists in passing to operators on spaces of functions of two variables, where the integration is carried out only with respect to one variable. Such operators are sometimes called *partial integral operators*; they provide a powerful tool in the theory and applications of integro-differential equations and will be studied in great detail in the next chapter.

In this subsection we again employ the space $C_t^1(X)$ introduced in the preceding subsection for $X = L_p([-1, 1])$. The proofs of the following two lemmas are straightforward.

Lemma 9.3. *Let $c \in L_\infty([-1, 1])$. Then, for any $\phi \in X_+$, $\psi \in X_-$, and $f \in C_t(X)$, the problem*

$$
\begin{cases}
\dfrac{\partial g(t, s)}{\partial t} = c(s)g(t, s) + f(t, s) & \text{if } (t, s) \in Q, \\[2mm]
g(0, s) = \phi(s) & \text{if } 0 < s \le 1, \\[2mm]
g(T, s) = \psi(s) & \text{if } -1 \le s < 0
\end{cases}
$$

has a unique solution $g \in C_t^1(X)$. This solution is given, for almost *all* $(t, s) \in Q$, *by* $g(t, s) =$

$$(9.13) \quad \begin{cases} \int_0^t e^{(t-\tau)c(s)} f(\tau, s)\, d\tau + e^{tc(s)}\phi(s) & (0 < s \le 1), \\[2mm] \int_T^t e^{(t-\tau)c(s)} f(\tau, s)\, d\tau + e^{(t-T)c(s)}\psi(s) & (-1 \le s < 0). \end{cases}$$

Lemma 9.4. *If the integral operator* (9.12) *is bounded in* $X = L_p([-1, 1])$, *the partial integral operator*

$$\hat{K}x(t, s) = \int_{-1}^1 k(s, \sigma)x(t, \sigma)\, d\sigma$$

is bounded in $L_p(Q), C_t(X)$, *and* $C_t^1(X)$.

Let us denote by L the operator defined by

$$(9.14) \quad Lx(t, s) = \begin{cases} \int_0^t e^{(t-\tau)c(s)} \hat{K}x(\tau, s)\, d\tau & \text{if } 0 < s \le 1, \\[2mm] \int_T^t e^{(t-\tau)c(s)} \hat{K}x(\tau, s)\, d\tau & \text{if } -1 \le s < 0. \end{cases}$$

As a consequence of Lemma 9.4, we get the following

Lemma 9.5. *If* $c \in L_\infty([-1, 1])$ *and the operator* (9.12) *is regular in* $X = L_p([-1, 1])$, *then* L *is a bounded operator from* $L_p(Q)$ *into* $C_t(X)$.

Combining the preceding three lemmas we obtain still another reformulation of the problem (9.1)/(9.2):

Theorem 9.2. *Let* $c \in L_\infty([-1, 1])$, *and suppose that the operator* (9.12) *is regular in* $X = L_p([-1, 1])$. *Then every solution of the problem* (9.1)/(9.2) *solves the operator equation*

$$(9.15) \quad x(t, s) = Lx(t, s) + g(t, s),$$

where L is defined by (9.14) *and g is defined by* (9.13). *Conversely,*
every solution $x \in C_t(X)$ *of* (9.15), *with L given by* (9.14) *and g given*
by (9.13), *is a solution of the problem* (9.1)/(9.2).

By Theorem 9.2, we may reduce the solvability problem for the Barbashin equation (9.1) in the space $C_t^1(X) = C_t^1(L_p)$ to that of the operator equation (9.15) in the space $C_t(X)$. Let us introduce some notation. For $0 \le t, \tau \le T$ and $0 < s \le 1$ we put

$$u(t, s) = x(t, s), \quad v(t, s) = x(t, -s),$$

$$\xi(t, s) = g(t, s), \quad \eta(t, s) = g(t, -s).$$

Moreover, for $0 \le t, \tau \le T$ and $0 < s, \sigma \le 1$ we put

(9.16)
$$\begin{cases} a(t, \tau, s, \sigma) = e^{(t-\tau)c(s)}k(s, \sigma), \\[2mm] b(t, \tau, s, \sigma) = e^{(t-\tau)c(s)}k(s, -\sigma), \\[2mm] c(t, \tau, s, \sigma) = e^{(t-\tau)c(-s)}k(-s, \sigma), \\[2mm] d(t, \tau, s, \sigma) = e^{(t-\tau)c(-s)}k(-s, -\sigma), \end{cases}$$

The functions $a, b, c,$ and d give rise to four operators $A, B, C,$ and D defined by

(9.17)
$$\begin{cases} Au(t, s) = \displaystyle\int_0^t \int_0^1 a(t, \tau, s, \sigma)u(\tau, \sigma)\, d\sigma\, d\tau, \\[4mm] Bv(t, s) = \displaystyle\int_0^t \int_0^1 b(t, \tau, s, \sigma)v(\tau, \sigma)\, d\sigma\, d\tau, \\[4mm] Cu(t, s) = \displaystyle\int_T^t \int_0^1 c(t, \tau, s, \sigma)u(\tau, \sigma)\, d\sigma\, d\tau, \\[4mm] Dv(t, s) = \displaystyle\int_T^t \int_0^1 d(t, \tau, s, \sigma)v(\tau, \sigma)\, d\sigma\, d\tau, \end{cases}$$

respectively. The equation (9.15) may then be written as a system

$$\begin{cases} u(t, s) = Au(t, s) + Bv(t, s) + \xi(t, s), \\[2mm] v(t, s) = Cu(t, s) + Dv(t, s) + \eta(t, s), \end{cases}$$

or, in matrix form

$$\text{(9.18)} \qquad \begin{pmatrix} I - A & -B \\ -C & I - D \end{pmatrix} \begin{pmatrix} u \\ v \end{pmatrix} = \begin{pmatrix} \xi \\ \eta \end{pmatrix}.$$

Now, the matrix in (9.18) admits an inverse if the operators $(I - A)^{-1}$ and $(I - D)^{-1}$ exist, and the operators $I - A - B(I - D)^{-1}C$ and $I - D - C(I - A)^{-1}B$, or, equivalently, the operators $I - (I - A)^{-1}B(I - D)^{-1}C$ and $I - (I - D)^{-1}C(I - A)^{-1}B$ are invertible. This holds, of course, precisely if

$$\text{(9.19)} \qquad 1 \notin \sigma((I - A)^{-1}B(I - D)^{-1}C).$$

In fact, the following holds.

Theorem 9.3. *Let $c \in L_\infty([-1, 1])$, and suppose that the operator (9.12) is regular in $X = L_p([-1, 1])$. Then (9.19) implies that the operator $I - L$, with L given by (9.14), is invertible in $C_t(X)$.*

□ It suffices to show that 1 belongs neither to $\sigma(A)$ nor to $\sigma(D)$. For this in turn it is clearly sufficient to show that

$$r_\sigma(A) = r_\sigma(D) = 0,$$

where $r_\sigma(K)$ denotes the spectral radius

$$r_\sigma(K) = \lim_{n \to \infty} \sqrt[n]{\|K^n\|_{\mathcal{L}(X)}}$$

of an operator $K \in \mathcal{L}(X)$. Since the function c is bounded, we find an $M > 0$ such that $e^{Tc(s)} \leq M$ for almost all $s \in [-1, 1]$. This implies that

$$|a(t, \tau, s, \sigma)| \leq M|k(s, \sigma)|, \quad |d(t, \tau, s, \sigma)| \leq M|k(-s, -\sigma)|,$$

hence

$$|Au(t, s)| \leq \overline{A}u(t, s), \quad |Dv(t, s)| \leq \overline{D}v(t, s),$$

where the operators $\overline{A}$ and $\overline{D}$ are defined by

$$\text{(9.20)} \qquad \overline{A}u(t, s) = M \int_0^t \int_0^1 |k(s, \sigma)|u(\tau, \sigma) \, d\sigma \, d\tau,$$

and

$$(9.21) \qquad \overline{D}v(t,s) = M \int_t^T \int_0^1 |k(-s,-\sigma)| v(\tau,\sigma)\, d\sigma\, d\tau,$$

respectively. As usual, it is easy to prove that the iterates of the Volterra operators (9.20) and (9.21) satisfy the estimates

$$(9.22) \qquad \|\overline{A}^n\|, \|\overline{D}^n\| \leq \frac{1}{n!} M^n T^n \| |K| \|^n,$$

where

$$(9.23) \qquad |K| x(s) = \int_{-1}^1 |k(s,\sigma)| x(\sigma)\, d\sigma$$

as above. But (9.22) implies that the spectral radii of both $\overline{A}$ and $\overline{D}$, and hence also of A and D, are zero, and so we are done. ∎

The requirement (9.19) is essential in Theorem 9.3, as may be seen by the following very simple

Example 9.1. Let $c(s) \equiv 0$, and

$$k(s,\sigma) = \begin{cases} 0 & \text{if } s,\sigma > 0 \text{ or } s,\sigma < 0, \\ 1 & \text{if } s > 0 \text{ and } \sigma < 0, \\ -1 & \text{if } s < 0 \text{ and } \sigma > 0. \end{cases}$$

We consider the operators (9.17) for $T = \frac{\pi}{2}$, i.e. $(t,s) \in [0, \frac{\pi}{2}] \times [-1, 1]$. In this case we have $Au(t,s) = Dv(t,s) \equiv 0$,

$$Bv(t,s) = \int_0^t \int_0^1 v(\tau,\sigma)\, d\sigma\, d\tau$$

and

$$Cu(t,s) = -\int_{\pi/2}^t \int_0^1 u(\tau,\sigma)\, d\sigma\, d\tau.$$

Since the function $u(t,s) = \sin t$ $(0 \leq t \leq \frac{\pi}{2})$ satisfies $u = BCu$, we have $1 \in \sigma(BC) = \sigma((I - A)^{-1} B(I - D)^{-1} C)$. The assertion of Theorem 9.3 fails, since the function

$$x(t,s) = \begin{cases} \sin t & \text{if } 0 \leq t \leq \frac{\pi}{2} \text{ and } 0 < s \leq 1, \\ \cos t & \text{if } 0 \leq t \leq \frac{\pi}{2} \text{ and } -1 \leq s < 0 \end{cases}$$

belongs to $C_t(X)$ and satisfies $x - Lx = 0$.

We are now in the position to state our main existence result for the integro-differential equation (9.1) with boundary condition (9.2).

Theorem 9.4. *Let $c \in L_\infty([-1,1])$, and suppose that the operator (9.12) is regular in $X = L_p([-1,1])$. Assume that (9.19) holds, where the operators A, B, C, and D are given by (9.17). Then the problem (9.1)/(9.2) has a unique solution $x \in C_t^1(X)$ for any $\phi \in X_+, \psi \in X_-$, and $f \in C_t(X)$.*

□ By Theorem 9.2, every solution of (9.1)/(9.2) is a solution of the operator equation (9.15), and vice versa. By Theorem 9.3 in turn, the hypothesis (9.19) ensures the unique solvability of the operator equation (9.15) for any g. This proves the assertion. ■

When applying Theorem 9.4, a crucial point is of course the verification of (9.19), which may be very hard in practice. However, one may avoid this by proving directly an upper estimate for the spectral radius of the operator L given in (9.13). We illustrate this in the following simple, though useful

Theorem 9.5. *Let $c \in L_\infty([-1,1])$ with $c(s) \le 0$ for $s > 0$ and $c(s) \ge 0$ for $s < 0$. Suppose that the operator (9.23) is bounded in $X = L_p([-1,1])$, and its spectral radius satisfies*

$$(9.24) \qquad\qquad r_\sigma(|K|) < \frac{1}{T}.$$

Then the problem (9.1)/(9.2) has a unique solution $x \in C_t^1(X)$ for any $\phi \in X_+, \psi \in X_-$, and $f \in C_t(X)$.

□ For any positive function $x \in C_t(X)$ we have

$$|Lx(t,s)| \le \int_0^T \int_{-1}^1 |k(s,\sigma)| x(\tau,\sigma) \, d\sigma \, d\tau = (J \otimes |K|) x(t,s),$$

where

$$Jh(t) = \int_0^T h(\tau) \, d\tau$$

denotes the integral mean operator. Consequently, from (9.24) we conclude that

$$r_\sigma(L) \le r_\sigma(J \otimes |K|) = r_\sigma(J)r(|K|) = Tr_\sigma(|K|) < 1,$$

and the assertion follows from Theorem 9.2. ∎

Sometimes it is also possible to solve the operator equation (9.15) directly, for instance in the case of degenerate kernels:

Example 9.2. Let $Q = [0,1] \times [-1,1]$, $c(s) \equiv c$, $k(s,\sigma) = a(s)b(\sigma)$, and $f(t,s) \equiv 0$, i. e. we consider the problem

$$(9.25) \qquad \frac{\partial x(t,s)}{\partial t} = cx(t,s) + a(s) \int_{-1}^{1} b(\sigma)x(t,\sigma)\,d\sigma$$

for $(t,s) \in Q$. In this case we have

$$Lx(t,s) = \begin{cases} a(s) \displaystyle\int_0^t e^{c(t-\tau)} \int_{-1}^{1} b(\sigma)x(\tau,\sigma)\,d\sigma\,d\tau & \text{if } 0 < s \le 1, \\[2ex] a(s) \displaystyle\int_1^t e^{c(t-\tau)} \int_{-1}^{1} b(\sigma)x(\tau,\sigma)\,d\sigma\,d\tau & \text{if } -1 \le s < 0, \end{cases}$$

and

$$g(t,s) = \begin{cases} \phi(s)e^{ct} & \text{if } 0 < s \le 1, \\[1ex] \psi(s)e^{c(t-1)} & \text{if } -1 \le s < 0. \end{cases}$$

Put

$$\int_{-1}^{1} b(\sigma)x(t,\sigma)\,d\sigma = \xi(t), \qquad \int_0^1 a(\sigma)b(\sigma)\,d\sigma = \alpha,$$

$$\int_{-1}^{0} a(\sigma)b(\sigma)\,d\sigma = \beta, \qquad \int_0^1 \phi(\sigma)b(\sigma)\,d\sigma = \gamma,$$

$$\int_{-1}^{0} \psi(\sigma)b(\sigma)\,d\sigma = \delta,$$

and suppose that $\alpha, \beta > 0$. The operator equation (9.15) reduces then to the equation

$$\xi(t) = \alpha \int_0^t e^{c(t-\tau)}\xi(\tau)\,d\tau +$$

$$(9.26)$$

$$\beta \int_1^t e^{c(t-\tau)}\xi(\tau)\,d\tau + \gamma e^{ct} + \delta e^{c(t-1)}.$$

Putting $y(t) = \xi(t)e^{-ct}$ and $\omega = \gamma + \delta e^{-c}$, we may rewrite (9.26) in the form

$$(9.27) \qquad y(t) = \alpha \int_0^t y(\tau)\, d\tau + \beta \int_1^t y(\tau)\, d\tau + \omega.$$

Differentiating (9.27) yields

$$y' = (\alpha + \beta)y, \quad y(0) = \omega - \beta \int_0^1 y(\tau)\, d\tau$$

with solution

$$y(t) = \omega \frac{\alpha + \beta}{\alpha + \beta e^{\alpha + \beta}} e^{(\alpha + \beta)t},$$

hence

$$\xi(t) = \omega \frac{\alpha + \beta}{\alpha + \beta e^{\alpha + \beta}} e^{(\alpha + \beta + c)t}.$$

We conclude that the solution of (9.15) is given by

$$x(t,s) = \begin{cases} \dfrac{\omega a(s) e^{ct}(e^{(\alpha + \beta)t} - 1)}{\alpha + \beta e^{\alpha + \beta}} + \phi(s)e^{ct} & (0 < s \leq 1), \\[4mm] \dfrac{\omega a(s) e^{ct}(e^{(\alpha + \beta)t} - e^{\alpha + \beta})}{\alpha + \beta e^{\alpha + \beta}} + \psi(s)e^{c(t-1)} & (-1 \leq s < 0). \end{cases}$$

A straightforward calculation shows that this satisfies indeed the equation (9.25) with boundary conditions (9.2).

There are several other possibilities for obtaining existence and uniqueness results for the problem (9.1)/(9.2). For instance, if the kernel k is positive and the corresponding operator (9.12) is bounded in $X = L_p([-1,1])$, one may use well-known lower estimates and monotonicity properties of the spectral radius of an integral operator (see e.g. ZABREJKO [1967]) in order to get more precise information on the spectral radius of the operator $(I - A)^{-1}B(I - D)^{-1}C$ in X. Moreover, integro-differential operators may be studied success-fully by means of fixed point principles in so-called K-normed spaces; for the theory and several applications of K-normed spaces to differential equations, see e.g. ZABREJKO [1992]. We shall study the Barbashin equation (9.1) in the setting of K-normed spaces in the following subsection.

9.4. Application of K-normed spaces

Let X be an arbitrary linear space, and Z a real linear space which is ordered by some cone K (see e.g. KRASNOSEL'SKIJ-LIFSHITS-SOBOLEV [1985]). A functional $]| \cdot |[: X \to K$ is called K-*norm* on X if

(a) $]|x|[= 0$ if and only if $x = 0$;

(b) $]|\lambda x|[= |\lambda| \,]|x|[$;

(c) $]|x + y|[\leq]|x|[+]|y|[$

$(x, y \in X, \lambda \in \mathbb{R})$. The space $(X,]| \cdot |[)$ is then called K- *normed space*. Of course, every normed space $(X, \| \cdot \|)$ is a trivial example with $Z = \mathbb{R}$ and $K = [0, \infty)$. On the other hand, using various nontrivial K-normed spaces one may obtain interesting new results, or interesting new proofs of known results. For instance, studying fixed point theorems in K-normed and related spaces (see ZABREJ-KO [1990, 1992], ZABREJKO-MAKAREVICH [1987, 1987a] and below) leads to useful existence results for the Cauchy problem for differential equations with "badly behaved" right-hand side and, especially, partial differential equations (see BARKOVA-ZABREJKO [1991], EVKHUTA-ZABREJKO [1985], ZABREJKO [1989, 1992]).

Let X be a K-normed space. A sequence $(x_n)_n$ in X is called *order-convergent* to $x \in X$ (respectively *order-Cauchy*) if there is a sequence $(z_n)_n$ in the cone K which converges monotonically to zero and satisfies $]|x_n - x|[\leq z_n$ for $n = 1, 2, \ldots$ (respectively $]|x_m - x_n|[\leq z_n$ for $n = 1, 2, \ldots$ and $m > n$). A K-normed space X is called *order-complete* if every order-Cauchy sequence in X is order-convergent in X. For example, if we take $X = Z = \mathbb{R}^m, K = \{(\zeta_1, \ldots, \zeta_m) : \zeta_j \geq 0 \ (j = 1, 2, \ldots, m)\}$, and

$$(9.28) \qquad]|x|[=]|(\xi_1, \ldots, \xi_m)|[= (|\xi_1|, \ldots, |\xi_m|) \quad (x \in \mathbb{R}^m),$$

then $(X,]| \cdot |[)$ is of course order-complete.

Two important examples of infinite dimensional order-complete K-normed spaces are as follows. First, let Ω be an arbitrary measure space and X an ideal space of measurable real functions on Ω. If we denote by $|x|$ the function defined by $|x|(s) = |x(s)|$, we may define a K-norm on X by putting

$$(9.29) \qquad]|x|[= |x| \quad (x \in X).$$

Second, let B some Banach space, and denote by $S(\Omega, B)$ the set of all (Bochner-) measurable functions on Ω with values in B. Then

$$(9.30) \qquad]|x|[= ||x|| \quad (x \in S(\Omega, B))$$

defines a K-norm on $S(\Omega, B)$. Similarly, any linear subspace $X \subseteq S(\Omega, B)$ may be equipped with the K-norm (9.30). A particularly important example is the *Bochner-Lebesgue space* $L_p(\Omega, B)$ $(1 \le p \le \infty)$ defined by the norm

$$(9.31) \qquad ||x|| = \begin{cases} \left(\int_\Omega ||x(s)||_B^p \, ds \right)^{1/p} & \text{if } 1 \le p < \infty, \\ \text{ess sup } \{ ||x(s)||_B : s \in \Omega \} & \text{if } p = \infty. \end{cases}$$

The following is a natural extension of the well-known Banach-Caccioppoli fixed point principle to K-normed spaces. Recall (see Subsection 4.1) that an ideal space Z is called *regular* if every element $z \in Z$ has an absolutely continuous norm.

Theorem 9.6. *Let $(X,]| \cdot |[)$ be an order-complete K-normed space with K-norm $]| \cdot |[: X \to K \subset Z$, where Z is a regular ideal space. Let $Q : K \to K$ be a linear operator with spectral radius $r_\sigma(Q) < 1$. Suppose that $F : X \to X$ is a (linear or nonlinear) operator which satisfies a contraction type condition*

$$(9.32) \qquad]|Fx_1 - Fx_2|[\le Q(]|x_1 - x_2|[) \quad (x_1, x_2 \in X).$$

Then F has a unique fixed point in X; this fixed point may be obtained as limit of successive approximations $x_{n+1} = Fx_n$ $(n = 0, 1, 2, \dots; x_0 \in X$ arbitrary).

Now we shall apply Theorem 9.6 to the problem (9.1)/(9.2). To this end, we take $B = L_p([0, 1]) \times L_p([0, 1])$ $(1 \le p < \infty)$, equipped with the norm

$$||(u, v)||_B = ||u||_{L_p} + ||v||_{L_p}.$$

Moreover, let $X = L_p([0,T], B)$ be the Bochner-Lebesgue space of all B-valued functions $t \mapsto x(t,\cdot) = (u(t,\cdot), v(t,\cdot))$, equipped with the norm (9.31) and the K-norm

$$(9.33) \qquad]|x|[= (||u(t,\cdot)||_{L_p}, ||v(t,\cdot)||_{L_p}).$$

Thus, the K-norm (9.33) takes its values in the natural cone of the Banach space $Z = L_p([0,T], \mathbb{R}^2)$. It is easy to see that the norm (9.31) is here equivalent on X to the somewhat simpler norm

$$(9.34) \qquad ||x|| = \left\{ \int_a^b \int_0^1 [|u(t,s)| + |v(t,s)|]^p \, dt \, ds \right\}^{1/p}$$

which will be considered throughout. As we have seen in Subsection 9.3, the problem (9.1)/(9.2) may be reduced to the matrix operator equation

$$(9.35) \qquad \begin{pmatrix} I - A & -B \\ -C & I - D \end{pmatrix} \begin{pmatrix} u \\ v \end{pmatrix} = \begin{pmatrix} \xi \\ \eta \end{pmatrix},$$

where $u(t,s) = x(t,s)$ and $v(t,s) = x(t,-s)$ for $0 \le t \le T$ and $0 < s \le 1$,

$$\begin{cases} \xi(t,s) = \displaystyle\int_0^t e^{(t-\tau)c(s)} f(\tau,s) \, d\tau + e^{tc(s)} \phi(s), \\[2mm] \eta(t,s) = \displaystyle\int_T^t e^{(t-\tau)c(-s)} f(\tau,-s) \, d\tau + e^{(t-T)c(-s)} \psi(-s), \end{cases}$$

and the operators $A, B, C,$ and D are given by (9.17). With the help of the functions (9.16) which generate the operators (9.17), we may now define the operator Q occurring in the contraction condition (9.32). Suppose that

$$(9.36) \qquad \begin{cases} \left\| \displaystyle\int_0^1 a(t,\tau,\cdot,\sigma) u(\sigma) \, d\sigma \right\| \le \alpha ||u||, \\[4mm] \left\| \displaystyle\int_0^1 b(t,\tau,\cdot,\sigma) v(\sigma) \, d\sigma \right\| \le \beta ||v||, \\[4mm] \left\| \displaystyle\int_0^1 c(t,\tau,\cdot,\sigma) u(\sigma) \, d\sigma \right\| \le \gamma ||u||, \\[4mm] \left\| \displaystyle\int_0^1 d(t,\tau,\cdot,\sigma) v(\sigma) \, d\sigma \right\| \le \delta ||v|| \end{cases}$$

for some $\alpha, \beta, \gamma, \delta > 0$, where all norms in (9.36) are taken in the corresponding L_p spaces. (Estimates of the type (9.36) may be verified by applying well-known formulas or inequalities for the norm of an integral operator in Lebesgue spaces, see e.g. KRASNOSEL'SKIJ-ZA-BREJKO-PUSTYL'NIK-SOBOLEVSKIJ [1966].) If we define $F : X \rightarrow X$ by

(9.37)
$$Fx(t,s) = F\left(\begin{array}{c} u(t,s) \\ v(t,s) \end{array}\right)$$
$$= \left(\begin{array}{cc} A & B \\ C & D \end{array}\right)\left(\begin{array}{c} u(t,s) \\ v(t,s) \end{array}\right) + \left(\begin{array}{c} \xi(t,s) \\ \eta(t,s) \end{array}\right)$$

and $Q : Z \rightarrow Z$ by

(9.38)
$$Qz(t) = Q\left(\begin{array}{c} u(t) \\ v(t) \end{array}\right) = \left(\begin{array}{c} \int_0^t [\alpha u(\tau) + \beta v(\tau)]\, d\tau \\ \int_t^T [\gamma u(\tau) + \delta v(\tau)]\, d\tau \end{array}\right),$$

the contraction condition (9.32) is simply a consequence of the estimates (9.36). By the definition (9.37) of the operator F, every fixed point point of F is a solution of the operator equation (9.35), and vice versa. Thus, for applying Theorem 9.6 it remains to impose suitable conditions which ensure that the operator (9.38) has spectral radius $r_\sigma(Q) < 1$. Since Q is a positive operator in Z, by the classical Krejn-Rutman theorem (KREJN-RUTMAN [1948], see also KRASNOSEL'SKIJ-LIFSHITS-SOBOLEV [1985] and ZABREJKO-SMITSKIKH [1979]) we have to find $\rho > 0$ such that $Qz = \rho z$ for some nonnegative function $z = (u, v) \in Z$. Writing this out in components, we get the system

(9.39)
$$\begin{cases} \rho u(t) = \int_0^t [\alpha u(\tau) + \beta v(\tau)]\, d\tau, \\ \rho v(t) = \int_t^T [\gamma u(\tau) + \delta v(\tau)]\, d\tau. \end{cases}$$

Differentiating (9.39) yields

(9.40)
$$\begin{cases} \rho u' = \alpha u + \beta v, \\ \rho v' = -\gamma u - \delta v. \end{cases}$$

with boundary conditions $u(0) = v(T) = 0$. Since $\rho = 0$ for $(\beta, \gamma) = (0,0)$, we suppose that $(\beta, \gamma) \neq (0,0)$; for definiteness, let $\beta \neq 0$.

The solution behaviour of (9.40) depends, of course, on the sign of the discriminant $\Delta = (\alpha + \delta)^2 - 4\beta\gamma$. In fact, putting v from the first equation into the second equation in (9.40), we get the second order differential equation

$$u'' - \frac{\alpha - \delta}{\rho} u' - \frac{\alpha\delta - \beta\gamma}{\rho^2} u = 0.$$

Solving the corresponding characteristic equation

$$\lambda^2 - \frac{\alpha - \delta}{\rho}\lambda - \frac{\alpha\delta - \beta\gamma}{\rho^2} = 0,$$

and choosing the free constants in the solution in such a way that the boundary conditions $u(0) = v(T) = 0$ are fulfilled, we arrive at the formula

$$\rho = \begin{cases} \dfrac{T\sqrt{\Delta}}{\log \frac{\alpha+\delta+\sqrt{\Delta}}{\alpha+\delta-\sqrt{\Delta}}} & \text{if } \Delta > 0, \\[3ex] \dfrac{T(\alpha + \delta)}{2} & \text{if } \Delta = 0, \\[3ex] \dfrac{T\sqrt{-\Delta}}{2\arctan \frac{\sqrt{-\Delta}}{\alpha+\delta}} & \text{if } \Delta < 0 \text{ and } \alpha + \delta \neq 0, \\[3ex] \dfrac{T\sqrt{-\Delta}}{\pi} & \text{if } \Delta < 0 \text{ and } \alpha + \delta = 0. \end{cases}$$

Thus, we have now all the necessary information to apply Theorem 9.6 to the operator equation (9.35), and hence to the integro-differential equation (9.1) with boundary condition (9.2). We summarize with the following

Theorem 9.7. *Let $c \in L_\infty([-1, 1])$, and suppose that the integral operator defined by the kernel k is regular in $X = L_p([-1, 1])$. Assume that the estimates (9.36) hold, and that one of the following four conditions is satisfied:*

(a) $\Delta = (\alpha + \delta)^2 - 4\beta\gamma > 0$ and $T\sqrt{\Delta} < \log \frac{\alpha+\delta+\sqrt{\Delta}}{\alpha+\delta-\sqrt{\Delta}}$;

(b) $(\alpha + \delta)^2 = 4\beta\gamma$ and $T(\alpha + \delta) < 2$;

(c) $\Delta < 0, \alpha + \delta \neq 0$, and $\sqrt{-\Delta} < 2\arctan \frac{\sqrt{-\Delta}}{\alpha+\delta}$;

(d) $\Delta < 0, \alpha + \delta = 0$, and $\sqrt{-\Delta} < \pi$.

Then the problem (9.1)/(9.2) has a unique solution $x \in C_t^1(X)$ *for any* $\phi \in X_+, \psi \in X_-$, *and* $f \in C_t(X)$.

9.5. Unbounded multipliers

In Theorem 9.7 above, as well as in all existence and uniqueness results for (9.1)/(9.2) derived in the previous subsections, we supposed that $c \in L_\infty([-1,1])$. From now on we consider also unbounded multipliers c, and we require that

$$(9.41) \qquad c(s) \leq -1 \, (0 < s \leq 1), \quad c(s) \geq 1 \, (-1 \leq s < 0).$$

Under this hypothesis, the existence and uniqueness results given above may become false:

Example 9.3. Let $X = L_1([-1,1]), c(s) = -1/s, k(s,\sigma) \equiv 0, f(t,s) \equiv 0, \phi(s) \equiv 1$, and $\psi(s) \equiv 0$, i.e. we consider the problem

$$\begin{cases} \dfrac{\partial x(t,s)}{\partial t} = -\dfrac{x(t,s)}{s} & \text{if } (t,s) \in Q, \\[2mm] x(0,s) = 1 & \text{if } 0 < s \leq 1, \\[2mm] x(1,s) = 0 & \text{if } -1 \leq s < 0. \end{cases}$$

We certainly have then $\phi \in X_+, \psi \in X_-$, and $f \in C_t(X)$, but the solution of (9.1)/(9.2) in this case, viz.

$$(9.42) \qquad x(t,s) = \begin{cases} e^{-t/s} & \text{if } 0 < s \leq 1, \\[2mm] 0 & \text{if } -1 \leq s < 0, \end{cases}$$

does not belong to $C_t^1(X)$, since $\partial x(0,\cdot)/\partial t \notin X$.

This phenomenon is due to the fact that, even if the data ϕ, ψ, and f are very smooth, the corresponding solution x may become singular at the boundary points $t = 0$ or $t = T$. This difficulty may be overcome in two different ways: either we weaken the regularity requirement on the possible solutions near the boundary, or we take the data from *weighted function spaces*.

Given an ideal space X of functions over $[-1, 1]$, by $\check{C}_t(X)$ we denote the linear space of all functions of two variables $x : Q \to \mathbb{R}$ with the property that the map $t \mapsto x(t, \cdot)$ is continuous from the *open* interval $(0, T)$ into X; the space $\check{C}_t^1(X)$ is defined analogously. Thus, in contrast to the spaces $C_t(X)$ and $C_t^1(X)$ introduced in Subsection 9.2, the points $t = 0$ and $t = T$ are excluded in the regularity requirement. As before, a model case will always be $X = L_p([-1, 1])$.

If X is an ideal space over $[-1, 1]$ and w a nonvanishing measurable function ("weight function") on $[-1, 1]$, we denote by $X(w), X_+(w)$, and $X_-(w)$ the ideal spaces defined by the norms

$$(9.43) \qquad \|x\|_{X(w)} = \|wx\|_X, \quad \|x\|_{X_\pm(w)} = \|wx\|_{X_\pm},$$

respectively. Now, if c is a multiplier satisfying (9.41), the imbeddings

$$(9.44) \qquad X(c) \subseteq X \subseteq X(\tfrac{1}{c}), \quad X_\pm(c) \subseteq X_\pm \subseteq X_\pm(\tfrac{1}{c})$$

hold with imbedding constants 1.

The following two lemmas are completely analogous to Lemma 9.3.

Lemma 9.6. *Let* $X = L_p([-1, 1]), \phi \in X_+(c), \psi \in X_-(c),$ *and* $f \in \check{C}_t(X(c)),$ *where* c *satisfies* (9.41). *Then the problem*

$$(9.45) \qquad \begin{cases} \dfrac{\partial g(t, s)}{\partial t} = c(s)g(t, s) + f(t, s) & \text{if } (t, s) \in Q, \\[2mm] g(0, s) = \phi(s) & \text{if } 0 < s \le 1, \\[2mm] g(T, s) = \psi(s) & \text{if } -1 \le s < 0 \end{cases}$$

has a unique solution $g \in \check{C}_t^1(X)$; *this solution is given, for almost* all $(t, s) \in Q$, by $g(t, s) =$

(9.46)
$$\begin{cases} \displaystyle\int_0^t e^{(t-\tau)c(s)} f(\tau, s) \, d\tau + e^{tc(s)} \phi(s) & (0 < s \le 1), \\[2ex] \displaystyle\int_T^t e^{(t-\tau)c(s)} f(\tau, s) \, d\tau + e^{(t-T)c(s)} \psi(s) & (-1 \le s < 0). \end{cases}$$

Lemma 9.7. *Let $X = L_p([-1, 1])$, $\phi \in X_+$, $\psi \in X_-$, and $f \in C_t(X)$, where c satisfies (9.41). Then the problem (9.45) has a unique solution in $\check{C}_t^1(X)$; this solution is given by (9.46).*

For the following lemma, recall the definition of the operator L in (9.14) and the function g in (9.13). The following assertions are parallel to Lemma 9.5 and Theorem 9.2, respectively; the proof is a simple consequence of the imbeddings (9.44).

Lemma 9.8. *If c satisfies (9.41), and the integral operator defined by the kernel k is regular in $X = L_p([-1, 1])$, then (9.14) is a bounded operator from $L_p(Q)$ into $C_t(X)$.*

Theorem 9.8. *Suppose that c satisfies (9.41), and the integral operator defined by the kernel k is regular in $X = L_p([-1, 1])$. Then every solution of the problem (9.1)/(9.2) solves the equation*

(9.47)
$$x(t, s) = Lx(t, s) + g(t, s).$$

Conversely, every solution $x \in C_t(X)$ of (9.47) is a solution of (9.1)/(9.2), and the partial derivative $\partial x/\partial t$ belongs to $C_t(X(\frac{1}{c}))$.

Using weighted function spaces, Theorem 9.4 above reads now as follows.

Theorem 9.9. *Suppose that c satisfies (9.41), and the integral operator defined by the kernel k is regular in $X = L_p([-1, 1])$. Assume that*

(9.48)
$$1 \notin \sigma((I - A)^{-1} B(I - D)^{-1} C),$$

where the operators A, B, C, and D are given by (9.17). Then the operator $I - L$, with L given by (9.14), is invertible in $C_t(X)$. Consequently, the problem (9.1)/(9.2) has a unique solution $x \in C_t(X)$ with $\partial x/\partial t \in C_t(X(\frac{1}{c}))$ for any $\phi \in X_+, \psi \in X_-$, and $f \in C_t(X)$.

Of course, Example 9.3 at the beginning of this subsection is covered by Theorem 9.9, but not by any existence theorem derived in the preceding subsections. In fact, for the solution (9.42) we have

$$\left\| \frac{\partial x(t, \cdot)}{\partial t} \right\|_{L_1(1/c)} = \int_{-1}^1 \left| \frac{\partial x(t, s)}{\partial t} s \right| \, ds = \int_{-1}^1 e^{-t/s} \, ds,$$

and the last integral depends continuously on $t \in [0, 1]$. Thus, we have $\partial x/\partial t \in C_t(L_1(\frac{1}{c}))$, but $\partial x/\partial t \notin C_t(L_1)$.

Theorem 9.9 gives the existence of a solution $x \in C_t(X)$ of (9.1)/(9.2) with certain additional properties. A parallel existence theorem may be proved in the space $C_t^1(X)$; we do not present the details.

An application of the above results in the weighted space $X(\frac{1}{c})$ will be given in Subsection 19.5.

§ 10. Non-stationary boundary value problems

In this section we briefly discuss a parallel theory for the non-stationary equation

$$(10.1) \qquad \frac{\partial x(t,s)}{\partial t} = c(t,s)x(t,s) + \int_{-1}^{1} k(t,s,\sigma)x(t,\sigma)\, d\sigma + f(t,s)$$

$((t,s) \in Q = [0,T] \times [-1,1])$, with boundary conditions

$$(10.2) \qquad \begin{cases} x(0,s) = \phi(s) & \text{if } 0 < s \le 1, \\ x(T,s) = \psi(s) & \text{if } -1 \le s < 0. \end{cases}$$

The results of this section may be found in KALITVIN [1993, 1993a], APPELL-DIALLO [1995], and ZABREJKO-KALITVIN [1997].

10.1. Equations with variable operators

Since we are now interested in the non-stationary equation (10.1), we have to put

$$(10.3) \qquad A(t)x(s) = c(t,s)x(s) + \int_{-1}^{1} k(t,s,\sigma)x(\sigma)\, d\sigma$$

and to consider, instead of (9.3), the differential equation

$$(10.4) \qquad \frac{dx}{dt} = A(t)x + f(t).$$

We recall that the evolution operator $U = U(t,\tau)$ of (10.4) has the form

$$(10.5) \qquad U(t,\tau)x(s) = e(t,\tau,s)x(s) + \int_{-1}^{1} h(t,\tau,s,\sigma)x(\sigma)\, d\sigma,$$

where

$$(10.6) \qquad e(t,\tau,s) = \exp\left\{ \int_{\tau}^{t} c(\xi,s)\, d\xi \right\},$$

and $h = h(t,\tau,s,\sigma)$ is a measurable function on $[0,T] \times [0,T] \times [-1,1] \times [-1,1]$. It is therefore not surprising that the results of Subsection 9.2 carry over provided we replace throughout the function

$tc(s)$ by the integral $\int_0^t c(\xi, s)\, d\xi$, and the function $h(t, s, \sigma)$ by the function $h(t, 0, s, \sigma)$. In particular, we may write the solution of (10.4) with initial condition $x(0) = x_0$ in the form

$$
(10.7) \qquad x(t, s) = \begin{cases} m(t, s) + h_z(t, s), \\[2mm] n(t, s) + e(t, 0, s)z(s) + h_z(t, s) \end{cases}
$$

for $0 < s \leq 1$ and $-1 \leq s < 0$, respectively, where we have put

$$
m(t, s) = e(t, 0, s)\phi(s) + \int_0^1 h(t, 0, s, \sigma)\phi(\sigma)\, d\sigma
$$

$$
+ \int_0^t \left\{ e(t, \tau, s)f(\tau, s) + \int_{-1}^1 h(t, \tau, s, \sigma)f(\tau, \sigma)\, d\sigma \right\} d\tau,
$$

$$
h_z(t, s) = \int_{-1}^0 h(t, 0, s, \sigma)z(\sigma)\, d\sigma,
$$

and

$$
n(t, s) = \int_0^1 h(t, 0, s, \sigma)\phi(\sigma)\, d\sigma
$$

$$
+ \int_0^t \left\{ e(t, \tau, s)f(\tau, s) + \int_{-1}^1 h(t, \tau, s, \sigma)f(\tau, \sigma)\, d\sigma \right\} d\tau.
$$

For obtaining the functions $y \in X_+$ and $z \in X_-$ (see Subsection 9.2) we therefore have to solve the system

$$
(10.8) \qquad \begin{cases} y(s) = m(T, s) + h_z(T, s), \\[2mm] e(T, 0, s)z(s) + h_z(T, s) = \psi(s) - n(T, s) \end{cases}
$$

for $0 < s \leq 1$ and $-1 \leq s < 0$, respectively. Moreover, putting

$$
(10.9) \qquad p(s, \sigma) = \frac{h(T, 0, s, \sigma)}{e(T, 0, s)}
$$

and

$$
(10.10) \qquad r(s) = \frac{\psi(s) - n(T, s)}{e(T, 0, s)}
$$

(which is parallel to (9.9)), solving the system (10.8) is equivalent to solving the Fredholm integral equation

$$(10.11) \qquad z(s) + \int_{-1}^{0} p(s,\sigma)z(\sigma)\,d\sigma = r(s).$$

In this way, we arrive at the following generalization of Lemma 9.1:

Theorem 10.1. *Suppose that the functions* $c : Q \to \mathbb{R}$ *and* $k : Q \times [-1,1] \to \mathbb{R}$ *are measurable, the operator* $A(t)$ *is regular in the space* $X = L_p([-1,1])$, *the operator function* (10.3) *is strongly continuous, and* $-1 \notin \sigma(P)$, *where*

$$(10.12) \qquad Pz(s) = \int_{-1}^{0} p(s,\sigma)z(\sigma)\,d(\sigma).$$

Then the problem (10.1)/(10.2) *has a unique solution* $x \in C_t^1(X)$ *for any* $\phi \in X_+, \psi \in X_-$, *and* $f \in C(X)$. *This solution may be determined by formula* (10.7).

Of course, one may also formulate a parallel result to Lemma 9.2 in case $-1 \in \sigma(P)$.

10.2. Reduction to a two-dimensional integral equation

As in Subsection 9.3, we may reduce the boundary value problem (10.1)/(10.2) to the operator equation

$$(10.13) \qquad x(t,s) = Lx(t,s) + g(t,s),$$

where

$$(10.14) \qquad Lx(t,s) = \begin{cases} \displaystyle\int_{0}^{t} e(t,\tau,s)\hat{K}x(\tau,s)\,d\tau & \text{if } 0 < s \le 1, \\[4mm] \displaystyle\int_{T}^{t} e(t,\tau,s)\hat{K}x(\tau,s)\,d\tau & \text{if } -1 \le s < 0 \end{cases}$$

with

$$(10.15) \qquad \hat{K}x(t,s) = \int_{-1}^{1} k(t,s,\sigma)x(t,\sigma)\,d\sigma,$$

and

$$(10.16) \qquad g(t,s) = \begin{cases} \int_0^t e(t,\tau,s)f(\tau,s)\,d\tau + e(t,0,s)\phi(s), \\[2mm] \int_T^t e(t,\tau,s)f(\tau,s)\,d\tau + e(t,T,s)\psi(s) \end{cases}$$

for $0 < s \leq 1$ and $-1 \leq s < 0$, respectively. First we state the following analogue to Lemma 9.3, again for $X = L_p([-1,1])$, which may be proved by a straightforward calculation:

Lemma 10.1 *Let $c \in C_t(L_\infty)$. Then, for any $\phi \in X_+$, $\psi \in X_-$, and $f \in C_t(X)$, the equation*

$$\frac{\partial g(t,s)}{\partial t} = c(t,s)g(t,s) + f(t,s) \qquad ((t,s) \in Q)$$

with boundary condition (10.2) has a unique solution $g \in C_t^1(X)$ which is given by (10.16).

The following is parallel to Theorem 9.2:

Theorem 10.2 *Let $c \in C_t(L_\infty)$, and suppose that the operator function*

$$(10.17) \qquad K(t)x(s) = \int_{-1}^1 k(t,s,\sigma)x(\sigma)\,d\sigma$$

is regular in $X = L_p([-1,1])$ for each $t \in [0,T]$. Assume that the operator (10.15) acts in $C_t(X)$. Then every solution of the problem (10.1)/(10.2) solves the operator equation (10.13), where L is defined by (10.14) and g is defined by (10.16). Conversely, every solution $x \in C_t(X)$ of (10.13), with L given by (10.14) and g given by (10.16), belongs to $C_t^1(X)$ and is a solution of the problem (10.1)/(10.2).

□   Suppose that $x \in C_t^1(X)$ solves (10.1)/(10.2). The function $\hat{f}$ defined by

$$\hat{f}(t,s) := \hat{K}x(t,s) + f(t,s)$$

belongs then to $C_t(X)$, by Lemma 4.2. Consequently, from Lemma 10.1 we conclude that the unique C_t^1-solution of the equation

$$(10.18) \qquad \frac{\partial y(t,s)}{\partial t} = c(t,s)y(t,s) + \hat{f}(t,s)$$

is $y(t,s) = Lx(t,s) + g(t,s)$, i.e. $y = x$ solves (10.13).

Conversely, if $x \in C_t(X)$ solves (10.13), the problem (10.18) with boundary conditions

$$y(0,s) = \phi(s) \ (0 < s \le 1), \qquad y(T,s) = \psi(s) \ (-1 \le s < 0)$$

has the unique solution $y(t,s) = Lx(t,s) + g(t,s)$, i.e. $x = y$ solves (10.1)/(10.2). ∎

We point out that the assumption on the boundedness of the function c is essential in Theorem 10.2; in fact, otherwise the C_t-solution x of (10.13) need not belong to $C_t^1(X)$:

Example 10.1. Let $X := L_2([-1,1])$, $c(t,s) := 1/2\sqrt{t+|s|}$, $k(t,s,\sigma) := s\exp(-\sqrt{t+|s|})$, $f(t,s) := -2s$, $\phi(s) := \exp\sqrt{s}$, and $\psi(s) := \exp\sqrt{1-s}$, i.e. we consider on $Q = [0,1] \times [-1,1]$ the equation

$$\frac{\partial x(t,s)}{\partial t} = \frac{x(t,s)}{2\sqrt{t+|s|}} + s\int_{-1}^{1} e^{-\sqrt{t+|\sigma|}}x(t,\sigma)\,d\sigma - 2s,$$

subject to the boundary conditions

$$\begin{cases} x(0,s) = e^{\sqrt{s}} & \text{if } 0 < s \le 1, \\[2mm] x(1,s) = e^{\sqrt{1-s}} & \text{if } -1 \le s < 0. \end{cases}$$

In this case we have

$$Lx(t,s) = \begin{cases} se^{\sqrt{t+s}} \displaystyle\int_0^t \int_{-1}^1 \frac{x(\tau,\sigma)}{e^{\sqrt{\tau+s}}e^{\sqrt{t+|\sigma|}}}\,d\sigma\,d\tau & \text{if } 0 < s \le 1, \\[4mm] se^{\sqrt{t-s}} \displaystyle\int_1^t \int_{-1}^1 \frac{x(\tau,\sigma)}{e^{\sqrt{\tau-s}}e^{\sqrt{t+|\sigma|}}}\,d\sigma\,d\tau & \text{if } -1 \le s < 0, \end{cases}$$

and

$$g(t,s) = \begin{cases} -2se^{\sqrt{t+s}} \displaystyle\int_0^t \frac{d\tau}{e^{\sqrt{\tau+s}}} + e^{\sqrt{t+s}} & \text{if } 0 < s \leq 1, \\[3mm] -2se^{\sqrt{t-s}} \displaystyle\int_1^t \frac{d\tau}{e^{\sqrt{\tau-s}}} + e^{\sqrt{t-s}} & \text{if } -1 \leq s < 0. \end{cases}$$

An easy calculation shows that equation (10.13) has the solution $x(t,s) = \exp\sqrt{t + |s|}$; this solution belongs to $C_t(X)$, but not to $C_t^1(X)$, since $\partial x(0,\cdot)/\partial t \notin X$.

The fact that the assertion of Theorem 10.2 *fails* for unbounded functions c is not accidental as we have seen in Lemma 4.5: the strong continuity of the operator function $t \mapsto C(t)$ implies that $c \in L_\infty(Q)$ (even $C_t(L_\infty)$).

Introducing now as in the previous subsection the functions

$$u(t,s) = x(t,s), \quad v(t,s) = x(t,-s),$$

$$\xi(t,s) = g(t,s), \quad \eta(t,s) = g(t,-s)$$

$(0 \leq t \leq T, 0 < s \leq 1)$, replacing (9.16) by

$$(10.19) \quad \begin{cases} a(t,\tau,s,\sigma) = e(t,\tau,s)k(\tau,s,\sigma), \\[3mm] b(t,\tau,s,\sigma) = e(t,\tau,s)k(\tau,s,-\sigma), \\[3mm] c(t,\tau,s,\sigma) = e(t,\tau,-s)k(\tau,-s,\sigma), \\[3mm] d(t,\tau,s,\sigma) = e(t,\tau,-s)k(\tau,-s,-\sigma), \end{cases}$$

$(0 \leq t,\tau \leq T, 0 < s,\sigma \leq 1)$, and defining the operators A, B, C, and D as in (9.17), we arrive again at the matrix equation (9.18). The following is parallel to Theorem 9.3:

Theorem 10.3. *Let $c \in L_\infty(Q)$, and suppose that the operator function (10.17) is regular in $X = L_p([-1,1])$. Assume that the operator (10.15) acts in $C_t(X)$. Then the condition (9.19) implies that the operator $I - L$, with L given by (10.14), is invertible in $C_t(X)$.*

The verification of condition (9.19) amounts, in particular, to showing that $1 \notin \sigma(A)$ and $1 \notin \sigma(D)$. For example, the spectral radii of A and D are zero if

$$|a(t,\tau,s,\sigma)| \le a_1(t,\tau)a_2(s,\sigma), \quad |d(t,\tau,s,\sigma)| \le d_1(t,\tau)d_2(s,\sigma),$$

the integral operators

$$A_1x(t) = \int_0^t a_1(t,\tau)x(\tau)\,d\tau, \quad D_1x(t) = \int_t^T d_1(t,\tau)x(\tau)\,d\tau$$

are compact in $L_p([0,T])$, and the integral operators

$$A_2x(s) = \int_0^1 a_2(s,\sigma)x(\sigma)\,d\sigma, \quad D_2x(s) = \int_0^1 d_2(s,\sigma)x(\sigma)\,d\sigma$$

are bounded in $L_p([0,1])$.

We remark that some classes of kernels for which the crucial condition (9.19) holds are described in APPELL-DIALLO [1995] and ZABREJKO-KALITVIN [1997].

A special case which also occurs in applications (see e.g. Subsection 19.5) is the *sign condition*

$$(10.20) \qquad\qquad sc(t,s) \le 0 \qquad ((t,s) \in Q);$$

this condition implies that the function (10.6) satisfies $e(t,\tau,s) \le 1$ for $t \ge \tau$ and $s > 0$, as well as for $t \le \tau$ and $s < 0$. The definition (10.14) of the operator L shows then that the "smallness" of the spectral radius of L only depends on the "smallness" of the spectral radius of the operator $\hat{K}$. For example, the following holds:

Theorem 10.4. *Suppose that the hypotheses of Theorem 10.2 and the sign condition* (10.20) *are satisfied. Assume, moreover, that*

$$(10.21) \qquad || \, |K(t_1)| \cdots |K(t_n)| \, || < \frac{1}{T} \quad (t_1,\ldots,t_n \in [0,T])$$

for some n, where

$$|Kx(t)|x(s) = \int_{-1}^1 |k(t,s,\sigma)|x(\sigma)\,d\sigma.$$

Then the problem (10.1)/(10.2) has a unique solution.

□ For any $x \in C_t^1(X)$ we have

$$|Lx(t,s)| \leq \int_0^T \int_{-1}^1 |k(t,s,\sigma)| \, |x(\tau,\sigma)| \, d\sigma \, d\tau = \int_0^T |K(t)| \, |x(\tau,s)| \, d\tau.$$

By induction we get

$$|L^n x(t,s)| \leq \int_0^T \cdots \int_0^T |K(t_1)| \cdots |K(t_n)| \, |x(t_n)|(s) \, dt_n \ldots dt_1$$

which implies

$$\|L^n x(t,\cdot)\| \leq \left\| \int_0^T \cdots \int_0^T |K(t_1)| \cdots |K(t_n)| \, |x(t_n)(\cdot)| \, dt_n \ldots dt_1 \right\|$$

$$\leq \int_0^T \cdots \int_0^T \| \, |K(t_1)| \cdots |K(t_n)| \, |x(t_n)(\cdot)| \, \| \, dt_n \ldots dt_1$$

$$\leq \|x\|_{C_t(X)} \int_0^T \cdots \int_0^T \| \, |K(t_1)| \cdots |K(t_n)| \, \| \, dt_n \ldots dt_1,$$

hence $r_\sigma(L) \leq \|L^n\|^{1/n} < 1.$ ∎

We make some remarks on Theorem 10.4. First of all, the proof shows that condition (10.21) can be weakened to

$$\int_0^T \cdots \int_0^T \| \, |K(t_1)| \cdots |K(t_n)| \, \| \, dt_n \ldots dt_1 < 1.$$

In the stationary case $K(t) \equiv K$ condition (10.21) is equivalent to the condition $r_\sigma(|K|) < 1/T$ which we have imposed in Theorem 9.5.

Furthermore, we recall (see Subsection 1.1) that the sign condition (10.20) may be achived by introducing some new function $\tilde{c}$ which satisfies (10.20) (for example, $\tilde{c}(t,s) = -s\tilde{c}(t)$) and then passing from x to the new unknown function

$$(10.22) \qquad y(t,s) := x(t,s) \exp \left\{ - \int_0^t [c(\xi,s) - \tilde{c}(\xi,s)] \, d\xi \right\}.$$

After this substitution we get for y the equation

$$(10.23) \qquad \frac{\partial y(t,s)}{\partial t} = \tilde{c}(t,s)y(t,s) + \int_{-1}^{1} \tilde{k}(t,s,\sigma)y(t,\sigma)\,d\sigma + \tilde{f}(t,s)$$

with boundary conditions

$$(10.24) \qquad \begin{cases} y(0,s) = \tilde{\phi}(s) & \text{if } 0 < s \le 1, \\[2mm] y(T,s) = \tilde{\psi}(s) & \text{if } -1 \le s < 0, \end{cases}$$

where

$$\tilde{k}(t,s,\sigma) = k(t,s,\sigma)\exp\left\{ \int_{0}^{t} [\tilde{c}(\xi,s) - c(\xi,s) + c(\xi,\sigma) - \tilde{c}(\xi,\sigma)]\,d\xi \right\},$$

$$\tilde{f}(t,s) = f(t,s)\exp\left\{ \int_{0}^{t} [\tilde{c}(\xi,s) - c(\xi,s)]\,d\xi \right\},$$

and

$$\tilde{\phi}(s) = \phi(s), \quad \tilde{\psi}(s) = \psi(s)\exp\left\{ \int_{0}^{T} [c(\xi,s) - \tilde{c}(\xi,s)]\,d\xi \right\}.$$

10.3. Application of K-normed spaces

The fixed point principle given in Theorem 9.6 above applies also to the non-stationary equation (10.1). We summarize with the following

Theorem 10.5. *Let $c \in L_\infty(Q)$, and suppose that the operator function (10.17) is regular in $X = L_p([-1,1])$. Assume that the operator (10.15) acts in $C_t(X)$, and the estimates (9.36) hold, where the functions a, b, c and d are defined by (10.19). Finally, suppose that one of the following four conditions is satisfied:*

(a) $\Delta = (\alpha + \delta)^2 - 4\beta\gamma > 0$ and $T\sqrt{\Delta} < \log \frac{\alpha + \delta + \sqrt{\Delta}}{\alpha + \delta - \sqrt{\Delta}}$;

(b) $(\alpha + \delta)^2 = 4\beta\gamma$ and $T(\alpha + \delta) < 2$;

(c) $\Delta < 0, \alpha + \delta \ne 0$, and $T\sqrt{-\Delta} < 2\arctan \frac{\sqrt{-\Delta}}{\alpha + \delta}$;

(d) $\Delta < 0, \alpha + \delta = 0$, and $T\sqrt{-\Delta} < \pi$.

Then the operator equation (10.13) has a unique solution $x \in C_t(X)$ for any $g \in C_t(X)$.

We do not give the proof of this theorem since it is identical with that of Theorem 9.7. Instead, we illustrate this theorem with another example.

Example 10.2. Consider the equation

$$(10.25) \qquad \frac{\partial x(t, s)}{\partial t} = 4tx(t, s) - 6ts \int_{-1}^{1} \sigma x(t, \sigma) \, d\sigma$$

with boundary condition (10.2). Here we have

$$Lx(t, s) = \begin{cases} -6se^{2t^2} \int_{0}^{t} \int_{-1}^{1} e^{-2\tau^2} \tau \sigma x(\tau, \sigma) \, d\sigma \, d\tau & \text{if } 0 < s \le 1, \\[2ex] -6se^{2t^2} \int_{T}^{t} \int_{-1}^{1} e^{-2\tau^2} \tau \sigma x(\tau, \sigma) \, d\sigma \, d\tau & \text{if } -1 \le s < 0, \end{cases}$$

and

$$g(t, s) = \begin{cases} e^{2t^2} \phi(s) & \text{if } 0 < s \le 1, \\[2ex] e^{2(t^2 - T^2)} \psi(s) & \text{if } -1 \le s < 0. \end{cases}$$

A straightforward calculation shows that the operator equation (10.13) has the unique solution

$$x(t, s) = \begin{cases} -\dfrac{6A(1 + T^2)s(e^{2t^2} - 1)}{2T^2 + e^{-2T^2} + 1} + e^{2t^2} \phi(s) & (0 < s \le 1), \\[3ex] -\dfrac{6A(1 + T^2)s(e^{2t^2} - 1)}{2T^2 + e^{-2T^2} + 1} + e^{2(t^2 - T^2)} \psi(s) & (-1 \le s < 0), \end{cases}$$

where we have put

$$A := \frac{1}{2} \left[\int_{0}^{1} s\phi(s) \, ds + e^{-2T^2} \int_{-1}^{0} s\psi(s) \, ds \right].$$

Since the functions defined in (10.19) are here

$$a(t, \tau, s, \sigma) = d(t, \tau, s, \sigma) = -6s\tau\sigma e^{2(t^2 - \tau^2)}$$

and

$$b(t, \tau, s, \sigma) = c(t, \tau, s, \sigma) = 6s\tau\sigma e^{2(t^2 - \tau^2)},$$

the estimates (9.36) hold for $\alpha = \beta = \gamma = \delta = 6Te^{2T^2}$. Thus, by condition (b) of Theorem 10.5 we know that there is a unique solution of (10.13) (and hence also of (10.25), since the function c is bounded) if $36T^2 e^{4T^2} < 1$. Of course, this condition is far from being necessary.

Chapter III

Partial Integral Operators

§ 11. General properties

In this section we shall be concerned with general properties of so-called *partial integral operators*. For simplicity, we shall restrict ourselves throughout to spaces of functions of two variables. The operators we shall study in such spaces are of the form

$$(11.1) \qquad P = C + L + M + N,$$

where

$$(11.2) \qquad Cx(t,s) = c(t,s)x(t,s),$$

$$(11.3) \qquad Lx(t,s) = \int_T l(t,s,\tau)x(\tau,s)\,d\mu(\tau),$$

$$(11.4) \qquad Mx(t,s) = \int_S m(t,s,\sigma)x(t,\sigma)\,d\nu(\sigma),$$

and

$$(11.5) \qquad Nx(t,s) = \int_{T\times S} n(t,s,\tau,\sigma)x(\tau,\sigma)d(\mu\times\nu)(\tau,\sigma).$$

Here T and S are arbitrary nonempty sets equipped with σ-algebras $\mathfrak{A}(T)$ and $\mathfrak{A}(S)$, and separable measures μ and ν on $\mathfrak{A}(T)$ and $\mathfrak{A}(S)$, respectively; by $\mu\times\nu$ we mean the product measure on the σ-algebra $\mathfrak{A}(T)\otimes\mathfrak{A}(S)$. The coefficient $c = c(t,s)$, as well as the kernels $l = l(t,s,\tau), m = m(t,s,\sigma)$, and $n = n(t,s,\tau,\sigma)$ are measurable functions, and the integrals (11.3) - (11.5) are meant in the Lebesgue-Radon sense. The whole operator (11.1) is a partial integral operator in its most general form.

The main results of this paragraph have been obtained in KALITVIN-ZABREJKO [1991], see also KALITVIN [1983, 1986a, 1987].

11.1. Continuity properties

In the sequel we use the following notation. By $\mathfrak{S} = \mathfrak{S}(T \times S)$ we denote the *space of all* (classes of) *measurable real functions* on $T \times S$, and by X and Y ideal spaces of functions over $T \times S$. (For the definition and properties of ideal spaces, see Subsection 4.1.)

Theorem 11.1. *Suppose that P is a partial integral operator which acts from X into Y. Then P is continuous.*

□ Together with (11.1), consider the operator

$$(11.6) \qquad]P[\ = \ ]C[\ + \ ]L[\ + \ ]M[\ + \ ]N[,$$

where

$$(11.7) \qquad]C[x(t,s) = |c(t,s)|x(t,s),$$

$$(11.8) \qquad]L[x(t,s) = \int_T |l(t,s,\tau)|x(\tau,s)\,d\mu(\tau),$$

$$(11.9) \qquad]M[x(t,s) = \int_S |m(t,s,\sigma)|x(t,\sigma)\,d\nu(\sigma),$$

and

$$(11.10) \qquad]N[x(t,s) = \int_{T \times S} |n(t,s,\tau,\sigma)|x(\tau,\sigma)\,d(\mu \times \nu)(\tau,\sigma).$$

Let $x \in X$. By hypothesis, the function $y = Px$ belongs to Y. This implies, in particular, that the functions

$$\tau \mapsto l(t,s,\tau)x(\tau,s), \quad \sigma \mapsto m(t,s,\sigma)x(t,\sigma),$$

$$(\tau,\sigma) \mapsto n(t,s,\tau,\sigma)x(\tau,\sigma)$$

are integrable, for almost all $(t,s) \in T \times S$, on T, S, and $T \times S$, respectively. By well-known properties of the Lebesgue-Radon integral, the same is true for the functions $\tau \mapsto |l(t,s,\tau)||x(\tau,s)|, \sigma \mapsto |m(t,s,\sigma)||x(t,\sigma)|$ and $(\tau,\sigma) \mapsto |n(t,s,\tau,\sigma)||x(\tau,\sigma)|$; thus, we may define $]P[x$ and have $]P[x \in \mathfrak{S}$, by Fubini's theorem. In other words, the operator (11.6) acts from X into $\mathfrak{S}$.

We claim that the operator P is closed. Suppose that $(x_n)_n$ converges (in X) to $x^* \in X$, and $(Px_n)_n$ converges (in Y) to $y^* \in Y$; we have to show that $Px^* = y^*$. Choose a sequence n_k of natural numbers such that

$$\sum_{k=1}^{\infty} ||x_{n_k} - x^*|| < \infty.$$

Without loss of generality we may then assume that $x_{n_k}(t, s) \to x^*(t, s)$ almost everywhere on $T \times S$, and the function z defined by

$$z(t, s) = \sum_{k=1}^{\infty} |x_{n_k}(t, s) - x^*(t, s)|$$

belongs to X. Consequently, we have convergence

$$l(t, s, \cdot)x_{n_k}(\cdot, s) \to l(t, s, \cdot)x^*(\cdot, s) \quad (k \to \infty),$$

$$m(t, s, \cdot)x_{n_k}(t, \cdot) \to m(t, s, \cdot)x^*(t, \cdot) \quad (k \to \infty),$$

and
$$n(t, s, \cdot, \cdot)x_{n_k}(\cdot, \cdot) \to n(t, s, \cdot, \cdot)x^*(\cdot, \cdot) \quad (k \to \infty)$$

for almost all $(t, s) \in T \times S$. Moreover, these sequences are majorized by the integrable functions $|l(t, s, \cdot)|z(\cdot, s), |m(t, s, \cdot)|z(t, \cdot)$, and $|n(t, s, \cdot, \cdot)|z(\cdot, \cdot)$, respectively. By Lebesgue's dominated convergence theorem, we have $Lx_{n_k}(t, s) \to Lx^*(t, s), Mx_{n_k}(t, s) \to Mx^*(t, s)$, and $Nx_{n_k}(t, s) \to Nx^*(t, s)$, for almost all $(t, s) \in T \times S$, and hence also $Px_{n_k}(t, s) \to Px^*(t, s)$ almost everywhere on $T \times S$. But, by hypothesis, the sequence $(Px_{n_k})_k$ converges in Y to y^*, and thus $Px^* = y^*$ as claimed.

We have shown that the operator P is closed between X and Y. The assertion follows now from Banach's closed graph theorem. ∎

Theorem 11.1 is an extension of Banach's well-known theorem on the continuity of integral operators (see BANACH [1932]) to partial integral operators.

It is clear that the operator P maps X into Y if all the operators C, L, M, and N given by (11.2), (11.3), (11.4), and (11.5), respectively, do so. Interestingly, the converse is not true, at least if one of the sets T or S contains a countable number of atoms:

Example 11.1. Let $S = T = \mathbb{N}$ with $\mu = \nu$ being the counting measure, and take $X = Y = l_2(\mathbb{N} \times \mathbb{N})$ (the space of all square-summable sequences with double index). If we take

$$c(t, s) := -t \qquad (t \in \mathbb{N})$$

$$l(t, s, \tau) = m(t, s, \sigma) \equiv 0,$$

and

$$n(t, s, \tau, \sigma) := \begin{cases} t & \text{if } t = \tau \text{ and } s = \sigma, \\ 0 & \text{otherwise}, \end{cases}$$

the operator (11.1) is simply the zero operator. On the other hand, the operator (11.2) with $c(t, s) = -t$ certainly does not act in the space l_2.

The acting problem for the components of the operator (11.1) is related to that of the *uniqueness* of the representation of the operator P in the form (11.1). We point out that this representation is unique if the measures μ and ν are continuous (atom-free) on $\mathfrak{S}(T)$ and $\mathfrak{S}(S)$, respectively; this follows from regularity theorems which we shall give in the next subsection. Such regularity theorems allow us also to conclude the action of all the operators (11.2) - (11.5) from the action of the single operator (11.1) between X and Y.

11.2. Regularity properties

As in Subsection 4.1, the notation $x \leq y$ for $x, y \in \mathfrak{S}(T \times S)$ means that $x(t, s) \leq y(t, s)$ almost everywhere on $T \times S$.

Recall that a linear operator $P : X \to Y$ is called *regular* if P may be written as a difference of two positive linear operators between X

and Y (i.e. operators which preserve inequalities almost everywhere). In this case there exists a positive operator $\tilde{P} : X \to Y$ such that

$$|Px| \leq \tilde{P}|x| \qquad (x \in X).$$

The classical Kantorovich theorem (see e.g. KANTOROVICH [1956] or VULIKH [1961, 1977]) states that a linear operator is regular if and only if it preserves order-boundedness. Moreover, among all positiv majorants $\tilde{P}$ of P there exists a minimal one (in the sense of the induced ordering on the space of linear operators); this minimal positive majorant is usually called the *module* of P and denoted by $|P|$ (see Subsection 3.1).

Theorem 11.2. *Suppose that the measures μ and ν are continuous, and let $P : X \to Y$ be a partial integral operator. Then P is regular if and only if the operator $]P[$ given by (11.6) also maps X into Y. In this case,*

$$(11.11) \qquad\qquad |P| = \,]P[.$$

☐ The sufficiency follows from the obvious inequality

$$(11.12) \qquad\qquad |Px| \leq \,]P[\,|x| \qquad (x \in X),$$

which implies that $]P[$ is a positive majorant of P, and, hence $|P| \leq \,]P[$. To prove the necessity, let $P : X \to Y$ be regular. For any nonnegative function $x \in X$, we then get (see again VULIKH [1961])

$$|P|x = \sup\left\{|Pz| : |z| \leq x\right\} \in Y.$$

Since X and Y are K-spaces of countable type (KANTOROVICH-AKILOV [1977]), we find a countable set M which is dense (in measure) in the conic interval $[-x, x] = \{z : z \in X, |z| \leq x\}$ and such that

$$|P|x = \sup\left\{|Pz| : z \in M\right\}.$$

By the continuity of the measures μ and ν, we can choose sequences of sets $T_n \subseteq T$ and $S_n \subseteq S$ such that

$$T = \bigcup_{n=1}^{\infty} T_n, \quad S = \bigcup_{n=1}^{\infty} S_n, \quad \mu(T_n) \to 0, \quad \nu(S_n) \to 0,$$

and every point $t \in T$ and $s \in S$ belongs to infinitely many subsets T_n and S_n, respectively. Let

$$U_n = T \times S_n, \quad V_n = T_n \times S, \quad \Delta_n = U_n \cup V_n, \quad \nabla_n = U_n \cap V_n,$$

and denote by M^* the set of all functions

$$z_n = P_{\nabla_n} x \operatorname{sign} c + P_{\Delta_n'} u + P_{U_n' \cap V_n} v + P_{U_n \cap V_n'} w \quad (n \in \mathbb{N}),$$

where $u, v, w \in M$, P_D is the operator (3.19) for $D \subseteq T \times S$, and D' denotes the complement $(T \times S) \setminus D$. Since $M \subseteq M^* \subseteq [-x, x]$, we have

$$|P|x = \sup \{|Pz| : z \in M^*\};$$

moreover, since M^* is countable, we also have

$$(11.13) \qquad |P|x(t, s) = \sup \{|Pz(t, s)| : z \in M^*\}$$

for almost all $(t, s) \in T \times S$. Fix $(t, s) \in T \times S$ with (11.13), and choose sequences $(n_k)_k$ and $(m_k)_k$ such that $t \in T_{n_k}$ and $s \in S_{m_k}$ for all k. Moreover, let $u_k, v_k,$ and w_k be functions in M such that $u_k(\tau, \sigma) \to \operatorname{sign} l(t, s, \tau) x(\tau, \sigma), v_k(\tau, \sigma) \to \operatorname{sign} m(t, s, \sigma) x(\tau, \sigma)$, and $w_k(\tau, \sigma) \to \operatorname{sign} n(t, s, \tau, \sigma) x(\tau, \sigma)$ in measure. Finally, put

$$z_k = P_{\tilde{\nabla}_k} x \operatorname{sign} c + P_{\tilde{\Delta}_k'} u_k + P_{\tilde{U}_k' \cap \tilde{V}_k} v_k + P_{\tilde{U}_k \cap \tilde{V}_k'} w_k,$$

where

$$\tilde{U}_k = T \times S_{m_k}, \quad \tilde{V}_k = T_{n_k} \times S, \quad \tilde{\Delta}_k = \tilde{U}_k \cup \tilde{V}_k, \quad \tilde{\nabla}_k = \tilde{U}_k \cap \tilde{V}_k.$$

By Lebesgue's dominated convergence theorem, we conclude that

$$\lim_{k\to\infty} |P_{z_k}(t,s)| = |c(t,s)|x(t,s) + \int_T |l(t,s,\tau)|x(\tau,s)\,d\mu(\tau)$$

$$+ \int_S |m(t,s,\sigma)|x(t,\sigma)\,d\nu(\sigma)$$

$$+ \int_{T\times S} |n(t,s,\tau,\sigma))|x(\tau,\sigma)\,d(\mu\times\nu)(\tau,\sigma) =]P[x(t,s).$$

On the other hand, from $z_k \in M^*$ $(k = 1,2,3,\ldots)$ it follows that

$$]P[x(t,s) = \lim_{k\to\infty} |P_{z_k}(t,s)| = \sup\{|Pz| : z \in M^*\}.$$

This implies, together with (11.13), that $]P[x(t,s) \leq |P|x(t,s)$ and hence $]P[x \leq |P|x$. We have shown that the operator $]P[$ maps all nonnegative functions $x \in X$ into Y. But every function in X may be written as a difference of two nonnegative functions, and, thus, $]P[$ acts from X into Y as claimed.

Equality (11.11) follows from the just established inequality $]P[\leq |P|$ and from the inequality (11.12) which is always true. ∎

The hypothesis on the continuity of the measures μ and ν is essential not just for the proof but also for the statement of Theorem 11.2. Nevertheless, one may modify Theorem 11.2 in such a way that its statement remains true also for discrete (purely atomic) measures and even for arbitrary measures. Denote by T_d and S_d the "discrete part", and by T_c and S_c the "continuous part", respectively, of the sets T and S. We say that the representation (11.1) of the operator P is *normal* if $c(t,s) = 0$ for $(t,s) \in T_d \times S_d, l(t,s,\tau) = 0$ for $s \in S_d$, and $m(t,s,\sigma) = 0$ for $t \in T_d$. It is not hard to see that one can always find a normal representation for a partial integral operator, just by using "δ-type functions" defined by

$$\delta(t,\tau) = \begin{cases} \mu(T)^{-1} & \text{if } t = \tau, \\ 0 & \text{if } t \neq \tau, \end{cases}$$

and

$$\delta(s,\sigma) = \begin{cases} \nu(S)^{-1} & \text{if } s = \sigma, \\ 0 & \text{if } s \neq \sigma. \end{cases}$$

In fact, replacing the functions $c(t,s), l(t,s,\tau), m(t,s,\sigma)$, and $n(t,s,\tau,\sigma)$ in (11.1) by the functions

$$\overline{c}(t,s) = P_{T_c \times S_c} c(t,s),$$

$$\overline{l}(t,s,\tau) = \delta(t,\tau) P_{T_d \times S_c} c(t,s) + P_{T \times S_c} l(t,s,\tau),$$

$$\overline{m}(t,s,\sigma) = \delta(s,\sigma) P_{T_c \times S_d} c(t,s) + P_{T_c \times S} m(t,s,\sigma),$$

and

$$\overline{n}(t,s,\tau,\sigma) = \delta(t,\tau) \delta(s,\sigma) P_{T_d \times S_d} c(t,s)$$
$$+ \delta(s,\sigma) P_{T \times S_d} l(t,s,\tau) + \delta(t,\tau) P_{T_d \times S} m(t,s,\sigma) + n(t,s,\tau,\sigma),$$

respectively, one gets a normal representation for P. For instance, in Example 11.1 we could take

$$\delta(t,\tau) = \begin{cases} 1 & \text{if } t = \tau, \\ 0 & \text{if } t \neq \tau. \end{cases}$$

We formulate an analogue of Theorem 11.2 in this general case; the modification of the proof is straightforward.

Theorem 11.3. *Let $P : X \to Y$ be a partial integral operator with normal representation (11.1). Then P is regular if and only if the operator $]P[$ given by (11.6) also maps X into Y. In this case, equality (11.11) holds.*

11.3. The associate operator

In this subsection we are interested in finding an explicit representation of the *associate operator* $P' : Y' \to X'$ to the partial integral operator (11.1) which is supposed to act from X into Y. First let us define an operator $P^{\#}$ by

$$P^{\#}y(t,s) = c(t,s)y(t,s) + \int_T l^{\#}(t,s,\tau)y(\tau,s)\,d\mu(\tau)$$

(11.14)
$$+ \int_S m^{\#}(t,s,\sigma)y(t,\sigma)\,d\nu(\sigma)$$

$$+ \int_{T\times S} n^{\#}(t,s,\tau,\sigma)y(\tau,\sigma)\,d(\mu\times\nu)(\tau,\sigma)$$

where

$$l^{\#}(t,s,\tau) = l(\tau,s,t), \quad m^{\#}(t,s,\sigma) = m(t,\sigma,s),$$

$$n^{\#}(t,s,\tau,\sigma) = n(\tau,\sigma,t,s).$$

The following theorem may be regarded as analogue of Lemma 3.3.

Theorem 11.4. *Let P be a partial integral operator which acts between two ideal spaces X and Y. Then the associate operator P' of P is given by*

(11.15)
$$P'y = P^{\#}y$$

for any $y \in Y'$ with $P^{\#}y \in \mathfrak{S}$. In particular, if the operator (11.1) is regular, then (11.15) holds for all $y \in Y'$.

□   As was shown in the proof of Theorem 11.1, the operator $]P[$ defined by (11.6) acts from X into $\mathfrak{S}$ and is continuous. Consequently, the image N of the unit ball $||x||_X \leq 1$ of X under $]P[$ is a bounded subset of $\mathfrak{S}$, and, hence, so is the set

(11.16) $$\tilde{N} = \overline{\bigcup\{\{v : -]P[x \leq v \leq]P[x\} : ||x||_X \leq 1\}}$$

(closure in $\mathfrak{S}$). Denote by $\tilde{Y}$ the ideal space whose unit ball coincides with the set (11.16); such a space may be easily constructed. By construction, the operator $]P[$ acts from X into $\tilde{Y}$ and is continuous.

Let $x \in X$ and $y \in \tilde{Y}'$. By Fubini's theorem, we then have

$$\int_{T \times S} c(t,s)x(t,s)y(t,s)\,d(\mu \times \nu)(t,s)$$

$$+ \int_{T \times S} \left[\int_T l(t,s,\tau)x(\tau,s)d\mu(\tau)\right] y(t,s)\,d(\mu \times \nu)(t,s)$$

$$+ \int_{T \times S} \left[\int_S m(t,s,\sigma)x(t,\sigma)d\nu(\sigma)\right] y(t,s)\,d(\mu \times \nu)(t,s)+$$

$$\int_{T \times S} \left[\int_{T \times S} n(t,s,\tau,\sigma)x(\tau,\sigma)\,d(\mu \times \nu)(\tau,\sigma)\right] y(t,s)\,d(\mu \times \nu)(t,s)$$

$$= \int_{T \times S} c(t,s)x(t,s)y(t,s)\,d(\mu \times \nu)(t,s)$$

$$+ \int_{T \times S} x(\tau,s) \left[\int_T l(t,s,\tau)y(t,s)\,d\mu(t)\right] d(\mu \times \nu)(\tau,s)$$

$$+ \int_{T \times S} x(t,\sigma) \left[\int_S m(t,s,\sigma)y(t,s)\,d\nu(s)\right] d(\mu \times \nu)(t,\sigma)+$$

$$\int_{T \times S} x(\tau,\sigma) \left[\int_{T \times S} n(t,s,\tau,\sigma)y(t,s)\,d(\mu \times \nu)(t,s)\right] d(\mu \times \nu)(\tau,\sigma).$$

This shows that

$$(11.17) \qquad \langle Px, y \rangle = \langle x, P^{\#}y \rangle \qquad (x \in X, y \in Y'),$$

and thus the operator $P^{\#}$ coincides with the associate operator P' of P if we consider P as an operator from X into $\tilde{Y}$.

Now let v be a fixed unit in $\tilde{Y}'$. For any $y \in Y'$, we define y_n ($n = 1,2,3,...$) by

$$y_n(t,s) := \begin{cases} y(t,s) & \text{if } |y(t,s)| \leq nv(t,s), \\ nv(t,s)\,\text{sign}\,y(t,s) & \text{if } |y(t,s)| > nv(t,s). \end{cases}$$

By Fubini's theorem we get then

$$(11.18) \qquad \langle Px, y_n \rangle = \langle x, P^{\#}y_n \rangle \qquad (n = 1,2,...).$$

Since the limit, as $n \to \infty$, of the left-hand side of (11.18) exists for any $x \in X$, the sequence $(P^{\#}y_n)_n$ is X-weakly Cauchy in X'. But the space X' is X-weakly complete (see ZABREJKO [1966, 1968]), and hence $P^{\#}y_n \to P'y$ for some $y \in Y'$. Since this function y obviously satisfies

$$\langle Px, y \rangle = \langle x, P'y \rangle \qquad (x \in X, y \in Y'),$$

we have shown that the associate operator P' of P also exists if we consider P as an operator from X into Y.

Now let $y \in Y'$ and $P^{\#}y \in \mathfrak{S}$. Then $]P^{\#}[\,|y| \in \mathfrak{S}$ and, by Lebesgue's theorem, the sequence $(P^{\#}y_n)_n$ converges (in $\mathfrak{S}$) to $P^{\#}y$. But, by what has been proved before, the sequence $(P^{\#}y_n)_n$ converges as well X-weakly to $P'y$. We conclude that $P'y = P^{\#}y$, and so we are done. ∎

11.4. Algebras of partial integral operators

Given two ideal spaces X and Y, denote by $\mathfrak{L}(X,Y)$ the space of all bounded linear operators and by $\mathfrak{L}^r(X,Y)$ the space of all regular bounded linear operators between X and Y. Similarly, by $\mathfrak{L}_p(X,Y)$ and $\mathfrak{L}_p^r(X,Y)$ (where the subscript p stands for "partial integral operator") we denote the space of all operators and regular operators, respectively, of the form (11.1). Theorems 11.1 and 11.2 state that $\mathfrak{L}_p(X,Y) \subseteq \mathfrak{L}(X,Y)$ and $\mathfrak{L}_p^r(X,Y) \subseteq \mathfrak{L}^r(X,Y)$. If $\mathfrak{L}(X,Y)$ and $\mathfrak{L}^r(X,Y)$ are equipped with the usual operator norm, the subspaces $\mathfrak{L}_p(X,Y)$ and $\mathfrak{L}_p^r(X,Y)$ are not closed. However, if we consider $\mathfrak{L}^r(X,Y)$ with the norm

$$(11.19) \qquad \|P\|_{\mathfrak{L}^r(X,Y)} := \|\,|P|\,\|_{\mathfrak{L}(X,Y)},$$

where $|P|$ is the module of P as in Subsection 11.2, then $\mathfrak{L}^r(X,Y)$ becomes a Banach space in the norm (11.19), and $\mathfrak{L}_p^r(X,Y)$ is closed in $\mathfrak{L}^r(X,Y)$. In order to state more precise results, some auxiliary definitions are in order. These definitions are the same as in § 3 and § 4, but now for functions of two variables.

Recall that, given two ideal spaces X and Y over $T \times S$, by Y/X we denote the *multiplicator space* of all functions c such that $cx \in Y$ for all $x \in X$. This is an ideal space with norm

(11.20) $$\|c\|_{Y/X} = \sup \{\|cx\|_Y : \|x\|_X \leq 1\}.$$

Further, by $\mathfrak{R}_l(X,Y), \mathfrak{R}_m(X,Y)$ and $\mathfrak{R}_n(X,Y)$ we denote the sets of all measurable functions $l = l(t,s,\tau)$ on $T \times S \times T, m = m(t,s,\sigma)$ on $T \times S \times S$, and $n = n(t,s,\tau,\sigma)$ on $T \times S \times T \times S$, respectively, such that $l(t,s,\tau) = 0$ for $s \in S_d$ and $m(t,s,\sigma) = 0$ for $t \in T_d$. All these three sets are ideal spaces equipped with the norms

(11.21) $$\|l\|_{\mathfrak{R}_l(X,Y)} = \sup_{\|x\|_X \leq 1} \left\| \int_T |l(\cdot,\cdot,\tau)x(\tau,\cdot)| \, d\mu(\tau) \right\|_Y ,$$

(11.22) $$\|m\|_{\mathfrak{R}_m(X,Y)} = \sup_{\|x\|_X \leq 1} \left\| \int_S |m(\cdot,\cdot,\sigma)x(\cdot,\sigma)| \, d\nu(\sigma) \right\|_Y ,$$

and

$$\|n\|_{\mathfrak{R}_n(X,Y)} =$$

(11.23)

$$\sup_{\|x\|_X \leq 1} \left\| \int_{T \times S} |n(\cdot,\cdot,\tau,\sigma)x(\tau,\sigma)| \, d(\mu \times \nu)(\tau,\sigma) \right\|_Y ,$$

respectively. Of course, these definitions are formulated just in such a way that the kernel norms (11.21), (11.22) and (11.23) coincide with the operator norms of the corresponding (regular) integral operators, i.e.

$$\|l\|_{\mathfrak{R}_l(X,Y)} = \|L\|_{\mathcal{L}^r(X,Y)}, \quad \|m\|_{\mathfrak{R}_m(X,Y)} = \|M\|_{\mathcal{L}^r(X,Y)},$$

$$\|n\|_{\mathfrak{R}_n(X,Y)} = \|N\|_{\mathcal{L}^r(X,Y)}.$$

Consider now the direct sum

(11.24) $$\mathfrak{R}(X,Y) = \mathfrak{R}_c(X,Y) \oplus \mathfrak{R}_l(X,Y) \oplus$$
$$\mathfrak{R}_m(X,Y) \oplus \mathfrak{R}_n(X,Y),$$

where $\mathfrak{R}_c(X,Y)$ is the subspace of Y/X consisting of all functions $c = c(t,s)$ such that $c(t,s) = 0$ for $t \in T_d$ or $s \in S_d$. The space (11.24) will be equipped with the norm

$$\|(c,l,m,n)\|_{\mathfrak{R}(X,Y)} = \|c\|_{\mathfrak{R}_c(X,Y)} + \|l\|_{\mathfrak{R}_l(X,Y)}$$
$$+ \|m\|_{\mathfrak{R}_m(X,Y)} + \|n\|_{\mathfrak{R}_m(X,Y)}.$$

Theorem 11.5. *Let X and Y be two ideal spaces. Then $\mathfrak{L}_p^r(X,Y)$ is a closed subspace of $\mathfrak{L}^r(X,Y)$ which is isomorphic to the space $\mathfrak{R}(X,Y)$. More precisely, the two-sided estimate*

$$(11.25) \qquad \|P\|_{\mathfrak{L}^r(X,Y)} \leq \|(c,l,m,n)\|_{\mathfrak{R}(X,Y)} \leq 4\|P\|_{\mathfrak{L}^r(X,Y)}$$

holds, if P is a regular operator of the form (11.1).

☐ The statement is an immediate consequence of the completeness of the space $\mathfrak{R}(X,Y)$ and of Theorems 11.2 and 11.3. ∎

In view of applications, the following theorem on the superposition of partial integral operators is useful, which follows by a standard reasoning from Fubini's theorem.

Theorem 11.6. *Let X,Y and Z be three ideal spaces, and let*

$$P_j x(t,s) = c_j(t,s)x(t,s) + \int_T l_j(t,s,\tau)x(\tau,s)\,d\mu(\tau)$$

$$(11.26) \qquad\qquad + \int_S m_j(t,s,\sigma)x(t,\sigma)\,d\nu(\sigma)$$

$$+ \int_{T \times S} n_j(t,s,\tau,\sigma)x(\tau,\sigma)\,d(\mu \times \nu)(\tau,\sigma)$$

($j = 1,2$) be two partial integral operators such that $P_1 \in \mathfrak{L}^r(X,Y)$ and $P_2 \in \mathfrak{L}^r(Y,Z)$. Then the linear operator $P = P_2 P_1$ is also a partial integral operator with multiplier

$$c(t,s) = c_2(t,s)c_1(t,s)$$

and kernels

$$l(t, s, \tau) = c_2(t, s)l_1(t, s, \tau) + l_2(t, s, \tau)c_1(\tau, s)$$

$$+ \int_T l_2(t, s, \xi)l_1(\xi, s, \tau) \, d\mu(\xi),$$

$$m(t, s, \sigma) = c_2(t, s)m_1(t, s, \sigma) + m_2(t, s, \sigma)c_1(t, \sigma)$$

$$+ \int_S m_2(t, s, \eta)m_1(t, \eta, \sigma) \, d\nu(\eta),$$

and

$$n(t, s, \tau, \sigma) = c_2(t, s)n_1(t, s, \tau, \sigma)$$

$$+ n_2(t, s, \tau, \sigma)c_1(\tau, \sigma) + l_2(t, s, \tau)m_1(\tau, s, \sigma)$$

$$+ m_2(t, s, \sigma)l_1(t, \sigma, \tau) + \int_T l_2(t, s, \xi)n_1(\xi, s, \tau, \sigma) \, d\mu(\xi)$$

$$+ \int_S m_2(t, s, \eta)n_1(t, \eta, \tau, \sigma) \, d\nu(\eta)$$

$$+ \int_T n_2(t, s, \xi, \sigma)l_1(\xi, \sigma, \tau) \, d\mu(\xi)$$

$$+ \int_S n_2(t, s, \tau, \eta)m_1(\tau, \eta, \sigma) \, d\nu(\eta)$$

$$+ \int_{T \times S} n_2(t, s, \xi, \eta)n_1(\xi, \eta, \tau, \sigma) \, d(\mu \times \nu)(\xi, \eta).$$

Theorem 11.6 allows us to give some category-type inclusions between the classes introduced so far, which we state as

Theorem 11.7. *Let X, Y, and Z be three ideal spaces. Then the following inclusions hold:*
(a) $\mathfrak{R}_c(X, Y) \circ \mathfrak{R}_c(Y, Z) \subseteq \mathfrak{R}_c(X, Z)$,
(b) $\mathfrak{R}_c(X, Y) \circ \mathfrak{R}_l(Y, Z) \subseteq \mathfrak{R}_l(X, Z)$,
   $\mathfrak{R}_l(X, Y) \circ \mathfrak{R}_c(Y, Z) \subseteq \mathfrak{R}_l(X, Z)$,

(c) $\mathfrak{R}_c(X, Y) \circ \mathfrak{R}_m(Y, Z) \subseteq \mathfrak{R}_m(X, Z)$,
    $\mathfrak{R}_m(X, Y) \circ \mathfrak{R}_c(Y, Z) \subseteq \mathfrak{R}_m(X, Z)$,

(d) $\mathfrak{R}_l(X, Y) \circ \mathfrak{R}_m(Y, Z) \subseteq \mathfrak{R}_n(X, Z)$,
    $\mathfrak{R}_m(X, Y) \circ \mathfrak{R}_l(Y, Z) \subseteq \mathfrak{R}_n(X, Z)$,

(e) $\mathfrak{R}_n(X, Y) \circ \mathfrak{R}_n(Y, Z) \subseteq \mathfrak{R}_n(X, Z)$,
    $\mathfrak{R}_n(X, Y) \circ \mathfrak{R}_n(Y, Z) \subseteq \mathfrak{R}_n(X, Z)$.

In particular, the classes $\mathfrak{R}_c, \mathfrak{R}_l, \mathfrak{R}_m$ *and* $\mathfrak{R}_n$ *are subalgebras of the algebra* $\mathfrak{R}$, *and the classes* $\mathfrak{R}_n$ *and* $\mathfrak{R}_l \oplus \mathfrak{R}_m \oplus \mathfrak{R}_n$ *are ideals in* $\mathfrak{R}$.

The basic Theorem 11.5 is not only of theoretical interest. It implies, in fact, that showing that a partial integral operator (11.1) belongs to $\mathfrak{L}^r(X, Y)$ reduces to proving the four relations

(11.27)
$$c \in \mathfrak{R}_c(X, Y), \quad l \in \mathfrak{R}_l(X, Y),$$
$$m \in \mathfrak{R}_m(X, Y), \quad n \in \mathfrak{R}_n(X, Y).$$

The verification of the first relation in (11.27) reduces to a simple application of results on multiplicator spaces, while that of the last relation may be carried out by means of the theory of Zaanen spaces of kernel functions for linear integral operators (see e.g. ZABREJ-KO [1966, 1968] and ZABREJKO-KOSHELEV-KRASNOSEL'SKIJ-MIKH-LIN-RAKOVSHCHIK-STETSENKO [1968]). Both procedures have been studied extensively for general spaces, as well as for special (e.g., Lebesgue and Orlicz) spaces. The problem of verifying the second and third relation in (11.27), however, is harder and has not been given much attention yet in books.

In general, it seems to be difficult to give a fairly explicit description of the kernel classes $\mathfrak{R}_l, \mathfrak{R}_m$, and $\mathfrak{R}_n$. Such a description, apparently, depends heavily on specific properties of the spaces X and Y, and the problems are due to the lack of symmetry in the variables s and t. Some of the most important special cases, in which more information may be obtained, will be considered in the next sections.

§ 12. Operators in spaces with mixed norm

In this section we introduce and study so-called ideal spaces with mixed norm (of functions of two variables). These spaces give the natural setting for investigating partial integral operators. The most important examples are, as usual, Lebesgue spaces with mixed norm which we used (at least implicitly) already in Subsection 3.3, and Orlicz spaces with mixed norm which we will discuss in Subsection 12.3 below.

The results of Subsections 12.1 and 12.2 have been obtained in KALITVIN-ZABREJKO [1991], the main results of Subsection 12.3 may be found in APPELL-KALITVIN-ZABREJKO [1998], and the results of Subsections 12.4 and 12.5 are taken from FROLOVA-KALITVIN-ZABREJKO [1998]. We also mention the papers KALITVIN-MILOVIDOV [1981] and POVOLOTSKIJ-KALITVIN [1986] on interpolation of partial integral operators in spaces with mixed norm.

12.1. Ideal spaces with mixed norm

Let U be an almost perfect ideal space over a domain T with measure μ, and V an almost perfect ideal space over a domain S with measure ν. The *space with mixed norm* $[U \to V]$ consists, by definition, of all measurable functions $x : T \times S \to \mathbb{R}$ for which the norm

$$(12.1) \qquad \|x\|_{[U \to V]} = \|s \mapsto \|x(\cdot, s)\|_U\|_V$$

is finite. Similarly, the space with mixed norm $[U \leftarrow V]$ is defined by the norm

$$(12.2) \qquad \|x\|_{[U \leftarrow V]} = \|t \mapsto \|x(t, \cdot)\|_V\|_U.$$

Both $[U \to V]$ and $[U \leftarrow V]$ are ideal spaces. If they are regular they are also examples of *tensor products* of U and V; in fact, for any $u \in U$ and $v \in V$ the function w defined by $w(t, s) = u(t)v(s)$ belongs to both $[U \to V]$ and $[U \leftarrow V]$ and satisfies

$$(12.3) \qquad \|w\|_{[U \to V]} = \|w\|_{[U \leftarrow V]} = \|u\|_U \|v\|_V.$$

For example, the spaces $U(p,q)$ and $V(p,q)$ introduced in (3.22) and (3.23) are nothing else than the spaces with mixed norm $[L_q \leftarrow L_p]$ and $[L_p \rightarrow L_q]$, respectively. For further reference, we collect some properties of Lebesgue spaces with mixed norm in the following

Lemma 12.1. *For* $1 \leq p, p_1, p_2, q, q_1, q_2 \leq \infty$, *the following holds:*

(a) $||xy||_{[L_p \rightarrow L_q]} \leq ||x||_{[L_{p_1} \rightarrow L_{q_1}]} ||y||_{[L_{p_2} \rightarrow L_{q_2}]}$ *for* $\frac{1}{p} = \frac{1}{p_1} + \frac{1}{p_2}$ *and* $\frac{1}{q} = \frac{1}{q_1} + \frac{1}{q_2}$;

(b) $||xy||_{[L_p \leftarrow L_q]} \leq ||x||_{[L_{p_1} \leftarrow L_{q_1}]} ||y||_{[L_{p_2} \leftarrow L_{q_2}]}$ *for* $\frac{1}{p} = \frac{1}{p_1} + \frac{1}{p_2}$ *and* $\frac{1}{q} = \frac{1}{q_1} + \frac{1}{q_2}$;

(c) $||x||_{[L_p \rightarrow L_q]} \leq ||x||_{[L_p \leftarrow L_q]}$ *for* $p \leq q$;

(d) $||x||_{[L_p \leftarrow L_q]} \leq ||x||_{[L_p \rightarrow L_q]}$ *for* $p \geq q$;

(e) $[L_{p_1} \rightarrow L_{q_1}]' = [L_{p_2} \rightarrow L_{q_2}]$ *and* $[L_{p_1} \leftarrow L_{q_1}]' = [L_{p_2} \leftarrow L_{q_2}]$ *for* $\frac{1}{p_1} + \frac{1}{p_2} = \frac{1}{q_1} + \frac{1}{q_2} = 1$, *where* X' *denotes the associate space of* X.

□   The estimates (a) and (b) are proved simply by applying the classical Hölder inequality with respect to both variables. To prove (c) it suffices to apply the Minkowski inequality

$$\left|\left| \int_T u(t,\cdot)\,d\mu(t) \right|\right|_{L_r} \leq \int_T ||u(t,\cdot)||_{L_r}\,d\mu(t)$$

for $r = q/p$ to the function $u(t,s) := |x(t,s)|^p$; of course, (d) is proved similarly.

The estimate

$$||y||_{[L_{p_2} \rightarrow L_{q_2}]} \leq$$

$$\sup\left\{ \int_S \int_T |x(t,s)y(t,s)|\,d\mu(t)\,d\nu(s) : ||x||_{[L_{p_1} \rightarrow L_{q_1}]} \leq 1 \right\}$$

follows from (a) for $p = q = 1$. The reverse inequality may be proved by choosing

$$x(t,s) := \operatorname{sign} y(t,s)|y(\cdot,s)|^{p_2-1}||y(\cdot,s)||_{L_{p_2}}^{q_2-p_2}||y||_{[L_{p_2} \rightarrow L_{q_2}]}^{1-q_2}$$

and observing that

$$||y||_{[L_{p_2} \rightarrow L_{q_2}]} = \int_S \int_T |x(t,s)y(t,s)|\,d\mu(t)\,d\nu(s)$$

and $||x||_{[L_{p_1} \to L_{q_1}]} = 1.$ ∎

Observe that, in case of bounded domains S and T, Lemma 12.1 (c) and (d) imply the elementary though useful relations

$$L_{\max\{p,q\}} \subseteq [L_p \to L_q] \subseteq L_{\min\{p,q\}}$$

and

$$L_{\max\{p,q\}} \subseteq [L_p \leftarrow L_q] \subseteq L_{\min\{p,q\}}.$$

Moreover, Lemma 12.1 (e) carries over to more general spaces; in fact, the equalities

$$[U \to V]' = [U' \to V'], \qquad [U \leftarrow V]' = [U' \leftarrow V']$$

are true for all almost perfect ideal spaces U and V. Some more results on Lebesgue spaces with mixed norm may be found, for example, in BENEDEK-PANZONE [1961], on more general spaces with mixed norm in BUKHVALOV [1983], see also BUKHVALOV-KOROTKOV-KUSRAEV-KUTATELADZE-MAKAROV [1992].

Let us return to the partial integral operators

$$(12.4) \qquad Lx(t,s) = \int_T l(t,s,\tau)x(\tau,s)d\mu(\tau)$$

and

$$(12.5) \qquad Mx(t,s) = \int_S m(t,s,\sigma)x(t,\sigma)d\nu(\sigma).$$

In what follows, we shall describe three approaches to action conditions for these operators in spaces with mixed norm. For $t \in T$ and $s \in S$, consider the families $L(s)$ and $M(t)$ of linear integral operators defined by

$$(12.6) \qquad L(s)u(t) = \int_T l(t,s,\tau)u(\tau)d\mu(\tau) \qquad (s \in S)$$

and

$$(12.7) \qquad M(t)v(s) = \int_S m(t,s,\sigma)v(\sigma)d\nu(\sigma) \qquad (t \in T).$$

In the following Theorems 12.1-12.5 we assume for simplicity that both U_i and V_i $(i = 1, 2)$ are perfect ideal spaces (see Subsection 4.1). We point out, however, that these theorems are also true under more general assumptions on U_i and V_i $(i = 1, 2)$. For example, Theorem 12.1 is true for the operator (12.4) if both U_1 and U_2 are almost perfect ideal spaces. We recall that both Lebesgue and Orlicz spaces are perfect ideal spaces.

Theorem 12.1. *Let U_1 and U_2 be two ideal spaces over T, and V_1 and V_2 two ideal spaces over S. Suppose that the linear integral operator (12.6) maps U_1 into U_2, for each $s \in S$, and that the map $s \mapsto \|L(s)\|_{\mathfrak{L}(U_1,U_2)}$ belongs to V_2/V_1. Then the partial integral operator (12.4) acts between the spaces $X = [U_1 \to V_1]$ and $Y = [U_2 \to V_2]$ and satisfies*

$$(12.8) \qquad \|L\|_{\mathfrak{L}(X,Y)} \leq \|s \mapsto \|L(s)\|_{\mathfrak{L}(U_1,U_2)}\|_{V_2/V_1}.$$

Similarly, if the linear integral operator (12.7) maps V_1 into V_2, for each $t \in T$, and the map $t \mapsto \|M(t)\|_{\mathfrak{L}(V_1,V_2)}$ belongs to U_2/U_1, the partial integral operator (12.5) acts between the spaces $X = [U_1 \leftarrow V_1]$ and $Y = [U_2 \leftarrow V_2]$ and satisfies

$$(12.9) \qquad \|M\|_{\mathfrak{L}(X,Y)} \leq \|t \mapsto \|M(t)\|_{\mathfrak{L}(V_1,V_2)}\|_{U_2/U_1}.$$

☐ Without loss of generality, we prove only the first statement. Given $x \in X = [U_1 \to V_1]$, for almost all $s \in S$ we have

$$\|Lx(\cdot, s)\|_{U_2} \leq \|L(s)\|_{\mathfrak{L}(U_1,U_2)}\|x(\cdot, s)\|_{U_1},$$

hence, by the definition of the multiplicator space V_2/V_1,

$$\|Lx\|_Y = \|s \mapsto \|Lx(\cdot, s)_{U_2}\|_{V_2} \leq \|s \mapsto \|L(s)\|_{\mathfrak{L}(U_1,U_2)}\|x(\cdot, s)\|_{U_1}\|_{V_2}$$

$$\leq \|s \mapsto \|L(s)\|_{\mathfrak{L}(U_1,U_2)}\|_{V_2/V_1}\|x\|_X.$$

This shows that the operator (12.4) acts between X and Y and satisfies (12.8). ■

Interestingly, in the case $V_2/V_1 = L_\infty$ the conditions of Theorem 12.1 are also necessary for the operator (12.4) to act between $X = [U_1 \to V_1]$ and $Y = [U_2 \to V_2]$. In fact, considering the operator (12.4) on the "separated" functions $x(t,s) = u(t)v(s)$, where $u \in U_1$ and $v \in V_1$, we conclude that, by the obvious relation $Lx(t,s) = v(s)L(s)u(t)$,

$$(12.10) \qquad \sup_{\|u\|_{U_1} \leq 1} \|s \mapsto \|v(s)L(s)u\|_{U_2}\|_{V_2} \leq \|L\|_{\mathcal{L}(X,Y)} \|v\|_{V_2/V_1}.$$

In case $V_2/V_1 = L_\infty$ this means exactly that

$$(12.11) \qquad \|s \mapsto \|L(s)\|_{\mathcal{L}(U_1,U_2)} \|_{V_2/V_1} \leq \|L\|_{\mathcal{L}(X,Y)},$$

i.e. equality holds in (12.8). Analogous statements hold, of course, for the operator (12.5) in case $U_2/U_1 = L_\infty$. For example, the equalities

$$\|L\|_{\mathcal{L}([L_{p_1} \to L_q],[L_{p_2} \to L_q])} = \|s \mapsto \|L(s)\|_{\mathcal{L}(L_{p_1},L_{p_2})}\|_{L_\infty}$$

and

$$\|M\|_{\mathcal{L}([L_p \leftarrow L_{q_1}],[L_p \leftarrow L_{q_2}])} = \|t \mapsto \|M(t)\|_{\mathcal{L}(L_{q_1},L_{q_2})}\|_{L_\infty}$$

hold.

Generally speaking, the estimates (12.10) and (12.11) are not equivalent. Nevertheless, (12.10) implies that

$$(12.12) \qquad \sup_{\|u\|_{U_1} \leq 1} \|s \mapsto \|L(s)u\|_{U_2}\|_{V_2/V_1} < \infty,$$

which thus is necessary for the operator (12.4) to act between $X = [U_1 \to V_1]$ and $Y = [U_2 \to V_2]$. The analogous condition for the operator (12.5) reads

$$(12.13) \qquad \sup_{\|v\|_{V_1} \leq 1} \|t \mapsto \|M(t)v\|_{V_2}\|_{U_2/U_1} < \infty.$$

Observe that we did not suppose in Theorem 12.1 that the operators (12.4) and (12.5) be regular. Applying this theorem to the kernels $|l(t,s,\tau)|$ and $|m(t,s,\sigma)|$, rather than to $l(t,s,\tau)$ and $m(t,s,\sigma)$, we get a refinement of Theorem 12.1.

Let W_1 and W_2 be two ideal spaces over some set Ω with measure λ. We denote by $3(W_1, W_2)$ the space of all *Zaanen kernels* $z = z(\xi, \eta)$, defining regular linear integral operators from W_1 into W_2, with the norm

$$(12.14) \qquad \|z\|_{3(W_1,W_2)} = \sup_{\|w\|_{W_1} \leq 1} \left\| \int_\Omega |z(\cdot, \eta) w(\eta)| d\lambda(\eta) \right\|_{W_2}.$$

By definition of the norm of a linear operator, the norm $\|z\|_{3(W_1,W_2)}$ of the kernel z coincides then with the operator norm $\|Z\|_{\mathfrak{L}^r(W_1,W_2)}$ (see (3.10)) of the integral operator

$$Zw(\xi) = \int_\Omega z(\xi, \eta) w(\eta) \, d\lambda(\eta)$$

generated by z.

Theorem 12.2. *Let U_1 and U_2 be two ideal spaces over T, and V_1 and V_2 two ideal spaces over S. Suppose that $l(\cdot, s, \cdot) \in 3(U_1, U_2)$ for almost all $s \in S$, and that the map $s \mapsto \|l(\cdot, s, \cdot)\|_{3(U_1,U_2)}$ belongs to V_2/V_1. Then the partial integral operator (12.4) acts between the spaces $X = [U_1 \rightarrow V_1]$ and $Y = [U_2 \rightarrow V_2]$, is regular, and satisfies*

$$(12.15) \qquad \|l\|_{\mathfrak{R}_l(X,Y)} \leq \|s \mapsto \|l(\cdot, s, \cdot)\|_{3(U_1,U_2)}\|_{V_2/V_1}.$$

Similarly, if $m(t, \cdot, \cdot) \in 3(V_1, V_2)$ for almost all $t \in T$, and the map $t \mapsto \|m(t, \cdot, \cdot)\|_{3(V_1,V_2)}$ belongs to U_2/U_1, then the partial integral operator (12.5) acts between the spaces $X = [U_1 \leftarrow V_1]$ and $Y = [U_2 \leftarrow V_2]$, is regular, and satisfies

$$(12.16) \qquad \|m\|_{\mathfrak{R}_m(X,Y)} \leq \|t \mapsto \|m(t, \cdot, \cdot)\|_{3(V_1,V_2)}\|_{U_2/U_1}.$$

Consider now the two integral operators

$$(12.17) \qquad \tilde{L}u(t) = \int_T \tilde{l}(t, \tau) u(\tau) d\mu(\tau)$$

and

$$(12.18) \qquad \tilde{M}v(s) = \int_S \tilde{m}(s, \sigma) v(\sigma) d\nu(\sigma)$$

generated by the kernels $\tilde{l}(t,\tau) = ||l(t,\cdot,\tau)||_{V_2/V_1}$ and $\tilde{m}(s,\sigma) = ||m(\cdot,s,\sigma)||_{U_2/U_1}$, respectively.

Theorem 12.3. *Let U_1 and U_2 be two ideal spaces over T, and V_1 and V_2 two ideal spaces over S. Suppose that the linear integral operator (12.17) maps U_1 into U_2. Then the partial integral operator (12.4) acts between the spaces $X = [U_1 \leftarrow V_1]$ and $Y = [U_2 \leftarrow V_2]$, is regular, and satisfies*

(12.19)
$$||l||_{\mathfrak{R}_l(X,Y)} \leq ||\tilde{l}||_{\mathfrak{Z}(U_1,U_2)}$$
$$= ||(t,\tau) \mapsto ||l(t,\cdot,\tau)||_{V_2/V_1}||_{\mathfrak{Z}(U_1,U_2)}.$$

Similarly, if the linear integral operator (12.18) maps V_1 into V_2, then the partial integral operator (12.5) acts between the spaces $X = [U_1 \to V_1]$ and $Y = [U_2 \to V_2]$ is regular, and satisfies

(12.20)
$$||m||_{\mathfrak{R}_m(X,Y)} \leq ||\tilde{m}||_{\mathfrak{Z}(V_1,V_2)}$$
$$= ||(s,\sigma) \mapsto ||m(\cdot,s,\sigma)||_{U_2/U_1}||_{\mathfrak{Z}(V_1,V_2)}.$$

□ We prove again only the first statement. Obviously, for $x \in X$, we have

$$||Lx(t,\cdot)||_{V_2} = \left|\left|\int_T l(t,\cdot,\tau)x(\tau,\cdot)\,d\mu(\tau)\right|\right|_{V_2}$$
$$\leq \int_T \tilde{l}(t,\tau)||x(\tau,\cdot)||_{V_1}\,d\mu(\tau) = \tilde{L}u(t),$$

where $u(t) = ||x(t,\cdot)||_{V_1}$. Consequently,

$$||Lx||_Y = \left|\left|t \mapsto \left|\left|\int_T l(t,\cdot,\tau)x(\tau,\cdot)\,d\mu(\tau)\right|\right|_{V_2}\right|\right|_{U_2}$$
$$\leq ||\tilde{L}u||_{U_2} \leq ||\tilde{L}||_{\mathfrak{L}(U_1,U_2)}||u||_{U_1} = ||\tilde{L}||_{\mathfrak{L}(U_1,U_2)}||x||_X.$$

This shows that the operator (12.4) acts between X and Y and satisfies (12.19). ■

We conclude this subsection with two more acting conditions for partial integral operators in spaces with mixed norm; these conditions build on multiplicator spaces of functions of two variables.

Theorem 12.4. *Let U_1 and U_2 be two ideal spaces over T, and V_1 and V_2 two ideal spaces over S. Suppose that the function $(s, \tau) \mapsto \|l(\cdot, s, \tau)\|_{U_2}$ belongs to the multiplicator space $[L_1 \to V_2]/[U_1 \leftarrow V_1]$. Then the partial integral operator (12.4) acts between the spaces $X = [U_1 \leftarrow V_1]$ and $Y = [U_2 \to V_2]$, is regular, and satisfies*

$$(12.21) \quad \|l\|_{\mathfrak{R}_l(X,Y)} \leq \|(s, \tau) \mapsto \|l(\cdot, s, \tau)\|_{U_2}\|_{[L_1 \to V_2]/[U_1 \leftarrow V_1]}.$$

Similarly, if the function $(t, \sigma) \mapsto \|m(t, \cdot, \sigma)\|_{V_2}$ belongs to the multiplicator space $[U_2 \leftarrow L_1]/[U_1 \to V_1]$, then the partial integral operator (12.5) acts between the spaces $X = [U_1 \to V_1]$ and $Y = [U_2 \leftarrow V_2]$, is regular, and satisfies

$$(12.22) \quad \begin{aligned} &\|m\|_{\mathfrak{R}_m(X,Y)} \\ &\leq \|(t, \sigma) \mapsto \|m(t, \cdot, \sigma)\|_{V_2}\|_{[U_2 \leftarrow L_1]/[U_1 \to V_1]}. \end{aligned}$$

◻ Obviously,

$$\left\| \int_T l(\cdot, s, \tau) x(\tau, s) \, d\mu(\tau) \right\|_{U_2} \leq \int_T \|l(\cdot, s, \tau)\|_{U_2} |x(\tau, s)| \, d\mu(\tau);$$

consequently,

$$\|Lx\|_Y = \left\| t \mapsto \left\| \int_T l(\cdot, s, \tau) x(\tau, s) d\mu(\tau) \right\|_{U_2} \right\|_{V_2}$$

$$\leq \|(s, \tau) \mapsto \|l(\cdot, s, \tau)\|_{U_2}\|_{[L_1 \to V_2]/[U_1 \leftarrow V_1]} \|x\|_{[U_1 \leftarrow V_1]}.$$

This proves the first statement. The second statement is proved analogously. ∎

Theorem 12.5. *Let U_1 and U_2 be two ideal spaces over T, and V_1 and V_2 two ideal spaces over S. Suppose that the function $(t, s) \mapsto$*

$||l(t, s, \cdot)||_{U'_1}$ *belongs to the multiplicator space* $[L_1 \to V'_1]/[U'_2 \leftarrow V'_2]$. *Then the partial integral operator* (12.4) *acts between the spaces* $X = [U_1 \to V_1]$ *and* $Y = [U_2 \leftarrow V_2]$ *is regular and satisfies*

$$(12.23) \qquad ||l||_{\mathfrak{R}_l(X,Y)} \leq ||(t, s) \mapsto ||l(t, s, \cdot)||_{U'_1}||_{[L_1 \leftarrow V'_1]/[U'_2 \leftarrow V'_2]}.$$

Similarly, if the function $(t, s) \mapsto ||m(t, s, \cdot)||_{V'_1}$ *belongs to the multiplicator space* $[L_1 \leftarrow U'_1]/[U'_2 \to V'_2]$, *then the partial integral operator* (12.5) *acts between the spaces* $X = [U_1 \leftarrow V_1]$ *and* $Y = [U_2 \to V_2]$ *is regular and satisfies*

$$(12.24) \qquad ||m||_{\mathfrak{R}_m(X,Y)} \leq ||(t, s) \mapsto ||m(t, s, \cdot)||_{V'_1}||_{[L_1 \leftarrow U'_1]/[U'_2 \to V'_2]}.$$

□ The proof is completely analogous to that of Theorem 12.4; one has just to pass to the corresponding associate operators. ∎

All sufficient conditions of Theorems 12.1 - 12.5 are different for the partial integral operator (12.4) and the partial integral operator (12.5) and refer to different kernel spaces with mixed norm. In this way, the above theorems contain eight statements which guarantee the acting (and, except for Theorem 12.1, also the regularity) of partial integral operators between four possible combinations of spaces with mixed norm.

12.2. Lebesgue spaces with mixed norm

In Subsection 11.5 we had to show that certain functions of two variables, constructed by means of the kernels $l = l(t, s, \tau)$ and $m = m(t, s, \sigma)$, belong to certain ideal spaces, constructed by means of the spaces U_1, U_2, V_1 and V_2. Since these ideal spaces are rather complicated, however, the natural problem arises to replace them by simpler and more tractable ones. One possibility to do so is to introduce ideal spaces of functions defined either on $T \times S \times T$, or on $T \times S \times S$, or on some permutation of these Cartesian products. Denote by $\theta = (\theta_1, \theta_2, \theta_3)$ an arbitrary permutation of the arguments $(t, s, \tau) \in T \times S \times T$, or $(t, s, \sigma) \in T \times S \times S$. Given three ideal spaces

W_1, W_2, and W_3, by $[W_1, W_2, W_3; \theta]$ we denote the ideal space of all functions w of three variables for which the norm

$$\|w\|_{[W_1,W_2,W_3;\theta]} :=$$

$$\left\|\theta_3 \mapsto \left\|\theta_2 \mapsto \left\|\theta_1 \mapsto w(\theta_1, \theta_2, \theta_3)\right\|_{W_1}\right\|_{W_2}\right\|_{W_3}$$

is defined and finite. Using classical results on linear integral operators, from Theorems 12.2 - 12.5 we get the following

Theorem 12.6. *Let U_1 and U_2 be two ideal spaces over T, and V_1 and V_2 two ideal spaces over S. Suppose that $l \in [U_2, V_2/V_1, U_1'; \theta]$ for some $\theta = (\theta_1, \theta_2, \theta_3)$. Then the partial integral operator (12.4) acts between X and Y, is regular, and satisfies*

$$(12.25) \qquad \|l\|_{\mathfrak{R}_l(X,Y)} \leq \|l\|_{[U_2,V_2/V_1,U_1';\theta]}.$$

Here the spaces X and Y have to be chosen according to the formula

$$X = [U_1 \leftarrow V_1], \ Y = [U_2 \leftarrow V_2] \quad \text{if} \quad \theta = (s,t,\tau) \text{ or } \theta = (s,\tau,t)$$

$$X = [U_1 \rightarrow V_1], \ Y = [U_2 \rightarrow V_2] \quad \text{if} \quad \theta = (t,\tau,s) \text{ or } \theta = (\tau,t,s),$$

$$X = [U_1 \leftarrow V_1], \ Y = [U_2 \rightarrow V_2] \quad \text{if} \quad \theta = (t,s,\tau),$$

$$X = [U_1 \rightarrow V_1], \ Y = [U_2 \leftarrow V_2] \quad \text{if} \quad \theta = (\tau,s,t).$$

Similarly, if $m \in [U_2/U_1, V_2, V_1'; \theta]$ for some $\theta = (\theta_1, \theta_2, \theta_3)$, then the partial integral operator (12.5) acts between X and Y, is regular and satisfies

$$(12.26) \qquad \|m\|_{\mathfrak{R}_m(X,Y)} \leq \|m\|_{[U_2/U_1,V_2,V_1';\theta]}.$$

Here the spaces X and Y have to be chosen according to the formula

$$X = [U_1 \leftarrow V_1], \ Y = [U_2 \leftarrow V_2] \quad \text{if} \quad \theta = (s,\sigma,t) \text{ or } \theta = (\sigma,s,t),$$

$$X = [U_1 \rightarrow V_1], \ Y = [U_2 \rightarrow V_2] \quad \text{if} \quad \theta = (t,s,\sigma) \text{ or } \theta = (t,\sigma,s),$$

$$X = [U_1 \leftarrow V_1], \ Y = [U_2 \rightarrow V_2] \quad \text{if} \quad \theta = (\sigma,t,s),$$

$$X = [U_1 \rightarrow V_1], \ Y = [U_2 \leftarrow V_2] \quad \text{if} \quad \theta = (s,t,\sigma).$$

Of course, the formulation of Theorem 12.6 is very clumsy. To illustrate its applicability, let us consider Lebesgue spaces again.

Theorem 12.7. *Let $p_1, p_2, q_1, q_2 \in [1, \infty]$. Suppose that $q_1 \geq q_2$, with $\nu(S) < \infty$ in case $q_1 > q_2$, and*

$$l \in [L_{p_2}, L_{q_1 q_2/(q_1 - q_2)}, L_{p_1/(p_1 - 1)}; \theta].$$

Then the partial integral operator (12.4) acts between the spaces X and Y, is regular, and satisfies

(12.27) $\|l\|_{\mathfrak{R}_l(X,Y)} \leq \|l\|_{[L_{p_2}, L_{q_1 q_2/(q_1 - q_2)}, L_{p_1/(p_1 - 1)}; \theta]},$

provided one of the conditions of Table 1 below holds. Similarly, suppose that $p_1 \geq p_2$, with $\mu(T) < \infty$ in case $p_1 > p_2$, and

$$m \in [L_{p_1 p_2/(p_1 - p_2)}, L_{q_2}, L_{q_1/(q_1 - 1)}; \theta].$$

Then the partial integral operator (12.5) acts between the spaces X and Y, is regular, and satisfies

(12.28) $\|m\|_{\mathfrak{R}_m(X,Y)} \leq \|m\|_{[L_{p_1 p_2/(p_1 - p_2)}, L_{q_2}, L_{q_1/(q_1 - 1)}; \theta]},$

provided one of the conditions of Table 2 below holds.

$X =$ $Y =$	$[L_{p_1} \to L_{q_1}]$ $[L_{p_2} \to L_{q_2}]$	$[L_{p_1} \to L_{q_1}]$ $[L_{p_2} \leftarrow L_{q_2}]$	$[L_{p_1} \leftarrow L_{q_1}]$ $[L_{p_2} \to L_{q_2}]$	$[L_{p_1} \leftarrow L_{q_1}]$ $[L_{p_2} \leftarrow L_{q_2}]$
(t, s, τ)	$p_1 \geq q_1$	$p_1 \geq q_1$ $p_2 \geq q_2$		$p_2 \geq q_2$
(t, τ, s)		$p_2 \geq q_2$	$p_1 \leq q_1$	$p_1 \leq q_1$ $p_2 \geq q_2$
(s, t, τ)	$p_1 \geq q_1$ $p_2 \leq q_2$	$p_1 \geq q_1$	$p_2 \leq q_2$	
(s, τ, t)	$p_1 \geq q_1$ $p_2 \leq q_2$	$p_1 \geq q_1$	$p_2 \leq q_2$	
(τ, s, t)	$p_2 \leq q_2$		$p_1 \leq q_1$ $p_2 \leq q_2$	$p_1 \leq q_1$
(τ, t, s)		$p_2 \geq q_2$	$p_1 \leq q_1$	$p_1 \leq q_1$ $p_2 \geq q_2$

Table 1

$X =$ $Y =$	$[L_{p_1} \to L_{q_1}]$ $[L_{p_2} \to L_{q_2}]$	$[L_{p_1} \to L_{q_1}]$ $[L_{p_2} \leftarrow L_{q_2}]$	$[L_{p_1} \leftarrow L_{q_1}]$ $[L_{p_2} \to L_{q_2}]$	$[L_{p_1} \leftarrow L_{q_1}]$ $[L_{p_2} \leftarrow L_{q_2}]$
(t,s,σ)		$p_2 \geq q_2$	$p_1 \leq q_1$	$p_1 \leq q_1$ $p_2 \geq q_2$
(t,σ,s)		$p_2 \geq q_2$	$p_1 \leq q_1$	$p_1 \leq q_1$ $p_2 \geq q_2$
(s,t,σ)	$p_2 \leq q_2$		$p_1 \leq q_1$ $p_2 \leq q_2$	$p_1 \leq q_1$
(s,σ,t)	$p_1 \geq q_1$ $p_2 \leq q_2$	$p_1 \geq q_1$	$p_2 \leq q_2$	
(σ,t,s)	$p_1 \geq q_1$	$p_1 \geq q_1$ $p_2 \geq q_2$		$p_2 \geq q_2$
(σ,s,t)	$p_1 \geq q_1$ $p_2 \leq q_2$	$p_1 \geq q_1$	$p_2 \leq q_2$	

Table 2

The most interesting and important case in the preceding theorem is when the operators (12.4) and (12.5) are considered in the single Lebesgue space $L_p = L_p(T \times S)$. Since $[L_p \to L_p] = [L_p \leftarrow L_p] = L_p$, Theorem 12.7 reads in this case as follows:

Theorem 12.8. *Let* $1 \leq p \leq \infty$. *Suppose that*

$$l \in [L_p, L_\infty, L_{p/(p-1)}; \theta']$$

for some $\theta' = (\theta'_1, \theta'_2, \theta'_3)$, *and*

$$m \in [L_\infty, L_p, L_{p/(p-1)}; \theta'']$$

for some $\theta'' = (\theta''_1, \theta''_2, \theta''_3)$. *Then the partial integral operators* (12.4) *and* (12.5) *are regular in the space* L_p *and satisfy*

$$(12.29) \qquad \|l\|_{\Re_l(L_p, L_p)} \leq \|l\|_{[L_p, L_\infty, L_{p/(p-1)}; \theta']}$$

and

$$(12.30) \qquad \|m\|_{\Re_m(L_p, L_p)} \leq \|m\|_{[L_\infty, L_p, L_{p/(p-1)}; \theta'']}.$$

Apart from the acting and regularity conditions for the partial integral operators (12.4) and (12.5), many other statements may be formulated. For instance, one may obtain further statements by means of interpolation theory (applied to either classical or partial integral operators, see KALITVIN-MILOVIDOV [1981] and POVOLOTSKIJ-KALITVIN [1983, 1985]), of Kantorovich type theorems, or of other methods.

We point out that Theorem 12.1 and its partial converse considered above imply the following useful criterion: the partial integral operator (12.4) (respectively, (12.5)) acts in L_p if and only if all operators of the family (12.6) (respectively, (12.7)) act in L_p for each $s \in S$ (respectively, $t \in T$) and have uniformly bounded norms.

The statements of Theorems 12.7 and 12.8, referring to the Lebesgue type spaces $[L_p \to L_q]$ and $[L_p \leftarrow L_q]$, carry over as well to the Orlicz type spaces $[L_M \to L_N]$ and $[L_M \leftarrow L_N]$, as we shall show in the following subsection.

The following theorem which is taken from KALITVIN [1995a] gives an effective acting criterion for the operator (11.1) in $L_\infty = L_\infty(T \times S)$ and in $L_1 = L_1(T \times S)$.

Theorem 12.9. *Let* $D = T \times S$, *and define two functions* $\alpha, \beta : D \to$ ℝ *by*

(12.31)
$$\alpha(t, s) = |c(t, s)| + \int_T |l(t, s, \tau)| \, d\tau$$
$$+ \int_S |m(t, s, \sigma)| \, d\sigma + \int_T \int_S |n(t, s, \tau, \sigma)| \, d\tau \, d\sigma$$

and

(12.32)
$$\beta(t, s) = |c(t, s)| + \int_T |l(\tau, s, t)| \, d\tau$$
$$+ \int_S |m(t, \sigma, s)| \, d\sigma + \int_T \int_S |n(\tau, \sigma, t, s)| \, d\tau \, d\sigma,$$

respectively. Then the operator (11.1) *acts in* $L_\infty(D)$ *if and only if* $\alpha \in L_\infty(D)$, *and in* $L_1(D)$ *if and only if* $\beta \in L_1(D)$. *In this case we have* $\|P\|_{\mathcal{L}(L_\infty)} = \|\alpha\|_{L_\infty}$ *and* $\|P\|_{\mathcal{L}(L_1)} = \|\beta\|_{L_1}$.

□ We consider the case of the space L_∞. Since each continuous linear operator in L_∞ is regular, by Theorems 11.2 and 11.3 it suffices to show that the norm of the operator (11.1) coincides with $\|\alpha\|_{L_\infty}$. Let $[P] := \sup\{|Px(t,s)| : \|x\| \leq 1\}$ be the abstract norm of the operator (11.1) in L_∞. Then $[P] = [|P|]$, where $|P|$ denotes the module of P (see Subsection 3.1). By Theorems 11.2 and 11.3, this operator coincides with the operator (11.6). Therefore,

$$\|P\|_{L_\infty} = \|[P]\|_{L_\infty} = \||P|\|_{L_\infty} = \||P|e\|_{L_\infty},$$

where $e(t,s) \equiv 1$.

This proves the assertion for the space L_∞. In case of the space L_1 the proof follows from Theorem 11.4 and equality (12.32). ■

12.3. Orlicz spaces with mixed norm

We give now some parallel results for Orlicz spaces which we already introduced in Subsection 4.1. Recall that, given a Young function $M : \mathbb{R} \to [0,\infty)$, the Orlicz space $L_M = L_M(\Omega)$ is defined by one of the (equivalent) norms (4.3) or (4.4). To state the theorems of this subsection, some further notions and results on Orlicz spaces are in order. For simplicity, we only consider Orlicz spaces over bounded domains Ω in this subsection.

Given two Young functions M and N, we write $M \preceq N$ if there exist $k > 0$ and $u_0 \geq 0$ such that

$$M(u) \leq N(ku) \quad (u \geq u_0).$$

Moreover, we write $M \prec N$ if

$$\lim_{u \to \infty} \frac{M(u)}{N(ku)} = 0$$

for every $k > 0$. Of course, in case $M(u) = |u|^p$ and $N(u) = |u|^q$ $(1 < p, q < \infty)$ we have $M \preceq N$ if and only if $p \leq q$, and $M \prec N$ if and only if $p < q$. In general, one can show that $M \preceq N$ is equivalent to the fact that L_N is continuously imbedded in L_M, and $M \prec N$ is equivalent to the fact that L_N is absolutely continuously imbedded in L_M (i.e., the unit ball of L_N is an absolutely bounded subset of L_M). Moreover, the inclusions $L_\infty \subseteq L_M \subseteq L_1$ are always true.

Let $U = L_M(T)$ and $V = L_N(S)$ be two Orlicz spaces. We are interested in the Orlicz spaces with mixed norm $[U \to V]$ and $[U \leftarrow V]$ defined by the formulas (12.1) and (12.2), respectively. These spaces are perfect ideal spaces. They are regular if and only if the Young functions M and N satisfy a Δ_2-condition. If $M_2, N_2 \preceq M_1, N_1$ the inclusions

$$[L_{M_1} \to L_{N_1}] \subseteq [L_{M_2} \to L_{N_2}], \qquad [L_{M_1} \leftarrow L_{N_1}] \subseteq [L_{M_2} \leftarrow L_{N_2}]$$

are obvious. Moreover, the inclusions

$$L_M(T \times S) \subseteq [L_1 \to L_M], \qquad [L_M \leftarrow L_1] \subseteq L_1(T \times S)$$

follow from the Jensen integral inequality

$$M\left(\frac{1}{\mu(\Omega)} \int_\Omega x(\omega)\, d\mu(\omega)\right) \leq \frac{1}{\mu(\Omega)} \int_\Omega M(x(\omega))\, d\mu(\omega)$$

and the definition of the norm in L_M. In fact, for $x \in L_M$ and $k > 0$ sufficiently large we have

$$\int_S M\left[\frac{1}{k} \int_T |x(t,s)|\, dt\right] ds = \int_S M\left[\frac{1}{\mu(T)} \int_T \frac{\mu(T)|x(t,s)|}{k}\, dt\right] ds$$

$$\leq \frac{1}{\mu(T)} \int_S \int_T M\left[\frac{\mu(T)|x(t,s)|}{k}\right] dt\, ds < \infty.$$

Consequently, $x \in [L_1 \to L_M]$, and hence the left inclusion is proved. The right inclusion is proved analogously.

Lemma 12.2 *Let M_i and N_i $(i = 1, 2)$ be Young functions satisfying*

(12.33) $\qquad N_2(u)M_2(vw) \leq a + N_1(k_1 uv)M_1(k_2 w) \quad (v \geq v_0),$

where a, k_1, k_2, v_0 are positive constants, and let $L_{M_i} = L_{M_i}(T)$ and $L_{N_i} = L_{N_i}(S)$. Then

$$[L_{M_1} \leftarrow L_{N_1}] \subseteq [L_{M_2} \rightarrow L_{N_2}].$$

□ Put $X := [L_{M_1} \leftarrow L_{N_1}]$ and $[L_{M_2} \rightarrow L_{N_2}]$. We make all calculations in the norm (4.3). By virtue of (12.33) we can find a constant c such that $||z||_{L_{M_1}} \leq 1/k_2$ implies

$$(12.34) \qquad \int_T M_2(v_0 z(t)) \, dt \leq c.$$

Fix some positive function u_0 in the space $[L_{M_1} \leftarrow L_{N_1}] \cap [L_{M_2} \rightarrow L_{N_2}]$, and denote by E_{u_0} the linear space of all $x \in L_1(T \times S)$ with finite norm

$$||x||_{E_{u_0}} = \inf \{\lambda : |x(t,s)| \leq \lambda u_0(t,s)\}.$$

Now, for $x \in E_{u_0}$ with $||x||_X \leq (k_1 k_2)^{-1}$ and all $\lambda > 1$ we have

$$1 \leq \frac{1}{\lambda} \int_T M_2 \left[\lambda \frac{x(t,s)}{||x(\cdot,s)||_{L_{M_2}}} \right] dt.$$

Since $||x(t,\cdot)||_{L_{N_1}} \leq (k_1 k_2)^{-1}$, from (12.34) we get

$$\int_T M_2(k_1 v_0 ||x(t,\cdot)||_{L_{N_1}}) \, dt \leq c.$$

Consequently, our hypothesis (12.33) implies that

$$\int_S N_2 \left[\frac{1}{\lambda}\|x(\cdot,s)\|_{L_{M_2}}\right] ds \le$$

$$\frac{1}{\lambda}\int_S\int_T N_2\left[\frac{1}{\lambda}\|x(\cdot,s)\|_{L_{M_2}}\right] M_2\left[\lambda\frac{x(t,s)}{\|x(\cdot,s)\|_{L_{M_2}}}\right] dt\, ds$$

$$\le \frac{1}{\lambda}\int_S\int_T N_2\left[\frac{1}{\lambda}\|x(\cdot,s)\|_{L_{M_2}}\right] \times$$

$$\times M_2\left\{\max\left[v_0, \frac{\lambda x(t,s)}{k_1\|x(\cdot,s)\|_{L_{M_2}}\|x(t,\cdot)\|_{L_{N_1}}}\right] k_1\|x(t,\cdot)\|_{L_{N_1}}\right\} dt\, ds$$

$$\le \frac{1}{\lambda}\int_S\int_T N_2\left[\frac{1}{\lambda}\|x(\cdot,s)\|_{L_{M_2}}\right] M_2(v_0 k_1\|x(t,\cdot)\|_{L_{N_1}})\, dt\, ds$$

$$+\frac{a}{\lambda}\mu(T)\nu(S) + \frac{1}{\lambda}\int_S\int_T N_1\left[\frac{x(t,s)}{\|x(t,\cdot)\|_{L_{N_1}}}\right] M_1(k_1 k_2\|x(t,\cdot)\|_{L_{N_1}})\, dt\, ds$$

$$\le \frac{c}{\lambda}\int_S N_2\left[\frac{1}{\lambda}\|x(\cdot,s)\|_{L_{M_2}}\right] ds + \frac{1}{\lambda}(a\mu(T)\nu(S)+1).$$

Putting now in the last inequality $\lambda := a\mu(T)\nu(S) + c + 1$, we obtain

$$\int_S N_2\left[\frac{\|x(\cdot,s)\|_{L_{M_2}}}{a\mu(T)\nu(S)+c+1}\right] ds \le 1,$$

hence

$$\|x\|_Y \le a\mu(T)\nu(S) + c + 1$$

by the definition of the norm (4.3). Since the last inequality holds for all functions $x \in E_{u_0}$ with $\|x\|_X \le (k_1 k_2)^{-1}$, we conclude that

$$\|x\|_Y \le k_1 k_2(a\mu(T)\nu(S) + c + 1)\|x\|_X$$

for any $x \in E_{u_0}$. Furthermore, for arbitrary $n \in \mathbb{N}$ we have then

$$\|\min\{|x|, n u_0\}\|_Y \le k_1 k_2(a\mu(T)\nu(S) + c + 1)\|x\|_X$$

for $x \in X$. Finally, since the space Y is perfect we see that

$$\|x\|_Y \le k_1 k_2 (a\mu(T)\nu(S) + c + 1)\|x\|_X$$

for all $x \in X$ as claimed. ∎

Lemma 12.2 generalizes a classical statement on the inclusion $[L_p \to L_q] \subseteq [L_p \leftarrow L_q]$ $(p \ge q)$ to the case of Orlicz spaces.
From Lemma 12.2 it follows, in particular, that $[L_M \to L_M]$ is always isomorphic to $[L_M \leftarrow L_M]$. As was shown in RUTITSKIJ [1962], $L_M(T \times S)$ is isomorphic to $[L_M \to L_M]$ if and only if the inequalities

$$M(u)M(v) \le a_1 + b_1 M(k_1 uv),$$

(12.35)

$$M(k_2 uv) \le a_2 + b_2 M(u)M(v)$$

hold for the some constants $a_i, b_i, k_i > 0$ $(i = 1, 2)$.
Let us consider now some acting conditions for the operators (12.4) and (12.5) in Orlicz spaces with mixed norm. As in the case of Lebesgue spaces, the study of partial integral operators in these spaces is more convenient than in ordinary Orlicz spaces.
In the following Theorems 12.10, 12.12, and 12.15 we suppose that M_i and N_i are given Young functions, $U_i = L_{M_i}(T)$, and $V_i = L_{N_i}(S)$ $(i = 1, 2)$. First of all, from Theorem 12.1 we get the following

Theorem 12.10. *Let*

$$V := \begin{cases} L_\infty(S) & \text{if} \quad N_2 \preceq N_1 \text{ but } N_2 \not\prec N_1, \\ \\ L_R(S) & \text{if} \quad N_2 \prec N_1, \end{cases}$$

where the Young function R in the last case is given by

$$R(u) = \sup\{N_2(uv) - N_1(v) : 0 < v < \infty\}.$$

Suppose that the operator (12.6) acts between the spaces U_1 and U_2 for each $s \in S$, and the function $s \mapsto \|L(s)\|_{\mathcal{L}(U_1, U_2)}$ belongs to V. Then the operator (12.4) acts between the spaces $X = [U_1 \to V_1]$ and $Y = [U_2 \to V_2]$.

Similary, let

$$U := \begin{cases} L_\infty(T) & \text{if } M_2 \preceq M_1 \text{ but } M_2 \not\prec M_1, \\ \\ L_R(S) & \text{if } M_2 \prec M_1, \end{cases}$$

where the Young function R in the last case is given by

$$R(u) = \sup\{M_2(uv) - M_1(v) : 0 < v < \infty\}.$$

Suppose that the operator (12.7) acts between the spaces V_1 and V_2 for each $t \in T$, and the function $t \mapsto \|M(t)\|_{\mathcal{L}(V_1,V_2)}$ belongs to U. Then the operator (12.5) acts between the spaces $X = [U_1 \leftarrow V_1]$ and $Y = [U_2 \leftarrow V_2]$.

We recall (see Subsection 12.1) that the conditions $\|L(\cdot)\|_{\mathcal{L}(U_1,U_2)} \in V$ and $\|M(\cdot)\|_{\mathcal{L}(V_1,V_2)} \in U$ are necessary for the operators (12.4) and (12.5), respectively, to act in the indicated spaces.

If the hypotheses of Theorem 12.10 are satisfied, then as direct consequences of (12.8) and (12.9) we get the estimate

$$(12.36) \qquad \|L\|_{\mathcal{L}(X,Y)} \leq \begin{cases} \|s \to \|L(s)\|_{\mathcal{L}(U_1,U_2)}\|_{L_\infty}, \\ \\ a\|s \to \|L(s)\|_{\mathcal{L}(U_1,U_2)}\|_{L_R} \end{cases}$$

in case $V = L_\infty(S)$ and $V = L_R(S)$, respectively, as well as the estimate

$$(12.37) \qquad \|M\|_{\mathcal{L}(X,Y)} \leq \begin{cases} \|t \to \|M(t)\|_{\mathcal{L}(V_1,V_2)}\|_{L_\infty}, \\ \\ b\|t \to \|M(t)\|_{\mathcal{L}(V_1,V_2)}\|_{L_R} \end{cases}$$

in case $U = L_\infty(T)$ and $U = L_R(T)$, respectively; here a and b are the imbedding constants of the inclusions $V \subseteq V_2/V_1$ and $U \subseteq U_2/U_2$, respectively. In particular, the estimate (12.36) holds if $N_1 = N_2$, and the estimate (12.37) holds if $M_1 = M_2$. Moreover, the following theorem is true.

Theorem 12.11. *Suppose that the operator (12.6) acts in $L_M(T)$ for each $s \in S$ and $\|L(\cdot)\|_{\mathcal{L}(L_M)} \in L_\infty$, while the operator (12.7) acts in*

$L_M(S)$ *for each* $t \in T$ *and* $\|M(\cdot)\|_{\mathcal{L}(L_M)} \in L_\infty$. *Then the operators* (12.4) *and* (12.5) *act in each of the spaces* $X = [L_M \to L_M]$ *and* $Y = [L_M \leftarrow L_M]$. *Moreover, the estimates*

(12.38)
$$\|L\|_{\mathcal{L}(X)} \leq \|s \mapsto \|L(s)\|_{\mathcal{L}(L_M)}\|_{L_\infty},$$
$$\|M\|_{L(Y)} \leq \|t \mapsto \|M(t)\|_{\mathcal{L}(L_M)}\|_{L_\infty}$$

are true. If, in addition, the Young function M *satisfies the inequalities* (12.35), *then the operators* (12.4) *and* (12.5) *act in the Orlicz space* $L_M(T \times S)$.

□ For the proof it suffices to remark that X is isomorphic to Y and, under the additional hypothesis (12.35), X is isomorphic to $L_M(T \times S)$. The assertion follows then from Theorem 12.10. ∎

We suppose now that $M_i = N_i$ $(i = 1, 2)$ and $M_2 \preceq M_1$. Let $V = L_\infty(S)$ if $M_2 \not\prec M_1$, and $V = L_R(S)$ otherwise, where R is defined as in Theorem 12.10. Similarly, we define U with S replaced by T. The following theorem is a natural generalization of Theorem 12.11.

Theorem 12.12. *Suppose that the operator* (12.6) *acts between* $L_{M_1}(T)$ *and* $L_{M_2}(T)$ *for each* $s \in S$ *and* $\|L(\cdot)\| \in V$, *while the operator* (12.7) *acts between* $L_{M_1}(S)$ *and* $L_{M_2}(S)$ *for each* $t \in T$ *and* $\|M(\cdot)\| \in U$. *Then the operators* (12.4) *and* (12.5) *act between the spaces* $X \in \{[L_{M_1} \to L_{M_1}], [L_{M_1} \leftarrow L_{M_1}]\}$ *and* $Y \in \{[L_{M_2} \to L_{M_2}], [L_{M_2} \leftarrow L_{M_2}]\}$. *If, in addition, the Young functions* M_1 *and* M_2 *satisfy the inequalities* (12.35), *then the operators* (12.4) *and* (12.5) *act between the Orlicz spaces* $L_{M_1}(T \times S)$ *and* $L_{M_2}(T \times S)$.

We remark that the norms of the operators (12.4) and (12.5) may be estimated by the formulas (12.36) and (12.37).

A crucial hypothesis in the Theorems 12.10 - 12.12 is the acting of the linear integral operators (12.6) and (12.7) between suitable Orlicz spaces. Some simple and effectively verifiable acting conditions for such operators between Orlicz spaces are well known (see e.g. KRASNOSELSKIJ-RUTITSKIJ [1958] or RAO-REN [1991]). More general acting conditions may be found in ZABREJKO [1968].

The Theorems 12.10 and 12.11 do not contain regularity conditions for the operators (12.4) and (12.5). Such conditions follow from Theorems 12.2 - 12.5 if in these theorems we choose $U_i = L_{M_i}(T)$ and $V_i = L_{N_i}(S)$ $(i = 1, 2)$. The simplest conditions we can obtain are those achived by the general Theorem 12.6. In fact, according to Lemma 12.2 the inclusions $[L_{M_i} \leftarrow L_{N_i}] \subseteq [L_{M_i} \rightarrow L_{N_i}]$ and $[L_{M_i} \rightarrow L_{N_i}] \subseteq [L_{M_i} \leftarrow L_{N_i}]$ are true if the conditions

$$(A_i) \qquad N_i(u)M_i(v\omega) \leq a_i + N_i(b_i uv)M_i(c_i \omega) \qquad (v \geq v_i)$$

and

$$(B_i) \qquad M_i(u)N_i(v\omega) \leq \overline{a_i} + M_i(\overline{b_i} uv)N_i(\overline{c_i}\omega) \qquad (v \geq \overline{v_i}),$$

are satisfied, where $a_i, b_i, c_i, v_i, \overline{a_i}, \overline{b_i}, \overline{c_i},$ and $\overline{v_i}$ are positive constants $(i = 1, 2)$. Applying Theorem 12.6 to our choice of U_i and V_i $(i = 1, 2)$ and using the explicit formulas for the multiplicator spaces L_{M_2}/L_{M_1} and L_{N_2}/L_{N_1}, we arrive at the following results which is parallel to Theorem 12.7:

Theorem 12.13. *Suppose that $N_2 \preceq N_1$ and*

$$(12.39) \qquad l \in [L_{M_2}, V, L_{M_1^*}; \theta],$$

where

$$V = \begin{cases} L_\infty(S) & \text{if } N_2 \not\preceq N_1, \\ L_R(S) & \text{otherwise.} \end{cases}$$

Here the Young function R in the last case is given by

$$R(u) = \sup \{N_2(uv) - N_1(v) : 0 < v < \infty\}.$$

Then the partial integral operator (12.4) acts between the spaces X and Y according to Table 3 below and is regular.

Similarly, suppose that $M_2 \preceq M_1$ and

$$(12.40) \qquad m \in [U, L_{N_2}, L_{N_1^*}; \theta],$$

where

$$U = \begin{cases} L_\infty(T) & \text{if } M_2 \not\prec M_1, \\ L_R(T) & \text{otherwise.} \end{cases}$$

Here the Young function R in the last case is given by

$$R(u) = \sup \{ M_2(uv) - M_1(v) : 0 < v < \infty \}.$$

Then the partial integral operator (12.5) acts between the spaces X and Y according to Table 4 below and is regular.

$X =$ $Y =$	$[L_{M_1} \to L_{N_1}]$ $[L_{M_2} \to L_{N_2}]$	$[L_{M_1} \to L_{N_1}]$ $[L_{M_2} \leftarrow L_{N_2}]$	$[L_{M_1} \leftarrow L_{N_1}]$ $[L_{M_2} \to L_{N_2}]$	$[L_{M_1} \leftarrow L_{N_1}]$ $[L_{M_2} \leftarrow L_{N_2}]$
(t,s,τ)	B_1	B_1, B_2		B_2
(t,τ,s)		B_2	A_1	A_1, B_2
(s,t,τ)	B_1, A_2	B_1	A_2	
(s,τ,t)	B_1, A_2	B_1	A_2	
(τ,s,t)	A_2		A_1, A_2	A_1
(τ,t,s)		B_2	A_1	A_1, B_2

Table 3

$X =$ $Y =$	$[L_{M_1} \to L_{N_1}]$ $[L_{M_2} \to L_{N_2}]$	$[L_{M_1} \to L_{N_1}]$ $[L_{M_2} \leftarrow L_{N_2}]$	$[L_{M_1} \leftarrow L_{N_1}]$ $[L_{M_2} \to L_{N_2}]$	$[L_{M_1} \leftarrow L_{N_1}]$ $[L_{M_2} \leftarrow L_{N_2}]$
(t,s,σ)		B_2	A_1	A_1, B_2
(t,σ,s)		B_2	A_1	A_1, B_2
(s,t,σ)	A_2		A_1, A_2	A_1
(s,σ,t)	B_1, A_2	B_1	A_2	
(σ,t,s)	B_1	B_1, B_2		B_2
(σ,s,t)	B_1, A_2	B_1	A_2	

Table 4

We point out that the inclusions (12.39) and (12.40) are usually checked by majorant techniques. For example, (12.39) is certainly satisfied if

$$|l(t,s,\tau)| \le \sum_{i=1}^{n} a_i(t) b_i(s) c_i(\tau),$$

where $a_i \in L_{M_2}(T)$, $b_i \in V$, and $c_i \in L_{M_1'}$. Finally, since $[L_M \to L_M]$ is isomorphic to $[L_M \leftarrow L_M]$ from Theorem 12.13 we get the following

Theorem 12.14. *Suppose that*

$$l \in [L_M, L_\infty, L_{M^*}; \theta']$$

for some $\theta' = (\theta_1', \theta_2', \theta_3')$ *and*

$$m \in [L_\infty, L_M, L_{M^*}; \theta'']$$

for some $\theta'' = (\theta_1'', \theta_2'', \theta_3'')$. *Then the partial integral operators* (12.4) *and* (12.5) *are regular in the spaces* $[L_M \to L_M]$ *and* $[L_M \leftarrow L_M]$. *If, in addition, the Young function M satisfies the inequality* (12.35), *then the operators* (12.4) *and* (12.5) *are regular in the Orlicz space* $L_M(T \times S)$.

12.4. Operator functions with values in $\mathfrak{L}_p(L_\infty)$ and $\mathfrak{L}_p(L_1)$

Let J be a bounded or unbounded interval in R. We suppose now that the functions c, l, m, and n in (11.2) - (11.5) depend on a parameter $\varphi \in J$. This means that we replace the operator (11.1) by the operator function

$$(12.41) \qquad P(\varphi) = C(\varphi) + L(\varphi) + M(\varphi) + N(\varphi),$$

with

$$(12.42) \qquad C(\varphi)x(t, s) = c(\varphi, t, s)x(t, s),$$

$$(12.43) \qquad L(\varphi)x(t, s) = \int_a^b l(\varphi, t, s, \tau)x(\tau, s) \, d\tau,$$

$$(12.44) \qquad M(\varphi)x(t, s) = \int_c^d m(\varphi, t, s, \sigma)x(t, \sigma) \, d\sigma,$$

and

$$(12.45) \qquad N(\varphi)x(t,s) = \int_a^b \int_c^d n(\varphi,t,s,\tau,\sigma)x(\tau,\sigma) \, d\sigma \, d\tau.$$

We recall that an operator function $\varphi \mapsto P(\varphi)$ with values in the space $\mathfrak{L}_p(X)$ is called strongly continuous at φ_0 if

$$\lim_{\varphi \to \varphi_0} \|P(\varphi)x - P(\varphi_0)x\|_X = 0 \quad (x \in X),$$

and norm-continuous at φ_0 if

$$\lim_{\varphi \to \varphi_0} \|P(\varphi) - P(\varphi_0)\|_{\mathfrak{L}(X)} = 0.$$

The operator functions (12.42) - (12.45) defines a family of operators in the space $\mathfrak{L}_p(X)$ operators, depending on the parameter φ. Each such operator is uniquely defined by the corresponding function $c(\varphi,\cdot,\cdot)$, $l(\varphi,\cdot,\cdot,\cdot)$, $m(\varphi,\cdot,\cdot,\cdot)$, and $n(\varphi,\cdot,\cdot,\cdot,\cdot)$. Therefore it is natural to expect that the norm continuity or strong continuity of the operator functions (12.42) - (12.45) is equivalent (in some sense to be made precise) to the continuous dependence of the functions c, l, m, and n on φ. However, as was shown in KRASNOSEL'SKIJ-KREJN [1955], this is not true even for simple operators arising in nonlinear mechanics (see also ZABREJKO-KOLESOV-KRASNOSEL'SKIJ [1969]). The Examples 13.2, 13.3, and 13.4 of the following section show, in particular, that the norm continuity or strong continuity of the operator function (12.41) is not guaranteed by the continuous dependence of the functions c, l, m, and n on φ.

In view of such examples, it is of interest to find conditions of the functions c, l, m, and n, possibly both necessary and sufficient, for the strong continuity or norm continuity of the operator function (12.41). This is still an open problem. The theorems we are going to present below are of somewhat different nature, although they contain conditions for the norm continuity or strong continuity of the operator function (12.41).

If one can calculate (or estimate) the norm of the operator $P(\varphi)$ in some Banach space X for each $\varphi \in J$, then finding conditions for the norm-continuity of the operator function (12.41) in $\mathfrak{L}(X)$ reduces to calculating the norm of $P(\varphi) - P(\varphi_0)$. Conditions for the strong continuity of the operator function (12.41) may be obtained by means of the Banach-Steinhaus theorem. To this end, one has to find conditions for the convergence $P(\varphi)x \to P(\varphi_0)x$, as $\varphi \to \varphi_0$, on some set of functions x whose linear hull is dense in X. As a matter of fact, different choices of functions x lead then to different conditions of strong continuity.

Usually these schemes are easily realized for particular functional spaces. The cases $X = C([a,b] \times [c,d])$ and $X = L_p([a,b] \times [c,d])$ $(1 \le p \le \infty)$ are considered below. Analogous results hold for ideal spaces with mixed norm and, in particular, for the Orlicz spaces with mixed norm considered in the last subsection.

Let $D = [a,b] \times [c,d]$ and let

(12.46)
$$\gamma(\varphi,t,s) = |c(\varphi,t,s)| + \int_a^b |l(\varphi,t,s,\tau)| \, d\tau$$
$$+ \int_c^d |m(\varphi,t,s,\sigma)| \, d\sigma + \int_a^b \int_c^d |n(\varphi,t,s,\tau,\sigma)| \, d\sigma \, d\tau.$$

Criteria for the strong continuity of the operator function (12.41) in $\mathfrak{L}(L_\infty)$ and in $\mathfrak{L}(L_1)$ are contained in the following two theorems.

Theorem 12.15. *The operator function (12.41) is strongly continuous in the space $\mathfrak{L}(L_\infty)$ if and only if the function (12.46) is essentially bounded on $J \times D$ for each bounded interval $J \subset R$, the vector function $\varphi \mapsto c(\varphi,\cdot,\cdot)$ is continuous as a function with values in $L_\infty(D)$, and the vector functions*

$$\varphi \mapsto \int_a^u l(\varphi,\cdot,\cdot,\tau) \, d\tau, \qquad \varphi \mapsto \int_c^v m(\varphi,\cdot,\cdot,\sigma) \, d\sigma,$$

$$\varphi \mapsto \int_a^u \int_c^v n(\varphi,\cdot,\cdot,\tau,\sigma) \, d\sigma \, d\tau$$

are continuous for each $u \in [a,b]$, $v \in [c,d]$, and $(u,v) \in D$, respectively, as functions with values in $L_\infty(D)$.

Theorem 12.16. *The operator function* (12.41) *is strongly conti-nuous in the space* $\mathfrak{L}(L_1)$ *if and only if the following conditions are satisfied:*

(a) *The function*

$$\varphi \mapsto |c(\varphi,t,s)| + \int_a^b |l(\varphi,\tau,s,t)|\, d\tau$$

$$+ \int_c^d |m(\varphi,t,\sigma,s)|\, d\sigma + \int_a^b \int_c^d |n(\varphi,\tau,\sigma,t,s)|\, d\tau\, d\sigma$$

is bounded, for almost all (t,s)*, on each bounded interval* $J \subset \mathbb{R}$*.*

(b) *The vector function* $\varphi \mapsto c(\varphi,\cdot,\cdot)$ *is continuous as a function with values in* $L_\infty(D)$*.*

(c) *The the vector functions*

$$\varphi \mapsto \int_a^u l(\varphi,\cdot,\cdot,\tau)\, d\tau, \quad \varphi \mapsto \int_c^v m(\varphi,\cdot,\cdot,\sigma)\, d\sigma,$$

$$\varphi \mapsto \int_a^u \int_c^v n(\varphi,\cdot,\cdot,\tau,\sigma)\, d\sigma\, d\tau$$

are continuous for each $u \in [a,b]$*,* $v \in [c,d]$*, and* $(u,v) \in D$*, respecti-vely, as functions with values in* $L_1(D)$*.*

□ For the proof of Theorem 12.15 and Theorem 12.16 it suffices to remark that the strong continuity of the operator function (12.41) in these spaces is equivalent to the strong continuity of the operator functions (12.42) - (12.45). Now, as dense set in X we choose the linear hull of functions $\kappa(u,t)\kappa(v,s)$, where $a \le u \le b$, $c \le v \le d$,

$$\kappa(u,t) = \begin{cases} 1 & \text{if} \quad a \le t \le u, \\ 0 & \text{if} \quad u < t \le b, \end{cases} \qquad \kappa(v,s) = \begin{cases} 1 & \text{if} \quad c \le s \le v, \\ 0 & \text{if} \quad v < s \le d. \end{cases}$$

Applying now Theorem 12.15 we get the required conclusions. ∎

Let $U = L_1([a,b])$, $V = L_1([c,d])$, and $L_p = L_p(D)$ $(1 \le p \le \infty)$. As above, we denote by $[L_p \leftarrow U]$, $[L_p \leftarrow V]$, and $[L_p \leftarrow L_1]$ the spaces with mixed norms

$$\|x\|_{[L_p \leftarrow U]} = \|(t,s) \mapsto \|x(t,s,\cdot)\|_U\|_{L_p},$$

$$\|y\|_{[L_p \leftarrow V]} = \|(t, s) \mapsto \|y(t, s, \cdot)\|_V\|_{L_p},$$

and

$$\|z\|_{[L_p \leftarrow L_1]} = \|(t, s) \mapsto \|z(t, s, \cdot, \cdot)\|_{L_1}\|_{L_p}.$$

The following criterion for the norm continuity of the operator function (12.41) in $\mathcal{L}(L_\infty)$ and $\mathcal{L}(L_1)$ follows from Theorem 12.5.

Theorem 12.17. *The operator function* (12.41) *is norm continuous in* $\mathcal{L}(L_\infty)$ *(in* $\mathcal{L}(L_1)$, *respectively), if and only if the functions* $\varphi \mapsto c(\varphi, \cdot, \cdot)$, $\varphi \mapsto l(\varphi, \cdot, \cdot, \cdot)$, $\varphi \mapsto m(\varphi, \cdot, \cdot, \cdot)$, *and* $\varphi \mapsto n(\varphi, \cdot, \cdot, \cdot, \cdot)$ *(* $\varphi \mapsto c(\varphi, \cdot, \cdot)$, $\varphi \mapsto l(\varphi, \cdot, \cdot, \cdot)$, $\varphi \mapsto m(\varphi, \cdot, \cdot, \cdot)$, *and* $\varphi \mapsto n(\varphi, \cdot, \cdot, \cdot, \cdot)$, *respectively), are continuous as vector functions with the values in* L_∞, $[L_\infty \leftarrow U]$, $[L_\infty \leftarrow V]$, *and* $[L_\infty \leftarrow L_1]$, *respectively.*

12.5. Operator functions with values in $\mathcal{L}_p(L_p)$

Unfortunately, necessary and sufficient conditions for the strong continuity or norm continuity of the operator function (12.41) with values in $\mathcal{L}_p(L_p)$ for $1 < p < \infty$ are unknown. However, if we replace $\mathcal{L}_p(L_p)$ by the space $\mathcal{L}_p^r(L_p)$ of regular operators (11.1), one can give some conditions.

Recall that the sets $\mathfrak{R}_l(L_p)$, $\mathfrak{R}_m(L_p)$, and $\mathfrak{R}_n(L_p)$ are defined by the norms (11.21), (11.22), and (11.23), respectively, for $X = Y = L_p$. All these sets are ideal Banach spaces. The regularity of the operator (11.1) in L_p is equivalent to the four inclusions $c \in L_\infty$, $l \in \mathfrak{R}_l(L_p)$, $m \in \mathfrak{R}_m(L_p)$ and $n \in \mathfrak{R}_n(L_p)$. Checking the last three inclusions is difficult because of the complexity of spaces $\mathfrak{R}_l(L_p)$, $\mathfrak{R}_m(L_p)$, and $\mathfrak{R}_n(L_p)$.

We will say that the operator function (12.41) with values in $\mathcal{L}_p^r(L_p)$ is *absolutely strongly continuous* at φ_0 if $|P(\varphi) - P(\varphi_0)|x \to 0$ for $\varphi \to \varphi_0$ and for any $x \in L_p$, and *absolutely norm continuous* at φ_0 if $\||P(\varphi) - P(\varphi_0)|\| \to 0$ for $\varphi \to \varphi_0$).

Of course, the strong continuity (norm continuity) of the operator function (12.41) in $\mathcal{L}_p(L_p)$ follows from its absolute strong continuity (absolute norm continuity). The converse is not true for $1 < p < \infty$.

Since $|P(\varphi) - P(\varphi_0)| =]P(\varphi) - P(\varphi_0)[$ for the operator function (12.41) with values in $\mathfrak{L}_p^r(L_p)$, its absolute strong continuity (absolute norm continuity) is equivalent to the absolute strong continuity (absolute norm continuity) of the operator functions (12.43) - (12.46).

The following criterion for the absolute strong continuity of the operator function (12.41) with values in $\mathfrak{L}(L_p)$ follows from the last statement and the of Banach-Steinhaus theorem.

Theorem 12.18. *Let* $1 < p < \infty$. *The operator function* (12.41) *with values in* $\mathfrak{L}_p(L_p)$ *is absolutely strongly continuous if and only if the following conditions are satisfied:*

(a) *the function* $\varphi \mapsto c(\varphi, \cdot, \cdot)$ *takes values in* $L_\infty(D)$, *is bounded on each bounded subset of* J *as function with values in* $L_\infty(D)$, *and is continuous as function with values in* $L_1(D)$;

(b) *the functions*

$$\varphi \mapsto l(\varphi, \cdot, \cdot, \cdot), \quad \varphi \mapsto m(\varphi, \cdot, \cdot, \cdot), \varphi \mapsto n(\varphi, \cdot, \cdot, \cdot, \cdot)$$

take values in $\mathfrak{R}_l(L_p)$, $\mathfrak{R}_m(L_p)$, *and* $\mathfrak{R}_n(L_p)$, *respectively, and are bounded on each bounded subset of* J;

(c) *for all* $y \in [a, b]$ *and* $z \in [c, d]$, *the functions*

$$\varphi \mapsto l(\varphi, \cdot, \cdot, \cdot)\chi_{[a,y]}(\tau), \quad \varphi \mapsto m(\varphi, \cdot, \cdot, \cdot)\chi_{[c,z]}(\sigma),$$

$$\varphi \mapsto n(\varphi, \cdot, \cdot, \cdot, \cdot)\chi_{[a,y] \times [c,z]}(\tau, \sigma)$$

are continuous from J *into* $[L_p \leftarrow X]$, $[L_p \leftarrow Y]$, *and* $[L_p \leftarrow L_1]$, *respectively.*

We remark that, under the conditions of Theorem 12.18, the operator function (12.41) takes its values in $\mathfrak{L}_p(L_p)$ and is strongly continuous. The same is true if we replace the condition (c) in Theorem 12.18 by the condition

(c) *for all* $y \in [a, b]$ *and* $z \in [c, d]$, *the functions*

$$\varphi \mapsto \int_a^y l(\varphi, \cdot, \cdot, \tau) \, d\tau, \quad \varphi \mapsto \int_c^z m(\varphi, \cdot, \cdot, \sigma) \, d\sigma,$$

$$\varphi \mapsto \int_a^y \int_c^z n(\varphi, \cdot, \cdot, \tau, \sigma) \, d\tau \, d\sigma$$

are continuous from J into L_p.

Simple sufficient conditions for the norm continuity of the operator function (12.41) with values in $\mathfrak{L}_p^r(L_p)$ are contained in the following theorem.

Theorem 12.19. *Let $1 < p < \infty$ and $p^{-1} + q^{-1} = 1$. The operator function (12.41) is norm continuous in $\mathfrak{L}_p^r(L_p)$ if the following conditions are satisfied:*

(a) *the function $\varphi \mapsto c(\varphi, \cdot, \cdot)$ is continuous as function with values in $L_\infty(D)$;*

(b) *the function $\varphi \mapsto l(\varphi, \cdot, \cdot, \cdot)$ is continuous as function with values in one of the spaces with mixed norm $[L_\infty \leftarrow [L_p \leftarrow L_q]]$ or $[L_\infty \leftarrow [L_q \leftarrow L_p]]$, where the norm in $L_p([a, b])$, $L_q([a, b])$, and $L_\infty([c, d])$ is calculated in the variables t, τ, and s, respectively;*

(c) *the function $\varphi \mapsto m(\varphi, \cdot, \cdot, \cdot)$ is continuous as function with values in one of the spaces with mixed norm $[L_\infty \leftarrow [L_p \leftarrow L_q]]$ or $[L_\infty \leftarrow [L_q \leftarrow L_p]]$, where the norm in $L_p([c, d])$, $L_q([c, d])$, and $L_\infty([a, b])$ is calculated in the variables s, σ, and t, respectively;*

(d) *the function $\varphi \mapsto n(\varphi, \cdot, \cdot, \cdot, \cdot)$ is continuous as function with values in $[L_p \leftarrow L_q]$ or $[L_q \leftarrow L_p]$, where the norm in $L_p(D)$ and $L_q(D))$ is calculated in the variables (t, s) and (τ, σ), respectively.*

□ The proof follows simply by applying the Hölder inequality and the generalized Minkowski inequality. ∎

§ 13. Partial integral operators in the space C

In this section we give acting and boundedness conditions for partial integral operators on the space $C = C([a, b] \times [c, d])$ which are both necessary and sufficient. Moreover, we discuss some algebras and ideals of such operators, considered as subspaces of the algebra $\mathcal{L}(C)$ of all bounded linear operators on C. We also study conditions for the strong or norm continuity of operator functions whose values are partial integral operators acting in C.

These problems are related to the solvability of the equation

$$(13.1) \qquad x(t, s) = Px(t, s) + f(t, s),$$

where $f : [a, b] \times [c, d] \to \mathbb{R}$ is a given function and the operator P is a sum

$$(13.2) \qquad P = C + L + M + N$$

of the four operators

$$(13.3) \qquad Cx(t, s) = c(t, s)x(t, s),$$

$$(13.4) \qquad Lx(t, s) = \int_a^b l(t, s, \tau)x(\tau, s) \, d\tau,$$

$$(13.5) \qquad Mx(t, s) = \int_c^d m(t, s, \sigma)x(t, \sigma) \, d\sigma,$$

and

$$(13.6) \qquad Nx(t, s) = \int_a^b \int_c^d n(t, s, \tau, \sigma)x(\tau, \sigma) \, d\tau \, d\sigma.$$

The operators (13.4) and (13.5) are partial integral operators, while (13.6) is an ordinary integral operator acting on functions over $[a, b] \times [c, d]$. As we have seen in § 11, the properties of the partial integral operator (13.2) heavily depend on the spaces in which it is studied, and may be very different from those of an ordinary integral operator.

Simple acting and boundedness conditions for the operator (13.2) have been obtained in KALITVIN-JANKELEVICH [1993] for scalar functions and in KALITVIN-JANKELEVICH [1994] for vector functions. Our purpose here is two-fold: First, we give acting and boundedness conditions for the operator (13.2) in the space $C(D)$ ($D = [a, b] \times [c, d]$) which are both necessary and sufficient, as well as some formulas for calculating or estimating its norm. Second, we discuss some algebras and ideals of operators of the form (13.2), considered as subspaces of the algebra $\mathfrak{L}(C)$ of bounded linear operators on $C = C(D)$. Here we follow the recent article APPELL-FROLOVA-KALITVIN-ZABREJKO [1996]. The main results of Subsection 13.4 are taken from FROLOVA-KALITVIN-ZABREJKO [1997] and KALITVIN-FROLOVA [1995]. Other criteria for the boundedness of the operator (13.2) may be found in FROLOVA [1995, 1996], for its compactness and weak compactness in KALITVIN [1996b].

13.1. Weakly continuous functions

A crucial role in the investigation of partial integral operators in spaces of continuous functions is played by the classical Radon theorem (GLIVENKO [1936], RADON [1919]) which states that any bounded linear operator A on the space $C(D)$ admits a representation as a Stieltjes integral

$$Af(t, s) = \int_D f(\tau, \sigma) \, dg(t, s, \tau, \sigma)$$

involving a function $g : D \times D \to \mathbb{R}$ such that $g(t, s, \cdot, \cdot)$ is of bounded variation and $g(\cdot, \cdot, \tau, \sigma)$ is weakly continuous.

Recall that the function $g(\cdot, \cdot, \tau, \sigma)$ is called weakly continuous on $D = [a, b] \times [c, d]$ if, for any sequence $(t_n, s_n) \in D$ with $(t_n, s_n) \to (t, s)$ and any continuous function $f : D \to \mathbb{R}$, the relation

$$\lim_{n \to \infty} \int_D f(\tau, \sigma) \, d_{\tau, \sigma} g(t_n, s_n, \tau, \sigma) = \int_D f(\tau, \sigma) \, d_{\tau, \sigma} g(t, s, \tau, \sigma)$$

holds. By the classical Riesz theorem, the weak continuity of $g(\cdot,\cdot,\tau,\sigma)$ is equivalent to the pointwise convergence of the sequence of bounded linear functionals

$$\gamma_n(f) = \int_D f(\tau,\sigma)\,d_{\tau,\sigma}g_n(\tau,\sigma) \quad (g_n(\tau,\sigma) := g(t_n,s_n,\tau,\sigma))$$

to the bounded linear functional

$$\gamma(f) = \int_D f(\tau,\sigma)\,d_{\tau,\sigma}g(\tau,\sigma) \quad (g(\tau,\sigma) := g(t,s,\tau,\sigma)).$$

For the pointwise convergence $\gamma_n \to \gamma$ in turn it is necessary and sufficient that the sequence $(\|\gamma_n\|)_n$ be bounded and $\gamma_n(f) \to \gamma(f)$ for $f \in F$, with F being an arbitrary subset of $C(D)$ whose linear hull is dense in $C(D)$.

Let $\hat{g}_n = \hat{g}_n(\tau,\sigma)$ and $\hat{g} = \hat{g}(\tau,\sigma)$ be functions of bounded variation corresponding to the functionals γ_n and γ, respectively, i.e. $\mathrm{Var}_D\,\hat{g}_n = \|\gamma_n\|$ and $\mathrm{Var}_D\,\hat{g} = \|\gamma\|$. Then the boundedness of the sequence $(\|\gamma_n\|)_n$ is of course equivalent to the fact that all functions $\hat{g}_n$ $(n = 1,2,3,...)$ have uniformly bounded variation. A suitable choice for the set $F \subset C(D)$ is as follows: Since the set of all linear combinations of $x(t) \equiv 1$ and all functions

$$x_\xi(t) = \begin{cases} \xi - t & \text{if} \quad a \le t \le \xi \le b, \\ 0 & \text{if} \quad a \le \xi < t \le b, \end{cases}$$

and

$$x_\eta(s) = \begin{cases} \eta - s & \text{if} \quad c \le s \le \eta \le d, \\ 0 & \text{if} \quad c \le \eta < s \le d, \end{cases}$$

is dense in $C([a,b])$ and $C([c,d])$ respectivelly, we may put

$$F := \{1\} \cup \{x_\xi : a \le \xi \le b\} \cup \{x_\eta : c \le \eta \le d\}$$

$$\cup \{x_\xi \otimes x_\eta : a \le \xi \le b, c \le \eta \le d\}.$$

It is then not hard to verify the relations

$$(13.7) \qquad \Gamma := \int_D 1\,dg(t,s) = g(b,d) - g(a,d) - g(b,c) + g(a,c),$$

$$\Gamma_\xi := \int_D x_\xi(t)\, dg(t,s) =$$

(13.8)

$$\int_a^\xi [g(t,d) - g(t,c)]\, dt + [g(a,c) - g(a,d)](\xi - a),$$

$$\Gamma_\eta := \int_D x_\eta(s)\, dg(t,s) =$$

(13.9)

$$\int_c^\eta [g(b,s) - g(a,s)]\, ds + [g(a,c) - g(b,c)](\eta - c),$$

and

$$\Gamma_{\xi,\eta} := \int_D x_\xi(t) x_\eta(s)\, dg(t,s) =$$

(13.10)

$$\int_a^\xi \int_c^\eta g(t,s)\, dt\, ds - (\eta - c)\int_a^\xi g(t,c)\, dt$$

$$-(\xi - a)\int_c^\eta g(a,s)\, ds + g(a,c)(\xi - a)(\eta - c).$$

Consequently, the convergence $\gamma_n(1) \to \gamma(1)$, $\gamma_n(x_\xi) \to \gamma(x_\xi)$, $\gamma_n(x_\eta) \to \gamma(x_\eta)$, and $\gamma_n(x_\xi \otimes x_\eta) \to \gamma(x_\xi \otimes x_\eta)$ means precisely that

(13.11) $$g_n(b,d) - g_n(a,d) - g_n(b,c) + g_n(a,c) \to \Gamma,$$

(13.12) $$\int_a^\xi [g_n(t,d) - g_n(t,c)]\, dt + [g_n(a,c) - g_n(a,d)](\xi - a) \to \Gamma_\xi,$$

(13.13)

$$\int_c^\eta [g_n(b,s) - g_n(a,s)]\, ds$$

$$+[g_n(a,c) - g_n(b,c)](\eta - c) \to \Gamma_\eta,$$

and

$$(13.14) \quad \int_a^\xi \int_c^\eta g_n(t,s)\, dt\, ds - (\eta - c) \int_a^\xi g_n(t,c)\, dt -$$

$$(\xi - a) \int_c^\eta g_n(a,s)\, ds + g_n(a,c)(\xi - a)(\eta - c) \to \Gamma_{\xi,\eta}$$

as $n \to \infty$, for all $\xi \in [a,b]$ and $\eta \in [c,d]$. In this way, we have proved the following

Lemma 13.1. *The function $g(t,s,\cdot,\cdot)$ is of bounded variation for each $(t,s) \in D$, and the function $g(\cdot,\cdot,\tau,\sigma)$ is weakly continuous for each $(\tau,\sigma) \in D$ if and only if*

$$\sup \{ \mathrm{Var}_D\, g(t,s,\cdot,\cdot) : (t,s) \in D \} < \infty$$

and the relations (13.11) - (13.14) hold.

13.2. Acting and boundedness conditions

Now we are going to derive an acting condition (both necessary and sufficient) for the operator (13.2) in the space $C(D)$ which also ensures its boundedness. To this end, we put

$$g(t,s,\tau,\sigma) := c(t,s)\kappa(t,s,\tau,\sigma)$$

$$(13.15) \quad\quad + \int_a^\tau l(t,s,\tilde\tau)\, d\tilde\tau\, \kappa(s,\sigma)$$

$$+ \int_c^\sigma m(t,s,\tilde\sigma)\, d\tilde\sigma\, \kappa(t,\tau) + \int_a^\tau \int_c^\sigma n(t,s,\tilde\tau,\tilde\sigma)\, d\tilde\sigma\, d\tilde\tau,$$

where

$$\kappa(t,\tau) := \begin{cases} 1 & \text{if } \tau \geq t > a \quad \text{or} \quad \tau > t = a, \\ 0 & \text{if } \tau < t \leq b \quad \text{or} \quad \tau = t = a, \end{cases}$$

$$\kappa(s,\sigma) := \begin{cases} 1 & \text{if } \sigma \geq s > c \quad \text{or} \quad \sigma > s = c, \\ 0 & \text{if } \sigma < s \leq d \quad \text{or} \quad \sigma = s = c, \end{cases}$$

and

$$\kappa(t,s,\tau,\sigma) := \begin{cases} 1 & \text{if } \tau \geq t > a \text{ and } \sigma \geq s > c, \\ & \text{or } \tau > t = a \text{ and } \sigma > s = c, \\ 0 & \text{if } \tau < t \leq b, \ \sigma < s \leq d, \ \tau = t = a, \\ & \text{or } \sigma = s = c. \end{cases}$$

Obviously, $g(t,s,a,d) = g(t,s,b,c) = g(t,s,a,c) \equiv 0$ and

$$g(t,s,b,d) = c(t,s) + \int_a^b l(t,s,\tau)\, d\tau +$$

$$\int_c^d m(t,s,\sigma)\, d\sigma + \int_a^b \int_c^d n(t,s,\tau,\sigma)\, d\sigma\, d\tau.$$

Putting now (13.15) into (13.7) - (13.10) we obtain

(13.16)
$$\Gamma(t,s) = c(t,s) + \int_a^b l(t,s,\tau)\, d\tau$$
$$+ \int_c^d m(t,s,\sigma)\, d\sigma + \int_a^b \int_c^d n(t,s,\tau,\sigma)\, d\sigma\, d\tau,$$

(13.17)
$$\Gamma_\xi(t,s) = \int_a^\xi \{[c(t,s) + \int_c^d m(t,s,\sigma)\, d\sigma]\kappa(t,\tau)$$
$$+(\xi - \tau)[l(t,s,\tau) + \int_c^d n(t,s,\tau,\sigma)\, d\sigma]\}\, d\tau,$$

(13.18)
$$\Gamma_\eta(t,s) = \int_c^\eta \{[c(t,s) + \int_a^b l(t,s,\tau)\, d\tau]\kappa(s,\sigma)$$
$$+(\eta - \sigma)[m(t,s,\sigma) + \int_a^b n(t,s,\tau,\sigma)\, d\tau]\}\, d\sigma,$$

and

(13.19)
$$\Gamma_{\xi,\eta}(t,s) = \int_a^\xi \int_c^\eta [c(t,s)\kappa(t,s,\tau,\sigma)$$
$$+(\xi - \tau)l(t,s,\tau)\kappa(s,\sigma) + (\eta - \sigma)m(t,s,\sigma)\kappa(t,\tau)$$
$$+(\xi - \tau)(\eta - \sigma)n(t,s,\tau,\sigma)]\, d\sigma\, d\tau.$$

Finally, we put

$$
\overline{\Gamma}(t,s) = |c(t,s)| + \int_a^b |l(t,s,\tau)|\, d\tau
$$

(13.20)

$$
+ \int_c^d |m(t,s,\sigma)|\, d\sigma + \int_a^b \int_c^d |n(t,s,\tau,\sigma)|\, d\sigma\, d\tau.
$$

Theorem 13.1. *The linear operator* (13.2) *acts in the space* $C(D)$ *if and only if the functions* (13.16) - (13.19) *are continuous on* D *for all fixed* (ξ,η), *and the function* (13.20) *is bounded on* D. *In this case the operator* (13.2) *is automatically bounded on* $C(D)$, *and its norm is given by*

(13.21) $\|P\|_{\mathfrak{L}(C)} = \sup \{\overline{\Gamma}(t,s) : (t,s) \in D\}.$

□ Suppose first that the operator (13.2) acts in $C(D)$. It is known that this operator is then automatically bounded (KALITVIN-JANKE-LEVICH [1994]) and, by Radon's theorem, admits a representation

(13.22) $Px(t,s) = \int_a^b \int_c^d x(\tau,\sigma)\, dg(t,s,\tau,\sigma),$

where the function g is given by (13.15) and has the property that $g(t,s,\cdot,\cdot)$ is of bounded variation and $g(\cdot,\cdot,\tau,\sigma)$ is weakly continuous. The weak continuity of $g(\cdot,\cdot,\tau,\sigma)$ is equivalent, as we have seen, to the continuity of the functions (13.16) - (13.19). On the other hand, the boundedness of the function (13.20) follows from the fact that $g(t,s,\cdot,\cdot)$ is of bounded variation; we claim that

(13.23) $\mathrm{Var}_D\, g(t,s,\cdot,\cdot) = \overline{\Gamma}(t,s).$

To prove (13.23), let (t,s) be a fixed interior point of D, choose points $u \in (a,t)$, $v \in (t,b)$, $p \in (c,s)$, and $q \in (s,d)$, and split the rectangle $D = [a,b] \times [c,d]$ into the 9 parts $D_1 := [a,u] \times [c,p]$, $D_2 := [a,u] \times [p,q]$, $D_3 := [a,u] \times [q,d]$, $D_4 := [u,v] \times [q,d]$, $D_5 := [u,v] \times [p,q]$, $D_6 := [u,v] \times [c,p]$, $D_7 := [v,b] \times [q,d]$, $D_8 := [v,b] \times [p,q]$, and

$D_9 := [v, b] \times [c, p]$. Then the total variation of $g(t, s, \cdot, \cdot)$ on D is equal to the sum of the variations of $g(t, s, \cdot, \cdot)$ on D_j $(j = 1, \ldots, 9)$. Now, calculating the variation of $g(t, s, \cdot, \cdot)$ on D_j $(j = 1, \ldots, 9)$ for $u, v \to t$ and $p, q \to s$ we get for the total variation the formula

$$\operatorname{Var}_D g(t, s, \cdot, \cdot) = \lim_{\substack{u, v \to t \\ p, q \to s}} \sum_{j=1}^{9} \operatorname{Var}_{D_j} g(t, s, \cdot, \cdot)$$

$$= \int_a^t \int_c^s |n(t, s, \tau, \sigma)| \, d\sigma \, d\tau + \int_a^t |l(t, s, \tau)| \, d\tau$$

$$+ \int_a^t \int_s^d |n(t, s, \tau, \sigma)| \, d\sigma \, d\tau + \int_s^d |m(t, s, \sigma)| \, d\sigma$$

(13.24)

$$+ |c(t, s)| + \int_c^s |m(t, s, \sigma)| \, d\sigma$$

$$+ \int_t^b \int_s^d |n(t, s, \tau, \sigma)| \, d\sigma \, d\tau + \int_t^b |l(t, s, \tau)| \, d\tau$$

$$+ \int_t^b \int_c^s |n(t, s, \tau, \sigma)| \, d\sigma \, d\tau,$$

which proves (13.23). To see in turn that (13.24) is true consider, for example, the "center rectangle" $D_5 = [u, v] \times [p, q]$. On this rectangle we have

$$g(t, s, \tau, \sigma) := c(t, s)\kappa(t, s, \tau, \sigma) + \int_u^\tau l(t, s, \tilde{\tau}) \, d\tilde{\tau} \, \kappa(s, \sigma)$$

$$+ \int_p^\sigma m(t, s, \tilde{\sigma}) \, d\tilde{\sigma} \, \kappa(t, \tau) + \int_u^\tau \int_p^\sigma n(t, s, \tilde{\tau}, \tilde{\sigma}) \, d\tilde{\sigma} \, d\tilde{\tau}.$$

Given $\varepsilon > 0$, choose a partition Π of D_5 such that the corresponding variation of $g(t, s, \cdot, \cdot)$ with respect to Π is less than ε. By the absolute continuity of the integral and the definition of the functions κ, we get $|\operatorname{Var}_\Pi g(t, s, \cdot, \cdot) - |c(t, s)|| < \varepsilon$, and hence $|\operatorname{Var}_{D_5} g(t, s, \cdot, \cdot) - |c(t, s)|| < 2\varepsilon$ for $u, v \to t$ and $p, q \to s$. The other eight equalities in (13.24) are proved similarly.

Conversely, suppose now that the functions (13.16) - (13.19) are continuous on D for all fixed (ξ, η), and the function (13.20) is bounded on D. This immediately implies that $g(t, s, \cdot, \cdot)$ is of bounded variation and $g(\cdot, \cdot, \tau, \sigma)$ is weakly continuous. By Radon's theorem, the operator (13.2) is bounded on $C(D)$.

Finally, let us prove the formula (13.21) for the norm of the operator (13.2). We have

$$(13.25) \qquad \|P\|_{\mathfrak{L}(C)} = \sup \{\|Px(t, s)\|_{C^*} : (t, s) \in D\},$$

where $Px(t, s)$ is considered, for fixed $(t, s) \in D$, as a bounded linear functional on $C(D)$. Therefore it suffices to prove that $\operatorname{Var}_D g(t, s, \cdot, \cdot) = \|Px(t, s)\|_{C^*}$ for fixed $(t, s) \in D$. Given an arbitrary functional $f \in C^*$, let

$$g(\tau, \sigma) := \begin{cases} f(x_{\tau, \sigma}) & \text{if} \quad a < \tau \leq b \text{ and } c < \sigma \leq d, \\ 0 & \text{if} \quad \tau = a \text{ or } \sigma = c, \end{cases}$$

where

$$x_{\tau, \sigma}(u, v) := \begin{cases} 1 & \text{for} \quad a \leq u \leq \tau \text{ and } c \leq v \leq \sigma, \\ 0 & \text{for} \quad \tau < u \leq b \text{ or } \sigma < v \leq d. \end{cases}$$

By the Riesz representation theorem for bounded linear functionals on $C(D)$ we have

$$f(x) = \int_D x(\tau, \sigma) \, dg(\tau, \sigma),$$

where g is of bounded variation with $\operatorname{Var}_D g = \|f\|_{C^*}$. In particular, the function g corresponding to the functional $f = Px(t, s) \in C^*$ has the form

$$
\begin{aligned}
(13.26) \quad & g_{t,s}(\tau, \sigma) := c(t, s)\tilde{\kappa}(t, s, \tau, \sigma) + \int_a^\tau l(t, s, \tilde{\tau}) \, d\tilde{\tau} \, \tilde{\kappa}(s, \sigma) \\
& + \int_c^\sigma m(t, s, \tilde{\sigma}) \, d\tilde{\sigma} \, \tilde{\kappa}(t, \tau) + \int_a^\tau \int_c^\sigma n(t, s, \tilde{\tau}, \tilde{\sigma}) \, d\tilde{\sigma} \, d\tilde{\tau},
\end{aligned}
$$

where

$$\tilde{\kappa}(t, \tau) := \begin{cases} 1 & \text{if} \quad a \leq t \leq \tau \leq b, \\ 0 & \text{if} \quad a \leq \tau < t \leq b, \end{cases}$$

$$\tilde{\kappa}(s,\sigma) := \begin{cases} 1 & \text{if } c \leq s \leq \sigma \leq d, \\ 0 & \text{if } c \leq \sigma < s \leq d, \end{cases}$$

and

$$\tilde{\kappa}(t,s,\tau,\sigma) := \begin{cases} 1 & \text{if } a \leq t \leq \tau \leq b \text{ and } c \leq s \leq \sigma \leq d, \\ 0 & \text{if } a \leq \tau < t \leq b \text{ or } c \leq \sigma < s \leq d. \end{cases}$$

A comparison of (13.15) and (13.26) shows that the functions $g(t,s,\cdot,\cdot)$ and $g_{t,s}$ have the same total variation on D. Altogether we get

$$\|Px(t,s)\|_{C^*} = \operatorname{Var}_D g_{t,s} = \operatorname{Var}_D g(t,s,\cdot,\cdot) = \overline{\Gamma}(t,s)$$

which together with (13.25) completes the proof. ∎

We remark that the Radon representation of a bounded linear operator on $C(D)$ and, in particular, an arbitrary partial integral operator on $C(D)$, is in general not unique. However, the representation (13.2) of the sum of operators (13.3) - (13.6) is unique.

Since the functions (13.16) - (13.19) are rather messy, to verify their continuity may be very hard in practice. It is therefore of practical interest to have easier acting (and boundedness) conditions for the operator (13.2) in the space $C(D)$ at hand, even if these conditions are only sufficient.

Here is one such condition. Let us call the kernel functions l, m and n in (13.4) - (13.6) *continuous in the whole* if

$$\lim_{\substack{|t_1-t_2|\to 0 \\ |s_1-s_2|\to 0}} \int_a^b |l(t_1,s_1,\tau) - l(t_2,s_2,\tau)|\, d\tau = 0,$$

$$\lim_{\substack{|t_1-t_2|\to 0 \\ |s_1-s_2|\to 0}} \int_c^d |m(t_1,s_1,\sigma) - m(t_2,s_2,\sigma)|\, d\sigma = 0,$$

and

$$\lim_{\substack{|t_1-t_2|\to 0 \\ |s_1-s_2|\to 0}} \int_a^b \int_c^d |n(t_1,s_1,\tau,\sigma) - n(t_2,s_2,\tau,\sigma)|\, d\sigma\, d\tau = 0.$$

Theorem 13.2. *Suppose that the function c is continuous, the kernel functions l, m and n are continuous in the whole, and*

$$\Delta := \sup_{(t,s)\in D} [\,||l(t,s,\cdot)|| + ||m(t,s,\cdot)|| + ||n(t,s,\cdot,\cdot)||\,] < \infty,$$

where all norms are taken in the corresponding L_1 space. Then the linear operator (13.2) is bounded in the space $C(D)$, and its norm is given by

$$(13.27) \qquad \begin{aligned} ||P||_{\mathfrak{L}(C)} &= \sup_{(t,s)\in D} [\,|c(t,s)| \\ &+ ||l(t,s,\cdot)|| + ||m(t,s,\cdot)|| + ||n(t,s,\cdot,\cdot)||\,]. \end{aligned}$$

Of course, all hypotheses of Theorem 13.2 are satisfied if the functions c, l, m and n are jointly continuous, but not vice versa. The hypotheses of Theorem 13.2 are also satisfied if $||l(t,s,\cdot)||_{L^p} \leq A_l < \infty$, $||m(t,s,\cdot)||_{L^q} \leq A_m < \infty$, $||n(t,s,\cdot,\cdot)||_{L^r} \leq A_n < \infty$ ($1 < p,q,r < \infty$), the function c is continuous, and the kernel functions l, m and n are discontinuous only along finitely many surfaces with parameter representation $\tau = \tau(t,s)$ and $\sigma = \sigma(t,s)$ (KALITVIN-JANKELEVICH [1994]).

13.3. Algebras of partial integral operators

In what follows, by $\mathfrak{L}_p(C)$ we denote the subspace of all operators of the form (13.2) in the space $\mathfrak{L}(C)$. Theorem 13.1 proved above allows us not only to characterize the operators which belong to $\mathfrak{L}_p(C)$, but also to describe structural properties of the subspace $\mathfrak{L}_p(C)$. We begin with the following

Theorem 13.3. *The subspace $\mathfrak{L}_p(C)$ is closed in the algebra $\mathfrak{L}(C)$.*

□ Let $(P_n)_n$ be a sequence of operators $P_n \in \mathfrak{L}_p(C)$, i.e.

$$P_n x(t,s) = c_n(t,s)x(t,s)$$

(13.28) $$+ \int_a^b l_n(t,s,\tau)x(\tau,s)\,d\tau + \int_c^d m_n(t,s,\sigma)x(t,\sigma)\,d\sigma$$

$$+ \int_a^b \int_c^d n_n(t,s,\tau,\sigma)x(\tau,\sigma)\,d\sigma\,d\tau,$$

which converges (in $\mathfrak{L}(C)$) to some operator $\check{P} \in \mathfrak{L}(C)$. Then the sequence $(P_n)_n$ is Cauchy and fulfills, by Theorem 13.1, the relation

$$\lim_{i,j\to\infty} \sup_{(t,s)\in D} \left[|c_i(t,s) - c_j(t,s)| + \int_a^b |l_i(t,s,\tau) - l_j(t,s,\tau)|\,d\tau \right.$$

$$+ \int_c^d |m_i(t,s,\sigma) - m_j(t,s,\sigma)|\,d\sigma$$

$$\left. + \int_a^b \int_c^d |n_i(t,s,\tau,\sigma) - n_j(t,s,\tau,\sigma)|\,d\sigma\,d\tau \right] = 0.$$

This is of course equivalent to the four relations

(13.29) $$\sup_{(t,s)\in D} |c_i(t,s) - c_j(t,s)| \to 0,$$

(13.30) $$\sup_{(t,s)\in D} \int_a^b |l_i(t,s,\tau) - l_j(t,s,\tau)|\,d\tau \to 0,$$

(13.31) $$\sup_{(t,s)\in D} \int_c^d |m_i(t,s,\sigma) - m_j(t,s,\sigma)|\,d\sigma \to 0,$$

and

(13.32) $$\sup_{(t,s)\in D} \int_a^b \int_c^d |n_i(t,s,\tau,\sigma) - n_j(t,s,\tau,\sigma)|\,d\sigma\,d\tau \to 0$$

as $i, j \to \infty$. By (13.29), the sequence $(c_n)_n$ is Cauchy in the space of all bounded functions on D (with the supremum norm), and hence

$$\lim_{n \to \infty} \sup_{(t,s) \in D} |c_n(t,s) - c(t,s)| = 0$$

for some bounded function c on D. The relation (13.30) in turn shows that the sequence $(l_n)_n$ is Cauchy in the complete space with mixed norm $[L_1 \to L_\infty]$ (see e. g. BENEDEK-PANZONE [1961]), and hence

$$\lim_{n \to \infty} ||l_n - \tilde{l}||_{[L_1 \to L_\infty]} = 0$$

for some measurable function $\tilde{l} \in [L_1 \to L_\infty]$; more precisely, this means that there is a nullset $D_0 \subset D$ such that

$$\lim_{n \to \infty} \sup_{(t,s) \in D \setminus D_0} \int_a^b |l_n(t,s,\tau) - \tilde{l}(t,s,\tau)| \, d\tau = 0.$$

But for fixed $(t,s) \in D_0$ the sequence $(l_n(t,s,\cdot))_n$ is Cauchy in the space $L_1 = L_1([a,b])$, and hence $||l_n(t,s,\cdot) - \hat{l}(t,s,\cdot)||_{L_1} \to 0$ for some function $\hat{l}(t,s,\cdot) \in L_1$, as $n \to \infty$, uniformly in $(t,s) \in D_0$, by (13.30). Consequently,

$$\lim_{n \to \infty} \sup_{(t,s) \in D} \int_a^b |l_n(t,s,\tau) - l(t,s,\tau)| \, d\tau = 0,$$

where

$$l(t,s,\tau) := \begin{cases} \tilde{l}(t,s,\tau) & \text{if } (t,s) \in D \setminus D_0, \\ \hat{l}(t,s,\tau) & \text{if } (t,s) \in D_0. \end{cases}$$

Likewise, one may construct functions m and n as limits of the sequences $(m_n)_n$ and $(n_n)_n$, respectively. Now we put

$$Px(t,s) := c(t,s)x(t,s)$$

$$(13.33) \qquad + \int_a^b l(t,s,\tau)x(\tau,s) \, d\tau + \int_c^d m(t,s,\sigma)x(t,\sigma) \, d\sigma$$

$$+ \int_a^b \int_c^d n(t,s,\tau,\sigma)x(\tau,\sigma) \, d\sigma \, d\tau$$

and show that $P \in \mathfrak{L}_p(C)$ and $P = \check{P}$. First of all, by the construction of the functions c, l, m and n, and the obvious estimate

$$|P_n x(t,s) - P x(t,s)| \le |c_n(t,s) - c(t,s)|\, |x(t,s)|$$

$$+ \int_a^b |l_n(t,s,\tau) - l(t,s,\tau)|\, |x(\tau,s)|\, d\tau$$

$$+ \int_c^d |m_n(t,s,\sigma) - m(t,s,\sigma)|\, |x(t,\sigma)|\, d\sigma$$

$$+ \int_a^b \int_c^d |n_n(t,s,\tau,\sigma) - n(t,s,\tau,\sigma)|\, |x(\tau,\sigma)|\, d\sigma\, d\tau$$

$(x \in C(D))$ we see that $P \in \mathfrak{L}_p(C)$. Moreover, we have

$$\|P_n - P\|_{\mathfrak{L}(C)} \le \sup_{(t,s) \in D} \bigl[|c_n(t,s) - c(t,s)|$$

$$+ \int_a^b |l_n(t,s,\tau) - l(t,s,\tau)|\, d\tau + \int_c^d |m_n(t,s,\sigma) - m(t,s,\sigma)|\, d\sigma$$

$$+ \int_a^b \int_c^d |n_n(t,s,\tau,\sigma) - n(t,s,\tau,\sigma)|\, d\sigma\, d\tau \bigr].$$

We conclude that $P_n \to P$ in the norm of $\mathfrak{L}(C)$, and hence $P = \check{P}$. ∎

The space $\mathfrak{L}(C)$ is an algebra with respect to the usual composition of operators. Therefore the natural question arises: *Is the subspace $\mathfrak{L}_p(C)$ a subalgebra in $\mathfrak{L}(C)$, or even an ideal?*

The answer to the second question is of course negative: To see this, it suffices to compose, for example, an operator $P \in \mathfrak{L}_p(C)$ with the operator which associates to each $x \in C(D)$ the constant function $y(t,s) \equiv x(a,c)$. The answer to the first question, however, is positive:

Theorem 13.4. *The subspace $\mathfrak{L}_p(C)$ is a subalgebra of the algebra*

$\mathfrak{L}(C)$. *More precisely, if* $P_i \in \mathfrak{L}_p(C)$ *is given by*

$$P_i x(t, s) = c_i(t, s)x(t, s) + \int_a^b l_i(t, s, \tau)x(\tau, s)\, d\tau$$

$$+ \int_c^d m_i(t, s, \sigma)x(t, \sigma)\, d\sigma + \int_a^b \int_c^d n_i(t, s, \tau, \sigma)x(\tau, \sigma)\, d\sigma\, d\tau,$$

$(i = 1, 2)$, *then the operator* $P = P_2 P_1$ *is given by* (13.33) *where*

$$c(t, s) := c_2(t, s)c_1(t, s),$$

$$l(t, s, \tau) := c_2(t, s)l_1(t, s, \tau) + l_2(t, s, \tau)c_1(\tau, s)$$

$$+ \int_a^b l_2(t, s, \xi)l_1(\xi, s, \tau)\, d\xi,$$

$$m(t, s, \sigma) := c_2(t, s)m_1(t, s, \sigma) + m_2(t, s, \sigma)c_1(t, \sigma)$$

$$+ \int_c^d m_2(t, s, \eta)m_1(t, \eta, \sigma)\, d\eta,$$

and

$$n(t, s, \tau, \sigma) := c_2(t, s)n_1(t, s, \tau, \sigma) + n_2(t, s, \tau, \sigma)c_1(\tau, \sigma)$$

$$+ l_2(t, s, \tau)m_1(\tau, s, \sigma) + m_2(t, s, \sigma)l_1(t, \sigma, \tau)$$

$$+ \int_a^b l_2(t, s, \xi)n_1(\xi, s, \tau, \sigma)\, d\xi + \int_c^d m_2(t, s, \eta)n_1(t, \eta, \tau, \sigma)\, d\eta$$

$$+ \int_a^b n_2(t, s, \xi, \sigma)l_1(\xi, \sigma, \tau)\, d\xi + \int_c^d n_2(t, s, \tau, \eta)m_1(\tau, \eta, \sigma)\, d\eta$$

$$+ \int_a^b \int_c^d n_2(t, s, \xi, \eta)n_1(\xi, \eta, \tau, \sigma)\, d\eta\, d\xi.$$

□ The proof of Theorem 13.4 is a straightforward consequence of Theorem 13.1 and Fubini's theorem. ∎

By construction, every operator $P \in \mathcal{L}_p(C)$ is a sum of four operators, namely the multiplication operator (13.3), the partial integral operators (13.4) and (13.5), and the ordinary integral operator (13.6). It is natural to expect that, if such an operator P acts in the space $C(D)$, also all its components (13.3) - (13.6) act in the space $C(D)$. Surprisingly, this is not true, as is shown by the following

Example 13.1. Consider the decomposition of the square $D := [0,1] \times [0,1]$ into the four subsquares $D_1 := [0,\frac{1}{2}] \times [0,\frac{1}{2}]$, $D_2 := (\frac{1}{2},1] \times [0,\frac{1}{2}]$, $D_3 := [0,\frac{1}{2}] \times (\frac{1}{2},1]$, and $D_4 := (\frac{1}{2},1] \times (\frac{1}{2},1]$, and define $P : C(D) \to C(D)$ by

$$
Px(t,s) := \begin{cases}
x(t,s) & \text{if } (t,s) \in D_1, \\[2mm]
\dfrac{2}{2t-1} \displaystyle\int_{1/2}^{t} x(\tau,s)\, d\tau & \text{if } (t,s) \in D_2, \\[4mm]
\dfrac{2}{2s-1} \displaystyle\int_{1/2}^{s} x(t,\sigma)\, d\sigma & \text{if } (t,s) \in D_3, \\[4mm]
\dfrac{4}{(2t-1)(2s-1)} \displaystyle\int_{1/2}^{t}\int_{1/2}^{s} x(\tau,\sigma)\, d\sigma\, d\tau & \text{if } (t,s) \in D_4.
\end{cases}
$$

A straightforward calculation shows that $P \in \mathcal{L}_p(C)$ with

$$
c(t,s) = \begin{cases} 1 & \text{if } (t,s) \in D_1, \\ 0 & \text{if } (t,s) \in D_2 \cup D_3 \cup D_4, \end{cases}
$$

$$
l(t,s,\tau) = \begin{cases} 0 & \text{if } (t,s) \in D_1 \cup D_3 \cup D_4, \\[1mm] \dfrac{2}{2t-1} & \text{if } \frac{1}{2} < \tau \leq t \leq 1 \text{ and } 0 \leq s \leq \frac{1}{2}, \\[2mm] 0 & \text{if } \frac{1}{2} \leq t < \tau \leq 1 \text{ and } 0 \leq s \leq \frac{1}{2}, \end{cases}
$$

$$
m(t,s,\sigma) = \begin{cases} 0 & \text{if } (t,s) \in D_1 \cup D_2 \cup D_4, \\[1mm] \dfrac{2}{2s-1} & \text{if } \frac{1}{2} < \sigma \leq s \leq 1 \text{ and } 0 \leq t \leq \frac{1}{2}, \\[2mm] 0 & \text{if } \frac{1}{2} \leq s < \sigma \leq 1 \text{ and } 0 \leq t \leq \frac{1}{2}, \end{cases}
$$

and

$$
n(t, s, \tau, \sigma) =
\begin{cases}
0 & \text{if } (t, s) \in D_1 \cup D_2 \cup D_3, \\[2mm]
\frac{4}{(2t-1)(2s-1)} & \text{if } \frac{1}{2} < \tau \le t \le 1,\ \frac{1}{2} < \sigma \le s \le 1, \\[2mm]
0 & \text{otherwise.}
\end{cases}
$$

But neither of the single operators (13.3) - (13.6) generated by these functions acts in the space $C(D)$. ∎

In view of this counterexample, it is of interest to characterize the operators $P \in \mathfrak{L}_p(C)$ with the additional property that all their components belong to $\mathfrak{L}(C)$ as well. Let us denote by $\mathfrak{L}_c(C)$, $\mathfrak{L}_l(C)$, $\mathfrak{L}_m(C)$, and $\mathfrak{L}_n(C)$ the sets of all operators of the form (13.3), (13.4), (13.5), and (13.6), respectively, acting in the space $C(D)$. Thus we are interested in describing the direct sum $\mathfrak{L}_\Sigma(C) := \mathfrak{L}_c(C) \oplus \mathfrak{L}_l(C) \oplus \mathfrak{L}_m(C) \oplus \mathfrak{L}_n(C)$ as a subspace of $\mathfrak{L}_p(C)$.

Theorem 13.5. *The subspace $\mathfrak{L}_\Sigma(C)$ is closed in the space $\mathfrak{L}_p(C)$.*

□ Suppose that the sequence (13.28) of operators $P_n \in \mathfrak{L}_\Sigma(C)$ converges in the norm of $\mathfrak{L}(C)$ to some operator $\check{P} \in \mathfrak{L}_p(C)$. Then the equalities (13.29) - (13.32) hold, by Theorem 13.1. Construct the functions c, l, m and n through the functions c_n, l_n, m_n and n_n in the same way as in the proof of Theorem 13.3. We claim that the corresponding operator (13.33) actually belongs to $\mathfrak{L}_\Sigma(C)$ and coincides with $\check{P}$.

First of all, the continuity of the functions c_n implies that of the limit function c, and thus the corresponding operator (13.3) acts in the space $C(D)$. Moreover, for $x \in C(D)$ and fixed $(t_0, s_0) \in D$ we have

$$\left| \int_a^b l(t,s,\tau) x(\tau,s)\, d\tau - \int_a^b l(t_0,s_0,\tau) x(\tau,s_0)\, d\tau \right|$$

$$\leq \int_a^b |l(t,s,\tau) - l_n(t,s,\tau)|\, |x(\tau,s)|\, d\tau$$

$$+ \int_a^b |l_n(t_0,s_0,\tau) - l(t_0,s_0,\tau)|\, |x(\tau,s_0)|\, d\tau$$

$$+ \left| \int_a^b [l_n(t,s,\tau) - l_n(t_0,s_0,\tau)] x(\tau,s_0)\, d\tau \right|$$

$$+ \int_a^b |l_n(t,s,\tau)|\, |x(\tau,s) - x(\tau,s_0)|\, d\tau.$$

Every term in the right-hand side of this estimate tends to zero as $(t,s) \to (t_0,s_0)$. In fact, for the first two terms this follows from (13.30) and the boundedness of the function x, for the third term from the continuity of the function $t \mapsto L_n x(t,s_0)$, and for the fourth term from the continuity of the function x and formula (13.21) for the norm of P. We conclude that $Lx \in C(D)$, hence $L \in \mathfrak{L}_l(C)$.

The relations $M \in \mathfrak{L}_m(C)$ and $N \in \mathfrak{L}_n(C)$ are proved by exactly the same reasoning. It follows that $P \in \mathfrak{L}_\Sigma(C)$, and so we are done. ∎

Theorem 13.6. *The subspace $\mathfrak{L}_n(C)$ is an ideal in both the space $\mathfrak{L}_\Sigma(C)$ and the space $\mathfrak{L}_p(C)$, and the subspace $\mathfrak{L}_\Sigma(C)$ is a subalgebra in the algebra $\mathfrak{L}_p(C)$.*

☐ Let $\tilde{N}$ be an integral operator of the form (13.6) generated by some kernel function $\tilde{n} = \tilde{n}(t,s,\tau,\sigma)$, and let P be a partial integral operator of the form (13.33) which belongs to $\mathfrak{L}_\Sigma(C)$. By Fubini's theorem we have then

$$\tilde{N}Px(t,s) = \int_a^b \int_c^d [c(\tau,\sigma)\tilde{n}(t,s,\tau,\sigma)$$

$$+ \int_a^b l(\xi,\sigma,\tau)\tilde{n}(t,s,\xi,\sigma)\,d\xi + \int_c^d m(\tau,\eta,\sigma)\tilde{n}(t,s,\tau,\eta)\,d\eta$$

$$+ \int_a^b \int_c^d \tilde{n}(t,s,\xi,\eta)n(\xi,\eta,\tau,\sigma)\,d\eta\,d\xi\Big] x(\tau,\sigma)\,d\sigma\,d\tau,$$

as well as

$$P\tilde{N}x(t,s) = \int_a^b \int_c^d [c(t,s)\tilde{n}(t,s,\tau,\sigma)$$

$$+ \int_a^b l(t,s,\xi)\tilde{n}(\xi,s,\tau,\sigma)\,d\xi + \int_c^d m(t,s,\eta)\tilde{n}(t,\eta,\tau,\sigma)\,d\eta$$

$$+ \int_a^b \int_c^d n(t,s,\xi,\eta)\tilde{n}(\xi,\eta,\tau,\sigma)\,d\eta\,d\xi\Big] x(\tau,\sigma)\,d\sigma\,d\tau.$$

This shows that both $\tilde{N}P$ and $P\tilde{N}$ belong to $\mathfrak{L}_n(C)$. The second assertion follows from the fact that $\mathfrak{L}_\Sigma(C)$ is stable under composition and and closed in $\mathfrak{L}_p(C)$. ■

We remark that the space $\mathfrak{L}_n(C)$ is not an ideal in the whole algebra $\mathfrak{L}(C)$.

13.4. Operator functions with values in $\mathfrak{L}_p(C)$

Let again $D = [a,b] \times [c,d]$, and let J be a bounded or unbounded interval in $\mathbb{R}$. In this subsection we study the strong and norm continuity of the operator function
(13.34)

$$P(\varphi)x(t,s) = c(\varphi,t,s)x(t,s) + \int_a^b l(\varphi,t,s,\tau)x(\tau,s)\,d\tau$$

$$+ \int_c^d m(\varphi,t,s,\sigma)x(t,\sigma)\,d\sigma + \int_a^b \int_c^d n(\varphi,t,s,\tau,\sigma)x(\tau,\sigma)\,d\sigma\,d\tau;$$

here $c : J \times D \to \mathbb{R}$, $l : J \times D \times [a, b] \to \mathbb{R}$, $m : J \times D \times [c, d] \to \mathbb{R}$, and $n : J \times D \times D \to \mathbb{R}$ are jointly measurable functions, and the integrals are meant in the Lebesgue sense.

For each fixed $\varphi \in J$, the formula (13.34) defines a partial integral operator of the form (13.2). We consider the operator function (13.34) with values in the space $\mathcal{L}_p(X)$ of partial integral operators which act in the space $X = C(D)$.

The following three examples show that the norm continuity or strong continuity of the operator function (13.34) is not equivalent to the continuous dependence of the functions $c(\varphi, \cdot, \cdot), l(\varphi, \cdot, \cdot, \cdot), m(\varphi, \cdot, \cdot, \cdot)$, and $n(\varphi, \cdot, \cdot, \cdot, \cdot)$ on the parameter φ. In these examples we assume throughout that $J = [a, b] = [c, d] = [0, 1]$ and $c(\varphi, t, s) = m(\varphi, t, s, \sigma) = n(\varphi, t, s, \tau, \sigma) \equiv 0$; thus, we work only with the kernel function $l = l(\varphi, t, s, \tau)$.

Example 13.2. Let

$$l(\varphi, t, s, \tau) = \begin{cases} \sin \dfrac{\varphi}{\sqrt{\tau}} & \text{if } 0 \leq \varphi, t, s \leq 1 \text{ and } 0 < \tau \leq 1, \\ \\ 0 & \text{if } \tau = 0. \end{cases}$$

The function l is discontinuous at the point $(0, 0, 0, 0)$, but $P(\varphi)$ is a norm continuous operator function with values in $\mathcal{L}_p(C)$. In fact, for each $\varphi \in [0, 1]$ the operator $P(\varphi)$ acts in the space $C(D)$. From this and Theorem 13.1 we have

$$\|P(\varphi) - P(\varphi_0)\| = \int_0^1 |\sin \dfrac{\varphi}{\sqrt{\tau}} - \sin \dfrac{\varphi_0}{\sqrt{\tau}}| \, d\tau \leq 2|\varphi - \varphi_0|,$$

which means that $\|P(\varphi) - P(\varphi_0)\| \to 0$ for $\varphi \to \varphi_0$.

Observe that in Example 13.2 the function $l(\cdot, t, s, \tau)$ is continuous at 0 for each $(t, s, \tau) \in [0, 1] \times [0, 1] \times [0, 1]$. In the following example, the function $l(\cdot, t, s, \tau)$ is discontinuous at 0 for each $(t, s, \tau) \in [0, 1] \times [0, 1] \times [0, 1]$; nevertheless, $P(\varphi)$ is a strongly continuous operator function with values in $\mathcal{L}_p(C)$.

Example 13.3. Let

$$l(\varphi, t, s, \tau) = \begin{cases} \sin \dfrac{\tau}{\varphi} & \text{if } 0 < \varphi \leq 1 \text{ and } 0 \leq t, s, \tau \leq 1, \\ 0 & \text{if } \varphi = 0. \end{cases}$$

It is obvious that the operator function $P(\cdot)$ takes values in $\mathfrak{L}_p(C)$ and is strongly continuous at every point $\varphi_0 \neq 0$. We show that $P(\cdot)$ is also strongly continuous at 0. For continuous functions of the form $x(t, s) = u(t)v(s)$ we have

$$\|[P(\varphi) - P(0)]x\|_C = \max_{0 \leq s \leq 1} |v(s) \int_0^1 \sin \frac{\tau}{\varphi} u(\tau) \, d\tau|$$

$$= \|v\|_C \, \| \int_0^1 \sin \frac{\tau}{\varphi} u(\tau) \, d\tau \|_C \to 0$$

as $\varphi \to 0$. Since $\|P(\varphi)\| \leq 1$ and the linear hull of the functions of the form $x(t, s) = u(t)v(s)$ is dense in $C(D)$, we conclude that $P(\varphi)x \to P(0)x$ for any $x \in C(D)$, and hence $P(\cdot)$ is strongly continuous. We remark that in this example the operator function $P(\cdot)$ is not norm continuous.

Example 13.4. Let $l(\varphi, t, s, \tau) = \delta(\varphi)\delta(\tau) + l_0(\varphi, t, s, \tau)$, where l_0 is a jointly continuous on $[0, 1] \times [0, 1] \times [0, 1] \times [0, 1]$ and δ is the Dirichlet function (i.e., the characteristic function of $[0, 1] \cap \mathbb{Q}$). Obviously, the function $l(\cdot, t, s, \tau)$ is discontinuous everywhere on $[0,1]$ for each t, s, τ. On the other hand, $P(\cdot)$ takes values in $\mathfrak{L}_p(C)$ and is norm continuous at any point $\varphi \in [0, 1]$.

Before stating conditions for the strong continuity or norm continuity of the operator function (13.34), we have to introduce some notation. Let

$$B(\varphi, t, s) = c(\varphi, t, s) + \int_a^b l(\varphi, t, s, \tau) \, d\tau$$

(13.35)

$$+ \int_c^d m(\varphi, t, s, \sigma) \, d\sigma + \int_a^b \int_c^d n(\varphi, t, s, \tau, \sigma) \, d\sigma \, d\tau,$$

$$B_\xi(\varphi, t, s) = \int_a^\xi [(c(\varphi, t, s)$$

(13.36)
$$+ \int_c^d m(\varphi, t, s, \sigma) \, d\sigma) \kappa(t, \tau) + (\xi - \tau)(l(\varphi, t, s, \tau)$$

$$+ \int_c^d n(\varphi, t, s, \tau, \sigma) \, d\sigma)] \, d\tau,$$

$$B_\eta(\varphi, t, s) = \int_c^\eta [(c(\varphi, t, s)$$

(13.37)
$$+ \int_a^b l(\varphi, t, s, \tau) \, d\tau) \kappa(s, \sigma) + (\eta - \sigma)(m(\varphi, t, s, \sigma)$$

$$+ \int_a^b n(\varphi, t, s, \tau, \sigma) \, d\tau)] \, d\sigma,$$

and

$$B_{\xi\eta}(\varphi, t, s) = \int_a^\xi \int_c^\eta [c(\varphi, t, s) \kappa(t, s, \tau, \sigma)$$

$$+ (\xi - \tau) l(\varphi, t, s, \tau) \kappa(s, \sigma)$$

(13.38)
$$+ (\eta - \sigma) m(\varphi, t, s, \sigma) \kappa(t, \tau)$$

$$+ (\xi - \tau)(\eta - \sigma) n(\varphi, t, s, \tau, \sigma) \, d\sigma \, d\tau].$$

Moreover, we put

$$\gamma(\varphi, t, s) = |c(\varphi, t, s)|$$

(13.39)
$$+ \int_a^b |l(\varphi, t, s, \tau)| \, d\tau + \int_c^d |m(\varphi, t, s, \sigma)| \, d\sigma$$

$$+ \int_a^b \int_c^d |n(\varphi, t, s, \tau, \sigma)| \, d\sigma \, d\tau,$$

where

$$\kappa(t, \tau) := \begin{cases} 1 & \text{if } \tau \geq t > a \text{ or } \tau > t = a, \\ 0 & \text{if } \tau < t \leq b \text{ or } \tau = t = a, \end{cases}$$

$$\kappa(s, \sigma) := \begin{cases} 1 & \text{if } \sigma \geq s > c \text{ or } \sigma > s = c, \\ 0 & \text{if } \sigma < s \leq d \text{ or } \sigma = s = c, \end{cases}$$

and

$$\kappa(t, s, \tau, \sigma) := \begin{cases} 1 & \text{if } \tau \geq t > a \text{ and } \sigma \geq s > c \\ & \text{or } \tau > t = a \text{ and } \sigma > s = c, \\ 0 & \text{if } \tau < t \leq b, \ \sigma < s \leq d, \ \tau = t = a, \\ & \text{or } \sigma = s = c. \end{cases}$$

With the use of Theorem 13.1 one may then prove the following

Theorem 13.7. *The operator function* (13.34) *is strongly continuous in the space* $\mathcal{L}(C)$ *if and only if the functions* (13.35) - (13.38) *are continuous, and the function* (13.39) *is bounded on each bounded subset of its domain of definition.*

$\square$ Suppose that the operator function $P(\cdot)$ is strongly continuous in $\mathcal{L}(C)$. Put

$$x_\xi(t) = \begin{cases} \xi - t & \text{if } a \leq t \leq \xi \leq b, \\ 0 & \text{if } a \leq \xi < t \leq c \end{cases}$$

and

$$x_\eta(s) = \begin{cases} \eta - s & \text{if } c \leq s \leq \eta \leq d, \\ 0 & \text{if } c \leq \eta < s \leq d \end{cases}$$

Then the equalities

$$B(\varphi, t, s) = P(\varphi)1(t, s), \quad B_\xi(\varphi, t, s) = P(\varphi)x_\xi(t)1(s),$$

$$B_\eta(\varphi, t, s) = P(\varphi)1(t)x_\eta(s), \quad B_{\xi\eta}(\varphi, t, s) = P(\varphi)x_\xi(t)x_\eta(s)$$

and the Banach-Steinhaus theorem imply the continuity of the functions (13.35) - (13.38) and the boundedness of the function (13.39) on each bounded subset of its domain of definition.

To prove the converse it suffices to show that the operator function $\varphi \mapsto P(\varphi)x$ is continuous on J for x in some dense subset $M \subset C$. As set M we choose the linear hull of the constant functions, the functions x_ξ and x_η as above, and their products. In fact, the continuity of $\varphi \mapsto P(\varphi)x$ for these function is a direct consequence of our assumption on the functions (13.35) - (13.38), and the assertion follows from the linearity of $P(\varphi)$. ∎

Since $C(D) \subset L_\infty(D)$, another criterion for the strong continuity of the operator function (13.34) in $C(D)$ follows from Theorems 12.9 and 13.7:

Theorem 13.8. *Suppose that the operator function* (13.34) *takes its values in* $\mathfrak{L}_p(C)$. *Then it is strongly continuous if and only if the function* (13.39) *is bounded on* $J \times D$ *for the each bounded interval* $J \subset \mathbb{R}$, *the function* $\varphi \mapsto c(\varphi, \cdot, \cdot)$ *is continuous as a function with values in* $C(D)$, *and the functions*

$$\varphi \mapsto \int_a^u l(\varphi, \cdot, \cdot, \tau) \, d\tau, \quad \varphi \mapsto \int_c^v m(\varphi, \cdot, \cdot, \sigma) \, d\sigma,$$

$$\varphi \mapsto \int_a^u \int_c^v n(\varphi, \cdot, \cdot, \tau, \sigma) \, d\sigma \, d\tau$$

are continuous for each $u \in [a, b]$, $v \in [c, d]$, *and* $(u, v) \in D$, *respectively, as functions with values in* $C(D)$.

The following theorem gives a criterion for the norm continuity of the operator function (13.34).

Theorem 13.9. *Suppose that the values of the operator function* (13.34) *belong to* $\mathfrak{L}_p(C)$. *Then* (13.34) *is norm continuous in* $\mathfrak{L}(C)$ *at* $\varphi_0 \in J$ *if and only if the function* $c(\cdot, t, s)$ *continuous at* φ_0, *uniformly with respect to* $(t, s) \in D$, *and the functions* l, m, n *have the following behaviour for* $\varphi \to \varphi_0$, *uniformly in* $(t, s) \in D$:

(13.40) $$\mu(\{\tau : |l(\varphi, t, s, \tau) - l(\varphi_0, t, s, \tau)| > \theta\}) \to 0,$$

(13.41) $$\lim_{\mu(A) \to 0} \int_A |l(\varphi, t, s, \tau) - l(\varphi_0, t, s, \tau)|\, d\tau \to 0,$$

(13.42) $$\mu(\{\sigma : |m(\varphi, t, s, \sigma) - m(\varphi_0, t, s, \sigma)| > \theta\}) \to 0,$$

(13.43) $$\lim_{\mu(B) \to 0} \int_B |m(\varphi, t, s, \sigma) - m(\varphi_0, t, s, \sigma)|\, d\sigma \to 0,$$

(13.44) $$\mu(\{(\tau, \sigma) : |n(\varphi, t, s, \tau, \sigma) - n(\varphi_0, t, s, \tau, \sigma)| > \theta\}) \to 0,$$

(13.45) $$\lim_{\substack{\mu(A) \to 0 \\ \mu(B) \to 0}} \int_A \int_B |n(\varphi, t, s, \tau, \sigma) - n(\varphi_0, t, s, \tau, \sigma)|\, d\sigma\, d\tau \to 0.$$

□ Let $P(\cdot)$ be norm continuous in $\mathfrak{L}_p(C)$ at $\varphi_0 \in J$. Then we have, by Theorem 13.1,

(13.46) $$\sup_{(t,s) \in D} |c(\varphi, t, s) - c(\varphi_0, t, s)| \to 0,$$

(13.47) $$\sup_{(t,s) \in D} \int_a^b |l(\varphi, t, s, \tau) - l(\varphi_0, t, s, \tau)|\, d\tau \to 0,$$

(13.48) $$\sup_{(t,s) \in D} \int_c^d |m(\varphi, t, s, \sigma) - m(\varphi_0, t, s, \sigma)|\, d\sigma \to 0,$$

and

$$(13.49) \quad \sup_{(t,s)\in D} \int_a^b \int_c^d |n(\varphi,t,s,\tau,\sigma) - n(\varphi_0,t,s,\tau,\sigma)|\, d\sigma\, d\tau \to 0$$

as $\varphi \to \varphi_0$. Thus, the statement of the theorem for the function c follows from the equality (13.46). Further, the equality (13.47) and the estimates

$$\theta\, \mu((\{\tau : |l(\varphi,t,s,\tau) - l(\varphi_0,t,s,\tau)| \geq \theta\}))$$

$$\leq \int_a^b |l(\varphi,t,s,\tau) - l(\varphi_0,t,s,\tau)|\, d\tau$$

and

$$\int_A |l(\varphi,t,s,\tau) - l(\varphi_0,t,s,\tau)|\, d\tau$$

$$\leq \int_a^b |l(\varphi,t,s,\tau) - l(\varphi_0,t,s,\tau)|\, d\tau$$

imply (13.40) and (13.41). The equalities (13.42) - (13.45) are proved analogously.

Conversely, suppose that $P(\varphi) \in \mathfrak{L}_p(C)$ for each $\varphi \in J$, the function $c(\cdot,t,s)$ is continuous at φ_0, uniformly with respect $(t,s) \in D$, and the equalities (13.40) - (13.45) are satisfied. By Theorem 13.1 it suffices again to show that the equalities (13.46) - (13.49) hold.

Equality (13.46) is evident; let us show (13.47). Choose $\varepsilon > 0$, $\theta = \varepsilon/2(b-a)$, and consider the inequality

$$\int_a^b |l(\varphi,t,s,\tau) - l(\varphi_0,t,s,\tau)|\, d\tau$$

$$\leq \theta(b-a) + \int_A |l(\varphi,t,s,\tau) - l(\varphi_0,t,s,\tau)|\, d\tau,$$

where $A = \{\tau : |l(\varphi,t,s,\tau) - l(\varphi_0,t,s,\tau)| \geq \theta\}$. By (13.40), the relation

$$\lim_{\varphi \to \varphi_0} \sup_{(t,s)\in D} \mu(A) = 0$$

is correct, and hence, by (13.41),

$$\lim_{\varphi \to \varphi_0} \sup_{(t,s) \in D} \int_A |l(\varphi, t, s, \tau) - l(\varphi_0, t, s, \tau)| \, d\tau = 0.$$

Consequently,

$$\sup_{(t,s) \in D} \int_a^b |l(\varphi, t, s, \tau) - l(\varphi_0, t, s, \tau)| \, d\tau$$

$$\leq \frac{\varepsilon}{2(b-a)}(b-a) + \frac{\varepsilon}{2} = \varepsilon$$

for φ sufficiently close to φ_0. Since $\varepsilon > 0$ is arbitrary, the equality (13.47) is proved. The equalities (13.48) - (13.49) are proved similarly. ∎

The hypotheses of Theorem 13.9 are of course satisfied if c, l, m, and n are jointly continuous. Other classes of functions c, l, m, n for which the hypotheses of Theorem 13.9 are satisfied will be given in the following Theorem 13.10.

Let us say that the functions l, m and n are L_1-continuous at (φ_0, t_0, s_0) if

$$(13.50) \qquad \lim_{r \to 0} \int_a^b |l(\varphi, t, s, \tau) - l(\varphi_0, t_0, s_0, \tau)| \, d\tau = 0,$$

$$(13.51) \qquad \lim_{r \to 0} \int_c^d |m(\varphi, t, s, \sigma) - m(\varphi_0, t_0, s_0, \sigma)| \, d\sigma = 0,$$

and

$$(13.52) \quad \lim_{r \to 0} \int_a^b \int_c^d |n(\varphi, t, s, \tau, \sigma) - n(\varphi_0, t_0, s_0, \tau, \sigma)| \, d\sigma \, d\tau = 0,$$

where $r = |\varphi - \varphi_0| + |t - t_0| + |s - s_0|$. Moreover, let us call these functions L_1-bounded at $\varphi \in J$ if

$$(13.53) \qquad \sup_{(t,s) \in D} \int_a^b |l(\varphi, t, s, \tau)| \, d\tau = L(\varphi) < \infty,$$

(13.54)
$$\sup_{(t,s)\in D} \int_c^d |m(\varphi,t,s,\sigma)|\, d\sigma = M(\varphi) < \infty,$$

and

(13.55)
$$\sup_{(t,s)\in D} \int_a^b \int_c^d |n(\varphi,t,s,\tau,\sigma)|\, d\sigma\, d\tau = N(\varphi) < \infty.$$

Theorem 13.10. *Suppose that the function c is jointly continuous, and the functions $l(\varphi,\cdot,\cdot,\cdot)$, $m(\varphi,\cdot,\cdot,\cdot)$, and $n(\varphi,\cdot,\cdot,\cdot,\cdot)$ are L_1-continuous and L_1-bounded for each φ. Then the operator function $P(\cdot)$ is norm continuous in the space $\mathfrak{L}(C)$.*

□ First of all, the continuity of the function c and the L_1-continuity and L_1-boundedness of the functions l, m and n at φ are obviously sufficient for the operator $P(\varphi)$ to act in $C(D)$. Given $\varphi_0 \in J$, we show that $\lim_{\varphi\to\varphi_0} \|P(\varphi) - P(\varphi_0)\| = 0$. We have

$$\|P(\varphi) - P(\varphi_0)\| = \sup_{\|x\|\leq 1} \|[P(\varphi) - P(\varphi_0)]x\|$$

$$= \sup_{\|x\|\leq 1} \max_{(t,s)\in D} |[P(\varphi) - P(\varphi_0)]x(t,s)|$$

$$\leq \max_{(t,s)\in D} |c(\varphi,t,s) - c(\varphi_0,t,s)|$$

(13.56)
$$+ \max_{(t,s)\in D} \int_a^b |l(\varphi,t,s,\tau) - l(\varphi_0,t,s,\tau)|\, d\tau$$

$$+ \max_{(t,s)\in D} \int_c^d |m(\varphi,t,s,\sigma) - m(\varphi_0,t,s,\sigma)|\, d\sigma$$

$$+ \max_{(t,s)\in D} \int_a^b \int_c^d |n(\varphi,t,s,\tau,\sigma) - n(\varphi_0,t,s,\tau,\sigma)|\, d\sigma\, d\tau.$$

Since the function $c(\cdot,t,s)$ is continuous at φ_0 and the functions l, m and n are L_1-continuous, for each $\varepsilon > 0$ we can find a $\delta > 0$ such that

$$|c(\varphi,t,s) - c(\varphi_0,t,s)| < \frac{\varepsilon}{4},$$

$$\int_a^b |l(\varphi, t, s, \tau) - l(\varphi_0, t, s, \tau)| \, d\tau < \frac{\varepsilon}{4},$$

$$\int_c^d |m(\varphi, t, s, \sigma) - m(\varphi_0, t, s, \sigma)| \, d\sigma < \frac{\varepsilon}{4},$$

and

$$\int_a^b \int_c^d |n(\varphi, t, s, \tau, \sigma) - n(\varphi_0, t, s, \tau, \sigma)| \, d\sigma \, d\tau < \frac{\varepsilon}{4}$$

for $|\varphi - \varphi_0| < \delta$. Combining these inequalities and using (13.56) we get $\|P(\varphi) - P(\varphi_0)\| < \varepsilon$ for $|\varphi - \varphi_0| < \delta$ which proves the assertion. ∎

The hypotheses of Theorem 13.10 and, in particular, the L_1-continuity of the kernels l, m and n hold if l, m and n are jointly continuous. However, as was shown in FROLOVA-KALITVIN-ZABREJKO [1997] and KALITVIN-FROLOVA [1995], the kernel functions l, m and n are still L_1-continuous if they are discontinuous only along finitely many surfaces with continuous parameter representations $\tau = \tau(\varphi, t, s)$ and $\sigma = \sigma(\varphi, t, s)$.

We point out that many partial integral equations which arise in applications have to be considered, for physical reasons, neither in ideal spaces nor in spaces of continuous functions. Therefore, partial integral equations in other spaces are of particular interest, as those studied in KALITVIN-JANKELEVICH [1994], KALITVIN-DEMANOVA [1995], KALITVIN-KOLESNIKOVA [1995], or KALITVIN-NASONOV [1996].

§ 14. Spectral properties

In this section we study some spectral properties of the partial integral operator (11.1). Unfortunately, very few is known on the spectrum of this operator in the general case. For special classes of partial integral operators, however, some information is available. We illustrate this in the present section for the operator

(14.1)

$$Kx(t,s) = \int_T l(t,\tau)x(\tau,s)\,d\mu(\tau)$$

$$+ \int_S m(s,\sigma)x(t,\sigma)\,d\nu(\sigma).$$

This operator is general enough to cover various classes of equations arising in applications, and, on the other hand, special enough to admit a precise description of its spectral properties.

14.1. Essential spectra of bounded linear operators

Let us recall some notions from the general spectral theory of bounded linear operators. Given such an operator A in a complex Banach space X, we denote, as usual, by $\rho(A)$ the *resolvent set*, by $\sigma(A)$ the *spectrum*, and by $r_\sigma(A)$ the *spectral radius* of A. Important subsets of $\sigma(A)$ are the *point spectrum*

$$\sigma_p(A) = \{\lambda : \lambda \in \mathbb{C}, (A - \lambda I)x = 0$$

$$\text{for some } x \text{ with } \|x\|_X = 1\}$$

(consisting of all *eigenvalues* of A), the *limit spectrum*

$$\sigma_l(A) = \{\lambda : \lambda \in \mathbb{C}, (A - \lambda I)x_n \to 0$$

$$\text{for some sequence } (x_n)_n \text{ with } \|x_n\|_X = 1\},$$

and the *defect spectrum*

$$\sigma_d(A) = \{\lambda : \lambda \in \mathbb{C}, (A - \lambda I)^* x_n^* \to 0$$

$$\text{for some sequence } (x_n^*)_n \text{ with } \|x_n^*\|_{X^*} = 1\}$$

of A. Thus, $\lambda \notin \sigma_l(A)$ means that

$$(14.2) \qquad\qquad ||Ax - \lambda x|| \geq c||x|| \qquad (x \in X)$$

for some $c > 0$. Moreover, by $\sigma_+(A)$ and $\sigma_-(A)$ we denote the sets of all $\lambda \in \mathbb{C}$ such that the dimension $n_+(A - \lambda I)$ of the nullspace of $A - \lambda I$ and the codimension $n_-(A - \lambda I)$ of the range of $A - \lambda I$, respectively, are infinite or the range of $A - \lambda I$ is not closed. Finally, $\text{ind}\,(A - \lambda I) = n_+(A - \lambda I) - n_-(A - \lambda I)$ is the *index* of A, and $\kappa(A, \lambda)$ is the *algebraic multiplicity* of an eigenvalue $\lambda \in \sigma_p(A)$.

We recall that there are various definitions of the *essential spectrum* of A. The most important ones are the sets

$$\sigma_{ek}(A) = \sigma_+(A) \cap \sigma_-(A),$$

$$\sigma_{ew}(A) = \sigma_+(A) \cup \sigma_-(A),$$

and

$$\sigma_{es}(A) = \sigma_{ew}(A) \cup \{\lambda : \lambda \in \sigma(A), 0 \neq \text{ind}\,(A - \lambda I) < \infty\};$$

they are called the essential spectrum of A in the sense of KATO [1966], WOLF [1959], and SCHECHTER [1965], respectively. Moreover, a point $\lambda \in \sigma(A)$ belongs to the essential spectrum $\sigma_{eb}(A)$ in the sense of BROWDER [1960] if either the range of $A - \lambda I$ is not closed, or λ is an accumulation point of $\sigma(A)$, or λ is an eigenvalue of A of infinite multiplicity. Thus, $\lambda \notin \sigma_{ew}(A)$ means that $A - \lambda I$ is Fredholm, $\lambda \notin \sigma_{es}(A)$ means that $A - \lambda I$ is Fredholm of index zero, and $\lambda \notin \sigma_{eb}(A)$ means that λ is an isolated eigenvalue of finite multiplicity.

All sets mentioned above are closed subsets of $\sigma(A)$, hence compact. For further reference, we recall some relations between these subsets (see e.g. KATO [1966], KREJN [1971] or ISHINOSE [1978, 1978a]).

Lemma 14.1. *The following inclusions are true, where ∂M denotes the boundary of a set $M \subset \mathbb{C}$:*

(a) $\sigma_{ek}(A) \subseteq \sigma_{ew}(A) \subseteq \sigma_{es}(A) \subseteq \sigma_{eb}(A)$;

(b) $\sigma_{ek}(A) \cup \sigma_p(A) = \sigma_l(A)$;

(c) $\partial\sigma_{eb}(A) \subseteq \partial\sigma_{es}(A) \subseteq \partial\sigma_{ew}(A) \subseteq \partial\sigma_{+}(A), \partial\sigma_{-}(A) \subseteq \partial\sigma_{ek}(A)$;

(d) $\partial\sigma(A) = [\sigma(A) \setminus \sigma_{eb}(A)] \cup \partial\sigma_{eb}(A) \subseteq \partial\sigma_{l}(A)$.

Moreover, the equalities

$$\sigma_{+}(A^*) = \{\overline{\lambda} : \lambda \in \sigma_{-}(A)\}, \quad \sigma_{-}(A^*) = \{\overline{\lambda} : \lambda \in \sigma_{+}(A)\},$$

and

$$\sigma_{ek}(A^*) = \{\overline{\lambda} : \lambda \in \sigma_{ek}(A)\}, \; \sigma_{es}(A^*) = \{\overline{\lambda} : \lambda \in \sigma_{es}(A)\},$$

$$\sigma_{ew}(A^*) = \{\overline{\lambda} : \lambda \in \sigma_{ew}(A)\}$$

hold.

It is not hard to see that $\sigma(A) = \sigma_l(A) \cup S_r(A)$, where $S_r(A) = \{\lambda : \lambda \in \sigma(A), \|(A - \lambda I)x\| \geq c\|x\|, x \in X, c > 0\}$ is an open set. We also mention the following perturbation result (KATO [1966], KREJN [1971]): if $\lambda \notin \sigma_a(A)$ ($a \in \{es, ew, +, -\}$) and B is a compact linear operator in X or an operator with sufficiently small norm, then $\lambda \notin \sigma_a(A + B)$.

In the case of a compact operator A we have $\sigma_a(A) = \{0\}$ ($a \in \{+, -, ew, es, eb\}$) and $\sigma_l(A) = \sigma_d(A) = \sigma(A)$. Spectral mapping results like $\sigma(A^2) = \sigma(A)^2$ and $\sigma_a(A^2) = \sigma_a(A)^2$, where $a = +, -, l$, ew, eb, or p may be found in ICHINOSE [1978].

14.2. Application to partial integral operators in L_2

Since the operator (14.1) contains the special kernels $l = l(t, \tau)$ and $m = m(s, \sigma)$, it is natural to study also the corresponding ordinary integral operators

(14.3)
$$\tilde{L}u(t) = \int_T l(t, \tau)u(\tau)\, d\mu(\tau)$$

and

(14.4)
$$\tilde{M}v(s) = \int_S m(s, \sigma)v(\sigma)\, d\nu(\sigma),$$

and to try to get information on the spectrum of the operator (14.1) in terms of the spectrum of these operators. In fact, one may write the operator (14.1) in the form

$$(14.5) \qquad K = \tilde{L}\overline{\otimes}I + I\overline{\otimes}\tilde{M},$$

where $A\overline{\otimes}B$ denotes the closure of the operator $A \otimes B$ which acts on all linear combinations of the functions $(u \otimes v)(t, s) = u(t)v(s)$, i.e.

$$(\tilde{L} \otimes I)\left(\sum_{i=1}^{n} u_i \otimes v_i\right) = \sum_{i=1}^{n} \tilde{L}u_i \otimes v_i,$$

$$(I \otimes \tilde{M})\left(\sum_{i=1}^{n} u_i \otimes v_i\right) = \sum_{i=1}^{n} u_i \otimes \tilde{M}v_i.$$

By (14.5), the space in which we study the operator K should therefore be chosen as completion of a *tensor product* of two function spaces which are "natural" for the operators $\tilde{L}$ and $\tilde{M}$. For example, this could be the Lebesgue space $L_p(T \times S)$ for $1 \leq p < \infty$ or, more generally, a separable ideal space over $T \times S$ with mixed norm.

The basic facts which will be proved in this subsection are due to KALITVIN [1984, 1985, 1985a, 1986a, 1988a, 1988b]. For $L_p(T \times S)$ ($1 \leq p < \infty$) and for spaces with mixed norms which are are quasi-uniform crossnorms they follow from results by ISHINOSE [1978, 1978a], while the case of general mixed norms was considered in KALITVIN [1986, 1986a]. Recall that a crossnorm $||\cdot||$ on a tensor product $X\otimes Y$ is called *quasi-uniform* if, for any two operators $A \in \mathfrak{L}(X)$ and $B \in \mathfrak{L}(Y)$, one has

$$||(A \otimes B)\left(\sum_{i=1}^{n} u_i \otimes v_i\right)|| \leq c||A||_{\mathfrak{L}(X)}||B||_{\mathfrak{L}(Y)}||\sum_{i=1}^{n} u_i \otimes v_i||$$

for some $c \geq 1$; if one may choose $c = 1$ in this estimate, the crossnorm is called *uniform*.

Some properties of the operator (14.1) in case of degenerate or Jordan kernels have been studied in VITOVA [1975, 1976, 1977, 1984], and in case of symmetric kernels in LIKHTARNIKOV [1974, 1975] and LIKHTARNIKOV-VITOVA [1975]. If the sets S and T are compact and

the kernels l and m in (14.1) are jointly continuous, the results of
ISHINOSE [1978] may be used to get analogous results for K in the
space $C(T \times S)$.

In this subsection we suppose throughout that the operator (14.3)
acts in the space $L_2(T)$, and the operator (14.4) acts in the space
$L_2(S)$. By Theorem 12.1, the operator (14.1) acts then in the space
$L_2(T \times S)$ which may be represented as completion of the tensor
product $L_2(T) \otimes L_2(S)$. On the dense subset $L_2(T) \otimes L_2(S) \subset L_2(T \times S)$ the operator K is given by (14.5). This allows us to describe the
spectrum of K in $L_2(T \times S)$ by means of the spectra of the operators
$\tilde{L}$ and $\tilde{M}$ in $L_2(T)$ and $L_2(S)$, respectively.

The papers ISHINOSE [1978, 1978a] contain a detailed account of spectral properties of tensor products of closed linear operators with respect to crossnorms with certain additional properties. Since these
properties are satisfied in our case here, we may apply the results of
ISHINOSE [1978, 1978a] to derive a number of theorems on the spectrum of the operator (14.1). Let us state three such theorems without
proof. Throughout this subsection, we use the abbreviations

$$(14.6) \quad \begin{cases} a_k(\alpha) := n_+((\tilde{L} - \alpha I)^k), \quad b_k(\beta) := n_+((\tilde{M} - \beta I)^k), \\ c_k(\alpha) := n_-((\tilde{L} - \alpha I)^k), \quad d_k(\beta) := n_-((\tilde{M} - \beta I)^k), \end{cases}$$

where $n_+(A)$ and $n_-(A)$ are as above, and $\alpha, \beta \in \mathbb{C}$ are fixed.

Theorem 14.1. *The equalities*

$$(14.7) \qquad \sigma(K) = \sigma(\tilde{L}) + \sigma(\tilde{M})$$

and

$$(14.8) \qquad \sigma_{eb}(K) = [\sigma_{eb}(\tilde{L}) + \sigma(\tilde{M})] \cup [\sigma(\tilde{L}) + \sigma_{eb}(\tilde{M})]$$

hold, and for any $\lambda \notin \sigma_{eb}(K)$ *we have*

$$\kappa(K, \lambda) = \sum_{\substack{\alpha+\beta=\lambda \\ \alpha \in A, \beta \in B}} \kappa(\tilde{L}, \alpha)\kappa(\tilde{M}, \beta),$$

where $A = \sigma(\tilde{L}) \setminus \sigma_{eb}(\tilde{L})$ and $B = \sigma(\tilde{M}) \setminus \sigma_{eb}(\tilde{M})$. Moreover, the equality

$$(14.9) \qquad \sigma_{ew}(K) = [\sigma_{ew}(\tilde{L}) + \sigma(\tilde{M})] \cup [\sigma(\tilde{L}) + \sigma_{ew}(\tilde{M})]$$

holds, and for any $\lambda \notin \sigma_{ew}(K)$ we have

$$\operatorname{ind}(K - \lambda I) =$$

$$(14.10) \qquad \sum_{\substack{\alpha+\beta=\lambda \\ \alpha \in A, \beta \in B'}} \operatorname{ind}(\tilde{M} - \beta I) \sum_{k=1}^{\infty}[a_k(\alpha) - a_{k-1}(\alpha)]$$

$$+ \sum_{\substack{\alpha+\beta=\lambda \\ \alpha \in A', \beta \in B}} \operatorname{ind}(\tilde{L} - \alpha I) \sum_{k=1}^{\infty}[b_k(\beta) - b_{k-1}(\beta)],$$

where $A' = \sigma(\tilde{L}) \setminus \sigma_{ew}(\tilde{L})$ and $B' = \sigma(\tilde{M}) \setminus \sigma_{ew}(\tilde{M})$. Finally, the equality

$$(14.11) \qquad \begin{aligned} \sigma_{es}(K) &= \sigma_{ew}(K) \cup \\ \{\lambda : \lambda \in C, \ \lambda \notin \sigma_{ew}(K), \ 0 \neq \operatorname{ind}(K - \lambda I) &< \infty\} \end{aligned}$$

holds.

Theorem 14.2. *The equalities*

$$(14.12) \qquad \sigma_l(K) = \sigma_l(\tilde{L}) + \sigma_l(\tilde{M})$$

and

$$(14.13) \qquad \sigma_+(K) = [\sigma_+(\tilde{L}) + \sigma_l(\tilde{M})] \cup [\sigma_l(\tilde{L}) + \sigma_+(\tilde{M})]$$

hold, and for any $\lambda \notin \sigma_+(K)$ we have

$$n_+(K - \lambda I) =$$

$$(14.14) \qquad \sum_{\substack{\alpha+\beta=\lambda \\ \alpha \in \sigma_l(\tilde{L}), \beta \in \sigma_l(\tilde{M})}} \sum_{k=1}^{\infty}[a_k(\alpha) - a_{k-1}(\alpha)][b_k(\beta) - b_{k-1}(\beta)].$$

Theorem 14.3. *The equalities*

(14.15) $$\sigma_d(K) = \sigma_d(\tilde{L}) + \sigma_d(\tilde{M})$$

and

(14.16) $$\sigma_-(K) = [\sigma_-(\tilde{L}) + \sigma_d(\tilde{M})] \cup [\sigma_d(\tilde{L}) + \sigma_-(\tilde{M})]$$

hold, and for any $\lambda \notin \sigma_-(K)$ *we have*

$$n_-(K - \lambda I) =$$

(14.17)
$$\sum_{\substack{\alpha+\beta=\lambda \\ \alpha \in \sigma_d(\tilde{L}), \beta \in \sigma_d(\tilde{M})}} \sum_{k=1}^{\infty} [c_k(\alpha) - c_{k-1}(\alpha)][d_k(\beta) - d_{k-1}(\beta)].$$

The preceding three theorems give a precise description of the essential spectra $\sigma_{eb}(K)$ and $\sigma_{ew}(K)$. Since the essential spectrum in Kato's sense $\sigma_{ek}(K)$ is given by $\sigma_{ek}(K) = \sigma_+(K) \cap \sigma_-(K)$, it may be "calculated" from the right-hand sides of (14.13) and (14.16).

We illustrate the above theorems by means of four typical examples which often arise in applications.

Example 14.1. Suppose that the operator (14.3) is compact in $L_2(T)$, and the operator (14.4) is compact in $L_2(S)$. Then their spectra have the form

$$\sigma(\tilde{L}) = \{0, \alpha_1, \alpha_2, \ldots, \alpha_n, \ldots\}, \qquad \sigma(\tilde{M}) = \{0, \beta_1, \beta_2, \ldots, \beta_n, \ldots\}.$$

We have then

$$\sigma(\tilde{L}) = \sigma_l(\tilde{L}) = \sigma_d(\tilde{L}), \quad \sigma(\tilde{M}) = \sigma_l(\tilde{M}) = \sigma_d(\tilde{M}),$$

$$\sigma_+(\tilde{L}) = \sigma_-(\tilde{L}) = \sigma_{ek}(\tilde{L}) = \sigma_{ew}(\tilde{L})$$

$$= \sigma_{es}(\tilde{L}) = \sigma_{eb}(\tilde{L}) = \{0\},$$

and

$$\sigma_+(\tilde{M}) = \sigma_-(\tilde{M}) = \sigma_{ek}(\tilde{M}) = \sigma_{ew}(\tilde{M})$$

$$= \sigma_{es}(\tilde{M}) = \sigma_{eb}(\tilde{M}) = \{0\}.$$

From Theorems 14.1 - 14.3 we conclude that

$$\sigma_+(K) = \sigma_-(K) = \sigma_{ek}(K) = \sigma_{ew}(K)$$

$$= \sigma_{es}(K) = \sigma_{eb}(K) = \sigma(\tilde{L}) \cup \sigma(\tilde{M}).$$

Obviously, $K - \lambda I$ is Fredholm (of index zero) if and only if λ is a point of resolvent set or λ is an isolated eigenvalue of finite multiplicity, i.e. $\lambda = \alpha + \beta$ with $\alpha \in \sigma(\tilde{L})$ and $\beta \in \sigma(\tilde{M})$, or $\lambda = \alpha + \beta$ with $\lambda \neq 0, \gamma, \delta$ for $\alpha, \gamma \in \sigma(\tilde{L})$ and $\beta, \delta \in \sigma(\tilde{M})$.

Example 14.2. Suppose that the operator (14.3) is self-adjoint in $L_2(T)$, and the operator (14.4) is self-adjoint in $L_2(S)$. For a self-adjoint bounded linear operator in a Hilbert space, all types of essential spectra coincide. Consequently, Theorems 14.1 - 14.3 imply that K has the essential spectrum $[\sigma_{eb}(\tilde{L}) + \sigma(\tilde{M})] \cup [\sigma(\tilde{L}) + \sigma_{eb}(\tilde{M})]$. Moreover, $K - \lambda I$ is Fredholm (of index zero) if and only if either $\lambda \neq \alpha + \beta$ for $\alpha \in \sigma(\tilde{L})$ and $\beta \in \sigma(\tilde{M})$, or λ satisfies

$$(14.18) \qquad \begin{aligned} \lambda \in \{[\sigma(\tilde{L}) \setminus \sigma_{eb}(\tilde{L})] + [\sigma(\tilde{M}) \setminus \sigma_{eb}(\tilde{M})]\} \\ \setminus \{[\sigma(\tilde{L}) + \sigma_{eb}(\tilde{M})] \cup [\sigma_{eb}(\tilde{L}) + \sigma(\tilde{M})]\}. \end{aligned}$$

Let us take a closer look to the description of the eigenvalues of the operator K in this example. Equality (14.8) implies that a number $\lambda \in \sigma(K)$ is an isolated eigenvalue of finite multiplicity if and only if λ satisfies (14.18). In particular, if one of the operators $\tilde{L}$ or $\tilde{M}$ is zero, then K has no eigenvalues of finite multiplicity at all.

It is clear that $\sigma_p(\tilde{L}) + \sigma_p(\tilde{M}) \subseteq \sigma_p(K)$. On the other hand, from equality (14.7) and our assumption $\sigma(\tilde{L}) = \sigma_p(\tilde{L}), \sigma(\tilde{M}) = \sigma_p(\tilde{M})$ it follows that

$$(14.19) \qquad \sigma_p(K) = \sigma_p(\tilde{L}) + \sigma_p(\tilde{M}).$$

Example 14.3. Suppose that both operators (14.3) and (14.4) have degenerate kernels $l = l(t, \tau)$ and $m = m(s, \sigma)$, respectively. In this case both $\tilde{L}$ and $\tilde{M}$ have a pure point spectrum, and hence also K has a pure point spectrum which is given by (14.19).

The following example is essentially due to ICHINOSE [1978a]:

Example 14.4. Let $S = T = [0, 1]$ and let $\{e_1, e_2, \ldots, e_n, \ldots\}$ be an orthonormal basis in $L_2([0, 1])$. Suppose that

$$l(t, \tau) = -\sum_{k=2}^{\infty} \frac{e_{k-1}(t)\overline{e_k(\tau)}}{(k-1)^2}, \quad m(s, \sigma) = \sum_{k=1}^{\infty} \frac{e_{k+1}(s)\overline{e_k(\sigma)}}{k^2(k+1)}.$$

Then both operators $\tilde{L}$ and $\tilde{M}$ are compact in $L_2([0, 1])$, and the formulas

$$\tilde{L}e_1 = 0, \ \tilde{L}e_2 = -e_1, \ \ldots, \ \tilde{L}e_n = -\frac{1}{(n-1)^2}e_{n-1}, \ldots$$

and

$$\tilde{M}e_1 = \frac{1}{2}e_2, \ \tilde{M}e_2 = \frac{1}{12}e_3, \ \ldots, \ \tilde{M}e_n = \frac{1}{n^2(n+1)}e_{n+1}, \ldots$$

hold. It is easy to see that $\sigma(\tilde{L}) = \sigma_p(\tilde{L}) = \sigma(\tilde{M}) = \{0\}$ and $\sigma_p(\tilde{M}) = \emptyset$. Thus, for the operator K we have $\sigma(K) = \sigma(\tilde{L}) + \sigma(\tilde{M}) = \{0\}$, by (14.7). To find the point spectrum of K, consider the function

$$x = \sum_{n=1}^{\infty} \frac{e_n \otimes e_n}{n!} \in L_2([0, 1] \times [0, 1])$$

which is not zero. We claim that x is an eigenfunction of K corresponding to the eigenvalue $\lambda = 0$. In fact, by (14.5) we have

$$Kx = K\left(\sum_{n=1}^{\infty} \frac{e_n \otimes e_n}{n!}\right) = \sum_{n=1}^{\infty} \frac{e_n}{n!} \otimes \tilde{M}e_n + \sum_{n=1}^{\infty} \tilde{L}e_n \otimes \frac{e_n}{n!}$$

$$= \sum_{n=1}^{\infty} \frac{e_n \otimes e_{n+1}}{n! n^2(n+1)} - \sum_{n=2}^{\infty} \frac{e_{n-1} \otimes e_n}{n!(n-1)^2}$$

$$= \sum_{n=1}^{\infty} \frac{e_n \otimes e_{n+1}}{(n+1)! n^2} - \sum_{n=1}^{\infty} \frac{e_n \otimes e_{n+1}}{(n+1)! n^2} = 0.$$

This shows that $\sigma_p(K) = \{0\}$, i.e. the equality (14.19) fails in this example.

In view of this example, the problem arises to find general classes of operators K for which the equality (14.19) is true. One such class is described in the following

Theorem 14.4. *Suppose that the operator* (14.3) *is bounded in $L_2(T)$, the operator* (14.4) *is bounded in $L_2(S)$, and the eigenfunctions of either $\tilde{L}$ or $\tilde{M}$ form an unconditional basis in the corresponding L_2 space. Then the equality* (14.19) *holds.*

$\square$ Since always $\sigma_p(K) \supseteq \sigma_p(\tilde{L}) + \sigma_p(\tilde{M})$, we only have to prove the reverse inclusion. Let $\sigma_p(\tilde{L}) = \{\alpha_1, \alpha_2, \ldots, \alpha_n, \ldots\}$ and suppose, for definiteness, that the corresponding eigenfunctions $\{\phi_1, \phi_2, \ldots, \phi_n, \ldots\}$ form an unconditional basis in $L_2(T)$. Moreover, let $\{\psi_1, \psi_2, \ldots, \psi_n, \ldots\}$ be an arbitrary orthonormal basis in $L_2(S)$. By Lorch's theorem (GOKHBERG-KREJN [1965]) there exists an isomorphism $J : L_2(T) \to L_2(T)$ such that the set $\{J\phi_1, J\phi_2, \ldots, J\phi_n, \ldots\}$ is an orthonormal basis in $L_2(T)$. This implies that $J\overline{\otimes}I : L_2(T \times S) \to L_2(T \times S)$ is an isomorphism which maps the set $\{\phi_n \otimes \psi_m : m, n = 1, 2, \ldots\}$ onto the set $\{J\phi_n \otimes \psi_m : m, n = 1, 2, \ldots\}$. Again by Lorch's theorem, $\{\phi_n \otimes \psi_m : m, n = 1, 2, \ldots\}$ is then an unconditional basis in $L_2(T \times S)$.

Now fix $\lambda \in \sigma_p(K)$. The equation

$$(14.20) \qquad (\tilde{L}\overline{\otimes}I + I\otimes\tilde{M})x = \lambda x$$

has then a nontrivial solution

$$(14.21) \qquad x_0(t, s) = \sum_{m,n=1}^{\infty} c_{mn}\phi_n(t)\psi_m(s).$$

Putting (14.21) into (14.20) and observing that $\{\phi_n \otimes \psi_m : m, n = 1, 2, \ldots\}$ is an unconditional basis we obtain

$$\sum_{n=1}^{\infty} \phi_n \otimes \left[(\lambda - \alpha_n)\sum_{m=1}^{\infty} c_{mn}\psi_m - \tilde{M}\left(\sum_{m=1}^{\infty} c_{mn}\psi_m\right)\right] = 0,$$

hence

$$(14.22) \qquad (\lambda - \alpha_n) \sum_{m=1}^{\infty} c_{mn}\psi_m - \tilde{M}\left(\sum_{m=1}^{\infty} c_{mn}\psi_m\right) = 0$$

for all n. Since the function x_0 is not zero, we have $c_{m_0 n_0} \neq 0$ for at least one pair (m_0, n_0). But this means that the equation

$$(\lambda - \alpha_{n_0}) \sum_{m=1}^{\infty} c_{mn_0}\psi_m - \tilde{M}\left(\sum_{m=1}^{\infty} c_{mn_0}\psi_m\right) = 0$$

has a nontrivial solution, i.e. the number $\beta = \lambda - \alpha_{n_0}$ is an eigenvalue of $\tilde{M}$. We conclude that $\lambda = \alpha_{n_0} + \beta \in \sigma_p(\tilde{L}) + \sigma_p(\tilde{M})$ as claimed. ■ Obviously, the assertion of Theorem 14.4 is true if the operator $\tilde{L}$ is compact and self-adjoint in $L_2(T)$. On the other hand, Example 14.3 shows that the conditions of Theorem 14.4 are not necessary for the equality (14.19) to be true.

Let us now show how to find the eigenfunctions of the operator K under the hypotheses of Theorem 14.4. If equation (14.20) has a non-trivial solution x, we have $\lambda = \alpha + \beta$ with appropriate $\alpha \in \sigma_p(\tilde{L})$ and $\beta \in \sigma_p(\tilde{M})$, by Theorem 14.4. Without loss of generality we may assume that α and β are unique in this representation. Let $P(\alpha) = \{n : \alpha_n = \alpha\}$; the system (14.22) may then be written in the form

$$(14.23) \qquad \beta \sum_{m=1}^{\infty} c_{mn}\psi_m - \tilde{M}\left(\sum_{m=1}^{\infty} c_{mn}\psi_m\right) = 0$$

if $n \in P(\alpha)$, and $c_{mn} = 0$ if $n \notin P(\alpha)$. Since β is an eigenvalue of the operator $\tilde{M}$, the set Γ of all nontrivial solutions $\gamma = \gamma(s)$ of (14.23) is nonempty. But then the set

$$(14.24) \qquad E(\lambda) = \{\phi_n \otimes \gamma : n \in P(\alpha), \gamma \in \Gamma\}$$

forms a complete system of eigenfunctions of K corresponding to the eigenvalue λ. Thus we have proved the following

Theorem 14.5. *Under the hypotheses of Theorem 14.4, the set of all eigenfunctions of K which correspond to the eigenvalue $\lambda = \alpha + \beta$ is given by (14.24).*

We remark that a particularly precise description of the set of all eigenfunctions (or associated eigenfunctions) is possible if the operators (14.3) and (14.4) have degenerate kernels. More information for this case may be found in VITOVA [1976a, 1977, 1988].

14.3. Application to other partial integral operators

In the proofs of Theorems 14.1 - 14.3 we have used the results of ISHINOSE [1978, 1978a] on the spectrum of a tensor product of two operators in a Banach space with a quasi-uniform crossnorm. Examples of tensor products may be constructed as follows. If U is a regular ideal space over T, and V is a regular ideal space over S (see Subsection 4.1), the tensor products $U \tilde{\otimes} V$ and $V \tilde{\otimes} U$ coincide (as sets) with the spaces with mixed norm $[U \leftarrow V]$ and $[U \rightarrow V]$ (see (12.1) and (12.2)). Unfortunately, the corresponding crossnorms are, even in the simplest case $U = L_p(T)$ and $V = L_q(S)$, not quasi-uniform (see BUKHVALOV [1983]). On the other hand, as we have seen in § 12, ideal spaces with mixed norms are quite natural for the investigation of partial integral operators. Therefore it seems useful to study the spectral properties of partial integral operators in such spaces from a general point of view.

First of all, suppose that the operator (14.3) acts in the space $U = L_p(T)$, and the operator (14.4) acts in the space $V = L_p(S)$ $(1 \le p < \infty)$. Then the operator (14.1) acts in the space $[U \leftarrow V] = [U \rightarrow V] = U \tilde{\otimes} V = V \tilde{\otimes} U = L_p(T \times S)$ with the norm

$$(14.25) \qquad ||x|| = \left\{ \int_S \int_T |x(t,s)|^p \, d\mu(t) \, d\nu(s) \right\}^{1/p}.$$

Since this crossnorm is quasi-uniform, the operator K may be written in the form (14.5), and the results of ISHINOSE [1978] apply.

In the following Theorems 14.6 and 14.8 we prove some formulas for the spectrum, the essential spectrum in Browder's sense, and the essential spectrum in Wolf's sense, without assuming that the crossnorms on the spaces $U \tilde{\otimes} V = [U \leftarrow V]$, $V \tilde{\otimes} U = [V \rightarrow U]$ are quasi-uniform. To this end, we denote by $\rho_{reg}(A)$ the set of all $\lambda \in \rho(A)$ such that $(A - \lambda I)^{-1}$ is regular.

Theorem 14.6. *Suppose that the operator (14.3) acts in some regular ideal space U over T, the operator (14.4) acts in some regular ideal space V over S, and one of the following two conditions is satisfied:*
(a) $\check{L}$ is regular in U and $\rho(\check{L}) = \rho_{reg}(\check{L})$;
(b) $\tilde{M}$ is regular in V and $\rho(\tilde{M}) = \rho_{reg}(\tilde{M})$.
Then

$$(14.26) \qquad \sigma(K) = \sigma(\check{L}) + \sigma(\tilde{M})$$

and

$$(14.27) \qquad \sigma_{eb}(K) = [\sigma_{eb}(\check{L}) + \sigma(\tilde{M})] \cup [\sigma(\check{L}) + \sigma_{eb}(\tilde{M})],$$

where the operator K is considered in the space $[U \leftarrow V]$ in case (a) *and in the space $[U \rightarrow V]$ in case* (b).

☐ We prove the assertion only for condition (a); the proof for condition (b) is similar. Suppose that condition (a) holds. First we show that the operator $L : [U \leftarrow V] \rightarrow [U \leftarrow V]$ given by

$$(14.28) \qquad Lx(t, s) = \int_T l(t, \tau) x(\tau, s) \, d\mu(\tau)$$

has the same spectrum as the operator $\check{L} : U \rightarrow U$ given by (14.3). Given $\alpha \in \rho(L)$, the operator $L - \alpha I$ is invertible in $[U \leftarrow V]$. Fix an element $e \in V \setminus \{0\}$; then the operator $L - \alpha I$ leaves the subspace $E = \{x \otimes e : x \in U\}$ invariant. Furthermore, since $L - \alpha I$ is invertible on E, the operator $\check{L} - \alpha I$ is invertible on U. We conclude that $\alpha \in \rho(\check{L})$ and hence $\sigma(\check{L}) \subseteq \sigma(L)$.

Conversely, let $\alpha \in \rho(\check{L})$. Since the operator $(\check{L} - \alpha I)^{-1}$ is regular, the operator $A = (\check{L} - \alpha I)^{-1} \otimes I : U \otimes V \rightarrow U \otimes V$ admits a continuous extension $\tilde{A}$ on $U \tilde{\otimes} V = [U \leftarrow V]$ (see Theorem 6 in LEVIN [1969]). But the operator $\tilde{A}$ is the inverse of the operator $L - \alpha I$ on $U \otimes V$, and hence also on $U \tilde{\otimes} V$. This shows that $\alpha \in \rho(L)$, and hence $\sigma(L) \subseteq \sigma(\check{L})$.

Of course, one may prove similarly that also the spectrum of the operator (14.4) on V and the spectrum of the operator

$$(14.29) \qquad M x(t, s) = \int_S m(s, \sigma) x(t, \sigma) \, d\nu(\sigma)$$

on $[U \leftarrow V]$ coincide.

Now we prove (14.26). Since $L = \tilde{L} \overline{\otimes} I$ and $M = I \overline{\otimes} \tilde{M}$, where $\tilde{L} \overline{\otimes} I$ and $I \overline{\otimes} \tilde{M}$ are the extensions of $\tilde{L} \otimes I$ and $I \overline{\otimes} \tilde{M}$, respectively, from $U \otimes V$ to $[U \leftarrow V]$, the operators L and M commute. Since the spectrum of the sum of two commuting operators is contained in the sum of the spectra of the single operators, we have

$$\sigma(K) \subseteq \sigma(L) + \sigma(M) = \sigma(\tilde{L}) + \sigma(\tilde{M}).$$

To see that the converse inclusion is also true, let $\lambda = \alpha + \beta$ with $\alpha \in \sigma(\tilde{L})$ and $\beta \in \sigma(\tilde{M})$. Now we distinguish four cases:

Case 1: $\alpha \in \sigma_l(\tilde{L}), \beta \in \sigma_l(\tilde{M})$;

Case 2: $\alpha \in \sigma_l(\tilde{L}), \beta \in \sigma(\tilde{M}) \setminus \sigma_l(\tilde{M})$;

Case 3: $\alpha \in \sigma(\tilde{L}) \setminus \sigma_l(\tilde{L}), \beta \in \sigma_l(\tilde{M})$;

Case 4: $\alpha \in \sigma(\tilde{L}) \setminus \sigma_l(\tilde{L}), \beta \in \sigma(\tilde{M}) \setminus \sigma_l(\tilde{M})$.

In Case 1 we find sequences $(x_n)_n$ in U and $(y_n)_n$ in V such that $\|x_n\| = \|y_n\| = 1$, $\tilde{L} x_n - \alpha x_n \to 0$, and $\tilde{M} y_n - \beta y_n \to 0$. For the sequence $z_n = x_n \otimes y_n \in U \tilde{\otimes} V = [U \leftarrow V]$ we have then

$$(K - \lambda I) z_n = (\tilde{L} \overline{\otimes} I + I \overline{\otimes} \tilde{M} - \lambda I)(x_n \otimes y_n)$$

$$= [(\tilde{L} - \alpha I) \overline{\otimes} I + I \overline{\otimes} (\tilde{M} - \beta I)](x_n \otimes y_n)$$

$$= ((\tilde{L} - \alpha I) x_n) \otimes y_n + x_n \otimes (\tilde{M} - \beta I) y_n \to 0$$

as $n \to \infty$. Since $\|z_n\| = 1$, we conclude that $\lambda \in \sigma_l(K) \subseteq \sigma(K)$.

Case 4 may be reduced to Case 1 by passing to the adjoint operator. In fact, in this case we have $\overline{\alpha} \in \sigma_l(\tilde{L}^*)$ and $\overline{\beta} \in \sigma_l(\tilde{M}^*)$. This means that we can find sequences $(x_n^*)_n$ in U^* and $(y_n^*)_n$ in V^* such that $\|x_n^*\| = \|y_n^*\| = 1$, $\tilde{L}^* x_n^* - \overline{\alpha} x_n^* \to 0$, and $\tilde{M}^* y_n^* - \overline{\beta} y_n^* \to 0$. Now, $\tilde{L}^*$ is regular in U^* (being the adjoint of a regular operator, see ZABREJKO

[1968]), and I is regular in V^*. So, by Levin's theorem mentioned above both operators $\tilde{L}^* \overline{\otimes} I$ and $I \overline{\otimes} \tilde{M}^*$ are bounded in the space $U^* \tilde{\otimes} V^*$. The sequence $z_n^* = x_n^* \otimes y_n^* \in U^* \tilde{\otimes} V^*$ satisfies $||z_n^*|| = 1$ and

$$(\tilde{L}^* \overline{\otimes} I + I \overline{\otimes} \tilde{M}^* - \overline{\lambda} I) z_n^* =$$

$$((\tilde{L}^* - \overline{\alpha} I) x_n^*) \otimes y_n^* + x_n^* \otimes (\tilde{M}^* - \overline{\beta} I) y_n^* \to 0$$

as $n \to \infty$. Therefore we have $\overline{\lambda} \in \sigma_l(\tilde{L}^* \overline{\otimes} I + I \overline{\otimes} \tilde{M}^*)$. Since $U^* \tilde{\otimes} V^* \subseteq [U^* \leftarrow V^*] = [U \leftarrow V]^*$ (ZABREJKO [1968]), and the operators $\tilde{L}^* \overline{\otimes} I + I \overline{\otimes} \tilde{M}^*$ and K^* coincide on $U^* \tilde{\otimes} V^*$, the operator $K^* : [U \leftarrow V]^* \to [U \leftarrow V]^*$ is an extension of the operator $\tilde{L}^* \overline{\otimes} I + I \overline{\otimes} \tilde{M}^* : U^* \tilde{\otimes} V^* \to U^* \tilde{\otimes} V^*$. We conclude that $\overline{\lambda} \in \sigma_l(K^*) \subseteq \sigma(K^*)$ and hence $\lambda \in \sigma(K)$.

Now we consider Case 2. If α is a boundary point of $\sigma(\tilde{L})$, then $\overline{\alpha} \in \sigma_l(L^*)$. Since $\overline{\beta} \in \sigma_l(B^*)$, by Case 4 we have $\lambda \in \sigma(K)$. Suppose now that α is an interior point of $\sigma_l(\tilde{L})$. Since the set $\sigma(\tilde{M}) \setminus \sigma_l(\tilde{M})$ is open, β is an interior point of $\sigma(\tilde{M})$. Obviously, we can find $\alpha_1 \in \sigma_l(\tilde{L})$ and $\beta_1 \in \sigma(\tilde{M})$ such that $\lambda = \alpha_1 + \beta_1$, with α_1 being a boundary point of $\sigma_l(\tilde{L})$ or β_1 being a boundary point of $\sigma(\tilde{M})$. The result $\lambda \in \sigma(K)$ follows now in the first situation from Case 4 and in the second situation from Case 1.

Case 3 is proved in the same way as Case 2.

Let us now prove equality (14.27). Since all sets in this equality are contained in $\sigma(K)$, we may assume that the numbers λ, α and β considered above are either isolated points of $\sigma(K), \sigma(\tilde{L})$ and $\sigma(\tilde{M})$, respectively, otherwise $\lambda \in \sigma_{eb}(K)$ and $\lambda = \alpha + \beta$, where or $\alpha \in \sigma_{eb}(\tilde{L})$ and $\beta \in \sigma(\tilde{M})$, or $\alpha \in \sigma(\tilde{L})$ and $\beta \in \sigma_{eb}(\tilde{M})$.

We claim that $\sigma(\tilde{L}) + \sigma_{eb}(\tilde{M}) \subseteq \sigma_{eb}(K)$. In fact, let $\lambda = \alpha + \beta$ with $\alpha \in \sigma(\tilde{L})$ and $\beta \in \sigma_{eb}(\tilde{M})$. Since β is an isolated point of $\sigma_{eb}(\tilde{M})$, either β has infinite multiplicity, or $\tilde{M} - \beta I$ has a non-closed range. If α belongs to $\sigma(\tilde{L}) \setminus \sigma_{eb}(\tilde{L})$ and β has infinite multiplicity, $\lambda = \alpha + \beta$ has infinite multiplicity as well. If α belongs to $\sigma(\tilde{L}) \setminus \sigma_{eb}(\tilde{L})$ and the range of $\tilde{M} - \beta I$ is not closed, we can find a noncompact sequence $(y_n)_n$ in V such that $||y_n|| = 1$ and $\tilde{M} y_n - \beta y_n \to 0$ as $n \to \infty$. Fix $x \in U$ with $||x|| = 1$ and $\tilde{L} x = \alpha x$. Then the noncompact sequence $x \otimes y_n \in [U \leftarrow V]$ satisfies $||x \otimes y_n|| = 1$ and

$$(K - \lambda I)(x \otimes y_n) = (\tilde{L} - \alpha I)(x \otimes y_n) + x \otimes (\tilde{M} - \beta I)y_n$$

$$= x \otimes (\tilde{M} - \beta I)y_n \to 0$$

as $n \to \infty$. Thus, $\lambda \in \sigma_{eb}(K)$ in both cases.

Suppose now that $\alpha \in \sigma_{eb}(\tilde{L})$. Here the following four cases are possible:

Case 1: α and β have both infinite multiplicity;

Case 2: the range of $\tilde{L} - \alpha I$ is not closed, and β has infinite multiplicity;

Case 3: α has infinite multiplicity, and the range of $\tilde{M} - \beta I$ is not closed;

Case 4: the ranges of both $\tilde{L} - \alpha I$ and $\tilde{M} - \beta I$ are not closed.

The cases 1 - 3 are treated as above. In the last case we can find noncompact sequences $(x_n)_n$ in U and $(y_n)_n$ in V such that $\|x_n\| = \|y_n\| = 1$, $\tilde{L}x_n - \alpha x_n \to 0$, and $\tilde{M}y_n - \beta y_n \to 0$. Then the noncompact sequence $z_n = x_n \otimes y_n \in [U \leftarrow V]$ satisfies $\|z_n\| = 1$ and $Kz_n - \lambda z_n \to 0$, i.e. $\lambda \in \sigma_{eb}(K)$, hence $\sigma(\tilde{L}) + \sigma_{eb}(\tilde{M}) \subseteq \sigma_{eb}(K)$.

The inclusion $\sigma_{eb}(\tilde{L}) + \sigma(\tilde{M}) \subseteq \sigma_{eb}(K)$ is proved similarly. Thus, we have shown that

$$(14.30) \qquad [\sigma_{eb}(\tilde{L}) + \sigma(\tilde{M})] \cup [\sigma(\tilde{L}) + \sigma_{eb}(\tilde{M})] \subseteq \sigma_{eb}(K).$$

To prove the reverse inclusion, consider the extension $\hat{K}$ of the operator K to the completion $U \hat{\otimes} V$ of $U \otimes V$ with respect to the weak crossnorm

$$\left\| \sum_{i=1}^{n} x_i \otimes y_i \right\| = \sup_{\substack{\|x_i^*\|_{U^*} = 1 \\ \|y_i^*\|_{V^*} = 1}} \left| \sum_{i=1}^{n} \langle x_i, x_i^* \rangle \langle y_i, y_i^* \rangle \right|.$$

Since this is a uniform crossnorm (see SCHATTEN [1950]), from the results of ISHINOSE [1978] we obtain the equalities

$$(14.31) \qquad \sigma(\hat{K}) = \sigma(\tilde{L}) + \sigma(\tilde{M})$$

and

(14.32) $\sigma_{eb}(\hat{K}) = [\sigma_{eb}(\tilde{L}) + \sigma(\tilde{M})] \cup [\sigma(\tilde{L}) + \sigma_{eb}(\tilde{M})].$

Together with (14.30) this gives

(14.33)

$$\sigma(K) \setminus \sigma_{eb}(K) \subseteq \sigma(K) \setminus \{[\sigma_{eb}(\tilde{L}) + \sigma(\tilde{M})]$$

$$\cup [\sigma(\tilde{L}) + \sigma_{eb}(\tilde{M})]\} = \sigma(\hat{K}) \setminus \sigma_{eb}(\hat{K}).$$

Now we prove that also

(14.34) $\sigma(\hat{K}) \setminus \sigma_{eb}(\hat{K}) \subseteq \sigma(K) \setminus \sigma_{eb}(K).$

For this it suffices of course to show that every eigenfunction of $\hat{K}$ which corresponds to an eigenvalue $\lambda = \alpha + \beta \in \sigma(\hat{K}) \setminus \sigma_{eb}(\hat{K})$ belongs to the space $[U \leftarrow V]$. For any such λ, the numbers α and β are isolated eigenvalues of finite multiplicity for $\tilde{L}$ and $\tilde{M}$, respectively, and hence may assume only finitely many values, i.e. $\lambda = \alpha_i + \beta_i$ $(i = 1, \ldots, p)$ with $\kappa(\tilde{L}, \alpha_i) < \infty$ and $\kappa(\tilde{M}, \beta_j) < \infty$. Then the decomposition

$$U = E_1 \oplus \ldots \oplus E_p \oplus R_1, \quad V = F_1 \oplus \ldots \oplus F_p \oplus R_2$$

holds, where E_i is the nullspace of $(\tilde{L} - \alpha_i I)^{\kappa(\tilde{L}, \alpha_i)}$, and F_j is the nullspace of $(\tilde{M} - \beta_j I)^{\kappa(\tilde{M}, \beta_j)}$ $(i, j = 1, \ldots, p)$. Consequently,

$$U \hat{\otimes} V = \sum_{i=1}^{p} \sum_{j=1}^{p} E_i \otimes F_j \oplus \sum_{i=1}^{p} E_i \otimes R_2 \oplus \sum_{j=1}^{p} R_1 \otimes F_j \oplus R_1 \hat{\otimes} R_2.$$

Since all spaces $E_i \otimes F_j$, $E_i \otimes R_2$, $R_1 \otimes F_j$ and $R_1 \hat{\otimes} R_2$ are invariant under $\hat{K}$, we obtain

$$\sigma(\hat{K}) = \bigcup_{i,j=1}^{p} \sigma(\hat{K} | E_i \otimes F_j) \cup \bigcup_{i=1}^{p} \sigma(\hat{K} | E_i \otimes R_2)$$

$$\cup \bigcup_{j=1}^{p} \sigma(\hat{K} | R_1 \otimes F_j) \cup \sigma(\hat{K} | R_1 \hat{\otimes} R_2),$$

where $\hat{K}|R$ denotes the restriction of $\hat{K}$ to R. But from the results of ISHINOSE [1978] it follows that

$$\begin{cases} \sigma(\hat{K}|E_i \otimes F_j) = \sigma(\tilde{L}|E_i) + \sigma(\tilde{M}|F_j), \\[2mm] \sigma(\hat{K}|E_i \otimes R_2) = \sigma(\tilde{L}|E_i) + \sigma(\tilde{M}|R_2), \\[2mm] \sigma(\hat{K}|R_1 \otimes F_j) = \sigma(\tilde{L}|R_1) + \sigma(\tilde{M}|F_j), \\[2mm] \sigma(\hat{K}|R_1 \hat{\otimes} R_2) = \sigma(\tilde{L}|R_1) + \sigma(\tilde{M}|R_2). \end{cases}$$

The relations $\alpha \in \sigma(\tilde{L}) \setminus \sigma_{eb}(\tilde{L})$ and $\beta \in \sigma(\tilde{M}) \setminus \sigma_{eb}(\tilde{M})$ imply that $\alpha \notin \sigma(\tilde{L}|R_1)$ and $\beta \notin \sigma(\tilde{M}|R_2)$, hence

$$\lambda \notin \bigcup_{i=1}^{p} \sigma(\hat{K}|E_i \otimes R_2) \cup \bigcup_{j=1}^{p} \sigma(\hat{K}|R_1 \otimes F_j) \cup \sigma(\hat{K}|R_1 \hat{\otimes} R_2).$$

Since $\lambda \neq \alpha_i + \beta_j$ for $i \neq j$, we have in addition $\lambda \notin \sigma(\hat{K}|E_i \otimes F_j)$ for $i \neq j$. On the other hand, $\lambda \in \sigma(\hat{K}|E_j \otimes F_j)$ for $j = 1, \ldots, p$, and thus $\lambda \in \sigma(K) \setminus \sigma_{eb}(K)$, i.e. (14.34) is true. Combining (14.30) - (14.34) we conclude now that

$$\sigma(K) \setminus \sigma_{eb}(K) = \sigma(\hat{K}) \setminus \sigma_{eb}(\hat{K})$$

$$= \sigma(K) \setminus \{[\sigma_{eb}(\tilde{L}) + \sigma(\tilde{M})] \cup [\sigma(\tilde{L}) + \sigma_{eb}(\tilde{M})]\}$$

which proves (14.27) and completes the proof. ∎

Theorem 14.6 implies, in particular, that the operators L given by (14.28) and M given by (14.29) satisfy the equalities

$$\sigma(L) = \sigma_{eb}(L) = \sigma_{ew}(L) = \sigma_{es}(L) = \sigma(\tilde{L})$$

and

$$\sigma(M) = \sigma_{eb}(M) = \sigma_{ew}(M) = \sigma_{es}(M) = \sigma(\tilde{M}).$$

If the operators $\tilde{L}$ and $\tilde{M}$ are compact in U and V, respectively, the operator K fulfills the statements of Example 14.1 in the space

$[U \leftarrow V]$ in case of hypothesis (a) in Theorem 14.6, and in the space $[U \rightarrow V]$ in case of hypothesis (b). The following is also a consequence of Theorem 14.6:

Theorem 14.7. *Suppose that the hypotheses of Theorem 14.6 are satisfied, and let $\lambda \in \sigma(K) \setminus \sigma_{eb}(K)$. Then*

$$(14.35) \qquad \kappa(K, \lambda) = \sum_{\substack{\lambda = \alpha + \beta \\ \alpha \in A, \beta \in B}} \kappa(\tilde{L}, \alpha) \kappa(\tilde{M}, \beta),$$

where $A = \sigma(\tilde{L}) \setminus \sigma_{eb}(\tilde{L})$ and $B = \sigma(\tilde{M}) \setminus \sigma_{eb}(\tilde{M})$.

☐ As in the proof of Theorem 14.6 one may show that λ necessarily belongs to $\sigma(K | E_j \otimes F_j)$ $(j = 1, \ldots, p)$. Let us denote by K_j the restriction of K to $E_j \otimes F_j$ $(j = 1, \ldots, p)$, and by K_0 the restriction of K to $(E_1 \otimes F_1) \oplus \ldots \oplus (E_p \otimes F_p)$. Fix $\lambda \in \sigma(K) \setminus \sigma_{eb}(K)$ and choose r so small that in the complex disc $|\lambda - \mu| \leq r$ there are no points from $\sigma(K)$ different from λ; this is possible, since λ is an isolated eigenvalue. Then the operators

$$P_j = \frac{1}{2\pi i} \int_{|\lambda - \mu| = r} (K_j - \mu I)^{-1} \, d\mu,$$

and

$$P_0 = \frac{1}{2\pi i} \int_{|\lambda - \mu| = r} (K_0 - \mu I)^{-1} \, d\mu$$

have rank $\kappa(K_j, \lambda)$ and $\kappa(K_0, \lambda)$, respectively, and $P_0 = P_1 \oplus \ldots \oplus P_p$. Moreover, we have

$$\kappa(K, \lambda) = \kappa(K_0, \lambda) = \sum_{j=1}^{p} \kappa(K_j, \lambda).$$

To prove (14.35) we therefore have to calculate $\kappa(K_j, \lambda)$ for $j = 1, \ldots, p$. Now, since

$$\dim (E_j \otimes F_j) = \kappa(\tilde{L}, \alpha_j) \kappa(\tilde{M}, \beta_j),$$

we have $\kappa(K_j, \lambda) \leq \kappa(\tilde{L}, \alpha_j) \kappa(\tilde{M}, \beta_j)$. Let $\{e_1^{(j)}, \ldots, e_{m_j}^{(j)}\}$ $(m_j = \kappa(\tilde{L}, \alpha_j))$ be a basis in the nullspace of the operator $(\tilde{L} - \alpha_j I)^{m_j}$, and $\{f_1^{(j)}, \ldots, f_{n_j}^{(j)}\}$ $(n_j = \kappa(\tilde{M}, \beta_j))$ be a basis in the nullspace of the operator $(\tilde{M} - \beta_j I)^{n_j}$. Then it is easy to see that $B_j = \{e_m^{(i)} \otimes f_n^{(j)} : m = 1, \ldots, m_j; n = 1, \ldots n_j\}$ is a basis in $E_j \otimes F_j$, hence also in the nullspace of the operator $(K_j - \lambda I)^{m_j n_j}$. It follows that $\kappa(K_j, \lambda) \geq \kappa(\tilde{L}, \alpha_j) \kappa(\tilde{M}, \beta_j)$, and thus we have proved the equality (14.35). ∎

Theorem 14.8. *Suppose that the operator* (14.3) *acts in some completely regular ideal space* U *over* T, *the operator* (14.4) *acts in some completely regular ideal space* V *over* S, *and one of the following two conditions is satisfied:*

(a) *the function* $\tau \mapsto \|l(\cdot, \tau)\|_U$ *belongs to* U', *and the function* $t \mapsto \|l(t, \cdot)\|_{U'}$ *belongs to* U;

(b) *the function* $\sigma \mapsto \|m(\cdot, \sigma)\|_V$ *belongs to* V', *and the function* $s \mapsto \|m(s, \cdot)\|_{V'}$ *belongs to* V.

Then

$$(14.36) \qquad \sigma_{ew}(K) = [\sigma_{ew}(\tilde{L}) + \sigma(\tilde{M})] \cup [\sigma(\tilde{L}) + \sigma_{ew}(\tilde{M})],$$

where the operator K *is considered in the space* $[U \leftarrow V]$ *in case* (a) *and in the space* $[U \rightarrow V]$ *in case* (b).

☐   We restrict ourselves to the proof in case (a). Suppose that the condition (a) holds. Then the operator $\tilde{L}^2$ is compact in U (see ZAB-REJKO [1968]). We first show that

$$(14.37) \qquad \sigma_{ew}(K) \supseteq [\sigma_{ew}(\tilde{L}) + \sigma(\tilde{M})] \cup [\sigma(\tilde{L}) + \sigma_{ew}(\tilde{M})].$$

Suppose that $\lambda \notin \sigma_{ew}(K)$, then the operator $\lambda I - K$ is Fredholm. Let $\lambda = \alpha + \beta$ with $\alpha \in \sigma_{ew}(\tilde{L})$ and $\beta \in \sigma(\tilde{M})$. By the compactness of $\tilde{L}^2$ we know that $\alpha = 0$ and one of the following two alternatives holds:

Case 1: the range of $\tilde{L}$ is not closed, and $\beta \in \sigma_l(\tilde{M})$;

Case 2: the range of $\tilde{L}$ is not closed, and $\beta \in \sigma(\tilde{M}) \setminus \sigma_l(\tilde{M})$;

Case 3: the range of $\tilde{L}$ is closed and $\dim R(\tilde{L}) < \infty$.

In the first case we find a noncompact sequence $(x_n)_n$ in U and a sequence $(y_n)_n$ in V such that $\|x_n\| = \|y_n\| = 1$, $\tilde{L}x_n \rightarrow 0$, and

$\tilde{M} y_n - \beta y_n \to 0$. Then the noncompact sequence $z_n = x_n \otimes y_n \in [U \leftarrow V]$ satisfies $\|z_n\| = 1$ and

$$(K - \lambda I) z_n = (\tilde{L} x_n) \otimes y_n + x_n \otimes (\tilde{M} - \beta I) y_n \to 0 \quad (n \to \infty).$$

From the closedness of the range of $K - \lambda I$ it follows (see KATO [1966]) that $n_+(K - \lambda I) = \infty$, and thus $K - \lambda I$ cannot be Fredholm.

In the second case we have $\overline{\beta} \in \sigma_l(\tilde{M}^*)$, and the range of $\tilde{L}^*$ is not closed. As in the first case we conclude that $n_+(K^* - \overline{\lambda} I) = \infty$, and thus $K - \lambda I$ again cannot be Fredholm.

Now let $\lambda = \alpha + \beta$ with $\alpha \in \sigma(\tilde{L})$ and $\beta \in \sigma_{ew}(\tilde{M})$. If $\alpha = 0$ or the range of $\tilde{M} - \beta I$ is not closed, we deduce as in the first case that the operator $K - \lambda I$ cannot be Fredholm. Suppose that $\alpha \neq 0$ and the range of $\tilde{M} - \beta I$ is closed. Since $\beta \in \sigma_{ew}(\tilde{M})$, we have either $n_+(\tilde{M} - \beta I) = \infty$ or $n_-(\tilde{M} - \beta I) = \infty$. Moreover, since α is an eigenvalue of $\tilde{L}$ and $\overline{\alpha}$ is an eigenvalue of $\tilde{L}^*$, we see that either $n_+(K - \lambda I) = \infty$ or $n_-(K - \lambda I) = \infty$. But this shows again that $K - \lambda I$ cannot be Fredholm.

Case 3 is treated analogously.

We have shown that $K - \lambda I$ is not Fredholm whenever λ belongs to the right-hand side of (14.37), which shows that the assumption $\lambda \notin \sigma_{ew}(K)$ is not true. Consequently, the inclusion (14.37) is proved.

Let us now prove the reverse inclusion

$$(14.38) \qquad \sigma_{ew}(K) \subseteq [\sigma_{ew}(\tilde{L}) + \sigma(\tilde{M})] \cup [\sigma(\tilde{L}) + \sigma_{ew}(\tilde{M})].$$

For any $\varepsilon > 0$ put

$$U_\varepsilon = \{\alpha : \alpha \in \mathbb{C}, d(\alpha, \sigma_{eb}(\tilde{L})) < \varepsilon\}$$

and

$$V_\varepsilon = \{\beta : \beta \in \mathbb{C}, d(\beta, \sigma_{eb}(\tilde{M})) < \varepsilon\},$$

where $d(z, M)$ denotes the distance of $z \in \mathbb{C}$ to $M \subset \mathbb{C}$. It suffices to show that from $\lambda \in \sigma_{ew}(K)$ and

$$\lambda \in [\sigma_{ew}(\tilde{L}) + (\sigma(\tilde{M}) \setminus V_\varepsilon)] \cup [(\sigma(\tilde{L}) \setminus U_\varepsilon) + \sigma_{ew}(\tilde{M})] \cup [U_\varepsilon + V_\varepsilon]$$

for sufficiently small $\varepsilon > 0$ it follows that $\lambda \in [\sigma_{ew}(\tilde{L}) + \sigma(\tilde{M})] \cup [\sigma(\tilde{L}) + \sigma_{ew}(\tilde{M})]$.

Obviously, it is sufficient to consider the case $\lambda \in U_\varepsilon + V_\varepsilon$ for sufficiently small $\varepsilon > 0$. In fact, choosing $\varepsilon = 1/n$ for n sufficiently large, by the compactness of the sets $\sigma_{eb}(\tilde{L})$ and $\sigma_{eb}(\tilde{M})$ we have then

$$\lambda = \alpha_n + \beta_n \to \alpha_0 + \beta_0 \in \sigma_{eb}(\tilde{L}) + \sigma_{eb}(\tilde{M})$$

as $n \to \infty$. The proof of the inclusion $\sigma(\tilde{L}) + \sigma_{eb}(\tilde{M}) \subseteq \sigma_{eb}(K)$ shows that

$$\sigma_{eb}(\tilde{L}) + \sigma_{eb}(\tilde{M}) \subseteq [\sigma_{ew}(\tilde{L}) + \sigma(\tilde{M})] \cup [\sigma(\tilde{L}) + \sigma_{ew}(\tilde{M})].$$

It follows then that $\lambda \in [\sigma_{ew}(\tilde{L}) + \sigma(\tilde{M})] \cup [\sigma(\tilde{L}) + \sigma_{ew}(\tilde{M})]$.

Now, since $U_\varepsilon \supset \sigma_{eb}(\tilde{L})$ and $V_\varepsilon \supset \sigma_{eb}(\tilde{M})$ are open sets, we know that their complements to $\sigma(\tilde{L})$ and $\sigma(\tilde{M})$, respectively, are finite, say

$$\sigma(\tilde{L}) \setminus U_\varepsilon = \{\alpha_1, \ldots, \alpha_p\}, \quad \sigma(\tilde{M}) \setminus V_\varepsilon = \{\beta_1, \ldots, \beta_q\}.$$

As in the proof of (14.34) we have

$$\begin{aligned}
\sigma_{ew}(K) = & \bigcup_{i=1}^{p} \bigcup_{j=1}^{q} \sigma_{ew}(K|E_i \otimes F_j) \\
& \cup \bigcup_{i=1}^{p} \sigma_{ew}(K|E_i \otimes R_2) \\
& \cup \bigcup_{j=1}^{q} \sigma_{ew}(K|R_1 \otimes F_j) \cup \sigma_{ew}(K|R_1 \tilde{\otimes} R_2),
\end{aligned}$$

(14.39)

where E_i is the nullspace of $(\tilde{L} - \alpha_i I)^{\kappa(\tilde{L}, \alpha_i)}$, F_j is the nullspace of $(\tilde{M} - \beta_j I)^{\kappa(\tilde{M}, \beta_j)}$ $(i = 1, \ldots, p; j = 1, \ldots, q)$, and

$$U = E_1 \oplus \ldots \oplus E_p \oplus R_1, \quad V = F_1 \oplus \ldots \oplus F_q \oplus R_2.$$

In fact, let P_i denote the projector of U to E_i, Q_j the projector of V to F_j, $P = P_1 \oplus \ldots \oplus P_p$, and $Q = Q_1 \oplus \ldots \oplus Q_j$. Since P and Q are projectors of finite rank, P and Q are obviously regular operators. Consequently, the operators $I - P$ and $I - Q$ are also regular. Thus the operators $P \tilde{\otimes} Q$, $P \tilde{\otimes} (I - Q)$, $(I - P) \tilde{\otimes} Q$, and $(I - P) \tilde{\otimes} (I - Q)$ are continuous in $[U \leftarrow V]$. Let A be one of these operators. Since $A^2 = A$, A is a continuous projector. Consequently,

(14.40)
$$[U \leftarrow V] = R(P \tilde{\otimes} Q) \oplus R(P \tilde{\otimes} (I - Q))$$
$$\oplus R((I - P) \tilde{\otimes} Q) \oplus R((I - P) \tilde{\otimes} (I - Q)).$$

Since P and Q are finite-dimensional operators, we have

(14.41)
$$\begin{cases} R(P \tilde{\otimes} Q) = \sum_{i=1}^{p} \sum_{j=1}^{q} \oplus E_i \otimes F_j, \\[2mm] R(P \tilde{\otimes} (I - Q)) = \sum_{i=1}^{p} \oplus F_i \otimes R_2, \\[2mm] R((I - P) \tilde{\otimes} Q) = \sum_{j=1}^{q} \oplus R_1 \otimes F_j. \end{cases}$$

Taking into account that $(I - P)U = R_1$, $(I - Q)V = R_2$, the set $R((I - P) \tilde{\otimes} (I - Q))$ is closed in $[U \leftarrow V]$, and $R_1 \otimes R_2$ is dense in $R((I - P) \tilde{\otimes} (I - Q))$, we get

(14.42)
$$R((I - P) \tilde{\otimes} (I - Q)) = R_1 \tilde{\otimes} R_2.$$

From the equalities (14.40) - (14.42) it follows that

(14.43)
$$[U \leftarrow V] = \sum_{i=1}^{p} \sum_{j=1}^{q} \oplus E_i \otimes F_j$$
$$\oplus \sum_{i=1}^{p} \oplus E_i \otimes R_2 \oplus \sum_{j=1}^{q} \oplus R_1 \otimes F_j \oplus R_1 \tilde{\otimes} R_2.$$

Since every term in the direct sum (14.43) is invariant under K, we see that (14.39) follows from (14.43).

Obviously, $\sigma_{ew}(K | E_i \otimes F_j) = \emptyset$, since $E_i \otimes F_j$ is finite-dimensional. We claim that

(14.44) . $\quad \sigma(K | R_1 \tilde{\otimes} R_2) \subseteq \sigma(\tilde{L} | R_1) + \sigma(\tilde{M} | R_2).$

In fact, since the operators L and M defined in (14.28) and (14.29), respectively, commute on $[U \leftarrow V]$, they also commute on $R_1 \tilde{\otimes} R_2$. Consequently,

$$(14.45) \qquad \sigma(K|R_1 \tilde{\otimes} R_2) \subseteq \sigma(L|R_1 \tilde{\otimes} R_2) + \sigma(M|R_1 \tilde{\otimes} R_2).$$

On the other hand, we have

$$(14.46) \qquad \sigma(L|R_1 \tilde{\otimes} R_2) = \sigma(\tilde{L}|R_1), \quad \sigma(M|R_1 \tilde{\otimes} R_2) = \sigma(\tilde{M}|R_2).$$

To see this for the operator $\tilde{M}$, let $\beta \notin \sigma(M|R_1 \tilde{\otimes} R_2)$, and choose any element $e \in R_1 \setminus \{0\}$. Then the operator $\tilde{M} - \beta I$ is invertible on R_2, since $(M - \beta I)(\{e\} \otimes R_2) = \{e\} \otimes (\tilde{M} - \beta I) R_2$. Consequently, $\beta \notin \sigma(\tilde{M}|R_2)$. Conversely, let $\beta \notin \sigma(\tilde{M}|R_2)$, i.e. the operator $\tilde{M} - \beta I$ is invertible on R_2. By Theorem 6 in LEVIN [1969], the operator $I\overline{\otimes}(\tilde{M} - \beta I)^{-1}$ is then bounded and inverse to the operator $M - \beta I$ on the space $U \tilde{\otimes} R_2$. Consequently, the operator $M - \beta I$ is invertible on $R_1 \tilde{\otimes} R_2$, and thus $\beta \notin \sigma(M|R_1 \tilde{\otimes} R_2)$. This proves the second equality in (14.46).

Let us now prove the first equality in (14.46). From Theorems 12.1 - 12.5 we know that the operator L is bounded from $R_1 \tilde{\otimes} R_2$ into $R_1 \tilde{\otimes} R_2$, from $R_1 \tilde{\otimes} R_2$ into $R_2 \tilde{\otimes} R_1$, from $R_2 \tilde{\otimes} R_1$ into $R_1 \tilde{\otimes} R_2$, and from $R_2 \tilde{\otimes} R_1$ into $R_2 \tilde{\otimes} R_1$. Therefore L is also bounded from $R_1 \tilde{\otimes} R_2 + R_2 \tilde{\otimes} R_1$ into $R_1 \tilde{\otimes} R_2 \cap R_2 \tilde{\otimes} R_1$. From Theorem 2 in ZABREJKO [1976] it follows that $\sigma(L|R_1 \tilde{\otimes} R_2) = \sigma(L|R_2 \tilde{\otimes} R_1)$. The equality $\sigma(L|R_2 \tilde{\otimes} R_1) = \sigma(\tilde{L}|R_1)$ may be proved like the equality $\sigma(M|R_1 \tilde{\otimes} R_2) = \sigma(\tilde{M}|R_2)$ above. Altogether, we have

$$\sigma(K|R_1 \tilde{\otimes} R_2) \subseteq \sigma(L|R_1 \tilde{\otimes} R_2) + \sigma(M|R_1 \tilde{\otimes} R_2)$$

$$= \sigma(\tilde{L}|R_1) + \sigma(\tilde{M}|R_2),$$

i.e. (14.44) holds. Consequently,

$$\sigma_{ew}(K|R_1 \tilde{\otimes} R_2) \subseteq \sigma(K|R_1 \tilde{\otimes} R_2)$$

$$\subseteq \sigma(\tilde{L}|R_1) + \sigma(\tilde{M}|R_2) \subseteq U_\varepsilon + V_\varepsilon.$$

It remains to prove that

$$(14.47) \qquad \sigma_{ew}(K|E_i \otimes R_2) = \sigma_{ew}(\tilde{M}|R_2) + \alpha_i \quad (i = 1, \ldots, p)$$

and

$$(14.48) \qquad \sigma_{ew}(K|R_1 \otimes F_j) = \sigma_{ew}(\tilde{L}|R_1) + \beta_j \quad (j = 1, \ldots, q).$$

To prove (14.47) we show first that $\sigma_{ew}(K|E_i \otimes R_2) \subseteq \sigma_{ew}(\tilde{M}|R_2) + \alpha_i$ for $i = 1, \ldots, p$. Let $\lambda \in \sigma_{ew}(K|E_i \otimes R_2)$ and $\beta := \lambda - \alpha_i$; we have to show that $\beta \in \sigma_{ew}(\tilde{M}|R_2)$. Suppose that $K - \lambda I$ has an infinite-dimensional nullspace or the range of $K - \lambda I$ is not closed. Then we find a noncompact sequence $(u_n)_n$ in $E_i \otimes R_2$ such that $||u_n|| = 1$ and

$$(K - \lambda I)u_n = [(\tilde{L} - \alpha_i I) \otimes I + I \otimes (\tilde{M} - \beta I)]u_n \to 0$$

as $n \to \infty$. Consequently,

$$[(\tilde{L} - \alpha_i I)^{p_i - 1} \otimes I](K - \lambda I)u_n$$

$$= [(\tilde{L} - \alpha_i I)^{p_i} \otimes I + (\tilde{L} - \alpha_i I)^{p_i - 1} \otimes (\tilde{M} - \beta I)]u_n \to 0$$

as $n \to \infty$, where by p_i we denote the minimal exponent p for which the nullspace of $(\tilde{L} - \alpha_i I)^p$ becomes stationary. This and the equality $(\tilde{L} - \alpha_i I)^{p_i} E_i = \{0\}$ imply that

$$[(\tilde{L} - \alpha_i I)^{p_i - 1} \otimes (\tilde{M} - \beta I)]u_n \to 0$$

as $n \to \infty$. Since E_i is finite-dimensional, we may write $E_i = N_1^{(i)} \oplus \ldots \oplus N_{p_i}^{(i)}$, where $N_1^{(i)} \oplus \ldots \oplus N_k^{(i)}$ coincides with the nullspace of $(\tilde{L} - \alpha_i I)^k$ for $k = 1, \ldots, p_i$. We have then

$$E_i \otimes R_2 = (N_1^{(i)} \otimes R_2) \oplus \ldots \oplus (N_{p_i}^{(i)} \otimes R_2)$$

and $u_n = u_{1,n} + \ldots + u_{p_i,n}$ with $u_{k,n} \in N_k^{(i)}$ for $k = 1, \ldots, p_i$. Evidently,

$$[(\tilde{L} - \alpha_i I)^{p_i - 1} \otimes (\tilde{M} - \beta I)]u_n$$

$$= [(\tilde{L} - \alpha_i I)^{p_i - 1} \otimes (\tilde{M} - \beta I)](u_{1,n} + \ldots + u_{p_i - 1,n})$$

$$+ [(\tilde{L} - \alpha_i I)^{p_i - 1} \otimes (\tilde{M} - \beta I)]u_{p_i,n}$$

$$= [(\tilde{L} - \alpha_i I)^{p_i - 1} \otimes (\tilde{M} - \beta I)]u_{p_i,n} \to 0$$

as $n \to \infty$. If r_i denotes the dimension of $N_{p_i}^{(i)}$ and $\{e_1^{(i)}, \ldots, e_{r_i}^{(i)}\}$ is a normalized basis in $N_{p_i}^{(i)}$, the set $\{(\tilde{L} - \alpha_i I)^{p_i-1} e_1^{(i)}, \ldots, (\tilde{L} - \alpha_i I)^{p_i-1} e_{r_i}^{(i)}\}$ has also dimension r_i. We may write $u_{p_i,n}$ in the form

$$u_{p_i,n} = (e_1^{(i)} \otimes x_{1,n}) + \ldots + (e_{r_i}^{(i)} \otimes x_{r_i,n}) \quad (x_{1,n}, \ldots, x_{r_i,n} \in R_2).$$

It follows that

$$[(\tilde{L} - \alpha_i I)^{p_i-1} \otimes (\tilde{M} - \beta I)] u_{p_i,n}$$

$$= (\tilde{L} - \alpha_i I)^{p_i-1} [e_1^{(i)} \otimes (\tilde{M} - \beta I) x_{1,n}] + \ldots$$

$$+ (\tilde{L} - \alpha_i I)^{p_i-1} [e_{r_i}^{(i)} \otimes (\tilde{M} - \beta I) x_{r_i,n}] \to 0$$

as $n \to \infty$. This shows that the nullspace of $\tilde{M} - \beta I$ is infinite-dimensional or the range of $K - \lambda I$ is not closed, hence $\beta \in \sigma_{ew}(\tilde{M})$ and $\lambda \in \sigma_{ew}(\tilde{M}) + \alpha_i$.

If $\lambda \in \sigma_{ew}(K)$ and the nullspace of $(K - \lambda I)^*$ is infinite-dimensional, by the above reasoning we deduce that also the nullspace of $(\tilde{M} - \beta I)^*$ is infinite-dimensional, hence $\beta \in \sigma_{ew}(\tilde{M})$ and $\lambda \in \sigma_{ew}(\tilde{M}) + \alpha_i$. But for any bounded linear operator A in a Banach space we have $\lambda \in \sigma_{ew}(A)$ if and only if $A - \lambda I$ or $(A - \lambda I)^*$ have infinite-dimensional nullspaces or the range of $A - \lambda I$ is not closed; thus we have proved that $\sigma_{ew}(K | E_i \otimes R_2) \subseteq \sigma_{ew}(\tilde{M}) + \alpha_i$. The reverse inclusion is obvious.

In this way, we have shown that (14.47) is true. The equality (14.48) is proved similarly, and so we are done. ∎

We point out that the hypotheses of Theorem 14.6 follow from the hypotheses of Theorem 14.8. In fact, suppose, for instance, that the hypothesis (a) in Theorem 14.8 holds. Then the operator $\tilde{L}$ is regular in U and, by Theorem 10.5 in ZABREJKO [1968], we have $\rho(\tilde{L}) = \rho_{reg}(\tilde{L})$. But this means that the corresponding hypothesis (a) in Theorem 14.6 is satisfied.

A careful analysis of the proofs of the preceding theorems shows that the inclusions

$$\sigma(\tilde{L}) + \sigma(\tilde{M}) \subseteq \sigma(K),$$

$$[\sigma_{eb}(\tilde{L}) + \sigma(\tilde{M})] \cup [\sigma(\tilde{L}) + \sigma_{eb}(\tilde{M})] \subseteq \sigma_{eb}(K),$$

and

$$[\sigma_{ew}(\tilde{L}) + \sigma(\tilde{M})] \cup [\sigma(\tilde{L}) + \sigma_{ew}(\tilde{M})] \subseteq \sigma_{ew}(K)$$

remain true if one drops the assumptions $\rho(\tilde{L}) = \rho_{reg}(\tilde{L})$ and $\rho(\tilde{M}) = \rho_{reg}(\tilde{M})$ in the hypotheses (a) and (b), respectively, of Theorem 14.6.

14.4. An index formula for partial integral operators

Now we pass to the problem of calculating the defect numbers and the index of the partial integral operator (14.1) in ideal spaces. We keep the terminology of the preceding subsections.

Theorem 14.9. *Suppose that the hypotheses of Theorem 14.8 are satisfied, and*

(14.49) $\qquad \lambda \in \sigma(K) \setminus \{[\sigma_{ew}(\tilde{L}) + \sigma(\tilde{M})] \cup [\sigma(\tilde{L}) + \sigma_{ew}(\tilde{M})]\}.$

Then the formulas

$$n_+(K - \lambda I) =$$

(14.50)
$$\sum_{\substack{\alpha+\beta=\lambda \\ \alpha \in \sigma_l(\tilde{L}), \beta \in \sigma_l(\tilde{M})}} \sum_{k=1}^{\infty} [a_k(\alpha) - a_{k-1}(\alpha)][b_k(\beta) - b_{k-1}(\beta)],$$

$$n_-(K - \lambda I) =$$

(14.51)
$$\sum_{\substack{\alpha+\beta=\lambda \\ \alpha \in \sigma_d(\tilde{L}), \beta \in \sigma_d(\tilde{M})}} \sum_{k=1}^{\infty} [c_k(\alpha) - c_{k-1}(\alpha)][d_k(\beta) - d_{k-1}(\beta)]$$

and

$$\text{ind}\,(K - \lambda I) =$$

$$(14.52) \qquad \sum_{\substack{\alpha+\beta=\lambda \\ \alpha\in A, \beta\in B'}} \text{ind}\,(\tilde{M} - \beta I) \sum_{k=1}^{\infty} [a_k(\alpha) - a_{k-1}(\alpha)]$$

$$+ \sum_{\substack{\alpha+\beta=\lambda \\ \alpha\in A, \beta\in B'}} \text{ind}\,(\tilde{L} - \alpha I) \sum_{k=1}^{\infty} [b_k(\beta) - b_{k-1}(\beta)]\,.$$

hold, where the numbers $a_k(\alpha)$, $b_k(\beta)$, $c_k(\alpha)$, *and* $d_k(\beta)$ *are defined as in* (14.6) *and the sets* A, B, A', B' *are defined as in Theorem 14.1.*

☐   We assume that the hypothesis (a) of Theorem 14.8 holds; in case of hypothesis (b) the proof is similar. We begin with the proof of formula (14.50). Since $\sigma_{ew}(K) \subseteq \sigma_{eb}(K) \subseteq \sigma(K)$, our hypothesis (14.49) implies that either $\lambda \in \sigma(K) \setminus \sigma_{eb}(K)$ or $\lambda \in \sigma_{eb}(K) \setminus \sigma_{ew}(K)$.

Suppose first that $\lambda \in \sigma(K) \setminus \sigma_{eb}(K)$. Using the same notation as in the proof of (14.34), we may write (14.50) in the form

$$n_+(K - \lambda I) =$$

$$(14.53) \qquad \sum_{\substack{\alpha_i+\beta_i=\lambda \\ i=1,\dots p}} \sum_{k=1}^{p} [a_k(\alpha_i) - a_{k-1}(\alpha_i)][b_k(\beta_i) - b_{k-1}(\beta_i)].$$

Moreover, as in the proof of (14.34) we have

$$(14.54) \qquad n_+(K - \lambda I) = \sum_{i=1}^{p} n_+(K - \lambda I|E_i \otimes F_i),$$

where $n_+(K - \lambda I|E_i \otimes F_i)$ denotes as above the dimension of the nullspace of the restriction of $K - \lambda I$ to $E_i \otimes F_i$. This dimension may be calculated as follows. Fix any $i_0 \in \{1,\dots,p\}$ with $\alpha_{i_0} + \beta_{i_0} = \lambda$; otherwise we have $\lambda \notin \sigma(K|E_{i_0} \otimes F_{i_0})$ and $n_+(K - \lambda I|E_{i_0} \otimes F_{i_0}) = 0$. To simplify the notation, we write $E_{i_0} =: E$, $F_{i_0} =: F$, $\alpha_{i_0} =: \alpha$,

$\beta_{i_0} =: \beta$, $\kappa(\tilde{L}, \alpha_{i_0}) =: s$, and $\kappa(\tilde{M}, \beta_{i_0}) =: t$. Since $\dim E = s < \infty$ and $\dim F = t < \infty$, we may suppose that the operators $\tilde{L}$ and $\tilde{M}$ (more precisely, their restrictions to E and F, respectively) are matrices in canonical Jordan form.

Assume first that $n_+(\tilde{L} - \alpha I) = n_+(\tilde{M} - \beta I) = 1$. Then $\tilde{L} - \alpha I$ and $\tilde{M} - \beta I$ are quadratic matrices of order s and t, respectively, of the form

$$\tilde{L} - \alpha I, \tilde{M} - \beta I \sim \begin{pmatrix} 0 & 0 & 0 & \cdot & \cdot & \cdot & 0 & 0 & 0 \\ 1 & 0 & 0 & \cdot & \cdot & \cdot & 0 & 0 & 0 \\ 0 & 1 & 0 & \cdot & \cdot & \cdot & 0 & 0 & 0 \\ \cdot & \cdot & \cdot & \cdot & \cdot & \cdot & \cdot & \cdot & \cdot \\ \cdot & \cdot & \cdot & \cdot & \cdot & \cdot & \cdot & \cdot & \cdot \\ \cdot & \cdot & \cdot & \cdot & \cdot & \cdot & \cdot & \cdot & \cdot \\ 0 & 0 & 0 & \cdot & \cdot & \cdot & 0 & 0 & 0 \\ 0 & 0 & 0 & \cdot & \cdot & \cdot & 1 & 0 & 0 \\ 0 & 0 & 0 & \cdot & \cdot & \cdot & 0 & 1 & 0 \end{pmatrix}$$

Fix a basis $\{e_1, \ldots, e_s\}$ in E and a basis $\{f_1, \ldots, f_t\}$ in F such that

$$(\tilde{L} - \alpha I)e_1 = e_2, \ \ldots, \ (\tilde{L} - \alpha I)e_{s-1} = e_s, \ (\tilde{L} - \alpha I)e_s = 0$$

and

$$(\tilde{M} - \beta I)f_1 = f_2, \ \ldots, \ (\tilde{M} - \beta I)f_{t-1} = f_t, \ (\tilde{M} - \beta I)f_t = 0.$$

Since $\{e_n \otimes f_m : n = 1, \ldots, s; m = 1, \ldots, t\}$ is a basis in $E \otimes F$, every $u \in E \otimes F$ has a unique representation

$$u = \sum_{n=1}^{s} \sum_{m=1}^{t} u_{n,m}(e_n \otimes f_m).$$

Consequently,

$$0 = (K - \lambda I)u = (\tilde{L} \otimes I + I \otimes \tilde{M} - \lambda I \otimes I)u$$

$$= [(\tilde{L} - \alpha I) \otimes I + I \otimes (\tilde{M} - \beta I)]u$$

$$= \sum_{n=1}^{s} \sum_{m=1}^{t} u_{n,m}[(\tilde{L} - \alpha I)e_n \otimes f_m + e_n \otimes (\tilde{M} - \beta I)f_m]$$

$$= \sum_{n=1}^{s-1} \sum_{m=1}^{t} u_{n,m}(e_{n+1} \otimes f_m) + \sum_{n=1}^{s} \sum_{m=1}^{t-1} u_{n,m}(e_n \otimes f_{m+1})$$

$$= \sum_{n=2}^{s} \sum_{m=1}^{t} u_{n-1,m}(e_n \otimes f_m) + \sum_{n=1}^{s} \sum_{m=2}^{t} u_{n,m-1}(e_n \otimes f_m)$$

$$= \sum_{n=2}^{s} \sum_{m=2}^{t} [u_{n-1,m} + u_{n,m-1}](e_n \otimes f_m) +$$

$$\sum_{n=2}^{s} u_{n-1,1}(e_n \otimes f_1) + \sum_{m=2}^{t} u_{1,m-1}(e_1 \otimes f_m).$$

Since $\{e_n \otimes f_m : n = 1, \ldots, s; m = 1, \ldots, t\}$ is a basis in $E \otimes F$, we get the following linear system for the unknown scalars $u_{n,m}$ ($n = 1, \ldots, s; m = 1 \ldots, t$):

$$\begin{cases} u_{n-1,m} + u_{n,m-1} = 0 & \text{if } 2 \leq n \leq s \text{ and } 2 \leq m \leq t, \\ u_{n-1,1} = 0 & \text{if } 2 \leq n \leq s \text{ and } m = 1, \\ u_{1,m-1} = 0 & \text{if } 2 \leq m \leq t \text{ and } n = 1, \\ u_{s,t} \text{ arbitrary} & \text{if } n = s \text{ and } m = t. \end{cases}$$

This system has rank $st - \min\{s,t\}$, and hence $\min\{s,t\}$ linearly independent solutions. If $n_+(\tilde{L} - \alpha I) \geq 1$ and $n_+(\tilde{M} - \beta I) \geq 1$, we have $2a_k(\alpha) - a_{k-1}(\alpha) - a_{k+1}(\alpha)$ and $2b_k(\beta) - b_{k-1}(\beta) - b_{k+1}(\beta)$ Jordan blocks of order k, respectively, in the above matrix, where we have used again the notation (14.6). Applying to each such block the same reasoning as before, we see that

$$(14.55) \quad n_+(K - \lambda I | E \otimes F) = \sum_{n=1}^{s} \sum_{m=1}^{t} \min\{n, m\} A_n(\alpha) B_m(\beta),$$

where we have put

$$A_n(\alpha) = 2a_n(\alpha) - a_{n-1}(\alpha) - a_{n+1}(\alpha)$$

and

$$B_m(\beta) = 2b_m(\beta) - b_{m-1}(\beta) - b_{m+1}(\beta).$$

We claim that

$$(14.56)$$
$$\sum_{n=1}^{s} \sum_{m=1}^{t} \min\{n, m\} A_n(\alpha) B_m(\beta)$$
$$= \sum_{k=1}^{\infty} [a_k(\alpha) - a_{k-1}(\alpha)][b_k(\beta) - b_{k-1}(\beta)].$$

In fact, a straightforward but cumbersome calculation shows that, for fixed n, we have

$$(14.57) \quad \sum_{m=1}^{t} \min\{n, m\} B_m(\beta) = b_n(\beta).$$

Assuming without loss of generality that $s \leq t$ we get from (14.57)

$$n_+(K - \lambda I | E \otimes F) = \sum_{n=1}^{s} [a_n(\alpha) - a_{n-1}(\alpha) - a_{n+1}(\alpha)] b_n(\beta).$$

Taking into account that all terms in (14.56) are zero for $k > s$ and

$a_{s+1}(\alpha) = a_s(\alpha)$ we further obtain

$$\sum_{n=1}^{s} [a_n(\alpha) - a_{n-1}(\alpha)][b_n(\beta) - b_{n-1}(\beta)]$$

$$= \sum_{n=1}^{s} a_n(\alpha)b_n(\beta) - \sum_{n=1}^{s} a_{n-1}(\alpha)b_n(\beta)$$

$$- \sum_{n=1}^{s} a_n(\alpha)b_{n-1}(\beta) + \sum_{n=1}^{s} a_{n-1}(\alpha)b_{n-1}(\beta)$$

$$= \sum_{n=1}^{s} a_n(\alpha)b_n(\beta) - \sum_{n=1}^{s} a_{n-1}(\alpha)b_n(\beta)$$

$$- \sum_{n=1}^{s-1} a_{n+1}(\alpha)b_n(\beta) + \sum_{n=1}^{s-1} a_n(\alpha)b_n(\beta)$$

$$= \sum_{n=1}^{s} [2a_n(\alpha) - a_{n-1}(\alpha) - a_{n+1}(\alpha)]b_n(\beta) = \sum_{n=1}^{s} A_n(\alpha)b_n(\beta).$$

This together with (14.57) implies (14.56), hence also (14.53). The proof is complete for the case $\lambda \in \sigma(K) \setminus \sigma_{eb}(K)$.

Suppose now that $\lambda \in \sigma_{eb}(K) \setminus \sigma_{ew}(K)$. Since $\sigma_{eb}(\tilde{L}) + \sigma_{eb}(\tilde{M}) \subseteq \sigma_{ew}(K)$, from (14.27) and (14.36) we get

$$\lambda \in \{[\sigma(\tilde{L}) \setminus \sigma_{eb}(\tilde{L})] + [\sigma(\tilde{M}) \setminus \sigma_{ew}(\tilde{M})]\}$$

(14.58) $$\cup \{[\sigma(\tilde{L}) \setminus \sigma_{ew}(\tilde{L})] + [\sigma(\tilde{M}) \setminus \sigma_{eb}(\tilde{M})]\}$$

$$\setminus \{[\sigma_{ew}(\tilde{L}) + \sigma(\tilde{M})] \cup [\sigma(\tilde{L}) + \sigma_{ew}(\tilde{M})]\}.$$

As above, the equality $\lambda = \alpha + \beta$ can hold only for finitely many α and β. In fact, if there were an infinite number of such α, say, the same would be true for β, and thus $n_+(K - \lambda I) = \infty$ or $n_+((K - \lambda I)^*) = \infty$, contradicting the fact that $K - \lambda I$ is Fredholm. So, let $\{\alpha_1, \ldots, \alpha_p\} \subseteq \sigma(\tilde{L}) \setminus \sigma_{eb}(\tilde{L})$ be the set of all possible α, and put $\beta_i = \lambda - \alpha_i$ $(i = 1, \ldots, p)$. We have then

$$U = E_1 \oplus \ldots \oplus E_p \oplus R_1$$

and
$$U \tilde{\otimes} V = (E_1 \otimes V) \oplus \ldots \oplus (E_p \otimes V) \oplus (R_1 \tilde{\otimes} V),$$

where E_i is the nullspace of $(\tilde{L} - \alpha_i I)^{\kappa(\tilde{L}, \alpha_i)}$ $(i = 1, \ldots, p)$, and $R_1 \tilde{\otimes} V$ is the closure of $R_1 \otimes V$ in $[U \leftarrow V]$. Since $\{\alpha_1, \ldots, \alpha_p\} \cap \sigma(\tilde{L}|R_1) = \emptyset$ and $\sigma(K|R_1 \tilde{\otimes} V) \subseteq \sigma(\tilde{L}|R_1) + \sigma(\tilde{M})$, we get as in the proof of the inclusion (14.44) that $\lambda \notin \sigma(K|R_1 \tilde{\otimes} V)$. Moreover, since the operator $K - \lambda I$ leaves the spaces $E_i \otimes V$ $(i = 1, \ldots, p)$ and $R_1 \tilde{\otimes} V$ invariant, we have

(14.59) $$n_+(K - \lambda I) = \sum_{i=1}^{p} n_+(K - \lambda I | E_i \otimes V).$$

We claim that

$$n_+(K - \lambda I | E_i \otimes V) =$$

(14.60)
$$\sum_{k=1}^{\infty} [a_k(\alpha_i) - a_{k-1}(\alpha_i)][b_k(\beta_i) - b_{k-1}(\beta_i)]$$

for $i = 1, \ldots, p$. But this follows from the fact that the nullspace of $K - \lambda I | E_i \otimes V$ is contained in the tensor product of E_i with the nullspace of $(\tilde{M} - \beta_i I)^{\kappa(\tilde{L}, \alpha_i)}$, and hence the restriction of the operator $K - \lambda I$ on these two sets has the same nullspace. Combining (14.59) and (14.60) we see that (14.50) is true also in case $\lambda \in \sigma_{eb}(K) \setminus \sigma_{ew}(K)$.

The proof of formula (14.51) is now rather easy. Suppose first that $\lambda \in \sigma(K) \setminus \sigma_{eb}(K)$. Then $\lambda = \alpha + \beta$ with $\alpha \in \sigma(\tilde{L}) \setminus \sigma_{eb}(\tilde{L})$ and $\beta \in \sigma(\tilde{M}) \setminus \sigma_{eb}(\tilde{M})$, i.e. $K - \lambda I$, $\tilde{L} - \alpha I$, and $\tilde{M} - \beta I$ are all Fredholm operators. Consequently, (14.51) follows directly from (14.50).

On the other hand, assume that $\lambda \in \sigma_{eb}(K) \setminus \sigma_{ew}(K)$. Using the same notation as above and observing that the operator $K - \lambda I$ leaves the spaces $E_i \otimes V$ $(i = 1, \ldots, p)$ and $R_1 \tilde{\otimes} V$ invariant and is invertible on $R_1 \tilde{\otimes} V$, we obtain the equality

(14.61) $$n_-(K - \lambda I) = \sum_{i=1}^{p} n_-(K - \lambda I | E_i \otimes V).$$

It is therefore sufficient to prove that

$$n_-(K - \lambda I | E_i \otimes V) =$$

(14.62)
$$\sum_{k=1}^{\infty} [c_k(\alpha_i) - c_{k-1}(\alpha_i)][d_k(\beta_i) - d_{k-1}(\beta_i)].$$

Since the operator $(K - \lambda I)^* | E_i \otimes V$ coincides on the space $(E_i \otimes V)^* = E_i^* \otimes V^*$ with the operator $(\tilde{L} - \alpha_i I)^* \otimes I + I \otimes (\tilde{M} - \beta_i I)^*$, we obtain

$$n_+((K - \lambda I)^* | E_i^* \otimes V^*) =$$

(14.63)
$$\sum_{k=1}^{\infty} [a_k^*(\overline{\alpha_i}) - a_{k-1}^*(\overline{\alpha_i})][b_k^*(\overline{\beta_i}) - b_{k-1}^*(\overline{\beta_i})]$$

for $i = 1, \ldots, p$, where the asterisk denotes the corresponding numbers (14.6) for $\tilde{L}^*$ and $\tilde{M}^*$. But all operators occurring in (14.62) are Fredholm, and thus the relations $a_k^*(\overline{\alpha}) \equiv c_k(\alpha)$ and $b_k^*(\overline{\beta}) \equiv d_k(\beta)$ hold which imply that

$$n_+((K - \lambda I)^* | E_i^* \otimes V^*) = n_-(K - \lambda I | E_i \otimes V).$$

for $i = 1, \ldots, p$. In this way, we have reduced the proof of (14.51) to that of (14.50) in case $\lambda \in \sigma_{eb}(K) \setminus \sigma_{ew}(K)$.

It remains to prove the index formula (14.52). Again let $\lambda \in \sigma(K) \setminus \sigma_{eb}(K)$, i.e. $\lambda = \alpha + \beta$ with $\alpha \in \sigma(\tilde{L}) \setminus \sigma_{eb}(\tilde{L})$ and $\beta \in \sigma(\tilde{M}) \setminus \sigma_{eb}(\tilde{M})$. Here the formula (14.52) is a trivial consequence of the fact that all operators $K - \lambda I$, $\tilde{L} - \alpha I$, and $\tilde{M} - \beta I$ have index zero.

Suppose, finally, that $\lambda \in \sigma_{eb}(K) \setminus \sigma_{ew}(K)$. By (14.58) we have then $\lambda = \alpha + \beta$ with $\alpha \in \sigma(\tilde{L}) \setminus \sigma_{eb}(\tilde{L})$ or $\beta \in \sigma(\tilde{M}) \setminus \sigma_{eb}(\tilde{M})$. In this case ind $(\tilde{L} - \alpha I) = 0$ or ind $(\tilde{M} - \beta I) = 0$ and we get from (14.50) and (14.51)

$$\text{ind}\,(K - \lambda I) = \sum_{\alpha + \beta = \lambda} \text{ind}\,(\tilde{M} - \beta I) \sum_{k=1}^{\infty} [a_k(\alpha) - a_{k-1}(\alpha)]$$

$$+ \sum_{\alpha + \beta = \lambda} \text{ind}\,(\tilde{L} - \alpha I) \sum_{k=1}^{\infty} [d_k(\beta) - d_{k-1}(\beta)].$$

Suppose that $\alpha \in \sigma(\tilde{L}) \setminus \sigma_{eb}(\tilde{L})$. Since ind $(\tilde{L} - \alpha I) = 0$, this reduces to

$$\text{ind}\,(K - \lambda I) = \sum_{\alpha + \beta = \lambda} \text{ind}\,(\tilde{M} - \beta I) \sum_{k=1}^{\infty} [a_k(\alpha) - a_{k-1}(\alpha)],$$

and thus (14.52) holds in this case. The proof for $\beta \in \sigma(\tilde{M}) \setminus \sigma_{eb}(\tilde{M})$ gives the parallel formula

$$\text{ind}\,(K - \lambda I) = \sum_{\alpha + \beta = \lambda} \text{ind}\,(\tilde{L} - \alpha I) \sum_{k=1}^{\infty} [d_k(\beta) - d_{k-1}(\beta)].$$

Theorem 14.9 is completely proved. ∎

Observe that Theorem 14.9 implies, in particular, that

$$(14.64) \qquad \sigma_{es}(K) \subseteq [\sigma_{es}(\tilde{L}) + \sigma(\tilde{M})] \cup [\sigma(\tilde{L}) + \sigma_{es}(\tilde{M})].$$

The proof of the following theorem is parallel to that of the preceding theorems, and therefore we drop it.

Theorem 14.10. *Suppose that the operator* (14.3) *acts in some completely regular ideal space U over T, the operator* (14.4) *acts in some completely regular ideal space V over S, and one of the following four conditions is satisfied:*

(a) $[U \to V] \subseteq [U \leftarrow V]$ and the function $\tau \mapsto ||l(\cdot, \tau)||_U$ belongs to U';

(b) $[U \to V] \subseteq [U \leftarrow V]$ and the function $s \mapsto ||m(s, \cdot)||_V$ belongs to V;

(c) $[U \leftarrow V] \subseteq [U \to V]$ and the function $t \mapsto ||l(t, \cdot)||_{U'}$ belongs to U;

(d) $[U \leftarrow V] \subseteq [U \to V]$ and the function $\sigma \mapsto ||m(\cdot, \sigma)||_V$ belongs to V'.

Then the equalities (14.26), (14.27) *and* (14.36) *hold, where the operator K is considered in the space $[U \leftarrow V]$ in the cases* (a) *and* (c), *and in the space $[U \to V]$ in the cases* (b) *and* (d). *In addition, for any λ as in* (14.49), *the defect numbers and the index of the operator*

$K - \lambda I$ *may be calculated by means of the formulas* (14.50) - (14.52).

To illustrate the hypotheses of the last theorem, let us consider Lebesgue spaces as an elementary example.

Example 14.5. Suppose that the operator (14.3) acts in the space $L_{p_1} = L_{p_1}(T)$, and the operator (14.4) acts in the space $L_{p_2} = L_{p_2}(S)$. We know that $[L_{p_1} \leftarrow L_{p_2}] \subseteq [L_{p_1} \rightarrow L_{p_2}]$ for $1 \leq p_1 \leq p_2 < \infty$, and $[L_{p_1} \rightarrow L_{p_2}] \subseteq [L_{p_1} \leftarrow L_{p_2}]$ for $1 \leq p_2 \leq p_1 < \infty$ (see Subsection 12.2). Putting $\frac{1}{p_1} + \frac{1}{q_1} = \frac{1}{p_2} + \frac{1}{q_2} = 1$ as usual, the four conditions of Theorem 14.10 read then

(a) $p_2 \leq p_1$, $\tau \mapsto \|l(\cdot, \tau)\|_{L_{p_1}} \in L_{q_1}$;

(b) $p_2 \leq p_1$, $s \mapsto \|m(s, \cdot)\|_{L_{q_2}} \in L_{p_2}$;

(c) $p_1 \leq p_2$, $t \mapsto \|l(t, \cdot)\|_{L_{q_1}} \in L_{p_1}$;

(d) $p_1 \leq p_2$, $\sigma \mapsto \|m(\cdot, \sigma)\|_{L_{p_2}} \in L_{q_2}$.

Under these assumptions, the equalities (14.26), (14.27) and (14.36) hold, where the operator K is considered in the space $[L_{p_1} \leftarrow L_{p_2}]$ in the cases (a) and (c), and in the space $[L_{p_1} \rightarrow L_{p_2}]$ in the cases (b) and (d). In addition, for any λ as in (14.49), the defect numbers and the index of the operator $K - \lambda I$ may be calculated by means of the formulas (14.50) - (14.52).

14.5. The case of positive kernels

In this subsection we suppose throughout that the kernels $l = l(t, \tau)$ and $m = m(s, \sigma)$ in the definition of the operators (14.3) and (14.4) are nonegative on $T \times T$ and $S \times S$, respectively. We also assume that the hypotheses of Theorem 14.6 are satisfied, and hence the corresponding operator (14.1) acts in the space $[U \rightarrow V]$ or $[U \leftarrow V]$. Recall that under these hypotheses the equality (14.7) is true. Moreover, the positivity of the operators $\tilde{L}$ and $\tilde{M}$ implies that $r_\sigma(\tilde{L}) \in \sigma(\tilde{L})$ and $r_\sigma(\tilde{M}) \in \sigma(\tilde{M})$ (see e.g. SCHAEFFER [1971]), and thus

$$(14.65) \qquad r_\sigma(K) = r_\sigma(\tilde{L}) + r_\sigma(\tilde{M}).$$

The equality (14.65) is not true for arbitrary kernels l and m. However, since $0 \in \sigma(\tilde{L})$ and $0 \in \sigma(\tilde{M})$, we always have the two-sided estimate

$$(14.66) \qquad \max\{r_\sigma(\tilde{L}), r_\sigma(\tilde{M})\} \le r_\sigma(K) \le r_\sigma(\tilde{L}) + r_\sigma(\tilde{M}).$$

Consequently, any lower [respectively upper] estimate for $r_\sigma(\tilde{L})$ and $r_\sigma(\tilde{M})$ leads to a lower [respectively upper] estimate for $r_\sigma(K)$. For example, if the kernels l and m are nonnegative, and we know that

$$(14.67) \qquad \tilde{L}x(s) \ge \alpha x(s), \quad \tilde{M}y(t) \ge \beta y(t) \quad (\alpha, \beta > 0)$$

for some positive functions $x \in U$ and $y \in V$, then necessarily $r_\sigma(K) \ge \alpha + \beta$.

Classical estimates for the spectral radii of integral operators may be found in KRASNOSEL'SKIJ-VAJNIKKO-ZABREJKO-RUTITSKIJ [1969] and ZABREJKO-KOSHELEV-KRASNOSEL'SKIJ-
MIKHLIN-RAKOVSHCHIK-STETSENKO [1968].

Let A be a positive linear operator in a Banach space E with cone C. Following KRASNOSEL'SKIJ [1962] or KRASNOSEL'SKIJ-LIFSHITS-SOBOLEV [1986] we call an eigenvalue μ of A *cone-positive* if A has an eigenvector $x \in C$ which corresponds to μ, i.e. $Ax = \mu x$.

Lemma 14.2. *Suppose that $r_\sigma(\tilde{L})$ is a cone-positive eigenvalue of the operators $\tilde{L}$ and $\tilde{L}^*$, and $r_\sigma(\tilde{M})$ is a cone-positive eigenvalue of the operators $\tilde{M}$ and $\tilde{M}^*$. Then $r_\sigma(K) = r_\sigma(\tilde{L}) + r_\sigma(\tilde{M})$ is a cone-positive eigenvalue of the operators K and K^*.*

☐ The proof is almost trivial: If x and y are eigenfunctions in the corresponding cones such that $\tilde{L}x = r_\sigma(\tilde{L})x$ and $\tilde{M}y = r_\sigma(\tilde{M})y$, then $x \otimes y$ belongs to the induced cone in $U \otimes V$ and is an eigenfunction for the eigenvalue $r_\sigma(\tilde{L}) + r_\sigma(\tilde{M})$ of K. ■

Given a linear operator A in a Banach space E, by

$$(14.68) \qquad r_{es}(A) := \sup\{|\lambda| : \lambda \in \sigma_{es}(A)\}$$

we denote the *Fredholm radius* (or *radius of the essential spectrum*) of A.

Theorem 14.11 *Suppose that the kernels l and m in (14.3) and (14.4) are nonnegative, and that*

$$(14.69) \qquad r_{es}(\tilde{L}) < r_\sigma(\tilde{L}), \qquad r_{es}(\tilde{M}) < r_\sigma(\tilde{M}).$$

Then $r_\sigma(K) = r_\sigma(\tilde{L}) + r_\sigma(\tilde{M})$ is a cone-positive eigenvalue of the operators K and K^*.

☐ From ZABREJKO-SMITSKIKH [1979] it follows that, under the above hypotheses, $r_\sigma(\tilde{L})$ is a cone-positive eigenvalue of $\tilde{L}$, and $r_\sigma(\tilde{M})$ is a cone-positive eigenvalue of $\tilde{M}$. By Lemma 14.2, $r_\sigma(K) = r_\sigma(\tilde{L}) + r_\sigma(\tilde{M})$ is then a cone-positive eigenvalue of K. The proof for K^* is of course the same. ■

Theorem 14.11 applies, for example, to the case of compact operators $\tilde{L}$ and $\tilde{M}$ which satisfy (14.67). In fact, in this case we have $0 = r_{es}(\tilde{L}) < \alpha \le r_\sigma(\tilde{L})$ and $0 = r_{es}(\tilde{M}) < \beta \le r_\sigma(\tilde{M})$, and thus (14.69) holds.

We remark that, under the hypotheses of Theorem 14.8 or Theorem 14.10, the strict inequality

$$(14.70) \qquad r_{es}(K) < r_\sigma(K)$$

follows from the strict inequalities (14.69). In fact, the inclusion (14.64) implies that

$$r_{es}(K) \le \sup\{|\lambda| : \lambda \in [\sigma_{es}(\tilde{L}) + \sigma(\tilde{M})] \cup [\sigma(\tilde{L}) + \sigma_{es}(\tilde{M})]\}$$

$$\le \max\{\sup\{|\alpha| : \alpha \in \sigma_{es}(\tilde{L})\} + r_\sigma(\tilde{M}),$$

$$r_\sigma(\tilde{L}) + \sup\{|\beta| : \beta \in \sigma_{es}(\tilde{M})\}\}$$

$$= \max\{r_{es}(\tilde{L}) + r_\sigma(\tilde{M}), r_\sigma(\tilde{L}) + r_{es}(\tilde{M})\}$$

$$< r_\sigma(\tilde{L}) + r_\sigma(\tilde{M}) = r_\sigma(K).$$

However, if one of the inequalitites in (14.69) is not strict, the inequality (14.70) may fail. For example, let $\tilde{L}$ and $\tilde{M}$ be compact operators with $r_\sigma(\tilde{L}) = 0$ and $r_{es}(\tilde{M}) < r_\sigma(\tilde{M})$. From (14.36) and the inclusion

$\sigma_{ew}(K) \subseteq \sigma_{es}(K)$ we get then $r_\sigma(K) = r_\sigma(\tilde{M}) \leq r_{es}(K) \leq r_\sigma(K)$, hence $r_{es}(K) = r_\sigma(K)$.

We point out that a cone-positive eigenvalue of a bounded linear operator A in a Banach space with a cone C need not be simple. However, if A is a so-called u-*bounded operator*, any cone-positive eigenvalue of A is simple (see Theorem 11.1 in KRASNOSEL'SKIJ-LIFSHITS-SOBOLEV [1986]). We recall that A is called u-bounded for some $u \in C$ if for all $x \in C \setminus \{0\}$ one can find numbers $\alpha, \beta > 0$ such that $\alpha u \leq A^n x \leq \beta u$ for some $n \in \mathbb{N}$. (Here the inequality $\leq$ is meant, of course, in the sense of the partial ordering induced by the cone C.)

Theorem 14.12. *Suppose that the operators $\tilde{L}$ and $\tilde{M}$ are u-bounded, and $r_\sigma(\tilde{L})$ and $r_\sigma(\tilde{M})$ are cone-positive eigenvalues of $\tilde{L}$ and $\tilde{M}$, respectively. Assume, in addition, that*

$$(14.71) \qquad \sigma_p(K) = \sigma_p(\tilde{L}) + \sigma_p(\tilde{M}).$$

Then all eigenvalues $\lambda \in \sigma_p(K) \setminus \{r_\sigma(K)\}$ satisfy $|\lambda| < r_\sigma(K)$, and $r_\sigma(K)$ is a simple cone-positive eigenvalue of K.

☐ Let $\lambda = \alpha + \beta \in \sigma_p(K) = \sigma_p(\tilde{L}) + \sigma_p(\tilde{M})$ with $\lambda \neq r_\sigma(K)$, hence $\alpha \neq r_\sigma(\tilde{L})$ or $\beta \neq r_\sigma(\tilde{M})$, let $\alpha \neq r_\sigma(L)$. From Theorem 11.4 of KRASNOSEL'SKIJ-LIFSHITS-SOBOLEV [1986] it follows that $|\alpha| < r_\sigma(\tilde{L})$, since $\tilde{L}$ is u-bounded. Consequently,

$$|\lambda| = |\alpha + \beta| \leq |\alpha| + |\beta| < r_\sigma(\tilde{L}) + r_\sigma(\tilde{M}) = r_\sigma(K).$$

Now we show that the eigenvalue $r_\sigma(K)$ is simple. First of all, we can find decompositions

$$(14.72) \qquad \begin{cases} U = E_1 \oplus F_1, \quad V = E_2 \oplus F_2, \\[2mm] [U \to V] = (V \tilde{\otimes} E_1) \oplus (V \tilde{\otimes} F_1), \\[2mm] [U \leftarrow V] = (U \tilde{\otimes} E_2) \oplus (U \tilde{\otimes} F_2), \end{cases}$$

where E_1 is the (one-dimensional) nullspace of the operator $\tilde{L} - r_\sigma(\tilde{L})$, and E_2 is the (one-dimensional) nullspace of the operator $\tilde{M} - r_\sigma(\tilde{M})$.

For the sake of definiteness, let us consider the operator K in the space $[U \leftarrow V]$.

Since every space occurring in the decompositions (14.72) is invariant under K, we have

$$\sigma_p(K) = \sigma_p(K|U\tilde{\otimes}E_2) \cup \sigma_p(K|U\tilde{\otimes}F_2).$$

From (14.71) and from the equalities

$$\sigma(K|U\tilde{\otimes}E_2) = \sigma(\tilde{L}) + \sigma(\tilde{M}|E_2)$$

and

$$\sigma(K|U\tilde{\otimes}F_2) = \sigma(\tilde{L}) + \sigma(\tilde{M}|F_2)$$

we deduce that also

$$\sigma_p(K|U\tilde{\otimes}E_2) = \sigma_p(\tilde{L}) + \sigma_p(\tilde{M}|E_2)$$

and

$$\sigma_p(K|U\tilde{\otimes}F_2) = \sigma_p(\tilde{L}) + \sigma_p(\tilde{M}|F_2).$$

Since $r_\sigma(\tilde{M}) \notin \sigma_p(\tilde{M}|F_2)$, we have $r_\sigma(K) \notin \sigma_p(K|U\tilde{\otimes}F_2)$, hence

$$r_\sigma(K) \in \sigma_p(K|U\tilde{\otimes}E_2) = \sigma_p(\tilde{L}) + \sigma_p(\tilde{M}|E_2).$$

Moreover, since E_2 is one-dimensional, we have $U\tilde{\otimes}E_2 = \{x \otimes e : x \in U\}$, where e is some positive eigenvector of $\tilde{M}$ corresponding to the eigenvalue $r_\sigma(\tilde{M})$. It follows that the equation

$$[K - r_\sigma(K)](x \otimes e) = [(\tilde{L} - r_\sigma(\tilde{L}))x] \otimes e = 0$$

has a nontrivial solution if and only if the equation $\tilde{L}x = r_\sigma(\tilde{L})x$ has a nontrivial solution. But $r_\sigma(\tilde{L})$ is a simple eigenvalue of $\tilde{L}$, and thus $r_\sigma(K)$ is a simple eigenvalue of K as claimed. ∎

We remark that, under the hypotheses of Theorem 14.12, the operator K is in general not u-bounded.

The most important situation where Theorem 14.12 applies is as follows. Suppose that the operators $\tilde{L}$ and $\tilde{M}$ are both u-bounded, and some powers of $\tilde{L}$ and $\tilde{M}$ are compact. Then the equality (14.71) holds; consequently, $r_\sigma(K)$ is a simple eigenvalue, and all other eigenvalues λ of K satisfy $|\lambda| < r_\sigma(K)$. The proof is precisely the same as that of Theorem 14.12.

The results discussed in this subsection have been obtained in KALITVIN [1984, 1985, 1985a, 1988a, 1996]; they hold for operators K in the space $C(T \times S)$ (see also KALITVIN [1988b] and KALITVIN-GALITSKAJA-SOKOLOVA [1988]).

Finally, we remark that in some existence results for partial integral equations (for example, those containing Hardy-Littlewood type operators), the notion of the so-called generalized spectral radius plays an important role. Some results and applications in this direction may be found in KALITVIN-LAVROVA [1993], KALITVIN-PROVOTOROVA [1995], and POVOLOTSKIJ-KALITVIN [1994].

§ 15. Linear partial integral equations

In this section we collect some basic facts about the (unique) solvability of linear equations involving partial integral operators (of both Fredholm and Volterra type). Such equations occur rather often in applications (as we shall see in Chapter IV), but have not been studied systematically yet in book form. The main results of Subsections 15.1, 15.2 and 15.4 have been obtained in KALITVIN-ZABREJKO [1991].

15.1. Fredholm equations

Consider the linear partial integral equation

$$x(t,s) = \int_T l(t,s,\tau)x(\tau,s)\,d\mu(\tau)$$

(15.1)
$$+ \int_S m(t,s,\sigma)x(t,\sigma)\,d\nu(\sigma)$$

$$+ \int_{T\times S} n(t,s,\tau,\sigma)x(\tau,\sigma)d(\mu\times\nu)(\tau,\sigma) + f(t,s)$$

which may be written in a suitable Banach space concisely as operator equation

(15.2) $$(I - L - M - N)x = f,$$

where L, M, and N are given by (11.3), (11.4), and (11.5), respectively. Some typical examples of such spaces will be given below. We will be interested in the question to what extent the basic results for classical integral equations carry over to equation (15.1). We shall discuss some results which are related to specific properties of the operator $K = L + M + N$. Obviously,

(15.3)
$$I - K = (I - L)(I - M) - (N + LM)$$
$$= (I - M)(I - L) - (N + ML);$$

hence, in case the operators $I - L$ and $I - M$ are invertible, equation (15.2) is equivalent to both

(15.4) $[I - (I - M)^{-1}(I - L)^{-1}(N + LM)]x = g$

with

$$g := (I - M)^{-1}(I - L)^{-1}f$$

and

(15.5) $[I - (I - L)^{-1}(I - M)^{-1}(N + ML)]x = h$

with

$$h := (I - L)^{-1}(I - M)^{-1}f).$$

Observe that the iterated integral operators LM and ML may be written in the form

$$LMx(t, s) = \int_{T \times S} p(t, s, \tau, \sigma)x(\tau, \sigma)\, d(\mu \times \nu)(\tau, \sigma)$$

with $p(t, s, \tau, \sigma) := l(t, s, \tau)m(\tau, s, \sigma)$ and

$$MLx(t, s) = \int_{T \times S} q(t, s, \tau, \sigma)x(\tau, \sigma)\, d(\mu \times \nu)(\tau, \sigma)$$

with $q(t, s, \tau, \sigma) := m(t, s, \sigma)l(t, \sigma, \tau)$, respectively. Consequently, the operators $N + LM$ and $N + ML$ occurring in (15.4) and (15.5) are usual integral operators over the domain $T \times S$.

The invertibility of the operators $I - L$ and $I - M$ is of course related to the solvability of the equation

(15.6) $u(t, s) = \int_T l(t, s, \tau)u(\tau, s)\, d\mu(\tau) + f(t, s)$

(which is a classical integral equation containing a parameter s), and the equation

(15.7) $v(t, s) = \int_S m(t, s, \sigma)v(t, \sigma)\, d\nu(\sigma) + f(t, s)$

(which is a classical integral equation containing a parameter t). The study of the equations (15.6) and (15.7), in turn, reduces to analyzing the families of operators (12.6) and (12.7). This is connected with the solvability of the integral equations

$$(15.8) \qquad\qquad u = L(s)u + g \qquad (s \in S)$$

in an ideal space U over T, and the solvability of the integral equations

$$(15.9) \qquad\qquad v = M(t)v + h \qquad (t \in T)$$

in an ideal space V over S. By classical results on integral equations, the solutions u and v of these equations (if they exist!) are given by

$$(15.10) \qquad u(t) = g(t) + \int_T \varphi(t,s,\tau)g(\tau)\,d\mu(\tau) \qquad (s \in S)$$

and

$$(15.11) \qquad v(s) = h(s) + \int_S \psi(t,s,\sigma)h(\sigma)\,d\nu(\sigma) \qquad (t \in T)$$

involving the resolvent kernels $\varphi(t,s,\tau)$ and $\psi(t,s,\sigma)$. Now, if the spectral radii of the operators $]L(s)[$ and $]M(t)[$ (see (12.6), (12.7), (11.8) and (11.9)) satisfy

$$(15.12) \quad r_\sigma(]L(s)[) < 1 \quad (s \in S), \qquad r_\sigma(]M(t)[) < 1 \quad (t \in T),$$

the resolvent kernels may be represented as series of iterated kernels

$$(15.13) \qquad \varphi(t,s,\tau) = \sum_{k=1}^{\infty} l^{(k)}(t,s,\tau) \qquad (s \in S,\ t,\tau \in T)$$

and

$$(15.14) \qquad \psi(t,s,\sigma) = \sum_{k=1}^{\infty} m^{(k)}(t,s,\sigma) \qquad (t \in T,\ s,\sigma \in S)$$

which converge in the Zaanen kernel spaces $\mathfrak{Z}(U)$ and $\mathfrak{Z}(V)$, by each $s \in S$ and $t \in T$ respectively (see Subsection 12.1). If, in addition, we have

$$(15.15) \qquad\qquad \varphi \in \mathfrak{R}_l(X), \quad \psi \in \mathfrak{R}_m(X),$$

where $X = [U \to V]$ or $X = [U \leftarrow V]$, it is natural to expect that

(15.16) $$x(t,s) = f(t,s) + \int_T \varphi(t,s,\tau)f(\tau,s)\,d\mu(\tau)$$

and

(15.17) $$x(t,s) = f(t,s) + \int_S \psi(t,s,\sigma)f(t,\sigma)\,d\nu(\sigma)$$

are solutions of (15.6) and (15.7), respectively. In particular, this is true if

(15.18) $$r_\sigma(]L(s)[) = \lim_{k \to \infty} \sqrt[k]{\||l^{(k)}(\cdot,s,\cdot)\|_{\Re_l(U)}} \le p < 1$$

for all $s \in S$, and

(15.19) $$r_\sigma(]M(t)[) = \lim_{k \to \infty} \sqrt[k]{\||m^{(k)}(t,\cdot,\cdot)\|_{\Re_m(V)}} \le q < 1$$

for all $t \in T$. We summarize with the following

Theorem 15.1. *Let $X = [U \to V]$ or $X = [U \leftarrow V]$, and suppose that the linear operator $K = L + M + N$ acts in X and is regular. Assume that the estimates (15.12) hold, that the corresponding resolvent kernels satisfy (15.15), and that one of the operators $N + LM$ or $N + ML$ is compact in X. Then the linear partial integral equation (15.1) satisfies the Fredholm alternative in the space X. In particular, (15.1) admits a unique solution $x \in X$ for any $f \in X$ if and only if the equation $x = Kx$ has only the trivial solution $x(t,s) \equiv 0$.*

The conditions of Theorem 15.1 are sufficient but not necessary. The crucial points in this theorem are the compactness of $N + LM$ or $N + ML$ and the invertibility of both $I - L(s)$ $(s \in S)$ and $I - M(t)$ $(t \in T)$, i.e. the hypothesis

(15.20) $$1 \notin \sigma(L(s)) \cup \sigma(M(t)) \qquad (t \in T,\, s \in S).$$

Even in the case $l(t,s,\tau) \equiv l(t,\tau)$, $m(t,s,\sigma) \equiv m(s,\sigma)$, and $n(t,\tau,s,\sigma) \equiv 0$ (which frequently occurs in applications), one cannot drop the fundamental assumption (15.20).

Suppose now that S and T are compact sets, and the kernels l : $T \times S \times T \to \mathbb{R}$ and $m : T \times S \times S \to \mathbb{R}$ are continuous. Put

$$\sigma_L := \bigcup_{s \in S} \sigma(L(s)), \qquad \sigma_M := \bigcup_{t \in T} \sigma(M(t)),$$

with $L(s)$ given by (12.6) and $M(t)$ given by (12.7). Then the condition

$$(15.21) \qquad\qquad 1 \notin \sigma_L \cup \sigma_M$$

obviously implies the invertibility of the operators $I - L$ and $I - M$, where L is given by (11.3) and M by (11.4). From this we get the following

Theorem 15.2. *Suppose that the sets S and T are compact, the kernels $l = l(t, s, \tau)$ and $m = m(t, s, \sigma)$ are continuous, and the integral operator N given by (11.5) is compact in the space $X = C(T \times S)$. Then (15.21) implies that the linear partial integral equation (15.1) satisfies the Fredholm alternative in the space X.*

Under the hypotheses of the preceding two theorems we know, in particular, that *the partial integral equation* (15.1) *admits a unique solution* (in $X = [U \to V]$ or $X = [U \leftarrow V]$ in case of Theorem 15.1, and in $X = C(T \times S)$ in case of Theorem 15.2) *if*

$$(15.22) \qquad 1 \notin \sigma[(I - M)^{-1}(I - L)^{-1}(N + LM)]$$

or

$$(15.23) \qquad 1 \notin \sigma[(I - L)^{-1}(I - M)^{-1}(N + ML)].$$

We illustrate this for degenerate kernels over $T = S = [0, 1]$ in the following

Example 15.1. Let

$$l(t, s, \tau) = \alpha(t, s)\beta(\tau), \quad m(t, s, \sigma) = \gamma(t, s)\delta(\sigma),$$

and

$$n(t, s, \tau, \sigma) = \epsilon(t, s)\zeta(\tau)\eta(\sigma),$$

be continuous functions. We consider the equation

$$x(t,s) = \dot{a}(t,s) \int_0^1 \beta(\tau)x(\tau,s)\,d\tau + \gamma(t,s)\int_0^1 \delta(\sigma)x(t,\sigma)\,d\sigma$$

$$+ \epsilon(t,s)\int_0^1 \int_0^1 \zeta(\tau)\eta(\sigma)x(\tau,\sigma)\,d\sigma\,d\tau + f(t,s),$$

where f is continuous on $T \times S$. Then the operators $N + LM$ and $N + ML$ are compact in the space X and

$$\sigma_L = \bigcup_{s \in S} \sigma(L(s)) =$$

$$\{0\} \cup \bigcup_{s \in S} \{\lambda : \int_0^1 \alpha(\tau,s)\beta(\tau)d\tau = \lambda\} \supseteq \sigma(L),$$

$$\sigma_M = \bigcup_{t \in T} \sigma(M(t)) =$$

$$\{0\} \cup \bigcup_{t \in T}\{\lambda : \int_0^1 \gamma(t,\sigma)\delta(\sigma)d\sigma = \lambda\} \supseteq \sigma(M).$$

Obviously, the above equation is Fredholm if for any $s \in S$ and $t \in T$ we have

$$\int_0^1 \alpha(\tau,s)\beta(\tau)d\tau \neq 1, \quad \int_0^1 \gamma(t,\sigma)\delta(\sigma)d\sigma \neq 1.$$

In this case the inverse operators

$$(I - L)^{-1}x(t,s) = x(t,s) - \frac{\alpha(t,s)}{a(s)}\int_0^1 \beta(\tau)x(\tau,s)d\tau$$

and

$$(I - M)^{-1}x(t,s) = x(t,s) - \frac{\gamma(t,s)}{b(t)}\int_0^1 \delta(\sigma)x(t,\sigma)d\sigma$$

exist, where we have put

$$a(s) = 1 - \int_0^1 \alpha(\tau,s)\beta(\tau)d\tau, \quad b(t) = 1 - \int_0^1 \gamma(t,\sigma)\delta(\sigma)d\sigma.$$

Thus, if 1 is not an eigenvalues of the compact integral operator $(I - M)^{-1}(I - L)^{-1} (N + LM)$ (which may be written out explicitly), then the above equation has a unique solution in X.

15.2. Volterra equations

Consider now the special case when the sets T and S are real intervals and (15.1) is a *partial integral equation of Volterra type*

$$x(t, s) = \int_0^t l(t, s, \tau) x(\tau, s) \, d\tau$$

(15.24)
$$+ \int_0^s m(t, s, \sigma) x(t, \sigma) \, d\sigma$$

$$+ \int_0^t \int_0^s n(t, s, \tau, \sigma) x(\tau, \sigma) \, d\tau \, d\sigma + f(t, s).$$

In this case, (15.12) holds true (see e.g. ZABREJKO-KOSHELEV-KRAS-NOSEL'SKIJ-MIKHLIN-RAKOVSHCHIK-STETSENKO [1968]) if the operators (12.6) and (12.7) are compact in the spaces U and V, respectively, and U and V are *regular* spaces (i.e. all elements in U and V have absolutely continuous norms, see Subsection 4.1). Moreover, if the resolvent kernels $\varphi = \varphi(t, s, \tau)$ and $\psi = \psi(t, s, \sigma)$ satisfy (15.15), the study of equation (15.24) reduces to that of (15.4) and (15.5) (which are then classical Volterra integral equations, of course). Finally, if at least one of the operators $N + LM$ or $N + ML$ is compact, then, again by the regularity of the space X (see ZABREJKO-KOSHELEV-KRAS-NOSEL'SKIJ-MIKHLIN-RAKOVSHCHIK-STETSENKO [1968]), the spectral radius of the corresponding operator is zero. Thus we arrive at the following

Theorem 15.3. *Let U and V be regular ideal spaces and $X = [U \rightarrow V]$ or $X = [U \leftarrow V]$. Suppose that the linear operator $K = L + M + N$ acts in X, and the operators $L(s)$ and $M(t)$ defined by*

(15.25)
$$L(s)u(t) := \int_0^t l(t, s, \tau) u(\tau) \, d\tau \qquad (s \in S)$$

and

(15.26) $$M(t)v(s) := \int_0^s m(t,s,\sigma)v(\sigma)\,d\sigma \qquad (t \in T)$$

are compact in U and V, respectively. Assume that the resolvent kernels $\varphi = \varphi(t,s,\tau)$ and $\psi = \psi(t,s,\sigma)$ satisfy (15.15), and that at least one of the operators $N + LM$ or $N + ML$ is compact in X. Then the linear partial integral equation (15.24) admits a unique solution $x \in X$ for any $f \in X$.

The example

$$x(t,s) = \frac{1}{3t}\int_0^t x(\tau,s)\,d\tau + \frac{1}{3s}\int_0^s x(t,\sigma)\,d\sigma$$

$$+\frac{1}{3ts}\int_9^t\int_0^s x(\tau,\sigma)\,d\sigma\,d\tau$$

shows that equation (15.24) is, in general, not uniquely solvable without additional conditions on the kernel functions involved.

A simple (though important) special case where the assertions of Theorem 15.1 and Theorem 15.3 may be made more explicit is $U = V = L_p([0,1])$ and $X = L_p([0,1] \times [0,1])$. Here the hypothesis (15.15) means precisely that the partial integral operators

$$\Phi x(t,s) = \int_0^1 \varphi(t,s,\tau)x(\tau,s)\,d\tau$$

and

$$\Psi x(t,s) = \int_0^1 \psi(t,s,\sigma)x(t,\sigma)\,d\sigma$$

(in case of Theorem 15.1), or

$$\Phi x(t,s) = \int_0^t \varphi(t,s,\tau)x(\tau,s)\,d\tau$$

and

$$\Psi x(t,s) = \int_0^s \psi(t,s,\sigma)x(t,\sigma)\,d\sigma$$

(in case of Theorem 15.3), belong to $\mathfrak{L}^r(L_p)$. In this way, verifying the corresponding hypotheses simply reduces to studying families of Fredholm or Volterra integral equations in Lebesgue spaces. In particular, (15.24) has a unique L_p-solution for any $f \in L_p$ if the kernels l, m and n are bounded, and a unique continuous solution for any continuous function f if the kernels l, m and n are continuous and bounded.

Under the hypotheses of Theorem 15.3, the spectral radius of the operator K is zero. Other sufficient conditions for this may be found in ZABREJKO-LOMAKOVICH [1987, 1990] and also in the note KALITVIN [1997a]. The relation $r_\sigma(K) = 0$ was established in case of a bounded kernel function first in MÜNTZ [1934] and GOURSAT [1943]. The classical work of VEKUA [1948] is devoted to holomorphic solutions of partial integral equations with Volterra operators and their application in elasticity theory and in the theory of elliptic boundary value problems.

15.3. Bounded and continuous solutions

In view of some applications (see e.g. AGOSHKOV [1988], BRACK [1985], GNEDENKO [1988], KLIMONTOVICH [1982]) it is interesting to find bounded, continuous, or even smooth solutions of equations (15.1) and (15.24).

Unfortunately, the situation here is quite different from that for usual Fredholm integral equations. It is well-known that any solution of the linear Fredholm integral equation

$$x(t) = \int_T k(t,\tau)x(\tau)\,d\tau + f(t)$$

is continuous (or bounded) if, for example, the function f and the kernel k are continuous (respectively bounded). As the following example shows, the equation (15.1) with constant functions l, m, n, and f can have both bounded and unbounded solutions.

Example 15.2. Let $T = S = [0,1]$ and $l(t,s,\tau) = m(t,s,\sigma) =$

$n(t, s, \tau, \sigma) \equiv 1$ and $f(t, s) \equiv -2$. Then the equation

$$x(t, s) = \int_0^1 x(\tau, s)\, d\tau + \int_0^1 x(t, \sigma)\, d\sigma + \int_0^1 \int_0^1 x(\tau, \sigma)\, d\tau d\sigma - 2$$

has the continuous solution $x(t, s) \equiv 1$, the bounded solution

$$x(t, s) = \begin{cases} 2 & \text{if } 0 \leq s \leq \frac{1}{2}, \\ 0 & \text{if } \frac{1}{2} < s \leq 1, \end{cases}$$

and the unbounded solution

$$x(t, s) = \begin{cases} 1 - \dfrac{1}{\sqrt{\frac{1}{2} - s}} & \text{if } 0 \leq s < \frac{1}{2}, \\[2ex] 0 & \text{if } s = \frac{1}{2}, \\[2ex] 1 + \dfrac{1}{\sqrt{s - \frac{1}{2}}} & \text{if } \frac{1}{2} < s \leq 1. \end{cases}$$

Thus, even the analyticity of the functions l, m, n, and f does not guarantee the continuity or boundedness of the solutions of equation (15.1). Nevertheless, if the operator K is bounded in the space $C(T \times S)$ or $L_\infty(T \times S)$, the operator $I - K$ is invertible, and the function f in equation (15.1) is continuous, then every bounded solution x of (15.1) is continuous.

15.4. Using tensor products

Let us now take a closer look to the case $l(t, s, \tau) \equiv l(t, \tau)$ and $m(t, s, \sigma) \equiv m(s, \sigma)$ in (15.1), i.e. we consider the equation

$$x(t, s) = \int_T l(t, \tau) x(\tau, s)\, d\mu(\tau)$$

(15.27)
$$+ \int_S m(s, \sigma) x(t, \sigma)\, d\nu(\sigma)$$

$$+ \int_{T \times S} n(t, s, \tau, \sigma) x(\tau, \sigma)\, d(\mu \times \nu)(\tau, \sigma) + f(t, s).$$

We rewrite this again as

$$(15.28) \qquad (I - L - M - N)x = f,$$

where N is the operator (11.5), and L and M are now given by

$$(15.29) \qquad Lx(t, s) = \int_T l(t, \tau) x(\tau, s) \, d\mu(\tau)$$

and

$$(15.30) \qquad Mx(t, s) = \int_S m(s, \sigma) x(t, \sigma) \, d\nu(\sigma).$$

In general, even in case of continuous kernels the equation (15.27) is not Fredholm, because it contains the partial integral operators L and M. However, if the operators L and M are continuous in some Banach space X, and N is compact in X, in some important cases one may prove that the integral equation (15.27) is Fredholm. This is possible if the operators L and M may be represented as tensor products $L = \tilde{L} \bar{\otimes} I$ and $M = I \bar{\otimes} \tilde{M}$, where $\tilde{L}$ and $\tilde{M}$ are defined by (14.3) and (14.4). If we use now the results of § 14 and the fact that the Fredholm property and the index of an operator is stable with respect to compact perturbations (see e.g. KATO [1966] or KREJN [1971]), we arrive at the following results.

Theorem 15.6. *Suppose that the integral operator $\tilde{L}$ acts in an ideal space $U = U(T)$, the integral operator $\tilde{M}$ acts in an ideal space $V = V(S)$, and the integral operator N is compact in the space $X = [U \to V]$ or $X = [U \leftarrow V]$. Assume that one of the following conditions is satisfied:*

(a) $U = L_p(T)$ and $V = L_p(S)$ $(1 \leq p < \infty)$;

(b) $V = L_1(S)$, U is almost perfect, and $X = [U \to V]$;

(c) $U = L_1(T)$, V is almost perfect, and $X = [U \leftarrow V]$.

Then the integral equation (15.27) is Fredholm of index zero if and only if $1 \notin \sigma_{es}(K)$, and Fredholm if and only if $1 \notin \sigma_{ew}(K)$; here $\sigma_{es}(K)$ and $\sigma_{ew}(K)$ are the sets defined in Theorem 14.1. In case $1 \notin \sigma_{ew}(K)$, the index of equation (15.27) may be calculated by formula (14.10).

Now let T be a compact set, $U = C = C(T)$ the space of continuous functions on T, and $V = V(S)$ some Banach space. In LEVIN [1969] it is shown that the space $U \overline{\otimes} V$ with the norm (11.42) is a Banach space, and the norm (11.42) coincides with the weakest crossnorm λ which is defined by

(15.31)
$$\lambda \left(\sum_{k=1}^{n} u_k \otimes v_k \right) =$$
$$\sup \left\{ \sum_{k=1}^{n} |\langle u_k, u' \rangle \langle v_k, v' \rangle| : \|u'\|_{U'} \le 1, \|v'\|_{V'} \le 1 \right\}.$$

This norm is uniform (see e.g. SCHATTEN [1950]). It is well known (GROTHENDIECK [1955]) that in this case $U \overline{\otimes} V$ is just the space $C(T, V)$ of continuous on T functions with values in V. Similarly, if S is a compact set, $V = C = C(S)$, and $U = U(T)$ is some Banach space, then $V \overline{\otimes} U = C(S, U)$.

The following theorem follows from the results of T. ICHINOSE [1978, 1978a] on the spectra of tensor products of operators in tensor products of Banach spaces with quasiuniform crossnorms (see § 14).

Theorem 15.7. *Let T and S be compact sets. Suppose that the integral operator $\tilde{L}$ is bounded on $C = C(T)$, the integral operator $\tilde{M}$ is bounded on some Banach space $V = V(S)$, and the operator N is compact on the spaces $X = C(T, V)$. Then the assertion of Theorem 15.6 holds true.*

In the previous two theorems, we have given conditions under which the integral equation (15.27) is Fredholm (of index zero) in the tensor product of Banach spaces which may be represented as spaces with mixed norm or space of vector functions with uniform crossnorms. The Theorems 14.8 - 14.10 below give such conditions in spaces with mixed norm whose crossnorms are not necessarily uniform or quasiuniform.

Theorem 15.8. *Suppose that the integral operators $\tilde{L}$ and $\tilde{M}$ act in completely regular ideal spaces $U = U(T)$ and $V = V(S)$, respectively. Assume that one of the following conditions is satisfied:*

(a) $\tau \mapsto \|l(\cdot, \tau)\|_U \in U'$, $t \mapsto \|l(t, \cdot)\|_{U'} \in U$, and $X = [U \leftarrow V]$;

(b) $\sigma \mapsto \|m(\cdot, \sigma)\|_V \in V'$, $s \mapsto \|m(s, \cdot)\|_{V'} \in V$, and $X = [U \rightarrow V]$;

(c) $\tau \mapsto \|l(\cdot, \tau)\|_U \in U'$, $[U \rightarrow V] \subseteq [U \leftarrow V]$, and $X = [U \leftarrow V]$;

(d) $t \mapsto \|l(t, \cdot)\|_{U'} \in U$, $[U \leftarrow V] \subseteq [U \rightarrow V]$, and $X = [U \leftarrow V]$;

(e) $\sigma \mapsto \|m(\cdot, \sigma)\|_V \in V'$, $[U \leftarrow V] \subseteq [U \rightarrow V]$, and $X = [U \rightarrow V]$;

(f) $s \mapsto \|m(s, \cdot)\|_{V'} \in V$, $[U \rightarrow V] \subseteq [U \leftarrow V]$, and $X = [U \rightarrow V]$.

Finally, suppose that N is a compact operator in X. Then the assertion of Theorem 15.6 holds.

Using Theorem 14.10 and Example 14.5 we may of course formulate Theorem 15.8, in particular, for the Lebesgue spaces $[L_p \rightarrow L_q]$ and $[L_q \leftarrow L_p]$ with mixed norms.

We consider now another important special case of the equation (15.27), namely

$$x(t, s) = \int_T l(t, \tau) x(\tau, s) \, d\mu(\tau)$$

(15.32)

$$+ \int_S m(s, \sigma) x(t, \sigma) \, d\nu(\sigma) + f(t, s).$$

In various classes of function spaces (see § 14) this equation can be written in the form

(15.33) $\qquad (I - K)x = (I \overline{\otimes} I - \tilde{L} \overline{\otimes} I - I \overline{\otimes} \tilde{M})x = f;$

here K is the operator (14.1), and $\tilde{L}$ and $\tilde{M}$ are the operators (14.3) and (14.4). Applying the results of § 14 to equation (15.33) allows us to give a fairly complete description of the spectral properties of the equation (15.32). We confine ourselves only to the invertibility condition

$$1 \notin \sigma(K) = \sigma(\tilde{L}) + \sigma(\tilde{M})$$

(15.34)

$$= \{\alpha + \beta : \alpha \in \sigma(\tilde{L}), \ \beta \in \sigma(\tilde{M})\},$$

which follows from equality (14.7).

Let us suppose that $1 \notin \sigma(K)$. Then we can find open neighborhoods G of $\sigma(\tilde{L})$ and F of $\sigma(\tilde{M})$ such that the function $h(\xi, \eta) = (1-\xi-\eta)^{-1}$ is holomorphic in $G \times F$. Applying classical results of the theory of operator calculus of tensor products (see e.g. ICHINOSE [1978b]) we get

$$(I - K)^{-1} = (I \overline{\otimes} I - \tilde{L} \overline{\otimes} I - I \overline{\otimes} \tilde{M})^{-1}$$

(15.35)
$$= \frac{1}{(2\pi i)^2} \int_{\Gamma_F} \int_{\Gamma_G} \frac{(\xi I - \tilde{L})^{-1} \overline{\otimes} (\eta I - \tilde{M})^{-1}}{1 - \xi - \eta} \, d\xi \, d\eta,$$

where $\Gamma_G \subset G$ and $\Gamma_F \subset F$ are some rectifiable curves which do not meet the spectra $\sigma(\tilde{L})$ and $\sigma(\tilde{M})$, respectively. Consequently, the solution of equation (15.33) is given by $x = (I-K)^{-1} f$ with $(I-K)^{-1}$ given by (15.35). In this way, we have proved the following theorem.

Theorem 15.9. *Let $U = U(T)$ and $V = V(S)$ be Banach spaces. Suppose that the operator $\tilde{L}$ is bounded in U, the operator $\tilde{M}$ is bounded in V, and one of the following conditions is satisfied:*

(a) $U = L_p(T)$, $V = L_p(S)$, and $X = L_p(T \times S)$ $(1 \le p < \infty)$;

(b) $U = L_1(T)$, V is an almost perfect ideal space, and $X = [U \leftarrow V]$;

(c) $V = L_1(S)$, U is an almost perfect ideal space, and $X = [U \rightarrow V]$;

(d) U and V are regular ideal spaces, $X = [U \leftarrow V]$, and $\tilde{L}$ is a regular operator in U;

(e) U and V are regular ideal spaces, $X = [U \rightarrow V]$, and $\tilde{M}$ is a regular operator in V;

(f) T and S are compact sets, $U = C(T)$, $V = C(S)$, and $X = C(T \times S)$.

Then the equation (15.32) has a unique solution in the space X, for any $f \in X$, if and only if the condition (15.34) is satisfied. Moreover, this solution is $x = (I - K)^{-1} f$ with $(I - K)^{-1}$ given by formula (15.35).

To illustrate these theorems, let us consider the linear integral equation

(15.36)
$$\lambda x(t,s) + \int_{-1}^{1} l(t-\tau)x(\tau,s)\,d\tau$$
$$+ \int_{0}^{s} m(s,\sigma)x(t,\sigma)\,d\sigma = f(t,s)$$

which occurs in the mechanics of continuous media (see e.g. ALEK-SANDROV-KOVALENKO [1980, 1986] and KOVALENKO [1981]). Here we have $T = [-1,1]$, $S = [0,a]$, and the kernel l has the form

$$l(\xi) = \frac{1}{2}\int_{-\infty}^{+\infty} \tilde{l}(z)\exp^{i\theta z\xi}\,dz,$$

where λ and θ are a parameters (which have a mechanical and geometric meaning), and $\tilde{l}$ is a positive continuous function satisfying

$$\tilde{l}(z) = A + O(z^2)\ (z \to 0),\quad |z|\tilde{l}(z) = B + O(z^{-1})\ (z \to \infty).$$

Moreover, the kernel m is either continuous or weakly singular, and the function f has the form

$$f(t,s) = f_1(t) + f_2(s) + tf_3(s)\quad (f_1 \in L_p(T),\ f_2,f_3 \in L_p(S)).$$

Equation (15.36) may be studied in the space $L_p(T \times S)$. Under the hypotheses given above, the operators $\tilde{L}$ and $\tilde{M}$ are compact in $L_p(T)$ and $L_p(S)$, respectively, and thus the results of § 14 apply. The spectrum of the compact operator $\tilde{L}$ consists of 0 and either a finite number, or a sequence converging to 0, of eigenvalues λ. On the other hand, the spectrum of $\tilde{M}$ contains only 0. The operator $\tilde{L}$ is selfadjoint and positive definite in $L_2(T)$. Therefore, its spectrum is contained in the nonnegative real axis. Since $0 \in \sigma(\tilde{L})$ in any of the spaces $L_p(T)$, and the other points of spectrum form a discrete set, by a well-known results of HALBERG-TAYLOR [1956], the spectrum of the operator $\tilde{L}$ in $L_p(T)$ is the same for any p, i.e. consists of 0 and a finite number or a sequence (converging to 0) of positive eigenvalues. Thus, the spectrum of the corresponding operator K coincides in this case with that of the operator $\tilde{L}$; in particular, all elements of this spectrum are not eigenvalues. Altogether, equation (15.36) has a unique solution $x \in L_p(T \times S)$ for any $f \in L_p(T \times S)$, provided that $-\lambda \notin \sigma(\tilde{L})$. Moreover, this solution is $x = (I - K)^{-1}f$ with $(I - K)^{-1}$ given by formula (15.35). Analogous results are true for the case of the space $C(L_p)$ $(1 \leq p < \infty)$.

We remark that some approximate methods for constructing the solutions of equation (15.36) (for $\lambda > 0$) have been studied in KOVALENKO [1981].

15.5. Using eigenfunction expansions

Suppose that the integral operators $\tilde{L}$ and $\tilde{M}$ are compact in $L_2(T)$ and $L_2(S)$, respectively, and have symmetric kernels. Let $\{\phi_1, \phi_2, \phi_3, \ldots\}$ be an orthonormal basis in $L_2(T)$ which consists of eigenfunctions of the operator $\tilde{L}$ with corresponding eigenvalues $\alpha_1, \alpha_2, \alpha_3, \ldots$, counted according to their multiplicity. Similarly, let $\{\psi_1, \psi_2, \psi_3, \ldots\}$ be an orthonormal basis in $L_2(S)$ which consists of eigenfunctions of the operator $\tilde{M}$ with corresponding eigenvalues $\beta_1, \beta_2, \beta_3, \ldots$, again counted according to their multiplicity. Then the system $\{\phi_i \otimes \psi_j : i, j = 1, 2, 3, \ldots\}$ forms an orthonormal basis in $L_2(T \times S)$, and the equation (15.32) may be written in the form

$$(15.37) \qquad \sum_{i,j=1}^{\infty} \phi_i \otimes \psi_j [x_{ij}(1 - \alpha_i - \beta_j) - f_{ij}] = 0,$$

where x_{ij} and f_{ij} are the Fourier coefficients of the functions x and f with respect to the system $\{\phi_i \otimes \psi_j : i, j = 1, 2, 3, \ldots\}$. By the results of § 14 we have then

$$\sigma(K) = \sigma(\tilde{L}) + \sigma(\tilde{M}), \quad \sigma_p(K) = \sigma_p(\tilde{L}) + \sigma_p(\tilde{M}),$$

and

$$\mathbb{C} \setminus \sigma_{es}(K) = (\mathbb{C} \setminus \sigma(\tilde{L})) \cap (\mathbb{C} \setminus \sigma(\tilde{M})).$$

As we can see from (15.37), if $1 \notin \sigma(K)$, then $x_{ij} = (1 - \alpha_i - \beta_j)^{-1} f_{ij}$. Therefore, the unique solution of equation (15.32) may be written in the form

$$(15.38) \qquad x(t,s) = \sum_{i,j=1}^{\infty} \frac{f_{ij}}{1 - \alpha_i - \beta_j} \phi_i(t)\psi_j(s).$$

On the other hand, if $1 \in \sigma(K)$, but $1 \notin \sigma(\tilde{L}) \cup \sigma(\tilde{M})$, then $1 = \alpha_i + \beta_j$, where i runs over some finite set $P \subset \mathbb{N}$ and j runs over some finite set $Q \subset \mathbb{N}$. It follows from (15.37) that $x_{ij} = (1 - \alpha_i - \beta_j)^{-1} f_{ij}$ if $i \notin P$ and $j \notin Q$, and $f_{ij} = 0$ if $i \in P$ or $j \in Q$. We conclude that the condition

$$(15.39) \qquad f_{ij} = 0 \quad (i \in P, j \in Q)$$

is necessary and sufficient for the solvability of equation (15.32). This condition simply means that f is orthogonal to the nullspace

$$(15.40) \qquad N(I - K) = \text{span}\{\phi_i \otimes \psi_j : i \in P, j \in Q\}$$

of the operator $I - K$. If (15.39) is satisfied, the equation (15.32) has a finite number of linearly independent solutions of the form

$$(15.41) \qquad \begin{aligned} x(t,s) = &\sum_{(i,j)\in P\times Q} c_{ij}\phi_i(t)\psi_j(s) \\ + &\sum_{(i,j)\in \mathbb{N}^2\setminus P\times Q} \frac{f_{ij}}{1 - \alpha_i - \beta_j} \phi_i(t)\,\psi_j(s), \end{aligned}$$

where c_{ij} $(i \in P, j \in Q)$ are arbitrary constants.

Now suppose that $1 \in \sigma(\tilde{L}) \cup \sigma(\tilde{M})$, but 1 is not an eigenvalue of K. Then it follows from (15.37) and (15.38) that equation (15.32) is solvable if and only if the series

$$(15.42) \qquad \sum_{i,j=1}^{\infty} \left(\frac{f_{ij}}{1 - \alpha_i - \beta_j} \right)^2$$

converges: If the condition (15.42) holds, then the equation (15.32) has unique solution which may be found by formula (15.38). On the other hand, if $1 \in \sigma(\tilde{L}) \cup \sigma(\tilde{M})$ and 1 is an eigenvalue of K, then it follows from (15.37) that equation (15.31) is solvable if and only if the condition (15.39) holds; in this condition P and Q may be finite or countable subsets of $\mathbb{N}$ such that

$$(15.43) \qquad \sum_{(i,j)\in R}^{\infty} \left(\frac{f_{ij}}{1 - \alpha_i - \beta_j} \right)^2 < \infty.$$

The solution of equation (15.31) may then be written in the form

$$(15.44) \qquad x(t,s) = x_0(t,s) + \sum_{(i,j)\in R}^{\infty} \frac{f_{ij}}{1 - \alpha_i - \beta_j} \phi_i(t)\psi_j(s),$$

where $x_0 \in N(I - K)$ is arbitrary.

Suppose that the kernels l and m in the integral equation (15.32) are degenerate, i.e.

$$(15.45) \qquad l(t,\tau) = \sum_{i=1}^{p} \phi_i(t)\overline{\psi_i(\tau)}, \quad m(s,\sigma) = \sum_{j=1}^{q} \nu_j(s)\overline{\mu_j(\sigma)},$$

where $\{\phi_1,\ldots,\phi_p\}$, $\{\psi_1,\ldots,\psi_p\}$, $\{\nu_1,\ldots,\nu_q\}$, and $\{\mu_1,\ldots,\mu_q\}$ are linearly independent systems. Let

$$\begin{cases} E_1 := \operatorname{span}\left(\{\phi_1,\ldots,\phi_p\} \cup \{\psi_1,\ldots,\psi_p\}\right), \\[2mm] E_2 := \operatorname{span}\left(\{\nu_1,\ldots,\nu_q\} \cup \{\mu_1,\ldots,\mu_q\}\right), \end{cases}$$

and denote by F_1 and F_2 the orthogonal complement of E_1 and E_2 in $L_2(T)$ and $L_2(S)$, respectively. Then

$$L_2(T) = E_1 \oplus F_1, \qquad L_2(S) = E_2 \oplus F_2$$

and

$$L_2(T \times S) = E_1 \otimes E_2 \oplus E_1 \otimes F_2 \oplus F_1 \otimes E_2 \oplus F_1 \overline{\otimes} F_2.$$

Obviously, the spaces $E_1 \otimes E_2$, $E_1 \otimes F_2$, and $F_1 \otimes E_2$ are invariant under the operator (14.1) with the kernel (15.45), and $F_1 \overline{\otimes} F_2$ is mapped by this operator to the trivial functions. Consequently, equation (15.32) may be replaced by the equations

$$(15.46) \qquad x^{(1)}(t,s) = \sum_{i=1}^{p} \phi_i(t)a_i^{(1)}(s) + \sum_{j=1}^{q} \nu_j(s)b_j^{(1)}(t) + f^{(1)}(t,s),$$

$$(15.47) \qquad x^{(2)}(t,s) = \sum_{i=1}^{p} \phi_i(t) a_i^{(2)}(s) + f^{(2)}(t,s),$$

$$(15.48) \qquad x^{(3)}(t,s) = \sum_{j=1}^{q} \nu_j(s) b_j^{(3)}(t) + f^{(3)}(t,s),$$

and

$$(15.49) \qquad x^{(4)}(t,s) = f^{(4)}(t,s).$$

Here $x^{(1)}, f^{(1)} \in E_1 \otimes E_2$, $x^{(2)}, f^{(2)} \in E_1 \otimes F_2$, $x^{(3)}, f^{(3)} \in F_1 \otimes E_2$, $x^{(4)}, f^{(4)} \in F_1 \overline{\otimes} F_2$, $x = x^{(1)} + \ldots + x^{(4)}$, and $f = f^{(1)} + \ldots + f^{(4)}$. Moreover, the coefficients $a_1^{(k)}, \ldots, a_p^{(k)}$ $(k = 1, 2)$ and $b_1^{(k)}, \ldots, b_q^{(k)}$ $(k = 1, 3)$ are given by

$$(15.50) \qquad a_i^{(k)}(s) = \int_T x^{(k)}(\tau, s) \overline{\psi_i(\tau)} \, d\mu(\tau) \quad (i = 1, \ldots, p)$$

and

$$(15.51) \qquad b_j^{(k)}(t) = \int_S x^{(k)}(t, \sigma) \overline{\mu_j(\sigma)} \, d\nu(\sigma) \quad (j = 1, \ldots, q).$$

Let us consider first the equation (15.46). Putting (15.46) into (15.50) for $k = 1$ and into (15.51) for $k = 1$ we obtain the equations

$$(15.52) \qquad a_i^{(1)}(s) = \sum_{l=1}^{p} a_l^{(1)}(s) c_{il} + \sum_{j=1}^{q} \nu_j(s) x_{ij}^{(1)} + f_i^{(1)}(s)$$

for $i = 1, \ldots, p$, and

$$(15.53) \qquad b_j(t) = \sum_{k=1}^{q} d_{jk} b_k^{(1)}(t) + \sum_{l=1}^{p} \phi_l(t) y_{jl}^{(1)} + g_j^{(1)}(t)$$

for $j = 1, \ldots, q$, where the constants c_{il} and d_{jk} and the functions $f_i^{(1)}$ and $g_j^{(1)}$ are defined by

$$(15.54) \qquad \begin{cases} c_{il} = \displaystyle\int_T \phi_l(\tau) \overline{\psi_i(\tau)} \, d\mu(\tau), \\[2mm] d_{jk} = \displaystyle\int_S \nu_k(\sigma) \overline{\mu_j(\sigma)} \, d\nu(\sigma) \end{cases}$$

and

$$
(15.55) \quad
\begin{cases}
f_i^{(1)}(s) = \displaystyle\int_T f^{(1)}(\tau, s)\overline{\psi_i(\tau)}\, d\mu(\tau), \\[2mm]
g_j^{(1)}(t) = \displaystyle\int_S f^{(1)}(t, \sigma)\overline{\mu_j(\sigma)}\, d\nu(\sigma),
\end{cases}
$$

and the unknown constants $x_{ij}^{(1)}$ and $y_{jl}^{(1)}$ may be determined by

$$
(15.56) \quad
\begin{cases}
x_{ij}^{(1)} = \displaystyle\int_T b_j^{(1)}(\tau)\overline{\psi_i(\tau)}d\mu(\tau), \\[2mm]
y_{jl}^{(1)} = \displaystyle\int_S a_l^{(1)}(\sigma)\overline{\mu_j(\sigma)}d\nu(\sigma).
\end{cases}
$$

Multiplying the equations (15.52) and (15.53) by $\overline{\mu_j(s)}$ $(j = 1, \ldots, q)$ and $\overline{\psi_i(t)}$ $(i = 1, \ldots, p)$, and integrating the resulting equations over S and T, respectively, from (15.56) we get the systems

$$
\begin{cases}
x_{ij}^{(1)} = \displaystyle\sum_{l=1}^{p} c_{il} y_{jl}^{(1)} + \sum_{k=1}^{q} d_{jk} x_{ik}^{(1)} + f_{ij}^{(1)}, \\[3mm]
y_{ji}^{(1)} = \displaystyle\sum_{l=1}^{p} c_{il} y_{jl}^{(1)} + \sum_{k=1}^{q} d_{jk} x_{ik}^{(1)} + f_{ij}^{(1)}
\end{cases}
$$

for $i = 1, \ldots, p$ and $j = 1, \ldots, q$, where

$$
f_{ij}^{(1)} = \int_S \int_T f(\tau, \sigma)\overline{\psi_i(\tau)\mu_j(\sigma)}d\mu(\tau)\, d\nu(\sigma).
$$

From the definition it is clear that

$$
(15.57) \qquad y_{ji}^{(1)} = x_{ij}^{(1)} \quad (i = 1, \ldots, p,\ j = 1, \ldots, q).
$$

Consequently,

$$
x_{ij}^{(1)} = \sum_{l=1}^{p} c_{il} x_{lj}^{(1)} + \sum_{k=1}^{q} d_{jk} x_{ik}^{(1)} + f_{ij}^{(1)}
$$

for $i = 1, \ldots, p$ and $j = 1, \ldots, q$. The last system may be written as a matrix Ljapunov equation

$$
(15.58) \qquad X = CX + XD^* + f,
$$

where the matrices $C = (c_{ij})$, $D^* = (d_{ji})$, and $f = (f_{ij})$ are known, while the matrix $X = (x_{ij})$ is unknown $(i = 1, \ldots, p, \; j = 1, \ldots, q)$. Thus, the solution of equation (15.46) is reduced to the solution of the systems (15.52), (15.53), and (15.58). If these systems are solvable, then $x_{ij}^{(1)}$ may be found from (15.58), and $y_{jl}^{(1)} = x_{lj}^{(1)}$ may be found from (15.57). Putting $x_{ij}^{(1)}$ and $y_{jl}^{(1)}$ into (15.52) and (15.53) and solving the last systems we find then $a_i^{(1)}$ and $b_j^{(1)}$. Putting now $a_i^{(1)}$ and $b_j^{(1)}$ into equation (15.46) we get the solution $x^{(1)}$ of this equation. Obviously, if the systems (15.52), (15.53) and (15.58) are solvable, then equation (15.46) has a unique solution or a finite number of linearly independent solutions.

Analogously, putting (15.47) into (15.50) for $k = 2$ and (15.48) into (15.51) for $k = 3$, we end up with the systems

$$(15.59) \qquad a_i^{(2)}(s) = \sum_{l=1}^{p} c_{il} a_l^{(2)}(s) + f_i^{(2)}(s) \quad (i = 1, \ldots, p)$$

and

$$(15.60) \qquad b_j^{(3)}(t) = \sum_{k=1}^{q} d_{jk} b_k^{(3)}(t) + g_j^{(3)}(t) \quad (j = 1, \ldots, q).$$

From these systems (if they are solvable, of course) we may find the unknown functions $a_i^{(2)}$ and $b_j^{(3)}$ $(i = 1, \ldots, p, \; j = 1, \ldots, q)$. Putting these functions into (15.47) and (15.48) we get $x^{(2)}$ and $x^{(3)}$, respectively. However, if the systems (15.59) and (15.60) are solvable, the equations (15.47) and (15.48) have a unique solution or an infinite number of linearly independent solutions. In fact, if 1 is not an eigenvalue of the matrices C and D (i.e. $1 \notin \sigma(\tilde{L})$ and $1 \notin \sigma(\tilde{M})$), then the systems (15.59) and (15.60), and hence also the equations (15.47) and (15.48) have a unique solution. However, if 1 is an eigenvalue of C and D, then $1 \in \sigma_p(\tilde{L})$ and $\sigma_p(\tilde{M})$ and, obviously, the homogeneous equations

$$x^{(2)}(t, s) = \sum_{i=1}^{p} \phi_i(t) \int_T x^{(2)}(\tau, s) \overline{\psi_i(\tau)} \, d\mu(\tau)$$

and

$$x^{(3)}(t, s) = \sum_{j=1}^{q} \nu_j(s) \int_T x^{(3)}(t, \sigma)\overline{\mu_j(\sigma)}\, d\nu(\sigma)$$

have infinitely many solutions of the form

$$x^{(2)}(t, s) = h(t)u(s), \quad x^{(3)}(t, s) = g(s)v(t)$$

with $h \in N(I - \tilde{L}) \cap E_1$, $u \in F_2$, $g \in N(I - \tilde{M}) \cap E_2$, and $v \in F_1$.

All these arguments do not only allow us to give explicit criteria for the solvability of equation (15.32) in the case of degenerate kernels, but they also provide a constructive method for solving this equation. We summarize with the following

Theorem 15.10. *Suppose that the kernels l and m in equation (15.32) are of the form (15.45). Then the following solvability criteria hold true:*

(a) If $1 \notin \sigma(\tilde{L}) + \sigma(\tilde{M})$, the equation (15.32) has a unique solution for any function $f \in L_2(T \times S)$.

(b) If $1 \in \sigma(\tilde{L}) + \sigma(\tilde{M})$ but $1 \notin \sigma(\tilde{L}) \cup \sigma(\tilde{M})$, then the equation (15.32) is solvable if and only if the equation (15.58) is solvable; moreover, equation (15.32) has then a finite number of linearly independent solutions.

(c) If $1 \in \sigma(\tilde{L}) + \sigma(\tilde{M})$ and one of the following conditions is satisfied:
(c1) $1 \in \sigma(\tilde{L})$ but $1 \notin \sigma(\tilde{M})$,
(c2) $1 \notin \sigma(\tilde{L})$ but $1 \in \sigma(\tilde{M})$,
(c3) $1 \in \sigma(\tilde{L})$ and $1 \in \sigma(\tilde{M})$,
then the equation (15.31) is solvable if and only if both (15.58) and (15.59) (in case (c1)), (15.58) and (15.60) (in case (c2)), or (15.58), (15.59) and (15.61) (in case (c3)) are solvable; moreover, the equation (15.32) has then an infinite number of linearly independent solutions.

We remark that a fairly complete description equation (15.32) in the case of degenerate kernels and matrix Ljapunov equations was established in VITOVA [1975, 1976, 1976a, 1977]. As we have seen, the study of equation (15.32) may be reduced to the study of the equations (15.46) - (15.48). A detailed analysis of these equations may be carried out with the help of the methods which we used in the proof of Theorem 14.9.

We close this section with detailed bibliographical comments. Partial integral equations with continuous kernel functions have been studied in OKOLELOV [1967, 1967a, 1967b, 1968], with bounded kernel functions in FENYÖ [1955], and with symmetric kernel functions in BOLTJANSKIJ [1979, 1980, 1981], BOLTJANSKIJ-LIKHTARNIKOV [1982], LIKHTARNIKOV-MOROZOVA [1983], and LIKHTARNIKOV-VITOVA [1976]. We also mention the papers KAKICHEV-KOVALENKO [1973] and VITOVA [1988] on equations with degenerate or Jordan-type kernel functions, and LIKHTARNIKOV [1974, 1975] on equations with two parameters and selfadjoint operators. In KALITVIN [1989, 1989b] one may find conditions under which a partial integral operator is Fredholm (of index zero). We also mention the classical work by KANTOROVITZ [1957, 1958], KOVALENKO [1968, 1971, 1975], and MAURO [1976].

Singular equations have been studied in ABHUANKAR-FYMAT [1969], GAKHOV [1977], GAKHOV-KAKICHEV [1967], KAKICHEV [1967, 1968, 1969, 1978], PILIDI [1971, 1971a, 1982, 1988, 1989], PILIDI-SAZONOV [1983], SIMONENKO [1971], SKORIKOV [1980], and TOKILASHVILI [1985, 1987]. Partial integral operators containing difference-type kernel functions are considered in BÖTTCHER [1982, 1983], BÖTTCHER-PASENCHUK [1982], GOVORUKHINA-KOVALENKO-PARADOKSOVA [1985, 1986, 1987, 1987a, 1989], KOVALENKO [1975], MORARU [1969, 1975], SIMONENKO [1968], and MALYSHEV [1970].

We did not consider here another important class of partial integral operators, namely that of operators of so-called Romanovskij type. A classical reference is ROMANOVSKIJ [1932] which uses some kind of analogue to Fredholm determinants. This method is also employed in SHCHELKUNOV [1972, 1974]. Spectral properties of such operators have been studied in LIKHTARNIKOV [1976, 1980, 1981, 1987], LIKHTARNIKOV-MOROZOVA [1988], LIKHTARNIKOV-SPEVAK [1976, 1976a], MOROZOVA [1984, 1984a, 1984b, 1986], and KALITVIN [1983]. Other equations of Romanovskij type are treated in KALITVIN [1983a, 1984, 1985, 1986a, 1987a, 1987b, 1988, 1988a1996c], KALITVIN-GALITSKAJA-

SOKOLOVA [1988], KALITVIN-ROMANOVA [1993], and LIKHTARNI-KOV-KALITVIN [1989]. The paper KALITVIN [1985a] is concerned with so-called multi-spectrum operators with partial integrals.

Approximate and numerical methods for solutions of linear partial integral equations may be found in BELOTSERKOVSKIJ-LIFANOV [1985], DZJADYK-OSTROVETSKIJ [1981], GABDULKHAEV [1980], KOVALEN-KO [1981], LENHARDT [1986], MOROZOVA [1984, 1984b], OKOLELOV [1967b], and TIVONCHUK [1971].

Several special classes of partial integral equations of Volterra or Volterra-Fredholm type coming from applied problems will be discussed in detail in Section 20 below.

Chapter IV

Generalizations and

Applications

§ 16. Generalized equations of Barbashin type

In this section we bridge the gap between the integro-differential equations of Barbashin type we considered in Chapter I and Chapter II, and the theory of partial integral operators we studied so far in the present chapter. We also introduce and study a class of generalized Barbashin equations (for functions of three variables) which occur in certain problems of mathematical physics. The scheme of reducing integro-differential equations and generalized equations of Barbashin type to integral equations is outlined in KALITVIN [1991a, 1995, 1996a].

16.1. Reduction to partial integral equations

Suppose we are interested in finding a solution x of the integro-differential equation of Barbashin type

$$(16.1) \qquad \frac{\partial x(t,s)}{\partial t} = c(t,s)x(t,s) + \int_a^b k(t,s,\sigma)x(t,\sigma)\,d\sigma + f(t,s)$$

satisfying the initial condition

$$(16.2) \qquad x(t_0,s) = x_0(s) \qquad (a \le s \le b),$$

where $t_0 \in J$ is fixed and $x_0 : [a,b] \rightarrow \mathbb{R}$ is a given continuous function. Putting $\partial x(t,s)/\partial t := y(t,s)$ we arrive at the equation

$$
\begin{aligned}
y(t,s) = g(t,s) &+ \int_{t_0}^t c(t,s)y(\tau,s)\,d\tau \\
&+ \int_{t_0}^t \int_a^b k(t,s,\sigma)y(\tau,\sigma)\,d\sigma\,d\tau,
\end{aligned}
$$

(16.3)

where

$$(16.4) \qquad g(t,s) = f(t,s) + c(t,s)x_0(s) + \int_a^b k(t,s,\sigma)x_0(\sigma)\,d\sigma.$$

This is a partial integral equation of type (13.1) with

$$l(t,s,\tau) \equiv c(t,s), \quad m(t,s,\sigma) \equiv 0, \quad n(t,s,\tau,\sigma) \equiv k(t,s,\sigma).$$

More generally, one may study parameter-dependent integro-differential equations of the form

$$\frac{\partial u(\varphi, t, s)}{\partial \varphi} = c(\varphi, t, s)u(\varphi, t, s)$$

$$+ \int_T l(\varphi, t, s, \tau)u(\varphi, \tau, s)\, d\tau$$

(16.5)

$$+ \int_S m(\varphi, t, s, \sigma)u(\varphi, t, \sigma)\, d\sigma$$

$$+ \int_T \int_S n(\varphi, t, s, \tau, \sigma)u(\varphi, \tau, \sigma)\, d\sigma\, d\tau + f(\varphi, t, s),$$

subject to the initial condition

(16.6) $$u(\varphi_0, t, s) = u_0(t, s),$$

where $t \in T := [a, b]$, $s \in S := [c, d]$, and $\varphi_0 \in J$. Putting $\partial u(\varphi, t, s)/\partial \varphi =: v(\varphi, t, s)$ we obtain

$$v(\varphi, t, s) = g(\varphi, t, s) + \int_{\varphi_0}^{\varphi} c(\varphi, t, s)v(\rho, t, s)\, d\rho$$

$$+ \int_{\varphi_0}^{\varphi} \int_T l(\varphi, t, s, \tau)v(\rho, \tau, s)\, d\tau\, d\rho$$

(16.7)

$$+ \int_{\varphi_0}^{\varphi} \int_S m(\varphi, t, s, \sigma)v(\rho, t, \sigma)\, d\sigma\, d\rho$$

$$+ \int_{\varphi_0}^{\varphi} \int_T \int_S n(\varphi, t, s, \tau, \sigma)v(\rho, \tau, \sigma)\, d\sigma\, d\tau\, d\rho,$$

where

$$g(\varphi, t, s) = f(\varphi, t, s) + c(\varphi, t, s)u_0(t, s)$$

$$+ \int_T l(\varphi, t, s, \tau)u_0(\tau, s)\, d\tau$$

(16.8)

$$+ \int_S m(\varphi, t, s, \sigma)u_0(t, \sigma)\, d\sigma$$

$$+ \int_T \int_S n(\varphi, t, s, \tau, \sigma)u_0(\tau, \sigma)\, d\sigma\, d\tau.$$

We point out that the equations (16.1) and (16.5) may be transformed into partial integral equations by applying Fourier or Laplace transforms. For example, suppose that we are interested in bounded solutions x of the Barbashin equation (16.1) with the property that both x and $\partial x/\partial t$ belong to the space with mixed norm $[L_1 \leftarrow X]$, where X is some regular ideal space over $[a, b]$ and $L_1 = L_1(\mathbb{R})$. This means that we want x to satisfy

$$\int_{-\infty}^{+\infty} \|x(t, \cdot)\|_X \, dt < \infty, \qquad \int_{-\infty}^{+\infty} \left\|\frac{\partial x(t, \cdot)}{\partial t}\right\|_X \, dt < \infty.$$

Assume that both the multiplier c and the kernel k in (16.1) do not depend on t, the function $s \mapsto c(s)$ is bounded, and the function $(s, \sigma) \mapsto k(s, \sigma)$ belongs to $[X \leftarrow X']$, where X' is the associate space to X (see Subsection 4.1). Taking now the Fourier transform of the corresponding equation

$$\frac{\partial x(t, s)}{\partial t} = c(s)x(t, s) + \int_a^b k(s, \sigma)x(t, \sigma) \, d\sigma + f(t, s)$$

with respect to the variable t leads to the equation

$$i\xi \hat{x}(\xi, s) = c(s)\hat{x}(\xi, s) + \int_a^b k(s, \sigma)\hat{x}(\xi, \sigma) \, d\sigma + \hat{f}(\xi, s),$$

where

$$\hat{x}(\xi, s) = \int_{-\infty}^{+\infty} x(t, s)e^{-i\xi t} \, dt, \quad \hat{f}(\xi, s) = \int_{-\infty}^{+\infty} f(t, s)e^{-i\xi t} \, dt.$$

This is an integral equation for the unknown function $\hat{x} : \mathbb{R} \times [a, b] \to \mathbb{C}$.

Similarly, if both u and $\partial u/\partial \varphi$ in (16.5) belong to $[L_1 \leftarrow X]$, where X is some regular ideal space over $T \times S$, the coefficient c in (16.5) is bounded, and the kernels l, m and n do not depend on φ and generate regular integral operators in the space X, one may transform the corresponding equation

$$\frac{\partial u(\varphi, t, s)}{\partial \varphi} = c(t, s)u(\varphi, t, s)$$

$$+ \int_T l(t, s, \tau)u(\varphi, \tau, s)\, d\tau$$

$$+ \int_S m(t, s, \sigma)u(\varphi, t, \sigma)\, d\sigma$$

$$+ \int_T \int_S n(t, s, \tau, \sigma)u(\varphi, \tau, \sigma)\, d\sigma\, d\tau + f(\varphi, t, s)$$

into the form

$$i\xi\hat{u}(\xi, t, s) = c(t, s)\hat{u}(\xi, t, s)$$

$$+ \int_T l(t, s, \tau)\hat{u}(\xi, \tau, s)\, d\tau$$

$$+ \int_S m(t, s, \sigma)\hat{u}(\xi, t, \sigma)\, d\sigma$$

$$+ \int_T \int_S n(t, s, \tau, \sigma)\hat{u}(\xi, \tau, \sigma)\, d\sigma\, d\tau + \hat{f}(\xi, t, s),$$

where

$$\hat{u}(\xi, t, s) = \int_{-\infty}^{+\infty} u(\varphi, t, s)e^{-i\xi\varphi}\, d\varphi$$

and

$$\hat{f}(\xi, t, s) = \int_{-\infty}^{+\infty} f(\varphi, t, s)e^{-i\xi\varphi}\, d\varphi.$$

This is a partial integral equation for the unknown function $\hat{u} : \mathbb{R} \times T \times S \to \mathbb{C}$.

16.2. Volterra operators and Barbashin equations

As we have seen, integrating the Barbashin equation (16.1) with initial condition (16.2) leads to the Volterra equation (16.3). The corresponding Volterra operator

$$V y(t, s) = \int_{t_0}^{t} c(t, s) y(\tau, s) \, d\tau$$

(16.9)

$$+ \int_{t_0}^{t} \int_{a}^{b} k(t, s, \sigma) y(\tau, \sigma) \, d\sigma \, d\tau$$

has the pleasant property that, under natural assumptions on the functions c and k, its spectral radius is equal to zero. This fact will be used in the following theorems which refer to the cases $X = C(J \times [a, b])$ and $L_p(J \times [a, b])$. (Here we consider $C(J \times [a, b])$ as locally convex metric linear space if J is a noncompact interval; see Subsection 7.1.)

Theorem 16.1. *Suppose that the operator* (16.9) *is bounded in the space* $C(J \times [a, b])$. *Then the Volterra equation* (16.3) *has, for each* $g \in C(J \times [a, b])$, *a unique solution* $y \in C(J \times [a, b])$.

☐ Let t be an arbitrary point in J. By Theorem 13.1, the number

$$M = M(D) := \sup_{(t,s) \in D} \left[|c(t, s)| + \int_{a}^{b} |k(t, s, \sigma)| \, d\sigma \right],$$

is finite for every bounded D with $[t_0, t] \times [a, b] \subseteq D \subseteq J \times [a, b]$. Let $z \in C(J \times [a, b])$ be fixed. For any $(t, s) \in D$ we have then

$$|V z(t, s)| \leq 2M(t - t_0) \sup_{(t,s) \in D} |z(t, s)|,$$

and hence, by induction,

$$|V^n z(t, s)| \leq (2M)^n \frac{(t - t_0)^n}{n!} \sup_{(t,s) \in D} |z(t, s)|$$

$$\leq \frac{[2M(b - a)]^n}{n!} \sup_{(t,s) \in D} |z(t, s)|.$$

This implies that

(16.10) $$y(t, s) := \sum_{n=0}^{\infty} V^n g(t, s)$$

is a well-defined continuous function on D, being the uniform limit (on D) of the continuous functions $y_1 := Vy + g, y_2 := Vy_1 + g, \ldots, y_n := Vy_{n-1} + g$. It is clear that (16.10) is a solution of equation (16.3). To see that this solution is unique, let $\tilde{y}$ be another continuous solution and put $\eta(t) := y(t) - \tilde{y}(t)$. Since $\eta = V\eta$ and the sequence $(V^n\eta)_n$ converges uniformly to zero, we conclude that $\eta \equiv 0$, hence $\tilde{y} \equiv y$. ∎

We remark that it suffices to verify the boundedness of the operator

$$(16.11) \qquad V_1 x(t, s) = c(t, s)x(t, s) + \int_a^b k(t, s, \sigma)x(t, \sigma)\, d\sigma,$$

since V is the composition $V = V_2 V_1$ with

$$(16.12) \qquad\qquad V_2 y(t, s) = \int_{t_0}^t y(\tau, s)\, d\tau$$

which is always a bounded operator.

Let us now study the operator (16.9) in the space $L_p(J \times [a, b])$ $(1 \le p \le \infty)$.

Theorem 16.2. *Suppose that the operator* (16.9) *is bounded in the space* $L_p(J \times [a, b])$, *the functions* c *and* k *satisfy estimates of the form*

$$|c(t, s)| \le \tilde{c}(s), \quad |k(t, s, \sigma)| \le \tilde{k}(s, \sigma) \qquad (t \in J),$$

and the operator $\tilde{V}_1$ *defined by*

$$(16.13) \qquad \tilde{V}_1 x(s) := \tilde{c}(s)x(s) + \int_a^b \tilde{k}(s, \sigma)x(\sigma)\, d\sigma$$

is bounded in $L_p([a, b])$. *Then the Volterra equation* (16.3) *has, for each* $g \in L_p(J \times [a, b])$, *a unique solution* $y \in L_p(J \times [a, b])$.

☐ Since the operators (16.11) and (16.12) both act in $L_p(J \times [a, b])$ and commute, we have for the spectral radius $r_\sigma(V)$ of V in $L_p(J \times [a, b])$ the estimate

$$r_\sigma(V) = r_\sigma(V_2 V_1) \le r_\sigma(V_2 \tilde{V}_1) \le r_\sigma(V_2)r_\sigma(\tilde{V}_1) = 0,$$

with V_1 given by (16.11) and V_2 given by (16.12). From this the assertion follows immediately. ∎

Theorem 16.2 may be strengthened in several directions. First, if we suppose that the operator (16.13) acts in some ideal space X over $[a, b]$, it is not hard to see that the operator (16.11) acts in the space $[L_p \leftarrow X]$ for $1 \leq p \leq \infty$. Therefore the statement of Theorem 16.2 remains true if we replace $L_p(J \times [a, b])$ by $[L_p \leftarrow X]$.

Second, under the hypotheses of Theorem 16.2 the operator V in (16.9) is automatically regular. One may give different results without requiring V to be regular; for example, the following holds true.

Theorem 16.3. *Suppose that $c(t, s) \equiv c(s)$ and $k(t, s, \sigma) \equiv k(s, \sigma)$, and the operator*

$$(16.14) \qquad \hat{V}_1 x(s) = c(s)x(s) + \int_a^b k(s, \sigma)x(\sigma)\, d\sigma$$

is bounded in $L_p([a, b])$. Then the Volterra equation (16.3) has, for each $g \in L_p(J \times [a, b])$, a unique solution $y \in L_p(J \times [a, b])$.
□ The assertion follows from the equality $r_\sigma(V) = r_\sigma(V_2\hat{V}_1) = 0$ and the fact that the operators $\hat{V}_1$ and $\hat{V}_2$ commute. ∎

16.3. Generalized Barbashin equations

Analogous results may be obtained also for the generalized integro-differential equation (16.5) with initial condition (16.6). To this end, we define operators W_1 and W_2 by

$$W_1 u(\varphi, t, s) = c(\varphi, t, s)u(\varphi, t, s)$$

$$+ \int_T l(\varphi, t, s, \tau)u(\varphi, \tau, s)\, d\tau$$

(16.15)

$$+ \int_S m(\varphi, t, s, \sigma)u(\varphi, t, \sigma)\, d\sigma$$

$$+ \int_T \int_S n(\varphi, t, s, \tau, \sigma,)u(\varphi, \tau, \sigma)\, d\sigma\, d\tau$$

and

(16.16) $$W_2 v(\varphi, t, s) = \int_{\varphi_0}^{\varphi} v(\psi, t, s)\, d\psi.$$

We may write (16.7) then as an operator equation

(16.17) $$v = g + W_2 W_1 v,$$

with g given by (16.8). The following theorems are contained in KA-LITVIN [1991] and may be proved in the same way as Theorem 16.1 and Theorem 16.2, respectively.

Theorem 16.4. *Suppose that the operator* (16.15) *is bounded in the space* $C(J \times T \times S)$. *Then the Volterra equation* (16.7) *has, for each* $g \in C(J \times T \times S)$, *a unique solution* $y \in C(J \times T \times S)$.

Theorem 16.5. *Suppose that the operator* (16.15) *is bounded in the space* $L_p(J \times T \times S)$, *the functions* c, l, m *and* n *satisfy estimates of the form*

$$\begin{cases} |c(\varphi, t, s)| \le \tilde{c}(t, s), \\[2mm] |l(\varphi, t, s, \tau)| \le \tilde{l}(t, s, \tau), \\[2mm] |m(\varphi, t, s, \sigma)| \le \tilde{m}(t, s, \sigma), \\[2mm] |n(\varphi, t, s, \tau, \sigma)| \le \tilde{n}(t, s, \tau, \sigma) \end{cases}$$

$(\varphi \in J)$, *and the operator* $\tilde{W}_1$ *defined by*

(16.18)
$$\begin{aligned}
\tilde{W}_1 u(t, s) &:= \tilde{c}(t, s) u(t, s) \\[2mm]
&+ \int_T \tilde{l}(t, s, \tau) u(\tau, s)\, d\tau \\[2mm]
&+ \int_S \tilde{m}(t, s, \sigma) u(t, \sigma)\, d\sigma \\[2mm]
&+ \int_T \int_S \tilde{n}(t, s, \tau, \sigma) u(\tau, \sigma)\, d\sigma\, d\tau
\end{aligned}$$

is bounded in $L_p(T \times S)$. *Then the Volterra equation* (16.7) *has, for each* $g \in L_p(J \times T \times S)$, *a unique solution* $v \in L_p(J \times T \times S)$.

Theorem 16.6. *Suppose that the functions* c, l, m *and* n *in* (16.159 *do not depend on the variable* φ, *and the operator* (11.1) *acts in* $L_p(T \times S)$. *Then the Volterra equation* (16.7) *has, for each* $g \in L_p(J \times T \times S)$, *a unique solution* $v \in L_p(J \times T \times S)$.

Theorem 16.5 and Theorem 16.6 also extend to more general spaces than Lebesgue spaces. For example, if the operator (16.18) acts in some ideal space U over $T \times S$, the operator (16.16) acts in the space $[L_p \leftarrow U]$, and hence $L_p(J \times T \times S)$ may be replaced by $[L_p \leftarrow U]$ in Theorem 16.5.

16.4. Differential equations in Banach spaces

A large part of Chapter I was devoted to studying the connections between integro-differential equations of Barbashin type and differential equations in Banach spaces. Of course, we may try the same for the generalized Barbashin equation (16.5). In this and the following subsection we confine ourselves to just some simple remarks.

Let X be an ideal space of real functions on $T \times S$. As in § 1, we identify the function $u : J \times T \times S \to \mathbb{R}$ with the function $\varphi \mapsto u(\varphi, \cdot, \cdot)$ from J into X, and the function $f : J \times T \times S \to \mathbb{R}$ with the function $\varphi \mapsto f(\varphi, \cdot, \cdot)$ from J into X. Putting $u_0 = u(\varphi_0, \cdot, \cdot)$ and defining $P(\varphi) : X \to X$ by

$$P(\varphi)u(t, s) := c(\varphi, t, s)u(t, s)$$

$$+ \int_T l(\varphi, t, s, \tau)u(\tau, s)\, d\tau$$

(16.19)

$$+ \int_S m(\varphi, t, s, \sigma)u(t, \sigma)\, d\sigma$$

$$+ \int_T \int_S n(\varphi, t, s, \tau, \sigma,)u(\tau, \sigma)\, d\tau\, d\sigma,$$

we may write (16.5)/(16.6) as differential equation in X

(16.20)
$$\frac{du}{d\varphi} = P(\varphi)u + f(\varphi)$$

with initial condition

(16.21)
$$u(\varphi_0) = u_0.$$

Denote by $C_\varphi(X)$ the space of all functions $u : J \times T \times S \to \mathbb{R}$ such that $\varphi \mapsto u(\varphi, \cdot, \cdot)$ is continuous from J into X. Likewise, by $C_\varphi^1(X)$ we denote the space of all $u \in C_\varphi(X)$ such that $u(\varphi, t, s)$ is absolutely continuous on J for $(t, s) \in T \times S$ and $\partial u / \partial \varphi \in C_\varphi(X)$ as well. We may then formulate the following analogue of Lemma 1.2: *If $P(\varphi)$ is strongly continuous operator function in $\mathfrak{L}(C)$ or is absolutely strongly continuous in $\mathfrak{L}(X)$, then the generalized $C_\varphi^1(X)$-solutions $u = u(\varphi, t, s)$ of* (16.5) *coincide with the $C^1(J, X)$-solutions of the differential equation* (16.20). The most important special case is of course $X = C(T \times S)$ and $X = L_p(T \times S)$ $(1 \leq p \leq \infty)$.

We recall that nessesary and sufficient conditions for the strong continuity of the operator function (16.19) in $\mathfrak{L}(C)$ and its absolute strong continuity in $\mathfrak{L}_p(L_p)$ $(1 \leq p \leq \infty)$ have been obtained in Theorem 13.7 and in Theorems 12.17, 12.18, and 12.20, respectively.

16.5. Representation of the evolution operator

This subsection is parallel to Subsection 2.7 and Subsection 3.6. Our goal is to compare the evolution operator $U(\varphi, \varphi_0)$ of equation (16.20) with the evolution operator $U_0(\varphi, \varphi_0)$ of the "reduced" equation

(16.22)
$$\frac{du}{d\varphi} = C(\varphi)u + f(\varphi),$$

where

(16.23)
$$C(\varphi)u(t, s) := c(\varphi, t, s)u(t, s).$$

As in Subsection 1.4, we give first an explicit formula for $U_0(\varphi, \varphi_0)$.

Lemma 16.1. *Suppose that* $c : J \times T \times S \to \mathbb{R}$ *is bounded on each bounded subset in* $J \times T \times S$, *and the corresponding operator function* (16.23) *is strongly continuous on* J. *Then the evolution operator* $U_0 = U_0(\varphi, \varphi_0)$ *of equation* (16.22) *is given by*

(16.24) $U_0(\varphi, \varphi_0)u(t, s) = e(\varphi, \varphi_0, t, s)u(t, s),$

where

(16.25) $e(\varphi, \varphi_0, t, s) := \exp \left\{ \int_{\varphi_0}^{\varphi} c(\xi, t, s) \, d\xi \right\}.$

Similarly, the following two theorems are proved as Theorem 2.3 (in case $X = C$) and Theorem 3.4 (in case $X = L_p$).

Theorem 16.7. *Suppose that the functions* (13.35)-(13.38) *are continuous, and the function* (13.39) *is bounded on each bounded subset its domain of definition. Then the evolution operator* $U(\varphi, \varphi_0)$ *for the differential equation* (16.20) *admits a representation*

(16.26) $U(\varphi, \varphi_0)u(t, s) = e(\varphi, \varphi_0, t, s)u(t, s) + H(\varphi, \varphi_0)u(t, s),$

where $e = e(\varphi, \varphi_0, t, s)$ *is given by* (16.25) *and* H *is defined by*

$$H(\varphi, \varphi_0)u(t, s) := \int_T \hat{l}(\varphi, \varphi_0, t, s, \tau)u(\tau, s) \, d\tau$$

$$+ \int_S \hat{m}(\varphi, \varphi_0, t, s, \sigma)u(t, \sigma) \, d\sigma$$

$$+ \int_T \int_S \hat{n}(\varphi, \varphi_0, t, s, \tau, \sigma)u(\tau, \sigma) \, d\sigma \, d\tau,$$

where $\hat{l}$, $\hat{m}$ *and* $\hat{n}$ *are measurable functions on* $J \times J \times T \times S \times T$, $J \times J \times T \times S \times S$, *and* $J \times J \times T \times S \times T \times S$, *respectively.*

Theorem 16.8. *Suppose that the operator function* (16.19) *is absolutely strongly continuous in the space* $\mathfrak{L}(L_p)$. *Then the evolution operator* $U(\varphi, \varphi_0)$ *for the differential equation* (16.20) *admits a representation* (16.26) *where* e, $\hat{l}$, $\hat{m}$ *and* $\hat{n}$ *are as in Theorem 16.7.*

The assertion of Theorem 16.8 is true also for more general spaces than L_p, see CHEN [1995, 1997].

As we have done in Section 10, one may also study boundary value problems for generalized integro-differential equations of Barbashin type. We do not do this here, but just refer to CHEN [1996, 1997].

§ 17. Nonlinear equations and operators

In this section we illustrate the applicability of topological methods, monotonicity methods, and variational methods to the solution of various types of nonlinear integro-differential equations of Barbashin type and nonlinear partial integral equations. We show first how to solve initial and boundary value problems for nonlinear Barbashin equations by means of the fixed point theorems considered in Subsection 7.1 and Subsection 9.4. An important step consists here in transforming the boundary value problem into an equivalent operator equation involving Uryson-type integral operators. Finally, we show how to use Minty's monotonicity principle to prove (unique) solvability of a Barbashin equation containing Hammerstein-type integral operators. For studying the boundary value problem (17.13)/(17.14) we apply a scheme which has recently been proposed in APPELL-KA-LITVIN-ZABREJKO [1994, 1996], see also KALITVIN [1993a].

17.1. Barbashin equations with Uryson operators

Consider the nonlinear integro-differential equation of Barbashin type

$$(17.1) \qquad \frac{\partial x(t,s)}{\partial t} = c(t,s)x(t,s) + \int_a^b k(t,s,\sigma,x(t,\sigma))\,d\sigma + f(t,s)$$

with initial condition

$$(17.2) \qquad\qquad x(t_0,s) = x_0(s),$$

where $x_0 : [a,b] \to \mathbb{R}$, $c : J \times [a,b] \to \mathbb{R}$, $k : J \times [a,b] \times [a,b] \times \mathbb{R} \to \mathbb{R}$, and $f : J \times [a,b] \to \mathbb{R}$ are given measurable functions, and J is a compact interval in $\mathbb{R}$. Imposing appropriate growth conditions on the functions c and k, one may apply "brute-force-estimates" to prove the solvability of the initial value problem (17.1)/(17.2) by means of the classical fixed point principles of Banach-Caccioppoli and Schauder. Instead, we are going to apply the more sophisticated fixed point principle by DARBO [1955] (see Theorem 7.1). To this end, we first give a general result on the solvability of nonlinear initial value problems in Banach spaces:

Proposition 17.1. *Let X be a Banach space, $t_0 \in \mathbb{R}$, $x_0 \in X$, and $Q = \{(t, x) \in \mathbb{R} \times X : |t - t_0| \le a, \|x - x_0\| \le b\}$. Suppose that $g_1 : Q \to X$ is continuous and also Lipschitz-continuous with respect to the second variable, i.e.*

(17.3) $$\|g_1(t, x) - g_1(t, y)\| \le L(t)\|x - y\|,$$

and $g_2 : Q \to X$ is compact and continuous. Choose $c \in (0, a]$ such that both

$$\int_{t_0-c}^{t_0+c} L(t)\, dt < 1, \qquad \sup_Q \|g_1(t, x)\| + \sup_Q \|g_2(t, x)\| \le \frac{b}{c}.$$

Then the initial value problem

(17.4) $$x'(t) = g_1(t, x(t)) + g_2(t, x(t)), \quad x(t_0) = x_0$$

has a continuously differentiable solution x on the compact interval $[t_0 - c, t_0 + c]$ with values in X.

□ We apply Theorem 7.1. Let $E = C([t_0 - c, t_0 + c], X)$, $M = \{x : x \in E, \|x - x_0\| \le b\}$, and

$$G_i x(t) = \int_{t_0}^{t} g_i(\tau, x(\tau))\, d\tau \quad (i = 1, 2).$$

Define $\Phi : M \to E$ by $\Phi x = G_1 x + G_2 x$. By assumption, Φ is continuous and leaves the set M invariant. Moreover, for any set $N \subseteq M$ we have

$$\gamma(\Phi(N)) \le \gamma(G_1(N)) + \gamma(G_2(N)) = \gamma(G_1(N))$$

$$\le \left(\int_{t_0-c}^{t_0+c} L(t)\, dt \right) \gamma(N),$$

where γ is the measure of noncompactness (4.50). This shows that Φ is condensing on M. The assertion follows now from Theorem 7.1. ■

Of course, Proposition 7.1 holds in the special situation when $g_1(t, x) \equiv 0$ or $g_2(t, x) \equiv 0$. In this case we get classical existence results for the initial value problem (17.4).

Consider now the nonlinear integro-differential equation (17.1), subject to the initial condition (17.2). Given an ideal space X (see Subsection 4.1) of measurable functions over $[a, b]$, in the same way as in the linear case (Subsection 1.4) we may write equation (17.1) as differential equation

$$(17.5) \qquad \frac{dx}{dt} = C(t)x + K(t)x + f(t)$$

in X, where

$$(17.6) \qquad C(t)x(s) = c(t, s)x(s)$$

is the (linear) multiplication operator generated by the multiplier c,

$$(17.7) \qquad K(t)x(s) = \int_a^b k(t, s, \sigma, x(\sigma))\, d\sigma$$

is the (nonlinear) *Uryson integral operator* generated by the kernel k, and $f(t)(s) = f(t, s)$. By means of Proposition 17.1, we get then a local existence result for the problem (17.1)/(17.2) in a suitable function space. As in Subsection 9.2, let us denote by $C_t^1 = C_t^1(X)$ the set of all functions $x : J \times [a, b] \to \mathbb{R}$ such that the map $t \mapsto x(t, \cdot)$ is continuously differentiable from J into X, equipped with the natural norm

$$(17.8) \qquad \|x\|_{C_t^1} = \max_{t \in J} \left[\|x(t, \cdot)\| + \left\| \frac{d}{dt} x(t, \cdot) \right\| \right].$$

A standard choice is of course $X = L_p$ for $1 \leq p < \infty$.

Theorem 17.1. *Suppose that the map $t \mapsto C(t)$ is strongly continuous in X, and the map $(t, x) \mapsto K(t)x$ is continuous and compact. Then the initial value problem (17.1)/(17.2) has, for each $f \in C(J, X)$, a local solution $x \in C_t^1(X)$.*

□ Let $g_1(t, x) = C(t)x + f(t)$ and $g_2(t, x) = K(t)x$, and let Q be as in Proposition 17.1. By assumption, $g_1 : Q \to X$ is continuous and satisfies a Lipschitz condition (17.3) with $L(t) = \|C(t)\|_{\mathcal{L}(X)}$, and $g_2 : Q \to X$ is continuous and compact. By Proposition 17.1, the initial value problem (17.4) has a continuously differentiable solution x on $[t_0 - c, t_0 + c]$ which is a C_t^1-solution of (17.1)/(17.2). ∎

Since the operator family $t \mapsto C(t)$ is supposed to be strongly continuous in Theorem 17.1, the smallness condition

$$\int_{t_0-c}^{t_0+c} \|C(\tau)\| \, d\tau < 1$$

may always be achieved for sufficiently small $c > 0$. The crucial hypothesis is therefore the compactness of the map of two variables $(t, x) \mapsto K(t)x$. The following lemma reduces the problem of verifying the compactness of this map to that of verifying the compactness of each operator $K(t)$ (t fixed), as well as the equicontinuity of the family of functions $t \mapsto K(t)x$ (x fixed):

Lemma 17.1. *Let $B \subset X$ be bounded, and suppose that the following conditions are satisfied:*
(a) *each operator $K(t) : B \to X$ is compact and continuous;*
(b) *the function set $\{K(\cdot)x : x \in B\}$ is equi-continuous.*
Then the map g defined by $g(t, x) = K(t)x$ is continuous and compact from $J \times B$ into X.

□ Continuity follows from the inequality

$$\|g(t, x) - g(t_0, x_0)\| \leq$$

$$\|K(t)x - K(t_0)x\| + \|K(t_0)x - K(t_0)x_0\|$$

and our hypotheses on K. For compactness, it suffices to prove that $g(J \times B)$ has a finite ε-net for any $\varepsilon > 0$. By hypothesis (b), there exists a $\delta(\varepsilon) > 0$ such that $\|K(t)x - K(t')x\| < \varepsilon/2$ whenever $|t - t'| < \delta(\varepsilon)$ and $x \in B$. Consider a partition $\{t_0, \ldots, t_n\}$ of J such that $|t_i - t_{i-1}| < \delta(\varepsilon)$ for $i = 1, \ldots, n$. Since $K(t_i)B$ is precompact, it contains a finite $\varepsilon/2$-net $\mathcal{N}_i$. We claim that $\mathcal{N}_1 \cup \ldots \cup \mathcal{N}_n$ is an ε-net for $g(J \times B)$. In fact, let $(t, x) \in J \times B$ be arbitrary. Then $t_{i-1} < t \leq t_i$ for some i. Choose $y \in \mathcal{N}_i$ such that $\|g(t_i, x) - y\| < \varepsilon/2$. This implies that

$$\|g(t, x) - y\| \leq \|g(t, x) - g(t_i, x)\| + \|g(t_i, x) - y\| < \varepsilon$$

as claimed. ■

As a typical application of Lemma 17.1, let us consider the case $k(t, s, \sigma, u) = k_0(t, s, \sigma)h(\sigma, u)$, i.e. the special family of *Hammerstein integral operators*

$$(17.9) \qquad K(t)x(s) = \int_a^b k_0(t, s, \sigma)h(\sigma, x(\sigma))\, d\sigma.$$

The operator (17.9) may be written as composition $K(t) = K_0(t)H$ of the nonlinear *Nemytskij operator*

$$(17.10) \qquad\qquad Hx(s) = h(s, x(s))$$

and the family of linear integral operators

$$(17.11) \qquad K_0(t)y(s) = \int_a^b k_0(t, s, \sigma)y(\sigma)\, d\sigma.$$

Theorem 17.2. *Suppose that the map $t \mapsto C(t)$ is strongly continuous in X, there exists some normed linear space Y such that $H : X \to Y$ is bounded and continuous, and each operator $K_0(t) : Y \to X$ is compact. Assume that the map $t \mapsto K_0(t)$ is continuous in the operator norm. Then the initial value problem $(17.1)/(17.2)$ has, for each $f \in C(J, X)$, a local solution $x \in C_t^1(X)$.*

□ It suffices to prove that $g : Q \to X$ is compact and continuous; to this end, we check the conditions of Lemma 17.1. If $B \subset X$ is bounded, $H(B) \subset Y$ is bounded as well. Since $K(t)$ is a compact operator, the set $K(t)B = K_0(t)(H(B))$ is precomapct. Moreover, since the map $t \mapsto K_0(t)$ is continuous in the operator norm, we find some $\delta > 0$ such that $|t_1 - t_2| \le \delta$ implies $\|K_0(t_1)y - K(t_2)y\| \le \varepsilon$ for $y \in H(B)$. Consequently,

$$\|K(t_1)x - K(t_2)x\| = \|K_0(t_1)Hx - K_0(t_2)Hx\| \le \varepsilon$$

for all $x \in B$, which shows that the set $\{K(\cdot)x : x \in B\}$ is equicontinuous. ■

17.2. Generalized Barbashin equations

In the preceding subsection we have seen that, under some natural hypotheses, the initial value problem for the integro-differential equation (17.1) has always a local solution; this is completely analogous to ordinary differential equations. As usual, the existence of solutions of *boundary value problems* for (17.1) is more difficult to prove (and, as a matter of fact, is not always true). This will be illustrated in this subsection. Our main tool is the fixed point principle for operators in K-normed spaces which turned out to be useful already in the linear case (Subsection 9.4).

Let E be a Banach space with K-norm $]| \cdot |[: E \to K$ which takes its values in the positive cone K of some ordered Banach space Z, and $F : E \to E$ a bounded continuous operator which satisfies a contraction type condition

$$(17.12) \qquad]|Fx - Fy|[\leq G(]|x - y|[) \qquad (x, y \in E).$$

Here $G : K \to K$ is some positive linear operator in Z. As we have seen (Theorem 9.6), the operator F has a unique fixed point in E, provided that G has spectral radius $r_\sigma(G) < 1$.

To apply Theorem 9.6, let us replace equation (17.1) by the more general integro-differential equation

$$\frac{\partial x(\varphi, t, s)}{\partial \varphi} = c(\varphi, t, s) x(\varphi, t, s)$$

$$+ \int_{-1}^{1} l(\varphi, t, s, \tau, x(\varphi, \tau, s)) \, d\tau$$

(17.13)

$$+ \int_{c}^{d} m(\varphi, t, \sigma, x(\varphi, t, \sigma)) \, d\sigma$$

$$+ \int_{c}^{d} \int_{-1}^{1} n(\varphi, t, s, \tau, \sigma, x(\varphi, \tau, \sigma)) \, d\tau \, d\sigma + f(\varphi, t, s)$$

$((\varphi, t, s) \in Q)$, subject to the boundary condition

$$(17.14) \qquad \begin{cases} x(a, t, s) = \phi(t, s) & \text{if } (t, s) \in Q_+, \\ x(b, t, s) = \psi(t, s) & \text{if } (t, s) \in Q_-; \end{cases}$$

here we have put $Q = [a, b] \times [-1, 1] \times [c, d]$, $Q_+ = (0, 1] \times [c, d]$, and $Q_- = [-1, 0) \times [c, d]$. Equations like (17.13) occur as Fourier transforms of a certain type of Schrödinger equations. As above, $c : Q \to \mathbb{R}$, $\phi : Q_+ \to \mathbb{R}$, $\psi : Q_- \to \mathbb{R}$, and $f : Q \to \mathbb{R}$ are given measurable functions, and $l : Q \times [-1, 1] \times \mathbb{R} \to \mathbb{R}$, $m : Q \times S \times \mathbb{R} \to \mathbb{R}$, and $n : Q \times [-1, 1] \times S \times \mathbb{R} \to \mathbb{R}$ are supposed to satisfy a Carathéodory condition.

We are going to study the boundary value problem (17.13)/(17.14) in the space $W^1_{\varphi p} = W^1_{\varphi p}(Q)$ $(1 \le p < \infty)$ of all measurable functions $x : Q \to \mathbb{R}$ for which the norm

$$\|x\|_{W^1_{\varphi p}} =$$

(17.15)
$$\left\{ \int_a^b \int_{-1}^1 \int_c^d \left[|x(\varphi, t, s)| + \left| \frac{\partial x(\varphi, t, s)}{\partial \varphi} \right| \right]^p ds\, dt\, d\varphi \right\}^{1/p}$$

$(1 \le p < \infty)$ is finite; this norm was introduced in NIKOL'SKIJ [1969] in a different context.

Consider the kernel of four variables

(17.16)
$$e(\varphi, \varphi_0, t, s) = \exp \int_{\varphi_0}^{\varphi} c(\xi, t, s) d\xi$$

which generates a partial integral operator (see Subsection 11.1)

(17.17)
$$Pf(\varphi, t, s) = \begin{cases} \displaystyle\int_a^\varphi e(\varphi, \theta, t, s) f(\theta, t, s)\, d\theta \\[2mm] \displaystyle\int_b^\varphi e(\varphi, \theta, t, s) f(\theta, t, s)\, d\theta \end{cases}$$

on Q_+ and Q_-, respectively. To begin with, we show how to solve the boundary value problem (17.13)/(17.14) explicitly in case $l = m = n \equiv 0$. The proof of the following lemma consists in a straightforward calculation:

Lemma 17.2. *For $c \in L_\infty(Q)$, $f \in L_p(Q)$, $\phi \in L_p(Q_+)$ and $\psi \in L_p(Q_-)$, the equation*

(17.18)
$$\frac{\partial z(\varphi, t, s)}{\partial \varphi} = c(\varphi, t, s) z(\varphi, t, s) + f(\varphi, t, s),$$

$((\varphi, t, s) \in Q)$, *subject to the boundary conditions*

$$\begin{cases} z(a, t, s) = \phi(t, s) & \text{if } (t, s) \in Q_+, \\ z(b, t, s) = \psi(t, s) & \text{if } (t, s) \in Q_- \end{cases}$$

has a unique solution $z \in W^1_{\varphi p}$. *This solution is given almost everywhere on* $[a, b] \times Q_+$ *and* $[a, b] \times Q_-$, *respectively, by*

(17.19)
$$z = \begin{cases} Pf + \tilde{\phi}, \\ Pf + \tilde{\psi}, \end{cases}$$

where

$$\tilde{\phi}(\varphi, t, s) = \phi(t, s)e(\varphi, a, t, s) \qquad ((t, s) \in Q_+)$$

and

$$\tilde{\psi}(\varphi, t, s) = \psi(t, s)e(\varphi, b, t, s) \qquad ((t, s) \in Q_-).$$

Lemma 17.2 allows us to transform the boundary value problem (17.13)/(17.14) into an operator equation which may be treated by Theorem 9.6. As in Subsection 11.1, we introduce the operators

$$Cx(\varphi, t, s) = c(\varphi, t, s)x(\varphi, t, s),$$

$$Lx(\varphi, t, s) = \int_{-1}^{1} l(\varphi, t, s, \tau, x(\varphi, \tau, s)) \, d\tau,$$

$$Mx(\varphi, t, s) = \int_{c}^{d} m(\varphi, t, s, \sigma, x(\varphi, t, \sigma)) \, d\sigma,$$

$$Nx(\varphi, t, s) = \int_{c}^{d} \int_{-1}^{1} n(\varphi, t, s, \tau, \sigma, x(\varphi, \tau, \sigma)) \, d\tau \, d\sigma,$$

and

$$K = L + M + N.$$

The equation (17.13) may then be written more concisely in the form

(17.20)
$$\frac{\partial x}{\partial \varphi} = [C + K]x + f.$$

Proposition 17.2. *Let the conditions of Lemma 17.2 be satisfied.*
Suppose that the nonlinear operator $K = L + M + N$ is continuous
and bounded in L_p. Then every solution $x \in W^1_{\varphi p}$ of the boundary
value problem (17.13)/(17.14) *solves the nonlinear operator equation*

$$(17.21) \qquad x = \begin{cases} P(Kx + f) + \tilde{\phi} & on \ \ [a,b] \times Q_+, \\ \\ P(Kx + f) + \tilde{\psi} & on \ \ [a,b] \times Q_-. \end{cases}$$

Conversely, every solution $x \in L_p$ of (17.21) *belongs to $W^1_{\varphi p}$ and*
solves the boundary value problem (17.13)/(17.14).

□ The proof follows immediately from Lemma 17.2 with f replaced
by $Kx + f$. ■

We remark that partial integral operators of Uryson type have been
studied by POVOLOTSKIJ-KALITVIN [1985, 1991], KALITVIN-GLOTOV
[1996], KALITVIN [1997], APPELL-DE PASCALE-KALITVIN-ZABREJ-
KO [1996], KALITVIN-KORENCHUK-EVTUKHINA [1993], and CHEN-
KALITVIN [1997, 1998].

We begin now to study the operator equation (17.21) from the viewpoint of fixed point theorems in K-normed spaces; the constructions
and results will be parallel to those in Subsection 9.4. For $a \leq \varphi \leq b$,
$0 < t \leq 1$, and $s \in S$, we put

$$\begin{cases} x(\varphi, t, s) = u(\varphi, t, s), \ x(\varphi, -t, s) = v(\varphi, t, s), \\ z(\varphi, t, s) = g(\varphi, t, s), \ z(\varphi, -t, s) = h(\varphi, t, s), \\ e(\varphi, \varphi_0, t, s) = i(\varphi, \varphi_0, t, s), \\ e(\varphi, \varphi_0, -t, s) = j(\varphi, \varphi_0, t, s). \end{cases}$$

Moreover, we define four operators A, B, C, and D by

$$Au(\varphi, t, s) = \int_a^\varphi \int_0^1 i(\varphi, \theta, t, s) l(\theta, t, s, \tau, u(\theta, \tau, s)) \, d\tau \, d\theta$$

$$+ \int_a^\varphi \int_c^d i(\varphi, \theta, t, s) m(\theta, t, s, \sigma, u(\theta, t, \sigma)) \, d\sigma \, d\theta$$

$$+ \int_a^\varphi \int_c^d \int_0^1 i(\varphi, \theta, t, s) n(\theta, t, s, \tau, \sigma, u(\theta, \tau, \sigma)) \, d\tau \, d\sigma \, d\theta,$$

$$Bv(\varphi, t, s) = \int_a^\varphi \int_0^1 i(\varphi, \theta, t, s) l(\theta, t, s, -\tau, v(\theta, \tau, s)) \, d\tau \, d\theta$$

$$+ \int_a^\varphi \int_c^d \int_0^1 i(\varphi, \theta, t, s) n(\theta, t, s, -\tau, \sigma, v(\theta, \tau, \sigma)) d\tau d\sigma \, d\theta,$$

$$Cu(\varphi, t, s) = \int_b^\varphi \int_0^1 j(\varphi, \theta, t, s) l(\theta, -t, s, \tau, u(\theta, \tau, s)) \, d\tau \, d\theta$$

$$+ \int_a^\varphi \int_c^d \int_0^1 j(\varphi, \theta, t, s) n(\theta, -t, s, \tau, \sigma, u(\theta, \tau, \sigma)) \, d\tau \, d\sigma \, d\theta,$$

and

$$Dv(\varphi, t, s) = \int_b^\varphi \int_0^1 j(\varphi, \theta, t, s) l(\theta, -t, s, -\tau, v(\theta, \tau, s)) \, d\tau \, d\theta$$

$$+ \int_b^\varphi \int_c^d j(\varphi, \theta, t, s) m(\theta, -t, s, \sigma, v(\theta, t, \sigma)) \, d\sigma \, d\theta$$

$$+ \int_b^\varphi \int_c^d \int_0^1 j(\varphi, \theta, t, s) n(\theta, -t, s, -\tau, \sigma, v(\theta, \tau, \sigma)) \, d\tau \, d\sigma \, d\theta.$$

The operator equation (17.21) may then be written as a system

(17.22)
$$\begin{pmatrix} u \\ v \end{pmatrix} = \begin{pmatrix} A & B \\ C & D \end{pmatrix} \begin{pmatrix} u \\ v \end{pmatrix} + \begin{pmatrix} g \\ h \end{pmatrix}.$$

Suppose now that the kernels l, m, and n satisfy Lipschitz conditions

$$|l(\varphi, t, s, \tau, u) - l(\varphi, t, s, \tau, v)| \le a_1(\varphi, t, s, \tau)|u - v|,$$

$$|m(\varphi, t, s, \sigma, u) - m(\varphi, t, s, \sigma, v)| \le a_2(\varphi, t, s, \sigma)|u - v|,$$

$$|n(\varphi, t, s, \tau, \sigma, u) - n(\varphi, t, s, \tau, \sigma, v)| \le a_3(\varphi, t, s, \tau, \sigma)|u - v|,$$

$$|l(\varphi, t, s, -\tau, u) - l(\varphi, t, s, -\tau, v)| \le b_1(\varphi, t, s, \tau)|u - v|,$$

$$|n(\varphi, t, s, -\tau, \sigma, u) - n(\varphi, t, s, -\tau, \sigma, v)| \le b_2(\varphi, t, s, \tau, \sigma)|u - v|,$$

$$|l(\varphi, -t, s, \tau, u) - l(\varphi, -t, s, \tau, v)| \le c_1(\varphi, t, s, \tau)|u - v|,$$

$$|n(\varphi, -t, s, \tau, \sigma, u) - n(\varphi, -t, s, \tau, \sigma, v)| \le c_2(\varphi, t, s, \tau, \sigma)|u - v|,$$

$$|l(\varphi, -t, s, -\tau, u) - l(\varphi, -t, s, -\tau, v)| \le d_1(\varphi, t, s, \tau)|u - v|,$$

$$|m(\varphi, -t, s, \sigma, u) - m(\varphi, -t, s, \sigma, v)| \le d_2(\varphi, t, s, \sigma)|u - v|,$$

and

$$|n(\varphi, -t, s, -\tau, \sigma, u) - n(\varphi, -t, s, -\tau, \sigma, v)| \le d_3(\varphi, t, s, \tau, \sigma)|u - v|.$$

Moreover, assume that

$$\left\| \int_0^1 i(\varphi, \theta, \cdot, \cdot) a_1(\theta, \cdot, \cdot, \tau) u(\tau, \cdot) \, d\tau \right.$$

$$+ \int_c^d i(\varphi, \theta, \cdot, \cdot) a_2(\theta, \cdot, \cdot, \sigma) u(\cdot, \sigma) \, d\sigma$$

$$+ \left. \int_c^d \int_0^1 i(\varphi, \theta, \cdot, \cdot) a_3(\theta, \cdot, \cdot, \tau, \sigma) u(\tau, \sigma) \, d\tau \, d\sigma \right\| \le \alpha ||u||,$$

$$\left\| \int_0^1 i(\varphi, \theta, \cdot, \cdot) b_1(\theta, \cdot, \cdot, \tau) v(\tau, \cdot) \, d\tau \right.$$

$$+ \left. \int_c^d \int_0^1 i(\varphi, \theta, \cdot, \cdot) b_2(\theta, \cdot, \cdot, \tau, \sigma) v(\tau, \sigma) \, d\tau \, d\sigma \right\| \le \beta ||v||,$$

$$\left\|\int_0^1 j(\varphi,\theta,\cdot,\cdot)c_1(\theta,\cdot,\cdot,\tau)u(\tau,\cdot)\,d\tau\right.$$

$$\left. +\int_c^d \int_0^1 j(\varphi,\theta,\cdot,\cdot)c_2(\theta,\cdot,\cdot,\tau,\sigma)u(\tau,\sigma)\,d\tau\,d\sigma\right\| \le \gamma\|u\|,$$

and

$$\left\|\int_0^1 j(\varphi,\theta,\cdot,\cdot)d_1(\theta,\cdot,\cdot,\tau)v(\tau,\cdot)\,d\tau\right.$$

$$+\int_c^d j(\varphi,\theta,\cdot,\cdot)d_2(\theta,\cdot,\cdot,\sigma)v(\cdot,\sigma)\,d\sigma$$

$$\left. +\int_c^d \int_0^1 j(\varphi,\theta,\cdot,\cdot)d_3(\theta,\cdot,\cdot,\tau,\sigma)v(\tau,\sigma)\,d\tau\,d\sigma\right\| \le \delta\|v\|,$$

where all norms are taken in $L_p(Q_+)$. We define a linear operator G by

$$(17.23)\qquad Gz(\varphi) = G\begin{pmatrix} u(\varphi) \\ v(\varphi) \end{pmatrix} = \begin{pmatrix} \int_a^\varphi[\dot\alpha u(\theta) + \beta v(\theta)]\,d\theta \\ \int_\varphi^b[\gamma u(\theta) + \delta v(\theta)]\,d\theta \end{pmatrix}.$$

Theorem 17.3. *Let the assumptions of Proposition 17.3 be satisfied. Assume, moreover, that the numbers α, β, γ and δ can be defined as above and satisfy one of the following four conditions:*

(a) $\Delta = (\alpha + \delta)^2 - 4\beta\gamma > 0$ *and* $(b - a)\sqrt{\Delta} < \log\frac{\alpha+\delta+\sqrt{\Delta}}{\alpha+\delta-\sqrt{\Delta}}$;

(b) $(\alpha + \delta)^2 = 4\beta\gamma$ *and* $(b - a)(d + \delta) < 2$;

(c) $\Delta < 0$, $\alpha + \delta \ne 0$, *and* $(b - a)\sqrt{-\Delta} < 2\arctan\frac{\sqrt{-\Delta}}{\alpha+\delta}$;

(d) $\Delta < 0$, $\alpha + \delta = 0$, *and* $(b - a)\sqrt{-\Delta} < \pi$.

Then the operator equation (17.21), *and hence also the boundary value problem* (17.13)/(17.14), *has a unique solution* $x \in W^1_{\varphi p}$; *this solution may be obtained by the usual method of successive approximations.*

□ We have to construct a suitable K-normed space E such that the operator

$$(17.24)\qquad F\begin{pmatrix} u \\ v \end{pmatrix} = \begin{pmatrix} A & B \\ C & D \end{pmatrix}\begin{pmatrix} u \\ v \end{pmatrix} + \begin{pmatrix} g \\ h \end{pmatrix}$$

satisfies the contraction condition (17.12). We take $X = L_p(Q_+) \times L_p(Q_-)$, equipped with the norm $||(u,v)||_X = ||u||_{L_p} + ||v||_{L_p}$. Moreover, let $E = L_p([a,b], X)$ be the Bochner-Lebesgue space of all X-valued functions $\varphi \mapsto x(\varphi, \cdot, \cdot) = (u(\varphi, \cdot, \cdot), v(\varphi, \cdot, \cdot))$, equipped with the norm

$$||x||_E = \left\{ \int_a^b [||u(\varphi, \cdot, \cdot)||_p + ||v(\varphi, \cdot, \cdot)||_p]^p \, d\varphi \right\}^{1/p}$$

and the K-norm

$$]|x|[= (||u(\varphi, \cdot, \cdot)||_p, ||v(\varphi, \cdot, \cdot)||_p).$$

Thus, the K-norm takes its values in the natural cone of the Banach space $Z = L_p([a,b], \mathbb{R}^2)$. Our assumptions ensure that the estimate (17.12) is true for the operators (17.23) and (17.24). As in Theorem 9.7 we see that the spectral radius of G is less than 1. Consequently, Theorem 9.6 applies. ∎

17.3. Equations with Hammerstein operators

In the preceding subsections we have proved existence and uniqueness theorems for various types of integro-differential equations of Barbashin type in Lebesgue spaces by means of topological methods. Now we are going to discuss partial integral equations of the type

$$y(t,s) = \int_T l(t,s,\tau) h(\tau, s, y(\tau, s)) \, d\tau$$

(17.25)
$$+ \int_S m(t,s,\sigma) h(t, \sigma, y(t, \sigma)) \, d\sigma$$

$$+ \int_S \int_T n(t,s,\tau,\sigma) h(\tau, \sigma, y(\tau, \sigma)) \, d\tau \, d\sigma + g(t,s)$$

in the space $C(D)$ of continuous functions over $D := T \times S$. Here $l : D \times T \to \mathbb{R}$, $m : D \times S \to \mathbb{R}$, $n : D \times D \to \mathbb{R}$, and $g : D \to \mathbb{R}$ are given measurable functions, while $f : D \times \mathbb{R} \to \mathbb{R}$ is supposed to satisfy a Carathéodory condition. We restrict ourselves to the model case $T = [a,b]$ and $S = [c,d]$ in the sequel.

Let us define operators L, M, and N by

(17.26)
$$Lx(t,s) = \int_a^b l(t,s,\tau)x(\tau,s)\,d\tau,$$

(17.27)
$$Mx(t,s) = \int_c^d m(t,s,\sigma)x(t,\sigma)\,d\sigma,$$

and

(17.28)
$$Nx(t,s) = \int_c^d \int_a^b n(t,s,\tau,\sigma)x(\tau,\sigma)\,d\tau\,d\sigma,$$

and denote by

(17.29)
$$Hy(t,s) = h(t,s,y(t,s))$$

the Nemytskij operator generated by the function f. We may then rewrite equation (17.25) as an operator equation

(17.30)
$$y = KHy + g.$$

From standard theorems on Nemytskij operators in spaces of continuous functions (see e.g. APPELL-ZABREJKO [1990]) it follows that, by the continuity of the function f, the operator KH maps $C(D)$ into itself if the operator $K = L + M + N$ does. Moreover, Theorem 13.1 implies the following

Theorem 17.4. *Suppose that the function f is continuous, and the functions*

$$\Gamma(t,s) := \int_a^b l(t,s,\tau)\,d\tau + \int_c^d m(t,s,\sigma)\,d\sigma$$

$$+ \int_c^d \int_a^b n(t,s,\tau,\sigma)\,d\tau\,d\sigma,$$

$$\Gamma_\xi(t,s) := \int_a^\xi \kappa(t,\tau) \left\{ \int_c^d m(t,s,\sigma)\,d\sigma + (\xi - \tau)[l(t,s,\tau) \right.$$

$$\left. + \int_c^d n(t,s,\tau,\sigma)\,d\sigma] \right\} d\tau,$$

$$\Gamma_\eta(t,s) := \int_c^\eta \kappa(s,\sigma) \left\{ \int_a^b l(t,s,\tau)\,d\tau + (\eta - \sigma)[m(t,s,\sigma) \right.$$

$$\left. + \int_a^b n(t,s,\tau,\sigma)\,d\tau] \right\} d\sigma,$$

and

$$\Gamma_{\xi,\eta}(t,s) :=$$

$$\int_a^\xi \int_c^\eta \{(\xi - \tau)l(t,s,\tau)\kappa(s,\sigma) + (\eta - \sigma)m(t,s,\sigma)\kappa(t,\tau)$$

$$+(\xi - \tau)(\eta - \sigma)n(t,s,\tau,\sigma)]\}\,d\tau\,d\sigma$$

are continuous on D for all $(\xi, \eta) \in D$; *here* $\kappa(t,\tau)$ *and* $\kappa(s,\sigma)$ *are defined as in Subsection 13.2. Assume, moreover, that the function*

$$\overline{\Gamma}(t,s) = \int_a^b |l(t,s,\tau)|\,d\tau + \int_c^d |m(t,s,\sigma)|\,d\sigma$$

$$+ \int_c^d \int_a^b |n(t,s,\tau,\sigma)|\,d\tau\,d\sigma$$

is bounded on D. Then the operator KH *acts in the space* $C(D)$, *is bounded, continuous, and uniformly continuous on bounded sets.*

The following example shows that the acting condition $KH : C(D) \to C(D)$ does not imply the acting conditions $K : C(D) \to C(D)$ or $H : C(D) \to C(D)$:

Example 17.1. Let $g : \mathbb{R} \to \mathbb{R}$ be a continuous function, not identical zero, $p := \frac{1}{2}(a + b)$,

$$h(t, s, u) := \begin{cases} g(u) & \text{if } a \leq t \leq p, \\ -g(u) & \text{if } p < t \leq b, \end{cases}$$

$$l(t, s, \tau) := \begin{cases} 1 & \text{if } a \leq t \leq p, \\ -1 & \text{if } p < t \leq b, \end{cases}$$

and $m(t, s, \sigma) = n(t, s, \tau, \sigma) \equiv 0$. Obviously, the operators M and H do not map the space $C(D)$ into itself, but the operator

$$(17.31) \qquad LHx(t, s) = \int_a^b g(x(\tau, s)) \, d\tau$$

does.

Let U be some real interval (bounded or unbounded), and either $\Omega = T$, $\Omega = S$, or $\Omega = D$. We say that a measurable function $a : D \times \Omega \times U \to \mathbb{R}$ is *integral-bounded* if for each $r > 0$ we have

$$\sup_{|u| \leq r} \sup_{(t,s) \in D} \int_\Omega |a(t, s, \omega, u)| \, d\omega < \infty,$$

and *continuous in the whole* if for each $r > 0$ we have

$$\lim_{d \to 0} \sup_{|u_1|, |u_2| \leq r} \int_\Omega |a(t_1, s_1, \omega, u_1) - a(t_2, s_2, \omega, u_2)| \, d\omega = 0,$$

where $d = |t_1 - t_2| + |s_1 - s_2| + |u_1 - u_2|$.

Theorem 17.5. *Suppose that the functions*

$$(t, s, \tau, u) \mapsto l(t, s, \tau) h(\tau, s, u),$$

$$(t, s, \sigma, u) \mapsto m(t, s, \sigma) h(t, \sigma, u),$$

and

$$(t, s, \tau, \sigma, u) \mapsto n(t, s, \tau, \sigma) h(\tau, \sigma, u)$$

are both integral-bounded and continuous in the whole. Then the operator KH acts in the space $C(D)$, is bounded, continuous, and uniformly continuous on bounded sets.

□ We prove the assertion for the operator LH, the proof for MH
and NH is similar. First we show that LH maps the space $C(D)$ into
itself. Given $x \in C(D)$, we have $|x(t,s)| \le r$ for some $r > 0$, hence

$$(17.32) \qquad \sup_{|u| \le r} \sup_{(t,s) \in D} \int_a^b |l(t,s,\tau)h(\tau,s,u)|\, d\tau < \infty,$$

by assumption. This implies that

$$(17.33) \qquad \sup_{(t,s) \in D} |LHx(t,s)| < \infty,$$

which shows that LH is defined on $C(D)$. Further, the continuity in
the whole of the integrand in (17.32) implies that

$$\sup_{|u_1|,|u_2| \le r} \int_a^b |l(t_1,s_1,\tau)h(\tau,s_1,u_1) - l(t_2,s_2,\tau)h(\tau,s_2,u_2)|\, d\tau \to 0,$$

as $d := |t_1 - t_2| + |s_1 - s_2| + |u_1 - u_2| \to 0$. But this and the continuity
of x on D imply that

$$|LHx(t_1,s_1) - LHx(t_2,s_2)| \to 0 \qquad (|t_1 - t_2|, |s_1 - s_2| \to 0).$$

We have shown that the operator LH maps $C(D)$ into itself and
is bounded. Now, choose $x, y \in C(D)$ such that $\|x\|, \|y\| \le r$ and
$\|x - y\| \le \delta$, with r as in (17.32) and δ small. Then $\|LHx - LHy\| \le \varepsilon$,
which shows that LH is uniformly continuous on bounded sets. ■

We remark that the hypotheses of Theorem 17.5 are fulfilled if the
functions

$$(t,s,\tau,u) \mapsto l(t,s,\tau)h(\tau,s,u),$$

$$(t,s,\sigma,u) \mapsto m(t,s,\sigma)h(t,\sigma,u),$$

and

$$(t,s,\tau,\sigma,u) \mapsto n(t,s,\tau,\sigma)h(\tau,\sigma,u)$$

are all continuous. In particular, this true for the functions in Example 17.1, since $l(t, s, \tau)h(\tau, s, u) = g(u)$.

Now we will be interested in finding conditions for the Lipschitz continuity of the Hammerstein operator KH. Since

$$\|KHx - KHy\| = \|K(Hx - Hy)\| \leq \|K\| \, \|Hx - Hy\|,$$

we have only to care for the Lipschitz continuity of the Nemytskij operator (17.29). As was shown in APPELL [1981] (see also APPELL-MASSABÒ-VIGNOLI-ZABREJKO [1988]), the Lipschitz condition for H, i.e.

$$\|Hx - Hy\| \leq k(r)\|x - y\| \qquad (\|x\|, \|y\| \leq r)$$

is equivalent to the corresponding Lipschitz condition for the function $h(t, s, \cdot)$, i.e.

$$(17.34) \qquad |h(t, s, u) - h(t, s, v)| \leq k(r)|u - v| \qquad (|u|, |v| \leq r).$$

Consequently, the following result is true:

Theorem 17.6. *Suppose that both operators K and H act in the space $C(D)$. Then the operator KH satisfies a Lipschitz condition in $C(D)$ if (17.34) holds.*

Observe that Theorem 17.6 does not apply to the operator given in Example 17.1, since the operator H there does not act in the space $C(D)$. On the other hand, the operator (17.31) satisfies a Lipschitz condition on $C(D)$ if the function g is Lipschitz continuous.

Let us now give an elementary differentiability result for the operator KH. Since the Fréchet derivative of this operator has the form $[KH]' = KH'$ (if it exists), we have again to care only for the differentiability of the Nemytskij operator (17.29) in $C(D)$. From well-known differentiability conditions of the Nemytskij operator (see e.g. KRASNOSEL'SKIJ-ZABREJKO-PUSTYL'NIK-SOBOLEVSKIJ [1966]) we get therefore the following

Theorem 17.7. *Suppose that both operators* K *and* H *act in the space* $C(D)$, *and the function* h *has a continuous derivative* $\partial h / \partial u$. *Then the operator* KH *is differentiable at each point* $x_0 \in C(D)$ *and*

(17.35) $[KH]'(x_0) = \hat{L}(x_0) + \hat{M}(x_0) + \hat{N}(x_0),$

where

$$\hat{L}(x_0)h(t, s) := \int_a^b l(t, s, \tau) \frac{\partial h}{\partial u}[\tau, s, x_0(\tau, s)]h(\tau, s)\, d\tau,$$

$$\hat{M}(x_0)h(t, s) := \int_c^d m(t, s, \sigma) \frac{\partial h}{\partial u}[t, \sigma, x_0(t, \sigma)]h(t, \sigma)\, d\sigma,$$

and

$$\hat{N}(x_0)h(t, s) := \int_c^d \int_a^b n(t, s, \tau, \sigma) \frac{\partial h}{\partial u}[\tau, \sigma, x_0(\tau, \sigma)]h(\tau, \sigma)\, d\tau\, d\sigma.$$

We illustrate Theorem 17.7 by an elementary example:

Example 17.2. Using the notation of Example 17.1, let $g(u) = u$. As we have seen, the operator LH is then

$$LHx(t, s) = \int_a^b x(\tau, s)\, d\tau.$$

Being linear, this operator is clearly differentiable with $(LH)'(x_0) = LH$ for every $x_0 \in C(D)$. On the other hand, neither of the operators L or H acts in the space $C(D)$. A similar fact is true in case $g(u) = u^2$, say: here

$$(LH)'(x_0)h(t, s) = 2 \int_a^b x_0(\tau, s)h(\tau, s)\, d\tau,$$

but again neither of the operators L or H acts in the space $C(D)$.

17.4. Surjectivity results for monotone operators

In Subsection 17.2 we have studied the Barbashin equation (17.1) in Lebesgue spaces by means of topological methods. Since most Lebesgue spaces are reflexive, and many nonlinearities arising in applications are monotonically increasing or decreasing, it is a useful device to apply also monotonicity methods, rather than topological methods. This will be illustrated in this and the following subsection by means of the equation (17.25).

The existence results given below essentially build on Minty's celebrated *monotonicity principle* (MINTY [1962], see also BROWDER [1968]). Recall that a subset Z of the product space $X \times X^*$ is said to be *monotone* if

$$(17.36) \qquad \langle y_1 - y_2, x_1 - x_2 \rangle \geq 0 \qquad ((x_1, y_1), (x_2, y_2) \in Z).$$

A monotone set Z is *maximal monotone* if it is not properly contained in any other monotone set. A (multivalued) mapping $A : D(A) \to 2^{X^*}$ (with $D(A) \subseteq X$) is called *monotone* (respectively *maximal monotone*) if its graph $G(A) = \{(x, y) \in X \times X^* : x \in D(A), y \in Ax\}$ is a monotone (respectively maximal monotone) set. A mapping A is called *weakly coercive* if either $D(A)$ is bounded or $D(A)$ is unbounded and

$$(17.37) \qquad \lim_{\substack{x \in D(A) \\ \|x\| \to \infty}} \inf_{y \in Ax} \|y\| = \infty.$$

Here $\langle \cdot, \cdot \rangle$ denotes the natural pairing between X^* and X. Minty's monotonicity principle may be stated as follows:

Theorem 17.8. *Let X be a reflexive Banach space, and let $A : D(A) \to 2^{X^*}$ with $D(A) \subseteq X$ be maximal monotone and weakly coercive. Then $R(A) = X^*$, i.e. A is onto.*

17.5. Equations with monotone operators

As in Subsection 17.3, we rewrite equation (17.25) as operator equation (17.30) with $K = L + M + N$. Since all operators occuring here

act on functions of two variables, it is reasonable to study equation (17.30) in spaces with mixed norm (see Subsection 12.1). To this
end, we first reduce the operators (17.26) and (17.27) to families of
operators acting on functions of one variable by putting

$$(17.38) \qquad L(s)u(t) = \int_T l(t,s,\tau)u(\tau)\,d\tau \quad (s \in S)$$

and

$$(17.39) \qquad M(t)v(s) = \int_S m(t,s,\sigma)v(\sigma)\,d\sigma \quad (t \in T).$$

Lemma 17.3. *Let* $2 \le p,q < \infty$, $\frac{1}{p} + \frac{1}{p'} = 1$, *and* $\frac{1}{q} + \frac{1}{q'} = 1$. *Suppose
that the following three conditions are satisfied:*
(a) *the linear integral operators* (17.38) *are bounded from* L_p *into*
$L_{p'}$ *for each* $s \in S$, *and the map* $s \mapsto \|L(s)\|_{\mathcal{L}(L_p,L_{p'})}$ *belongs to*
$L_{qq'/(q-q')} = L_{q/(q-2)}$;
(b) *the linear integral operators* (17.39) *are bounded from* L_q *into*
$L_{q'}$ *for each* $t \in T$, *and the map* $t \mapsto \|M(t)\|_{\mathcal{L}(L_q,L_{q'})}$ *belongs to*
$L_{pp'/(p-p')} = L_{p/(p-2)}$;
(c) *the linear integral operator* (17.28) *is bounded from* $[L_p \to L_q]$ *into*
$[L_{p'} \to L_{q'}]$ *if* $p \ge q$, *or from* $[L_p \leftarrow L_q]$ *into* $[L_{p'} \leftarrow L_{q'}]$ *if* $p \le q$.
Then the operator $K = L + M + N$ *is bounded from* $[L_p \to L_q]$ *into*
$[L_{p'} \to L_{q'}]$ *if* $p \ge q$, *or from* $[L_p \leftarrow L_q]$ *into* $[L_{p'} \leftarrow L_{q'}]$ *if* $p \le q$.
□ By Theorem 12.1, the operator L is bounded from $[L_p \to L_q]$
into $[L_{p'} \to L_{q'}]$ and the operator M is bounded from $[L_p \leftarrow L_q]$
into $[L_{p'} \leftarrow L_{q'}]$. Moreover, by the Minkowski inequality, we have
the continuous embeddings $[L_p \to L_q] \subseteq [L_p \leftarrow L_q]$ if $p \ge q$, and
$[L_p \to L_q] \supseteq [L_p \leftarrow L_q]$ if $p \le q$. Thus, we get the conclusion. ∎

With p,q,p',q' as given above, let us put for shortness $X = [L_p \to L_q]$
if $p \ge q$, and $X = [L_p \leftarrow L_q]$ if $p \le q$. Since $[L_p \to L_q]^* = [L_p \to
L_q]' = [L_{p'} \to L_{q'}]$ and $[L_p \leftarrow L_q]^* = [L_p \leftarrow L_q]' = [L_{p'} \leftarrow L_{q'}]$ for
these p and q, Lemma 17.3 gives a set of sufficient conditions under
which the operator $K = L+M+N$ maps X into its dual X^*. Observe,
moreover, that the space X is reflexive, by our choice of p and q.

In order to apply Theorem 17.8, we make the following further assumptions:

(A1) *there exists $c > 0$ such that $\langle -Kx, x \rangle \geq c \, \|Kx\|^2$ for all $x \in X$;*

(A2) *there exist a function $a \in X$ and $b \geq 0$ such that*

$$(17.40) \qquad |h(t, s, w)| \leq a(t, s) + b|w|^{r(p,q)},$$

where $r(p, q) = p'/p$ if $p \geq q$, and $r(p, q) = q'/q$ if $p \leq q$;

(A3) *the function $h(t, s, \cdot)$ is monotonically increasing for almost all $(t, s) \in D$;*

(A4) *there exist a function $\hat{h} \in L_1$ and $d > 0$ such that*

$$(17.41) \qquad h(t, s, w)w \geq d|w|^{s(p,q)} + \hat{h}(t, s),$$

where $s(p, q) = q'$ if $p \geq q$, and $s(p, q) = p'$ if $p \leq q$.

Theorem 17.9. *Suppose that the hypotheses of Lemma 17.3 as well as the assumptions* (A1) - (A3) *are satisfied. Then the operator equation* (17.30) *has, for each $g \in X^*$, a unique solution $y \in X^*$.*

□ We first show the existence of a solution. The operator equation (17.30), i.e. $y = KHy + g$ with $K = L + M + N$, is equivalent to the relation

$$(17.42) \qquad 0 \in (-K)^{-1}(y - g) + Hy$$

(see BROWDER-DE FIGUEIREDO-GUPTA [1970]). Let $z = y - g$, and let H_g be defined by $H_g z = H(z + g)$. Then equation (17.42) holds if and only if

$$(17.43) \qquad 0 \in [(-K)^{-1} + H_g]z.$$

Under the assumptions (A2) and (A3), it is easy to show that the Nemytskij operator H_g acts from X^* into X and is monotone and continuous. Since $(-K)^{-1}$ and H_g are both maximal monotone, by a result of ROCKAFELLAR [1970], the mapping $(-K)^{-1} + H_g$ is also maximal monotone. Moreover, the estimate

$$\inf_{x \in (-K)^{-1}z} ||x + H_g(z)|| \geq \inf_{x \in (-K)^{-1}z} \frac{\langle x + H_g(z), z \rangle}{||z||}$$

$$\geq c||z|| - ||H_g(0)||$$

shows that $(-K)^{-1} + H_g$ is weakly coercive. The solvability of equation (17.43) thus follows from Theorem 17.8.

For the proof of uniqueness, let $y_1, y_2 \in X^*$ such that $y_1 = KHy_1 + g$, $y_2 = KHy_2 + g$. Then

$$\langle y_1 - y_2, Hy_1 - Hy_2 \rangle +$$

$$\langle (-K)Hy_1 - (-K)Hy_2, Hy_1 - Hy_2 \rangle = 0,$$

and, by the monotonicity of H and the assumption (A1),

$$0 \geq \langle (-K)Hy_1 - (-K)Hy_2, Hy_1 - Hy_2 \rangle$$

$$\geq c||K(Hy_1 - Hy_2)||^2.$$

This implies that $KHy_1 = KHy_2$ and hence $y_1 = y_2$. ∎

In fact, under the assumptions (A2) and (A3), we only need that $(-K)$ is monotone, i.e. $\langle -Kx, x \rangle \geq 0$ for all $x \in X$, such that $(-K)^{-1} + H_g$ is maximal monotone. Furthermore, if the assumption (A4) is satisfied, by the Hölder inequality we have

$$\langle Hy, y \rangle \geq d'||y||^{s(p,q)} + d''$$

for some $d' > 0$ and $d'' \in \mathbb{R}$. Now let $g = 0$; since

$$\inf_{x \in (-K)^{-1}y} ||x + Hy|| \geq \inf_{x \in (-K)^{-1}y} \frac{\langle x + Hy, y \rangle}{||y||}$$

$$\geq \frac{d'||y||^{s(p,q)} + d''}{||y||},$$

$(-K)^{-1} + H$ is weakly coercive. In this way, we have proved the following

Theorem 17.10. *Suppose that the hypotheses of Lemma 17.3 as well as the assumptions* (A2) - (A4) *are satisfied. If* $(-K)$ *is monotone and* $g = 0$, *then the operator equation* (17.30) *has a solution* $y \in X^*$.

We remark that other solvability results for partial integral equations involving Hammerstein operators have been given in POVOLOTSKIJ-KALITVIN [1985a, 1987, 1989, 1989a, 1991].

17.6. Variational methods

Let the operators L, M, N and H be defined as in (17.26) - (17.29). We will show how to study the operator equation

$$(17.44) \qquad\qquad y = KHy,$$

with $K = L + M + N$ as before, by means of variational methods. For solving an operator equation $Ay = 0$, the main idea consists in looking for critical points of a *potential* for A. Recall that a real functional Ψ on a Banach space E is called *weakly lower semicontinuous* if

$$\Psi(u) \leq \varliminf_{n \to \infty} \Psi(u_n)$$

for every sequence $(u_n)_n$ converging weakly to u in E. The following is a standard result.

Proposition 17.3. *Let* E *be a reflexive real Banach space, and let* $\Psi : E \to \mathbb{R}$ *be weakly lower semicontinuous. Suppose that the gradient* $\nabla\Psi$ *of* Ψ *exists on* E *and is coercive, i.e.*

$$\lim_{\|u\| \to \infty} \frac{\langle \nabla\Psi(u), u \rangle}{\|u\|} = \infty.$$

Then there exists some $u_0 \in X$ *such that*

$$\Psi(u_0) = \inf_{u \in E} \Psi(u).$$

Moreover, u_0 *is a critical point of* Ψ, *i.e.* $\Psi'(u_0) = 0$.

Now let $1 < p, q < \infty$, and let $X = [L_p \to L_q]$ or $X = [L_p \leftarrow L_q]$. Our first step consists in finding potentials. For the Nemytskij operator H we can explicitly give its potential (KRASNOSEL'SKIJ [1956]), namely

$$(17.45) \qquad \Phi(y) = \int_T \int_S \left[\int_0^{y(t,s)} h(t,s,u)\, du \right] ds\, dt.$$

Lemma 17.4. *Suppose that the operator H maps X^* into X. Then (17.45) defines a continuous functional $\Phi : X^* \to \mathbb{R}$. Moreover, Φ is differentiable with gradient $\nabla \Phi = H$.*

□ By the classical Krasnosel'skij-Ladyzhenskij lemma (KRASNOSEL'-SKIJ-LADYZHENSKIJ [1954]) and Hölder's inequality we have

$$|\Phi(y)| \leq \int_T \int_S |y(t,s)|\, |h(t,s,y^*(t,s))|\, ds\, dt$$

$$\leq \|y\|_{X^*} \|Hy^*\|_X \qquad (y \in X^*),$$

where y^* is a suitable measurable function on $D = T \times S$ with $|y^*| \leq |y|$. Since the Nemytskij operator H maps X^* into X, we have $\|Hy^*\|_X < \infty$, and so Φ is defined on the whole of X^*. To prove the second assertion, let $y, \eta \in X^*$ and consider

$$w(y, \eta) := \Phi(y + \eta) - \Phi(y) - \langle Hy, \eta \rangle$$

$$= \int_T \int_S \left[\int_0^{y(t,s)+\eta(t,s)} h(t,s,u)\, du \right.$$

$$\left. - \int_0^{y(t,s)} h(t,s,u)\, du - h(t,s,y(t,s))\eta(t,s) \right] ds\, dt$$

$$= \int_T \int_S \left[\int_{y(t,s)}^{y(t,s)+\eta(t,s)} \{ h(t,s,u) - h(t,s,y(t,s)) \}\, du \right] ds\, dt.$$

Again by the Krasnosel'skij-Ladyzhenskij lemma, there exists a measurable function y_η with $|y| \leq |y_\eta| \leq |y + \eta|$ such that

$$|w(y, \eta)| \leq \int_T \int_S |\eta(t,s)|\, |h(t,s,y_\eta(t,s))|\, ds\, dt$$

$$\leq \|\eta\|_{X^*} \|Hy_\eta - Hy\|_X.$$

For $||\eta||_{X^*} \to 0$ we have $||y_\eta - y||_{X^*} \to 0$, and hence $||Hy_\eta - Hy||_X \to 0$, by the continuity of H. This implies that

$$\lim_{||\eta|| \to 0} \frac{|w(y,\eta)|}{||\eta||} = 0,$$

i.e. $\nabla \Phi = H$ as claimed. ∎

A standard way to apply Proposition 17.3 is to consider a square root of K. The usual assumption for this is that there is some Hilbert space H with $X^* \subseteq H \subseteq X^{**} = X$. However, for the "natural choice" $H = L_2$ this implies that $p, q \leq 2$. On the other hand, Lemma 17.3 holds only for $p, q \geq 2$. Thus we may cover only the Hilbert space case $p = q = 2$ in this way.

To overcome this difficulty, we apply the following result of BROW-DER-GUPTA [1969]:

Theorem 17.11. *Let E be a real Banach space, and let $K : E \to E^*$ be bounded, linear, monotone, and selfadjoint. Then there exists a Hilbert space H and a bounded linear operator $B : E \to H$ such that $K = B^*B$. Moreover, $B^* : H \to E^*$ is injective, and $||B||^2 \leq ||K||$.*

To apply Proposition 17.3 to the equation (17.44) we make now the following assumptions:

(H1) the operator $K : X \to X^*$ is monotone and selfadjoint;

(H2) the growth estimate

$$(17.46) \qquad . \qquad |h(t,s,v)| \leq a(t,s) + b|v|^{r(p,q)}$$

holds for some $a \in X$ and $b \geq 0$, where $r(p,q) := p'/p$ if $p \geq q$, and $r(p,q) := q'/q$ if $p \leq q$;

(H3) the monotonicity condition

$$(17.47) \qquad \langle Hy_1 - Hy_2, y_2 - y_2 \rangle \leq c\,||y_1 - y_2||_{X^*}^2$$

holds, where $c\,||K|| < 1$.

Theorem 17.12. *Suppose that the hypotheses of Lemma 17.4 as well as the assumptions* (H1) - (H3) *are satisfied. Then the operator equation* (17.44) *has a unique solution* $y \in X^*$.

$\square$ As mentioned before, we have $H(X^*) \subseteq X$, by (H2), and $K(X) \subseteq X^*$, by Lemma 17.4. Using the assumption (H1) we can construct a Hilbert space H with scalar product $\langle \cdot, \cdot \rangle$ and an operator $B : X \to H$ as given in Theorem 17.11. To solve the equation (17.44) it suffices to find a solution $h \in H$ of

$$(17.48) \qquad\qquad h - BHB^*h = 0.$$

Indeed, $h \in H$ solves (17.48) if and only if $y = B^*h$ solves (17.44). Consider the functional $\Psi : H \to \mathbb{R}$ defined by

$$\Psi(h) = \frac{1}{2}\langle h, h \rangle - \Phi(B^*h),$$

where Φ is given by (17.45), i.e. $\nabla \Phi = H$. We verify the conditions of Proposition 17.3 for $E = H$. For fixed $h_0, h \in H$ we have

$$\lim_{t \to 0} \frac{1}{t}[\Psi(h_0 + th) - \Psi(h_0)] =$$

$$\langle h_0, h \rangle - \lim_{t \to 0} \frac{1}{t}[\Phi(B^*h_0 + tB^*h) - \Phi(B^*h_0)]$$

$$= \langle h_0, h \rangle - \langle H(B^*h_0), B^*h \rangle = \langle h_0, h \rangle - \langle BH(B^*h_0), h \rangle.$$

This implies that $\nabla \Psi = I - BHB^*$ exists. Moreover, for $h_1, h_2 \in H$ we have, by assumption (H3),

$$\langle \nabla \Psi h_1 - \nabla \Psi h_2 \rangle = ||h_1 - h_2||_H^2$$

$$-\langle BHB^*h_1 - BHB^*h_2, h_1 - h_2 \rangle$$

$$= ||h_1 - h_2||_H^2 - \langle HB^*h_1 - HB^*h_2, B^*h_1 - B^*h_2 \rangle$$

$$(17.49) \qquad\qquad \geq ||h_1 - h_2||_H^2 - c\,||B^*h_1 - B^*h_2||_X^2$$

$$\geq ||h_1 - h_2||_H^2 - c\,||B^*||^2||h_1 - h_2||_H^2$$

$$\geq (1 - c\,||K||)||h_1 - h_2||_H^2 > 0.$$

This shows that $\nabla\Psi$ is monotone and

$$\frac{\langle \nabla\Psi(h), h \rangle}{||h||_H} \to \infty \qquad (||h||_H \to \infty).$$

The monotonicity of $\nabla\Psi$ implies the lower semicontinuity of Ψ (see e.g. VAJNBERG [1969]). Therefore, all conditions of Proposition 17.3 are satisfied, and thus (17.48) has a solution.

To prove uniqueness, suppose that y_1 and y_2 are solutions of (17.44). Then $h_1 = (B^*)^{-1}y_1$ and $h_2 = (B^*)^{-1}y_2$ are solutions of (17.48). From inequality (17.49) it follows that $(B^*)^{-1}y_1 = (B^*)^{-1}y_2$, hence $y_1 = y_2$. ∎

We point out that the assumption (H3) is in some sense too restrictive for the application of variational methods: As we have seen in the proof of Theorem 17.12, this assumption implies that $\nabla\Psi$ is monotone and weakly coercive. But $\nabla\Psi$ is even maximal monotone, being continuous. So, we get the same result by applying Minty's monotonicity principle.

Variational methods have been applied to the study of partial integral equations with Hammerstein operators in KALITVIN [1986a] and POVOLOTSKIJ-KALITVIN [1985a, 1989, 1991]. A parallel theory for Uryson operators has not been given systematically; however, some results may be found in KALITVIN [1986, 1986a, 1997], KALITVIN-KORENCHUK-EVKHUTINA [1993], POVOLOTSKIJ-KALITVIN [1985, 1991], and CHEN-KALITVIN [1997, 1998]. More existence theorems for nonlinear partial integral equations have been given in POVOLOTS-KIJ-KALITVIN [1994], MURESAN [1982, 1984], and PACHPATTE [1981, 1983, 1984, 1986]. Nemytskij operators in spaces with mixed norm have been studied, for example, in KALITVIN [1984a] and CHEN-VÄTH [1997].

§ 18. The Newton-Kantorovich method

In this section we discuss the applicability of the Newton-Kantorovich method to the nonlinear partial integral equation of Uryson type

$$x(t,s) = \int_T l(t,s,\tau,x(\tau,s))\,d\tau$$

(18.1)
$$+ \int_S m(t,s,\sigma,x(t,\sigma))\,d\sigma$$

$$+ \int_T \int_S n(t,s,\tau,\sigma,x(\tau,\sigma))\,d\sigma\,d\tau$$

by means of the Newton-Kantorovich method. A basic ingredient of this method consists in verifying a local Lipschitz condition for the Fréchet derivative of the the corresponding nonlinear partial integral operators. The abstract results are illustrated in the spaces C and L_p for $1 \leq p \leq \infty$. In particular, we show that a local Lipschitz condition for the derivative in L_p for $p < \infty$ leads to a strong degeneracy of the corresponding kernels. The basic results of this section are taken from APPELL-DE PASCALE-KALITVIN-ZABREJKO [1996].

18.1. The abstract Newton-Kantorovich method

The Newton-Kantorovich method is one of the basic tools for finding approximate solutions of the operator equation

(18.2) $$F(x) = 0,$$

where F is some nonlinear operator in a Banach space X. In the corresponding iterative scheme

(18.3) $$x_{n+1} = x_n - F'(x_n)^{-1}F(x_n) \qquad (n = 0,1,2,\ldots)$$

one has to require, in particular, that the Fréchet derivative of F at all points x_n exists and is invertible in the algebra $\mathfrak{L}(X)$ of all bounded linear operators in X. The direct verification of this requirement may cause essential difficulties in practice. However, in the last years several new ideas have been developed to overcome these difficulties. For the reader's ease, let us briefly sketch some of these ideas related to the Newton-Kantorovich method ZABREJKO-NGUYEN [1987, 1989].

Suppose that F is defined on the closure $\overline{B}_R(X)$ of some ball $B_R(X) = \{x : x \in X, \ ||x|| < R\}$ and admits a Fréchet derivative $F'(x)$ at each point of $B_R(X)$ such that F' satisfies a Lipschitz condition

$$(18.4) \qquad ||F'(x_1) - F'(x_2)|| \leq k(r)||x_1 - x_2||$$

for $x_1, x_2 \in B_r(X)$ with $0 \leq r \leq R$, considered as a map from $B_R(X)$ into $\mathfrak{L}(X)$. We also assume that the Fréchet derivative at zero is invertible and put

$$(18.5) \qquad a := ||F'(0)^{-1}F(0)||$$

and

$$(18.6) \qquad b := ||F'(0)^{-1}||.$$

Since the (minimal) Lipschitz constant $k = k(r)$ in (18.4) is positive, the function $\varphi : [0, R] \to [0, \infty)$ defined by

$$(18.7) \qquad \varphi(r) := a + b \int_0^r (r - t)k(t)\,dt \qquad (0 \leq r \leq R)$$

is convex. The scalar equation

$$(18.8) \qquad r = \varphi(r)$$

may have no solution, a unique solution, or many solutions in $[0, R]$. Let us suppose for the moment that (18.8) has a unique solution $r_* \in [0, R]$ and $\varphi(R) \leq R$. In this case the equation (18.2) has a solution $x_* \in \overline{B}_{r_*}(X)$, and this solution is unique in the open ball $B_R(X)$ (see ZABREJKO-NGUYEN [1987]). Moreover, the iterations (18.3) are defined for every n and converge to the solution x_*.

The usefulness of the Newton-Kantorovich method does not only consist in reducing the operator equation (18.2) in a Banach space to the scalar equation (18.8) on a real interval. It is also possible to give estimates for the convergence speed. In fact, let

$$U(r) := \frac{\varphi(r) - r}{1 - \varphi'(r)},$$

$$V(r) := U(r + U^{-1}(r)),$$

and

$$W(r) := \sum_{k=0}^{\infty} V^k(r).$$

Then the estimates

$$\|x_{n+1} - x_n\| \leq V^n(a)$$

and

$$\|x_* - x_n\| \leq W(V^n(a))$$

hold ZABREJKO-NGUYEN [1987]. So, in order to study equation (18.2) from the viewpoint of existence, uniqueness, and approximation it suffices in many cases to calculate (or estimate) the constants a and b, as well as to calculate (or estimate) the scalar function $k = k(r)$.

As already mentioned, the simplest case is when the scalar equation (18.8) has a unique solution r_* in $[0, R]$ and $\varphi(R) \leq R$. Other cases may be reduced to this case. For example, if (18.8) has another solution $r^* \in [0, R]$ with $r^* > r_*$, say, we simply take $R < r^*$. Likewise, if (18.8) has a whole continuum of solutions $[r_*, r^*] \subset (0, R]$, we can choose $R = r_*$.

18.2. Lipschitz conditions for derivatives

Throughout we denote the (partial) integral operator generated by some kernel function by the corresponding capital letter, i. e.

$$(18.9) \qquad L(x)(t,s) := \int_T l(t, s, \tau, x(\tau, s)) \, d\tau,$$

$$(18.10) \qquad M(x)(t,s) := \int_S n(t,s,\sigma,x(t,\sigma))\, d\sigma,$$

and

$$(18.11) \qquad N(x)(t,s) := \int_T \int_S n(t,s,\tau,\sigma,x(\tau,\sigma))\, d\sigma\, d\tau.$$

The nonlinear partial integral equation (18.1) may then be rewritten as operator equation (18.2) if we put

$$(18.12) \qquad F(x) = x - L(x) - M(x) - N(x) \qquad (x \in B_R(X)),$$

where X is some Banach space of real functions over $T \times S$.

Suppose now that the three kernels in (18.1) have partial derivatives in the last argument

$$\begin{cases} l_1(t,s,\tau,u) := \dfrac{\partial l}{\partial u}(t,s,\tau,u), \\[2mm] m_1(t,s,\sigma,u) := \dfrac{\partial m}{\partial u}(t,s,\sigma,u), \\[2mm] n_1(t,s,\tau,\sigma,u) := \dfrac{\partial n}{\partial u}(t,s,\tau,\sigma,u), \end{cases}$$

Consider the operators L^*, M^* and N^* defined by

$$(18.13) \qquad L^*(x)(t,s,\tau) := l_1(t,s,\tau,x(\tau,s)),$$

$$(18.14) \qquad M^*(x)(t,s,\sigma) := m_1(t,s,\sigma,x(t,\sigma)),$$

and

$$(18.15) \qquad N^*(x)(t,s,\tau,\sigma) := n_1(t,s,\tau,\sigma,x(\tau,\sigma)).$$

These operators are not Nemytskij operators in the usual sense, since they map functions of the two variables (t,s) into functions of three (or even four) variables; we call such operators "generalized

Nemytskij operators" in the sequel. The operators (18.13) - (18.15) may be considered between the space X and the kernel spaces $\mathfrak{R}_l(X)$, $\mathfrak{R}_m(X)$, and $\mathfrak{R}_n(X)$ defined by the norms

$$(18.16) \qquad \|p\|_{\mathfrak{R}_l(X)} := \sup_{\|h\|_X \leq 1} \|\int_T |p(\cdot,\cdot,\tau)h(\tau,\cdot)|\,d\tau\|_X,$$

$$(18.17) \qquad \|q\|_{\mathfrak{R}_m(X)} := \sup_{\|h\|_X \leq 1} \|\int_S |q(\cdot,\cdot,\sigma)h(\cdot,\sigma)|\,d\sigma\|_X,$$

and

$$(18.18) \quad \|r\|_{\mathfrak{R}_n(X)} := \sup_{\|h\|_X \leq 1} \|\int_T \int_S |r(\cdot,\cdot,\tau,\sigma)h(\tau,\sigma)|\,d\sigma\,d\tau\|_X,$$

respectively. Of course, the functionals (18.16) - (18.18) are nothing else than the operator norms of the moduli of the corresponding (regular) linear partial integral operators

$$(18.19) \qquad Ph(t,s) := \int_T p(t,s,\tau)h(\tau,s)\,d\tau,$$

$$(18.20) \qquad Qh(t,s) := \int_S q(t,s,\sigma)h(t,\sigma)\,d\sigma,$$

and

$$(18.21) \qquad Rh(t,s) := \int_T \int_S r(t,s,\tau,\sigma)h(\tau,\sigma)\,d\sigma\,d\tau$$

in the algebra $\mathfrak{L}(X)$ of bounded linear operators on X. We set

$$(18.22) \qquad L_1(x)h(t,s) := \int_T l_1(t,s,\tau,x(\tau,s))h(\tau,s)\,d\tau,$$

$$(18.23) \qquad M_1(x)h(t,s) := \int_S m_1(t,s,\sigma,x(t,\sigma))h(t,\sigma)\,d\sigma,$$

and

(18.24) $N_1(x)h(t,s) := \int_T \int_S n(t,s,\tau,\sigma,x(\tau,\sigma))h(\tau,\sigma)\,d\tau\,d\sigma.$

Lemma 18.1. *Suppose that the generalized Nemytskij operators* (18.13) *- (18.15) act from $B_R(X)$ into $\Re_l(X)$, $\Re_m(X)$, and $\Re_n(X)$, respectively, and satisfy a Lipschitz condition. Then the operators* (18.9) *- (18.11) are Fréchet differentiable with $L' = L_1$, $M' = M_1$, and $N' = N_1$. Consequently,*

(18.25) $F'(x) = I - L_1(x) - M_1(x) - N_1(x)$ $(x \in B_R(X)).$

□ The assertion has been proved in APPELL-DE PASCALE-ZABREJ-KO [1991] for integral operators of the form (18.11), so we have to prove it only for the partial integral operators (18.9) and (18.10). For $x \in B_R(X)$ and $h \in X$ we have

$$[L(x+h) - L(x) - L_1(x)h](t,s)$$

$$= \int_T [l(t,s,\tau,x(\tau,s) + h(\tau,s)) - l(t,s,\tau,x(\tau,s))$$

$$-l_1(t,s,\tau,x(\tau,s))h(\tau,s)]\,d\tau$$

$$= \int_T \int_0^1 [l_1(t,s,\tau,x(\tau,s) + \lambda h(\tau,s))$$

$$-l_1(t,s,\tau,x(\tau,s))]h(\tau,s)\,d\lambda\,d\tau$$

$$= \left\{ \int_0^1 [L_1(x+\lambda h) - L_1(x)]h\,d\lambda \right\}(t,s).$$

Since the operator $L_1 : B_R(X) \to \mathfrak{L}(X)$ satisfies a Lipschitz condition, by assumption, we conclude that

$$\|L(x+h) - L(x) - L_1(x)h\| \le$$

$$\int_0^1 \|L_1(x+\lambda h) - L_1(x)\|\,\|h\|\,d\lambda = o(\|h\|),$$

which means that $L'(x) = L_1(x)$. The equality $M'(x) = M_1(x)$ is proved similarly. ∎

Applying Lemma 18.1 allows us to "find" the constants a and b for the equation (18.2), where F is given by (18.12). In fact, the function $h := F'(0)^{-1}F(0)$ satisfies the *linear* partial integral equation

$$h(t,s) - \int_T l_1(t,s,\tau,0)h(\tau,s)\,d\tau$$

(18.26)
$$- \int_S m_1(t,s,\sigma,0)h(t,\sigma)\,d\sigma$$

$$- \int_T \int_S n_1(t,s,\tau,\sigma,0)h(\tau,\sigma)\,d\sigma\,d\tau = g(t,s),$$

where

$$g(t,s) := - \int_T l(t,s,\tau,0)\,d\tau$$

(18.27)
$$- \int_S m(t,s,\sigma,0)\,d\sigma - \int_T \int_S n(t,s,\tau,\sigma,0)\,d\sigma\,d\tau.$$

Suppose that the (unique!) solution of equation (18.26) may be written in the form

$$h(t,s) = g(t,s) + \int_T r_l(t,s,\tau)g(\tau,s)\,d\tau$$

(18.28)
$$+ \int_S r_m(t,s,\sigma)g(t,\sigma)\,d\sigma$$

$$+ \int_T \int_S r_n(t,s,\tau,\sigma)g(\tau,\sigma)\,d\sigma\,d\tau$$

with some resolvent kernels r_l, r_m, and r_n which are defined through the kernels l_1, m_1, and n_1. Then the constant a in (18.5) is of course nothing else than the norm $\|h\|_X$ of the function (18.28) in the space X. Since the explicit form of these resolvent kernels is in general hard to find, one usually looks for a representation of the form $F'(0) = T - E$, where T is boundedly invertible and the norm of E in $\mathfrak{L}(X)$

is small. The elementary equality $T - E = T(I - T^{-1}E)$ implies then that, under the hypotheses of Lemma 18.1, the estimates

$$b = ||F'(0)^{-1}|| \leq ||T^{-1}|| \, ||(I - T^{-1}E)^{-1}||$$

$$\leq ||T^{-1}|| \sum_{k=0}^{\infty} ||T^{-1}E||^k \leq \frac{||T^{-1}||}{1 - ||E|| \, ||T^{-1}||}$$

and

$$a = ||F'(0)^{-1}F(0)|| \leq \frac{||T^{-1}F(0)||}{1 - ||E|| \, ||T^{-1}||}$$

are true. In this way, we have proved the following

Lemma 18.2. *Suppose that the hypotheses of Lemma 18.1 are satisfied, and $F'(0) = T - E$, where T is invertible in $\mathfrak{L}(X)$ and $||E|| \leq \varepsilon$. Then the estimates*

$$a \leq \frac{||T^{-1}F(0)||}{1 - \varepsilon \, ||T^{-1}||}, \qquad b \leq \frac{||T^{-1}||}{1 - \varepsilon \, ||T^{-1}||}$$

hold.

Under the conditions of Lemma 18.2 it is natural to write the linear operators T and E also as sums of (partial) integral operators like (18.9) - (18.11). Various conditions for the "smallness" of the norm of E may then be found in §12 - 13. On the other hand, the invertibility of T is often not easy to verify, except for particular cases. Assume, for instance, that T has the special form

$$Th(t, s) = h(t, s) - \int_T \phi(t, \tau)h(\tau, s) \, d\tau$$

$$- \int_S \psi(s, \sigma)h(t, \sigma) \, d\sigma - c \int_T \int_S \phi(t, \tau)\psi(s, \sigma)h(\tau, \sigma) \, d\sigma \, d\tau,$$

where $\phi : T \times T \to \mathbb{R}$, $\psi : S \times S \to \mathbb{R}$, and $c \in \mathbb{R}$ are given. Then T is invertible if and only if

(18.29) $1 \notin \sigma(\Phi) + \sigma(\Psi) + c\sigma(\Phi)\sigma(\Psi),$

where Φ and Ψ are the integral operators generated by the kernels ϕ and ψ, respectively. Moreover, in some cases the operator T^{-1} may be expressed explicitly through the operators Φ and Ψ. For example, in case $c = 0$ the formula

$$T^{-1} = -\frac{1}{4\pi^2} \int_{\Gamma_\Psi} \int_{\Gamma_\phi} \frac{(\Phi - \xi I)\overline{\otimes}(\Psi - \eta I)}{1 - \xi - \eta} \, d\xi \, d\eta$$

holds, where Γ_Φ and Γ_Ψ are simple closed contours around $\sigma(\Phi)$ and $\sigma(\Psi)$, respectively. If the kernels ϕ and ψ are symmetric or degenerate, the relation (18.29) may be verified by standard schemes.

18.3. Lipschitz conditions for partial Uryson operators

The crucial assumption in Lemma 18.1 is of course the Lipschitz condition for the operators (18.13) - (18.15). In this subsection we take a closer look to this condition from a general point of view. More information in some specific function spaces which arise frequently in applications will be given in the next subsections.

Suppose that the three kernels in (18.1) also have second partial derivatives in the last argument

$$\begin{cases} l_2(t, s, \tau, u) := \dfrac{\partial^2 l}{\partial u^2}(t, s, \tau, u), \\[2mm] m_2(t, s, \sigma, u) := \dfrac{\partial^2 m}{\partial u^2}(t, s, \sigma, u), \\[2mm] n_2(t, s, \tau, \sigma, u) := \dfrac{\partial^2 n}{\partial u^2}(t, s, \tau, \sigma, u), \end{cases}$$

and that

$$(18.30) \qquad |l_1(t, s, \tau, u_1) - l_1(t, s, \tau, u_2)| \leq \tilde{l}_2(t, s, \tau, w) |u_1 - u_2|,$$

$$(18.31) \qquad |m_1(t, s, \sigma, u_1) - m_1(t, s, \sigma, u_2)| \leq \tilde{m}_2(t, s, \sigma, w) |u_1 - u_2|,$$

and

(18.32)
$$|n_1(t, s, \tau, \sigma, u_1) - n_1(t, s, \tau, \sigma, u_2)|$$
$$\leq \tilde{n}_2(t, s, \tau, \sigma, w) |u_1 - u_2|$$

for $|u_1|, |u_2| \leq w$. Here

$$\tilde{l}_2(t, s, \tau, w) = \sup_{|u| \leq w} |l_2(t, s, \tau, u)|,$$

$$\tilde{m}_2(t, s, \sigma, w) = \sup_{|u| \leq w} |l_2(t, s, \sigma, u)|,$$

and

$$\tilde{n}_2(t, s, \tau, \sigma, w) = \sup_{|u| \leq w} |l_2(t, s, \tau, \sigma, u)|,$$

respectively. It is then natural to state the Lipschitz condition (18.4) in terms of the generalized Nemytskij operators

(18.33) $$L^{**}(x)(t, s, \tau) := l_2(t, s, \tau, x(\tau, s)),$$

(18.34) $$M^{**}(x)(t, s, \sigma) := m_2(t, s, \sigma, x(t, \sigma)),$$

and

(18.35) $$N^{**}(x)(t, s, \tau, \sigma) := n_2(t, s, \tau, \sigma, x(\tau, \sigma)),$$

which are second order analogues to (18.13) - (18.15). As a matter of fact, we want to replace the generalized Nemytskij operators (18.33) - (18.35) by the usual Nemytskij operators

(18.36) $$\tilde{L}^{**}(x)(t, s, \tau) := l_2(t, s, \tau, x(t, s, \tau)),$$

(18.37) $$\tilde{M}^{**}(x)(t, s, \sigma) := m_2(t, s, \sigma, x(t, s, \sigma)),$$

and

(18.38) $\tilde{N}^{**}(x)(t,s,\tau,\sigma) := n_2(t,s,\tau,\sigma,x(t,s,\tau,\sigma))$

defined on functions over $T \times S \times T$, $T \times S \times S$, and $T \times S \times T \times S$, respectively. To make this precise, we need some special definitions. First, given a Banach space X of real functions over $T \times S$, we denote by X_l, X_m, and X_n the space of functions $p : T \times S \times T \to \mathbb{R}$, $q : T \times S \times S \to \mathbb{R}$, and $r : T \times S \times T \times S \to \mathbb{R}$ defined by the norms

$$||p||_{X_l} = \inf \{||\hat{p}||_X : |p(t,s,\tau)| \leq \hat{p}(\tau,s) \ (\hat{p} \in X)\},$$

$$||q||_{X_m} = \inf \{||\hat{q}||_X : |q(t,s,\sigma)| \leq \hat{q}(t,\sigma) \ (\hat{q} \in X)\},$$

and

$$||r||_{X_n} = \inf \{||\hat{r}||_X : |r(t,s,\tau,\sigma)| \leq \hat{r}(\tau,\sigma) \ (\hat{r} \in X)\},$$

respectively. (The inequalities in these definitions are considered almost everywhere on $T \times S \times T$, $T \times S \times S$, and $T \times S \times T \times S$, respectively.) Further, we define spaces $\tilde{\mathfrak{R}}_l(X)$, $\tilde{\mathfrak{R}}_m(X)$, and $\tilde{\mathfrak{R}}_n(X)$ of measurable functions $p : T \times S \times T \to \mathbb{R}$, $q : T \times S \times S \to \mathbb{R}$, and $r : T \times S \times T \times S \to \mathbb{R}$ by the norms

$$||p||_{\tilde{\mathfrak{R}}_l(X)} :=$$

(18.39)

$$\sup_{||x||_X, ||h||_X \leq 1} || \int_T |p(\cdot,\cdot,\tau)x(\tau,\cdot)h(\tau,\cdot)| \, d\tau ||_X,$$

$$||q||_{\tilde{\mathfrak{R}}_m(X)} :=$$

(18.40)

$$\sup_{||x||_X, ||h||_X \leq 1} || \int_S |q(\cdot,\cdot,\sigma)x(\cdot,\sigma)h(\cdot,\sigma)| \, d\sigma ||_X,$$

and

$$||r||_{\tilde{\mathfrak{R}}_n(X)} :=$$

(18.41)

$$\sup_{||x||_X, ||h||_X \leq 1} || \int_T \int_S |r(\cdot,\cdot,\tau,\sigma)x(\tau,\sigma)h(\tau,\sigma)| \, d\sigma \, d\tau ||_X,$$

respectively.

The construction described above is not as trivial as it seems to be. For example, if X is a Lebesgue space L_p with $p \geq 2$, one can easily see that the spaces $\tilde{\mathfrak{R}}_l(X)$, $\tilde{\mathfrak{R}}_m(X)$, and $\tilde{\mathfrak{R}}_n(X)$ consist of kernels of linear integral operators (18.19) - (18.21) acting from $L_{p/2}$ into L_p; in the case $1 \leq p < 2$ these spaces strongly degenerate (i.e. contain only the zero function). More generally, if X is an Orlicz space L_M, the spaces $\tilde{\mathfrak{R}}_l(X)$, $\tilde{\mathfrak{R}}_m(X)$, and $\tilde{\mathfrak{R}}_n(X)$ contain kernels of linear integral operators (18.19) - (18.21) acting from L_N into L_M, where the Orlicz space L_N is generated by any Young function N satisfying for some $c > 0$ the asymptotic condition

$$\varlimsup_{u \to \infty} \frac{N(cu^2)}{M(u)} < \infty.$$

Below we use the usual abbreviation

$$(u \vee v)(t,s) := \sup \{u(t,s), v(t,s)\}$$

and put

$$\gamma(r) := \sup \{\|u \vee v\| : \|u\|, \|v\| \leq r\} \quad (r > 0).$$

Moreover, given a bounded nonlinear operator F between two normed spaces, by

$$\mu(F; \rho) := \sup \{\|Fx\| : \|x\| \leq \rho\} \quad (\rho > 0)$$

we denote the *growth function* of F. This growth function may be used to state Lipschitz conditions and differentiability conditions, both necessary and sufficient, for Nemytskij operators in terms of the generating nonlinearity (APPELL [1981], APPELL-MASSABÒ-VIGNOLI-ZABREJKO [1988], see also APPELL-ZABREJKO [1990]).

Lemma 18.3. *Let X be a Banach space of real functions over $T \times S$. Then the following holds:*

(a) *if the Nemytskij operator (18.36) is bounded from X_l into $\tilde{\mathfrak{R}}_l(X)$, then the generalized Nemytskij operator (18.13) satisfies, for $\|x_1\|_X \leq r$ and $\|x_2\|_X \leq r$, the Lipschitz condition*

$$(18.42) \qquad \|L^*(x_1) - L^*(x_2)\|_{\mathfrak{R}_l(X)} \leq \mu(\tilde{L}^{**}; \gamma(r))\|x_1 - x_2\|_X;$$

(b) *if the Nemytskij operator* (18.37) *is bounded from* X_m *into* $\tilde{\mathfrak{R}}_m(X)$, *then the generalized Nemytskij operator* (18.14) *satisfies, for* $\|x_1\|_X \leq r$ *and* $\|x_2\|_X \leq r$, *the Lipschitz condition*

$$(18.43) \quad \|M^*(x_1) - M^*(x_2)\|_{\mathfrak{R}_m(X)} \leq \mu(\tilde{M}^{**}; \gamma(r))\|x_1 - x_2\|_X;$$

(c) *if the Nemytskij operator* (18.38) *is bounded from* X_n *into* $\tilde{\mathfrak{R}}_n(X)$, *then the generalized Nemytskij operator* (18.15) *satisfies, for* $\|x_1\|_X \leq r$ *and* $\|x_2\|_X \leq r$, *the Lipschitz condition*

$$(18.44) \quad \|N^*(x_1) - N^*(x_2)\|_{\mathfrak{R}_n(X)} \leq \mu(\tilde{N}^{**}; \gamma(r))\|x_1 - x_2\|_X.$$

□ Let us prove (18.42), the estimates (18.43) and (18.44) are proved in the same way. For any $x_1, x_2 \in X$ with $\|x_1\|_X, \|x_2\|_X \leq r$ we have, by (18.30),

$$\|L^*(x_1) - L^*(x_2)\|_{\mathfrak{R}_l(X)}$$

$$= \sup_{\|h\|_X \leq 1} \|(t,s) \mapsto \int_T |l_1(t,s,\tau,x_1(\tau,s))$$

$$-l_1(t,s,\tau,x_2(\tau,s))| \, |h(\tau,s)| \, d\tau\|_X$$

$$\leq \sup_{\|h\|_X \leq 1} \|(t,s) \mapsto \int_T |\tilde{l}_2(t,s,\tau,|x_1(\tau,s)| \vee |x_2(\tau,s)|))| \times$$

$$\times |x_1(\tau,s) - x_2(\tau,s)| \, |h(\tau,s)| \, d\tau\|_X.$$

By the Krasnosel'skij-Ladyzhenskij lemma (see e.g. KRASNOSEL'SKIJ-ZABREJKO-PUSTYL'NIK-SOBOLEVSKIJ [1976]) there exists a function $w: T \times S \times T \to \mathbb{R}$ such that $|w(t,s,\tau)| \leq |x_1(\tau,s)| \vee |x_2(\tau,s)|$ and

$$|l_2(t,s,\tau,w(t,s,\tau))| = \tilde{l}_2(t,s,\tau,|x_1(\tau,s)| \vee |x_2(\tau,s)|).$$

Thus,

$$\|L^*(x_1) - L^*(x_2)\|_{\mathfrak{R}_l(X)} \le$$

$$\sup_{\|h\|_X \le 1} \| \int_T |l_2(t, s, \tau, w(t, s, \tau))| \, |x_1(\tau, s) - x_2(\tau, s)| \, |h(\tau, s)| \, d\tau \|_X.$$

This implies (18.42), by the definition of $\gamma(r)$ and $\mu(\tilde{L}^{**}; \gamma(r))$. ∎

18.4. The case $X = C(T \times S)$

Lemma 18.3 is in many cases sufficient to find the constants a and b and the function $k = k(r)$ in the space C explicitly. Consider the scalar functions

$$k_l(r) :=$$

(18.45)
$$\sup_{(t,s) \in T \times S} \int_T \sup_{|u_1|, |u_2| \le r} \frac{|l_1(t, s, \tau, u_1) - l_1(t, s, \tau, u_2)|}{|u_1 - u_2|} \, d\tau,$$

$$k_m(r) :=$$

(18.46)
$$\sup_{(t,s) \in T \times S} \int_S \sup_{|u_1|, |u_2| \le r} \frac{|m_1(t, s, \sigma, u_1) - m_1(t, s, \sigma, u_2)|}{|u_1 - u_2|} \, d\sigma,$$

and

(18.47)
$$k_n(r) :=$$

$$\sup_{(t,s) \in T \times S} \int_T \int_S \sup_{|u_1|, |u_2| \le r} \frac{|n_1(t, s, \tau, \sigma, u_1) - n_1(t, s, \tau, \sigma, u_2)|}{|u_1 - u_2|} \, d\sigma \, d\tau.$$

These functions are finite for $r \le R$ if and only if the operators (18.13) - (18.15) satisfy a local Lipschitz condition in C. Moreover, the numbers given in (18.45) - (18.47) are then the minimal Lipschitz constants for the corresponding operators on $\overline{B}_r(C)$. This may be stated in a more precise and convenient way:

Theorem 18.1. *The operators* (18.22) - (18.24) *satisfy a Lipschitz condition on $\overline{B}_R(C)$ if and only if the three kernels in* (18.1) *have second partial derivatives in the last argument*

$$l_2(t, s, \tau, u) = \frac{\partial^2 l(t, s, \tau, u)}{\partial u^2},$$

$$m_2(t, s, \sigma, u) = \frac{\partial^2 m(t, s, \sigma, u)}{\partial u^2},$$

and

$$n_2(t, s, \tau, \sigma, u) = \frac{\partial^2 n(t, s, \tau, \sigma, u)}{\partial u^2}$$

for all $(t, s) \in T \times S$ *and almost all* $(\tau, u) \in T \times \mathbb{R}$, $(\sigma, u) \in S \times \mathbb{R}$, *and* $(\tau, \sigma, u) \in T \times S \times \mathbb{R}$, *respectively, and the three functions*

(18.48) $$\tilde{k}_l(r) := \sup_{(t,s) \in T \times S} \int_T \sup_{|u| \leq r} |l_2(t, s, \tau, u)| \, d\tau,$$

(18.49) $$\tilde{k}_m(r) := \sup_{(t,s) \in T \times S} \int_S \sup_{|u| \leq r} |m_2(t, s, \sigma, u)| \, d\sigma,$$

and

(18.50) $$\tilde{k}_n(r) := \sup_{(t,s) \in T \times S} \int_T \int_S \sup_{|u| \leq r} |n_2(t, s, \tau, \sigma, u)| \, d\sigma \, d\tau$$

are finite for $r \leq R$. *Moreover, the numbers* $\tilde{k}_l(r)$, $\tilde{k}_m(r)$, *and* $\tilde{k}_n(r)$ *are then the minimal Lipschitz constants for the operators* (18.22), (18.23), *and* (18.24), *respectively, on* $\overline{B}_r(C)$. *Finally, the minimal Lipschitz constant* $k(r)$ *for the operator* (18.25) *on* $\overline{B}_r(C)$ *satisfies the two-sided estimate*

$$\max \{\tilde{k}_l(r), \tilde{k}_m(r), \tilde{k}_n(r)\} \leq k(r)$$

(18.51)

$$\leq \tilde{k}_l(r) + \tilde{k}_m(r) + \tilde{k}_n(r).$$

□ The proof for the integral operator (18.24) is contained in APPELL-
DE PASCALE-ZABREJKO [1991]. Let us prove the assertion for the
partial integral operator (18.22), the proof for (18.23) is similar. By
what has been observed before, for this it is necessary and sufficient
to show that the function (18.45) is finite for $r \leq R$.

Suppose first that (18.45) is finite for all $r \leq R$. This means that

$$\int_T \sup_{|u_1|,|u_2| \leq r} \frac{|l_1(t,s,\tau,u_1) - l_1(t,s,\tau,u_2)|}{|u_1 - u_2|} d\tau \leq k_l(r) < \infty$$

for all $(t,s) \in T \times S$, and hence the function $\lambda_{t,s}$ given by

$$\lambda_{t,s}(\tau) := \sup_{|u_1|,|u_2| \leq r} \frac{|l_1(t,s,\tau,u_1) - l_1(t,s,\tau,u_2)|}{|u_1 - u_2|}$$

is finite almost everywhere on T. Since $|l_1(t,s,\tau,u_1) - l_1(t,s,\tau,u_2)| \leq \lambda_{t,s}(\tau)|u_1 - u_2|$ for $|u_1|, |u_2| \leq r$, the map $u \mapsto l_1(t,s,\tau,u)$ is absolutely
continuous. Consequently, the partial derivative $l_2 = \partial l_1 / \partial u$ exists for
almost all u and satisfies

$$\sup_{|u| \leq r} |l_2(t,s,\tau,u)| \leq \lambda_{t,s}(\tau).$$

But this implies that

$$\tilde{k}_l(r) \leq \sup_{(t,s) \in T \times S} \int_T \lambda_{t,s}(\tau) d\tau < \infty.$$

Conversely, suppose that (18.48) is finite for $r \leq R$. This implies that
the function $\tilde{\lambda}_{t,s}$ given by

$$\tilde{\lambda}_{t,s}(\tau) := \sup_{|u| \leq r} |l_2(t,s,\tau,u)|$$

is finite almost everywhere on T, for all $(t,s) \in T \times S$. Consequently,
for $|u_1|, |u_2| \leq r$ we have

$$|l_1(t,s,\tau,u_1) - l_1(t,s,\tau,u_2)| =$$

$$\left| \int_{u_2}^{u_1} l_2(t,s,\tau,u) du \right| \leq \tilde{\lambda}_{t,s}(\tau)|u_1 - u_2|.$$

We conclude that

$$k_l(r) \leq \sup_{(t,s)\in T\times S} \int_T \tilde{\lambda}_{t,s}(\tau)\,d\tau < \infty.$$

Of course, the proof shows also that $\tilde{k}_l(r) = k_l(r)$ for all $r \leq R$. ∎

Theorem 18.1 implies, in particular, that the estimate $k(r) \leq \tilde{k}_l(r) + \tilde{k}_m(r) + \tilde{k}_n(r)$ holds for the Lipschitz constant in (18.4) in case $X = C$. The problem of calculating the numbers a in (18.5) and b in (18.6) is quite easy. In fact, suppose that the partial integral equation (18.26) has a unique solution (18.28) in the space $X = C$. From the definition of the norm in the space C we obtain then the equality

$$\begin{aligned}
a = \sup_{(t,s)\in T\times S} \Bigg| & g(t,s) + \int_T r_l(t,s,\tau)g(\tau,s)\,d\tau \\
& + \int_S r_m(t,s,\sigma)g(t,\sigma)\,d\sigma \\
& + \int_T \int_S r_n(t,s,\tau,\sigma)g(\tau,\sigma)\,d\sigma\,d\tau \Bigg|,
\end{aligned}$$

(18.52)

and from explicit formulas for the norm of a linear partial integral operator in the space C (see § 13) the equality

(18.53)
$$\begin{aligned}
b = 1 + \sup_{(t,s)\in T\times S} \Bigg[& \int_T |r_l(t,s,\tau)|\,d\tau \\
& + \int_S |r_m(t,s,\sigma)|\,d\sigma + \int_T \int_S |r_n(t,s,\tau,\sigma)|\,d\sigma\,d\tau \Bigg].
\end{aligned}$$

Here g is defined by (18.27), and r_l, r_m, and r_n are the resolvent kernels corresponding to l_1, m_1, and n_1, respectively.

The resolvent kernels r_l, r_m, and r_n are in general difficult to compute explicitly. An exceptional case is that of degenerate kernels; we illustrate this by means of a very elementary example.

Example 18.1. Let $S = T = [0,1]$, $l(t,s,\tau,u) = \lambda(u)$, $m(t,s,\sigma,u) = \mu(u)$, and $n(t,s,\tau,\sigma,u)$

$\equiv 0$. Here λ and μ are real C^2-functions with $\lambda'(0) \neq 1, \mu'(0) \neq 1$, and $\lambda'(0) + \mu'(0) \neq 1$. For any $g \in C([0,1] \times [0,1])$ the equation (18.26) has then the unique solution

$$h(t,s) = g(t,s) + \frac{\lambda'(0)}{1 - \lambda'(0)} \int_0^1 g(\tau,s)\, d\tau$$

$$+ \frac{\mu'(0)}{1 - \mu'(0)} \int_0^1 g(t,\sigma)\, d\sigma$$

$$+ \frac{\lambda'(0)\mu'(0)(2 - \lambda'(0) - \mu'(0))}{(1 - \lambda'(0))(1 - \mu'(0))(1 - \lambda'(0) - \mu'(0))} \int_0^1 \int_0^1 g(\tau,\sigma)\, d\sigma\, d\tau.$$

In particular, since $g(t,s) \equiv -[\lambda(0) + \mu(0)]$ in this case, we get here the constant solution $h(t,s) \equiv g(t,s)/(1 - \lambda'(0) - \mu'(0))$. Putting this into (18.52) and (18.53) yields

$$a = \left| \frac{\lambda(0) + \mu(0)}{1 - \lambda'(0) - \mu'(0)} \right|$$

and

$$b = 1 + \left| \frac{\lambda'(0)}{1 - \lambda'(0)} \right| + \left| \frac{\mu'(0)}{1 - \mu'(0)} \right|$$

$$+ \left| \frac{\lambda'(0)\mu'(0)(2 - \lambda'(0) - \mu'(0))}{(1 - \lambda'(0))(1 - \mu'(0))(1 - \lambda'(0) - \mu'(0))} \right|.$$

The function $k = k(r)$ from (18.4) may in turn be estimated by

$$k(r) \leq \sup_{|u| \leq r} |\lambda''(u)| + \sup_{|u| \leq r} |\mu''(u)|.$$

This gives a sufficiently effective "recipe" for finding the scalar function (18.7), and hence for applying the Newton-Kantorovich method to equation (18.1) in this special case. To be more specific, suppose that the functions λ and μ are quadratic polynomials

(18.54) $\lambda(u) = \lambda_2 u^2 + \lambda_1 u + \lambda_0, \qquad \mu(u) = \mu_2 u^2 + \mu_1 u + \mu_0$

which is the simplest nonlinear case. A trivial calculation shows that then

$$a = \left| \frac{\lambda_0 + \mu_0}{1 - \lambda_1 - \mu_1} \right|,$$

$$b = 1 + \left| \frac{\lambda_1}{1 - \lambda_1} \right| + \left| \frac{\mu_1}{1 - \mu_1} \right| + \left| \frac{\lambda_1 \mu_1 (2 - \lambda_1 - \mu_1)}{(1 - \lambda_1)(1 - \mu_1)(1 - \lambda_1 - \mu_1)} \right|,$$

and

$$k(r) = 2(|\lambda_2| + |\mu_2|),$$

hence $\varphi(r) = a + cr^2$, where $c = b(|\lambda_2| + |\mu_2|)$. Consequently, the number of real solutions of the fixed point equation (18.8) depends on the sign of the discriminant $D = 1 - 4ac$.

We remark that this effective calculation also applies to the more general case

$$l(t, s, \tau, u) = a(t)b(s)c(\tau)\lambda(u), \quad m(t, s, \sigma, u) = d(t)e(s)f(\sigma)\mu(u),$$

and also to the case of degenerate kernels.

18.5. The case $X = L_\infty(T \times S)$

In rather the same way as in $X = C(T \times S)$, the Lipschitz conditions for the operators (18.25) and (18.22) - (18.24) are also equivalent in the space $X = L_\infty(T \times S)$. This may again be analyzed by imposing appropriate conditions on the kernels k_1, m_1, and n_1, and the corresponding operators (18.13) - (18.15).

For $r > 0$ and $\delta > 0$, let

$$k_l(r, \delta) :=$$

(18.55)
$$\operatorname*{ess\,sup}_{(t,s) \in T \times S} \int_T \sup_{\substack{|u_1|,|u_2| \le r \\ |u_1 - u_2| \le \delta}} |l_1(t, s, \tau, u_1) - l_1(t, s, \tau, u_2)| \, d\tau,$$

$$k_m(r, \delta) :=$$

(18.56)
$$\operatorname*{ess\,sup}_{(t,s) \in T \times S} \int_S \sup_{\substack{|u_1|,|u_2| \le r \\ |u_1 - u_2| \le \delta}} |m_1(t, s, \sigma, u_1) - m_1(t, s, \sigma, u_2)| \, d\sigma,$$

and

(18.57)
$$k_n(r, \delta) :=$$

$$\operatorname*{ess\,sup}_{(t,s)\in T\times S} \int_T \int_S \sup_{\substack{|u_1|,|u_2|\leq r \\ |u_1-u_2|\leq\delta}} |n_1(t, s, \tau, \sigma, u_1) - n_1(t, s, \tau, \sigma, u_2)| \, d\sigma \, d\tau.$$

Lemma 18.4. *The following three conditions are equivalent:*

(a) *the limits*

(18.58)
$$\begin{cases} k_l(r) := \lim_{\delta\to 0} \frac{k_l(r, \delta)}{\delta}, \\[2ex] k_m(r) := \lim_{\delta\to 0} \frac{k_m(r, \delta)}{\delta}, \\[2ex] k_n(r) := \lim_{\delta\to 0} \frac{k_n(r, \delta)}{\delta} \end{cases}$$

are finite for $r \leq R$;

(b) *the operators* (18.22) - (18.24) *satisfy a Lipschitz condition from $B_R(L_\infty)$ into $\mathfrak{L}(L_\infty)$;*

(c) *the operator* (18.25) *satisfies a Lipschitz condition from $B_R(L_\infty)$ into $\mathfrak{L}(L_\infty)$.*

□   We prove this again for the function (18.55) and the operator (18.22). Suppose that (a) holds. For $\varepsilon > 0$ choose $\hat\delta > 0$ such that $k_l(r, \delta) \leq (k_l(r) + \varepsilon)\delta$ for $0 < \delta \leq \hat\delta$. By the definition (18.55) of $k_l(r, \delta)$ we get

$$\|L_1(x_1)h - L_1(x_2)h\|_{L_\infty} \leq k_l(r, \delta)\|x_1 - x_2\|_{L_\infty}$$

for $\|h\| \leq 1$, $\|x_1\|, \|x_2\| \leq r$, and $\|x_1 - x_2\| \leq \delta$. In fact, for $\|x_1 -$

$x_2|| < \delta' \leq \delta$ and fixed $(t,s) \in T \times S$ we have

$$\left| \int_T [l_1(t,s,\tau,x_1(\tau,s)) - l_1(t,s,\tau,x_2(\tau,s))]h(\tau,s)\,d\tau \right|$$

$$\leq \delta' \int_{T(x_1,x_2)} \frac{|l_1(t,s,\tau,x_1(\tau,s)) - l_1(t,s,\tau,x_2(\tau,s))|}{|x_1(\tau,s) - x_2(\tau,s)|}\,d\tau$$

$$\leq \delta' \int_T \sup_{\substack{|u_1|,|u_2| \leq r \\ |u_1-u_2| \leq \delta}} \frac{|l_1(t,s,\tau,u_1) - l_1(t,s,\tau,u_2)|}{|u_1 - u_2|}\,d\tau,$$

where we have put $T(x_1,x_2) := \{\tau : x_1(\tau,s) \neq x_2(\tau,s)\}$. Since $\delta' > ||x_1 - x_2||$ is arbitrary, this implies that

$$||L_1(x_1) - L_1(x_2)||_{\mathfrak{L}(L_\infty)} \leq k_l(r,\delta)||x_1 - x_2||_{L_\infty}$$

for $||x_1 - x_2|| \leq \delta$. Now, for arbitrary $x_1, x_2 \in L_\infty$, fix $m \in \mathbb{N}$ such that $||x_1 - x_2|| \leq m\delta$. Then

$$||L_1(x_1) - L_1(x_2)||_{\mathfrak{L}(L_\infty)} \leq$$

$$\sum_{j=1}^m ||L_1[(1 - \tfrac{j}{m})x_1 + \tfrac{j}{m}x_2] - L_1[(1 - \tfrac{j-1}{m})x_1 + \tfrac{j-1}{m}x_2]||_{\mathfrak{L}(L_\infty)}$$

$$\leq m(k_l(r) + \varepsilon)\frac{||x_1 - x_2||}{m} = (k_l(r) + \varepsilon)||x_1 - x_2||,$$

and hence (b) is true. Conversely, suppose that (b) holds. As was shown in ZABREJKO-ZLEPKO [1983] (see also APPELL-DE PASCALE-ZABREJKO [1991]), the equality

$$\sup_{\substack{||x_1||,||x_2|| \leq r \\ ||x_1-x_2|| \leq \delta}} |L_1(x_1)h(t,s) - L_1(x_2)h(t,s)|$$

$$= \int_T \sup_{\substack{|u_1|,|u_2| \leq r \\ |u_1-u_2| \leq \delta}} |l_1(t,s,\tau,u_1) - l_1(t,s,\tau,u_2)|\,d\tau$$

holds in the space L_∞, and (a) follows by taking L_∞-norms. The equivalence of (b) and (c) is clear. ∎

The following is parallel to Theorem 18.1.

Theorem 18.2. *The operators* (18.22) - (18.24) *satisfy a Lipschitz condition on* $\overline{B}_R(L_\infty)$ *if and only if the three kernels in* (18.1) *have second partial derivatives in the last argument*

$$l_2(t, s, \tau, u) = \frac{\partial^2 l(t, s, \tau, u)}{\partial u^2},$$

$$m_2(t, s, \sigma, u) = \frac{\partial^2 m(t, s, \sigma, u)}{\partial u^2}$$

and

$$n_2(t, s, \tau, \sigma, u) = \frac{\partial^2 n(t, s, \tau, \sigma, u)}{\partial u^2}$$

for all $(t, s) \in T \times S$ *and almost all* $(\tau, u) \in T \times \mathbb{R}$, $(\sigma, u) \in S \times \mathbb{R}$, *and* $(\tau, \sigma, u) \in T \times S \times \mathbb{R}$, *respectively, and the three functions*

$$(18.59) \qquad \tilde{k}_l(r) := \operatorname*{ess\,sup}_{(t,s)\in T\times S} \int_T \sup_{|u|\le r} |l_2(t, s, \tau, u)| \, d\tau,$$

$$(18.60) \qquad \tilde{k}_m(r) := \operatorname*{ess\,sup}_{(t,s)\in T\times S} \int_S \sup_{|u|\le r} |m_2(t, s, \sigma, u)| \, d\sigma,$$

and

$$(18.61) \qquad \tilde{k}_n(r) := \operatorname*{ess\,sup}_{(t,s)\in T\times S} \int_T \int_S \sup_{|u|\le r} |n_2(t, s, \tau, \sigma, u)| \, d\sigma \, d\tau$$

are finite for $r \le R$. *Moreover, the numbers* $\tilde{k}_l(r)$, $\tilde{k}_m(r)$, *and* $\tilde{k}_n(r)$ *are then the minimal Lipschitz constants for the operators* (18.22), (18.23), *and* (18.24), *respectively, on* $\overline{B}_r(L_\infty)$. *Finally, the minimal Lipschitz constant* $k(r)$ *for the operator* (18.25) *on* $\overline{B}_r(L_\infty)$ *satisfies the two-sided estimate* (18.51).

The example of the operator

$$Fx(t, s) := x(t, s) - \int_0^1 tsx^2(\tau, s) \, d\tau - \int_0^1 (1 - t)(1 - s)x^2(t, \sigma) \, d\sigma$$

shows that, in general, the equality

$$k(r) = \tilde{k}_l(r) + \tilde{k}_m(r) + \tilde{k}_n(r)$$

is not true.

Theorem 18.2 gives an effective algorithm for estimating the function (18.4) in the space $X = L_\infty$. Analogously to what we have done in the preceding subsection for $X = C$, we may calculate the numbers (18.5) and (18.6) in the space $X = L_\infty$. The following theorem follows from the definition of the norm in L_∞ and formula (12.31):

Theorem 18.3. *Suppose that the equation* (18.26) *has a unique solution* (18.28) *in the space* $X = L_\infty$. *Then the numbers* (18.5) *and* (18.6) *may be calculated in* L_∞ *by means of the formulas*

(18.62)
$$\begin{aligned} a = \; & \operatorname*{ess\,sup}_{(t,s)\in T\times S} |g(t,s) + \int_T r_l(t,s,\tau)g(\tau,s)\,d\tau \\ & + \int_S r_m(t,s,\sigma)g(t,\sigma)\,d\sigma \\ & + \int_T\int_S r_n(t,s,\tau,\sigma)g(\tau,\sigma)\,d\sigma\,d\tau|, \end{aligned}$$

and

(18.63)
$$\begin{aligned} b = 1 + \; & \operatorname*{ess\,sup}_{(t,s)\in T\times S} \bigg[\int_T |r_l(t,s,\tau)|\,d\tau \\ & + \int_S |r_m(t,s,\sigma)|\,d\sigma + \int_T\int_S |r_n(t,s,\tau,\sigma)|\,d\sigma\,d\tau \bigg]. \end{aligned}$$

Here g is defined by (18.27), *and r_l, r_m, and r_n are the resolvent kernels corresponding to l_1, m_1, and n_1, respectively.*

We illustrate the results of this subsection again by means of Example 18.1. The constants a and b may be calculated precisely as in the space $X = C$. The functions (18.59) - (18.61) have the form $\tilde{k}_l(r) = \sup\{|\lambda''(u)| : |u| \le r\}$, $\tilde{k}_m(r) = \sup\{|\mu''(u)| : |u| \le r\}$, and $\tilde{k}_n(r) \equiv$

0. For the polynomials (54) this gives, in particular, $\tilde{k}_l(r) \equiv 2\lambda_2$ and $\tilde{k}_m(r) \equiv 2\mu_2$.

18.6. The case $X = L_p(T \times S)$ $(1 \le p < \infty)$

The analysis of the preceding two subsections becomes more difficult when passing to the case of the Lebesgue space L_p with $1 \le p < \infty$. One reason for this is that the unit ball in L_p contains lots of unbounded functions, and therefore one "cannot get rid of the function h" under the integrals in the right-hand sides of (18.16) - (18.18). But this is not just a technical problem: in fact, imposing a Lipschitz condition like (18.4) in L_p may lead to a strong degeneracy of the kernel functions involved! For the integral operator (18.11), for example, it was shown in APPELL-DE PASCALE-ZABREJKO [1991] that the derivative N' satisfies a Lipschitz condition in L_2 only if the corresponding kernel n_1 satisfies a Lipschitz condition in u, and in L_p for $1 \le p < 2$ only if n_1 does not depend on u, i.e. the kernel n is *linear* in u.

We shall show now that the situation is even worse for the partial integral operators (18.9) and (18.10): a Lipschitz condition for the derivatives L' and M' necessarily leads to linear kernels *for all values of p!*

Theorem 18.4. *The derivatives of the operators* (18.9) *and* (18.10) *satisfy a Lipschitz condition in* $X = L_p(T \times S)$ $(1 \le p < \infty)$ *if and only if the corresponding kernels l and m are linear in the last argument.*

□ We prove the assertion for the operator L' or, what is equivalent by Lemma 18.1, for the operator L_1 given by (18.22). Of course, if the kernel l is linear in u, the kernel l_1 is independent of u, and there is nothing to prove. Suppose that the operator (18.22) satisfies a Lipschitz condition in L_p, i.e.

$$
\int_S \int_T \left| \int_T [l_1(t, s, \tau, x_1(\tau, s)) \right.
$$

(18.64)

$$
\left. - l_1(t, s, \tau, x_2(\tau, s))] h(\tau, s) \, d\tau \right|^p \, dt \, ds
$$

$$
\leq k_l^p(r) \|x_1 - x_2\|^p \|h\|^p \qquad (\|x_1\|, \|x_2\| \leq r),
$$

where all norms are taken in $L_p(T \times S)$. Choosing, in particular, $x_i(t, s) := u_i \chi_D(t) \chi_E(s)$ and $h(t, s) := \chi_D(t) \chi_E(s)$, where $D \subset T$ and $E \subset S$ satisfy $u_i \operatorname{mes} D \operatorname{mes} E \leq r^p$ ($i = 1, 2$), and putting this into (18.64) yields

(18.65)

$$
\int_E \int_T \left| \int_D [l_1(t, s, \tau, u_1) - l_1(t, s, \tau, u_2)] \, d\tau \right|^p \, dt \, ds
$$

$$
\leq k_l^p(r) |u_1 - u_2|^p (\operatorname{mes} D)^2 (\operatorname{mes} E)^2.
$$

Dividing by $\operatorname{mes} E$ and letting $\operatorname{mes} E$ in (18.65) tend to zero, we get

(18.66)

$$
\int_T \left| \int_D [l_1(t, s, \tau, u_1) - l_1(t, s, \tau, u_2)] \, d\tau \right|^p \, dt = 0
$$

for almost all $s \in E$. From (18.67) it follows in turn that

$$
\int_D [l_1(t, s, \tau, u_1) - l_1(t, s, \tau, u_2)] \, d\tau = 0
$$

for almost all $(t, s) \in T \times E$. Since D is an arbitrary measurable set, we conclude that $l_1(t, s, \tau, u_1) - l_1(t, s, \tau, u_2) = 0$ for almost all $(t, s, \tau) \in T \times E \times D$, and the assertion follows. ∎

Theorem 18.4 is, of course, rather disappointing: the Newton-Kantorovich method applies to equation (18.1) in L_p ($1 \leq p < \infty$) only if the kernels l and m are linear in u. Only for the kernel n we have a larger choice in L_p, provided that $p \geq 2$. Taking into account this degeneracy, we close with another example in $X = L_2$.

Example 18.2. Let $S = T = [0, 1]$ and $p = 2$. By what has been observed before, this choice of p forces us to choose l and m linear in u.

For example, let $l(t, s, \tau, u) = \lambda_1(t)\lambda_2(s)u + \lambda_0(t, s, \tau)$, $m(t, s, \sigma, u) = \mu_1(t)\mu_2(s)u + \mu_0(t, s, \sigma)$, and $n(t, s, \tau, \sigma, u) \equiv 0$. The function (18.27) is here

$$g(t, s) = - \int_0^1 \lambda_0(t, s, \tau)\, d\tau - \int_0^1 \mu_0(t, s, \sigma)\, d\sigma,$$

and the equation (18.26) for h takes the form

(18.67) $h(t, s) = \lambda_1(t)\lambda_2(s)\phi(s) + \mu_1(t)\mu_2(s)\psi(t) + g(t, s),$

where we have put

(18.68) $\phi(s) := \int_0^1 h(\tau, s)\, d\tau, \qquad \psi(t) := \int_0^1 h(t, \sigma)\, d\sigma.$

Inserting (18.67) into (18.68) we arrive at the system

$$
\begin{cases}
\phi(s) = \displaystyle\int_0^1 \lambda_1(\tau)\lambda_2(s)\phi(s)\, d\tau \\[2mm]
\qquad + \displaystyle\int_0^1 \mu_1(\tau)\mu_2(s)\psi(\tau)\, d\tau + \int_0^1 g(\tau, s)\, d\tau, \\[4mm]
\psi(t) = \displaystyle\int_0^1 \lambda_1(t)\lambda_2(\sigma)\phi(\sigma)\, d\sigma \\[2mm]
\qquad + \displaystyle\int_0^1 \mu_1(t)\mu_2(\sigma)\psi(t)\, d\sigma + \int_0^1 g(t, \sigma)\, d\sigma
\end{cases}
$$

for the unknown functions ϕ and ψ. If we suppose that

$$\alpha(s) := 1 - \lambda_2(s)\int_0^1 \lambda_1(\tau)\, d\tau \neq 0$$

and

$$\beta(t) := 1 - \mu_1(t)\int_0^1 \mu_2(\sigma)\, d\sigma \neq 0,$$

and put

$$\gamma(s) := \int_0^1 g(\tau, s)\, d\tau, \quad \delta(t) := \int_0^1 g(t, \sigma)\, d\sigma,$$

we end up at a system of two scalar equations

$$
\begin{cases}
-\left(\displaystyle\int_0^1 \frac{\lambda_1(\tau)\mu_1(\tau)}{\beta(\tau)}\, d\tau\right)\xi + \eta = \displaystyle\int_0^1 \frac{\lambda_1(\tau)\delta(\tau)}{\beta(\tau)}\, d\tau \\[2ex]
\xi - \left(\displaystyle\int_0^1 \frac{\lambda_2(\sigma)\mu_2(\sigma)}{\alpha(\sigma)}\, d\sigma\right)\eta = \displaystyle\int_0^1 \frac{\lambda_2(\sigma)\gamma(\sigma)}{\alpha(\sigma)}\, d\sigma
\end{cases}
$$

for the unknown real numbers

$$
\xi := \int_0^1 \lambda_2(\sigma)\phi(\sigma)\, d\sigma, \qquad \eta := \int_0^1 \mu_1(\tau)\psi(\tau)\, d\tau.
$$

The last system has a unique solution $(\xi, \eta) \in \mathbb{R}^2$ if and only if

$$
\left(\int_0^1 \frac{\lambda_1(\tau)\mu_1(\tau)}{\beta(\tau)}\, d\tau\right)\left(\int_0^1 \frac{\lambda_2(\sigma)\mu_2(\sigma)}{\alpha(\sigma)}\, d\sigma\right) \neq 1,
$$

and this solution may be used to find $(\phi(s), \psi(t))$ and, consequently, the solution $h(t, s)$ of (18.26).

In this way, we may find the numbers (18.5) and (18.6) by means of well-known upper estimates for the L_2-norm of a linear integral operator. The function (18.4) is very easy to compute in this case, since $l_1(t, s, \tau, u) = \lambda_1(t)\lambda_2(s)$ and $m_1(t, s, \sigma, u) = \mu_1(t)\mu_2(s)$ do not depend on u.

§ 19. Applications of Barbashin equations

In this section we describe various problems whose mathematical modelling leads to integro-differential equations of Barbashin type. Such problems arise, for example, in probability theory, in the theory of elementary particles, in transport problems, in kinetic gas theory, in radiation propagation problems, and in some problems of mathematical biology.

19.1. Applications to probability theory

The mathematical description of some problems in physics and engineering leads to systems in which certain changes of states occur in "jumps", i.e. discontinuously. Some of these discontinuities are related to *random processes*. In what follows, we will study such processes without "after-effect", assuming that the *set of possible states* of the system is $\mathbb{R}$. In this case a random process is simply a *set of random variables* $\xi(t)$ $(-\infty < t < \infty)$. In this setting, the variable t is usually considered as time, and the variable $\xi(t)$ is interpreted as state of the system at the moment t.

To describe a process without after-effect, one usually introduces a function $F = F(t, x; \tau, y)$ which represents the probability that the random variable $\xi(\tau)$ assumes the value y at time τ if $\xi(t)$ assumes the value x at time $t < \tau$. The absence of an after-effect means that the knowledge of the state of the system at times $t' < t$ has no influence on the function $F(t, x; \tau, y)$ or, lossely speaking, that we are dealing with a system "without memory".

In this subsection we follow the paper GNEDENKO [1988]. We assume throughout that the function $y \mapsto F(t, x; \tau, y)$ is left-continuous, and that

(19.1) $\displaystyle\lim_{y \to -\infty} F(t, x; \tau, y) = 0, \quad \lim_{y \to +\infty} F(t, x; \tau, y) = 1.$

If, in addition, the function $y \mapsto F(t, x; \tau, y)$ is even continuous then the *generalized Markov equation*

(19.2) $\displaystyle F(t, x; \tau, y) = \int_{-\infty}^{+\infty} F(s, z; \tau, y) \, d_z F(t, x; s, z)$

holds true. Observe that the function F is defined so far only for $\tau > t$; for $\tau = t$ we set

$$\lim_{\tau \to t+0} F(t, x; \tau, y) = \lim_{t \to \tau - 0} F(t, x; \tau, y)$$

$$=: E(x, y) = \begin{cases} 0 & \text{if } y \le x, \\ 1 & \text{if } y > x. \end{cases}$$

If the function $(t, x; \tau) \mapsto F(t, x; \tau, y)$ is even differentiable the generalized Markov equation (19.2) takes the form

$$f(t, x; \tau, y) = \int_{-\infty}^{+\infty} f(s, z; \tau, y) f(t, x; s, z) \, dz,$$

where the density f satisfies the relations

$$\int_{-\infty}^{y} f(t, x; \tau, z) \, dz = F(t, x; \tau, y), \quad \int_{-\infty}^{+\infty} f(t, x; \tau, y) \, dy = 1.$$

Suppose that the random process is "purely discontinuous", i.e. on every time interval $(t, t + \Delta t)$ one has $\xi(t) = x$ with probability $1 - P(t, x)\Delta t + o(\Delta t)$, and $\xi(t) \ne x$ with probability $P(t, x)\Delta t - o(\Delta t)$. In addition, we suppose that the probability for more than one change of $\xi(t)$ on $(t, t + \Delta t)$ is also $o(\Delta t)$.

By $P = P(t, x, y)$ we denote the *conditional distribution function* for $\xi(t)$ under the hypothesis that a jump occurs at time t, and $\lim_{\tau \to t-0} \xi(\tau) = x$ (i.e. the value of $\xi(t)$ "immediately before t" was x). In this case we have

$$\begin{aligned} F(t, x; \tau, y) &= [1 - P(t, x)(\tau - t)] E(x, y) \\ &\quad + (\tau - t) P(t, x) P(t, x, y) + o(\tau - t). \end{aligned}$$

(19.3)

By definition, the functions $(t, x) \mapsto P(t, x)$ and $(t, x, y) \mapsto P(t, x, y)$ are nonnegative, and the latter function satisfies

(19.4) $$\lim_{y \to -\infty} P(t, x, y) = 0, \quad \lim_{y \to +\infty} P(t, x, y) = 1.$$

Let us suppose now that the functions $(t, x) \mapsto P(t, x)$ and $(t, x, y) \mapsto P(t, x, y)$ are bounded, the functions $t \mapsto P(t, x)$ and $t \mapsto P(t, x, y)$

are continuous, and the functions $x \mapsto P(t, x)$ and $(x, y) \mapsto P(t, x, y)$ are Borel measurable.

Theorem 19.1. *The distribution function F of a purely discontinuous process without after-effect satisfies the integro-differential equations*

(19.5)
$$\frac{\partial F(t, x; \tau, y)}{\partial t} = P(t, x) F(t, x; \tau, y)$$
$$- \int_{-\infty}^{+\infty} P(t, x) F(t, z; \tau, y) \, d_z P(t, x, z)$$

and

(19.6)
$$\frac{\partial F(t, x; \tau, y)}{\partial \tau} = - \int_{-\infty}^{y} P(t, z) \, d_z F(t, x; \tau, z)$$
$$+ \int_{-\infty}^{+\infty} P(\tau, x) P(\tau, z, y) \, d_z F(t, x; \tau, z).$$

□ By the generalized Markov equations we have

$$F(t, x; \tau, y) = \int_{-\infty}^{+\infty} F(t + \Delta t, z; \tau, y) \, d_z F(t, x; t + \Delta t, z).$$

Replacing in this formula the function $F(t, x; t + \Delta t, z)$ by the right-hand side of (19.3) we get

$$F(t, x; \tau, y) = \int_{-\infty}^{+\infty} F(t + \Delta t, z; \tau, y) \, d_z[1 - P(t, x)\Delta t] E(x, z)$$

$$+ \int_{-\infty}^{+\infty} F(t + \Delta t, z; \tau, y) \, d_z P(t, x) P(t, x, z) \Delta t$$

$$+ \int_{-\infty}^{+\infty} F(t + \Delta t, z; \tau, y) \, d_z o(\Delta t).$$

Since

$$\int_{-\infty}^{+\infty} F(t + \Delta t, z; \tau, y) \, d_z E(x, z) = F(t + \Delta t, x; \tau, y),$$

we further get

$$F(t, x; \tau, y) = [1 - P(t, x)\Delta t] F(t + \Delta t, x; \tau, y)$$

$$+ P(t, x)\Delta t \int_{-\infty}^{+\infty} F(t + \Delta t, z; \tau, y) \, d_z P(t, x, z) + o(\Delta t).$$

Consequently,

$$\frac{F(t + \Delta t, x; \tau, y) - F(t, x; \tau, y)}{\Delta t}$$

$$= P(t, x) F(t + \Delta t; \tau, y)$$

$$- P(t, x) \int_{-\infty}^{+\infty} F(t + \Delta t, z; \tau, y) \, d_z P(t, x, z) + \frac{o(\Delta t)}{\Delta t}$$

which proves (19.5) by passing to the limit $\Delta t \to 0$.

Using the generalized Markov equation (19.2), the relation (19.3), and the definition of $E(x, y)$ we obtain

$$F(t, x; \tau + \Delta\tau, y) = \int_{-\infty}^{+\infty} F(\tau, z; \tau + \Delta\tau, y) \, d_z F(t, x; \tau, z)$$

$$= \int_{-\infty}^{+\infty} [(1 - P(\tau, z)\Delta\tau) E(z, y)$$

$$+ \Delta\tau P(\tau, z) P(\tau, z, y) + o(\Delta)] d_z F(t, x; \tau, z)$$

$$= \int_{-\infty}^{y} d_z F(t, x; \tau, z) - \Delta\tau \int_{-\infty}^{y} P(\tau, z) \, d_z F(t, x; \tau, z)$$

$$+ \Delta\tau \int_{-\infty}^{+\infty} P(\tau, z) P(\tau, z, y) d_z F(t, x; \tau, z) + o(\Delta\tau)$$

$$= F(t, x; \tau, y) - \Delta\tau \int_{-\infty}^{y} P(\tau, z) \, d_z F(t, x; \tau, z)$$

$$+ \Delta\tau \int_{-\infty}^{+\infty} P(\tau, z) P(\tau, z, y) \, d_z F(t, x; \tau, z) + o(\Delta\tau).$$

This shows that (19.6) holds true. ∎

The equations (19.5) and (19.6) are called *Kolmogorov-Feller equations*. They are Barbashin equations containing a Lebesgue-Stieltjes integral. If the function $z \mapsto P(t, x, z)$ is absolutely continuous, we may write (19.2) as a pure Barbashin equation.

There is another way to relate the above problem to a Barbashin equation (KOLMOGOROV [1986]). Suppose that the parameter y preserves the preceding value in the "infinitesimal" interval $(t, t + dt)$ with probability $1 - a(t, y)dt$, and that y passes to $y' \in (z, z + dz)$ with probability $u(t, y, z)dtdz$, where

$$\int_{-\infty}^{+\infty} u(t, y, z)\, dz = a(t, y).$$

Consider the "infinitesimal" distribution function g defined by

$$g(t, y) = \int_{-\infty}^{+\infty} g(t_0, x)f(t_0, x, t, y)\, dx \quad (t > t_0),$$

where $g(t_0, y)$ is known at the time moment $t = t_0$, and

$$f(t_0, x, t, y) = \frac{\partial}{\partial y}F(t_0, x, t, y).$$

Then this function satisfies the Barbashin equation

$$(19.7) \qquad \frac{\partial g(t, y)}{\partial t} = -a(t, y)g(t, y) + \int_{-\infty}^{+\infty} g(t, z)u(t, y, z)\, dz.$$

We still introduce some other equations which arise in the mathematical description of processes without after-effect (PROKOFIEV-ROZANOV [1987]). Suppose that $\xi(t)$ is a random process without after-effect in some measure space (X, Σ) with Borel measure λ. If $\xi(t) = x$ at time t, we suppose that the state does not change in the interval $(t, t + \Delta t)$ with probability $1 - \lambda(t, x)\Delta t + o(\Delta t)$, but passes to some other state $y \in B \subseteq X$, where $x \notin B$, with probability $\lambda(t, x, B)\Delta t + o(\Delta t)$. We assume that $B \mapsto \lambda(t, x, B)$ is a finite measure for fixed t and x, while $t \mapsto \lambda(t, x, B)$ is uniformly continuous for fixed x and B. Under these assumptions, the *distribution function* of the process satisfies

$$P(t, x; t + \Delta t, B) = \lambda(t, x, B)\Delta t + o(\Delta t).$$

If this equality holds uniformly in t, x, and B, for every bounded function φ_0 on X we may define another function $\varphi : \mathbb{R} \times X \to X$ by

$$(19.8) \qquad \varphi(s, x) = \int_X \varphi_0(z) P(s, x; t, dz) \qquad (s \le t).$$

This function satisfies then the Barbashin equation

$$(19.9) \qquad \frac{\partial \varphi(s, x)}{\partial s} = - \int_X \varphi(s, y)\lambda(s, x, dy) \qquad (s < t)$$

with initial condition

$$(19.10) \qquad \varphi(t, x) = \varphi_0(x).$$

In particular, if φ_0 is the characteristic function of B, we have $\varphi(s, x) = P(s, x; t, B)$, by (19.8). Consequently, we may define the distribution function $P(s, x; t, B)$ through (19.9) and (19.10).

Now let μ_0 be some finite measure and

$$(19.11) \qquad \mu(t, B) = \int_X \mu_0(dx) P(s, x; t, B) \qquad (t \ge s).$$

Then μ satisfies the integro-differential equation

$$(19.12) \qquad \frac{\partial \mu(t, B)}{\partial t} = \int_X \lambda(t, x, B)\mu(t, dx) \qquad (t > s)$$

with initial condition

$$(19.13) \qquad \mu(s, B) = \mu_0(B).$$

In particular, if $\mu_0(B) = P(s, x; s, B)$, we may define the distribution function $P(s, x; t, B)$ through (19.12) and (19.13).

We point out that the problems (19.9)/(19.10) and (19.12)/(19.13) may be solved by means of the method of successive approximations. In fact, for the problem (19.9)/(19.10) we have

$$\varphi_0(s, x) = \varphi_0(x), \ldots, \varphi_{n+1}(s, x)$$

$$= \varphi_0(x) + \int_s^t \int_X \varphi_n(u, y)\lambda(u, x, dy) \, du,$$

where

$$\operatorname*{ess\,sup}_{x \in X} \ \sup_{t_0 \le s \le t} \ |\varphi_n(s, x) - \varphi(s, x)| \to 0 \qquad (n \to \infty).$$

Likewise, for the problem (19.12)/(19.13) we have

$$\mu_0(t, B) = \mu_0(B), \ldots, \mu_{n+1}(t, B)$$

$$= \mu_0(B) + \int_s^t \int_X \mu_n(u, dx)\lambda(u, x, B) \, du,$$

where

$$Var\,[\mu_n(t, B) - \mu(t, B)] \to 0 \qquad (n \to \infty),$$

uniformly in t on every interval $[s, \overline{t}]$.

19.2. Applications to evolution equations

Various classes of *nonlinear evolution equations* with applications in physics may be studied by means of the linear theory, see TAKHTA-DZHJAN-FADEEV [1986] or ABLOWITZ-SEGUR [1981]. We sketch some results in this direction.

Given two operators L and M, consider the problem

(19.14)
$$\begin{cases} Lv = v, \\ v_t = Mv. \end{cases}$$

The compatibility of the system (19.14) and its relation to specific evolution equations depend of course on the choice of the operators

L and M. For instance, in the *Korteweg-de Vries equation* one takes $L = \partial^2/\partial x^2 + u(x,t)$ (see e.g. ABLOWITZ-SEGUR [1981]). Another example is given by the system

$$(19.15) \qquad \begin{cases} \dfrac{\partial v(x,t)}{\partial x} = i\xi Dv + Nv, \\[2mm] \dfrac{\partial v(x,t)}{\partial t} = Qv, \end{cases}$$

where $v = (v_1, \ldots, v_n)$, $\xi_t = 0$, and D, N, and Q are $n \times n$-matrices such that the diagonal entries of N and the off-diagonal entries of D are zero. The system (19.15) describes various evolution processes with a natural physical interpretation (see ABLOWITZ-SEGUR [1981]). In particular, the *Boussinesq equation* and the *nonlinear vector Schrödinger equation* may be obtained in this way.

If one passes in (19.15) formally to infinitely many variables, i.e. to the limit $n \to \infty$, one is led to the Barbashin equations

$$(19.16) \qquad \begin{aligned} \frac{\partial v(x,y,t)}{\partial x} &= i\xi d(y)v(x,y,t) \\ &+ \int_{-\infty}^{+\infty} N(x,y,z,t)v(x,z,t)\,dz \end{aligned}$$

and

$$(19.17) \qquad \frac{\partial v(x,y,t)}{\partial t} = \int_{-\infty}^{+\infty} Q(x,y,z,t)v(x,z,t)\,dz.$$

Here the kernel $N(x,y,z,t)$ must satisfy the nonlinear integro-differential equation of Barbashin type

$$(19.18) \qquad \begin{aligned} \frac{\partial N(x,y,z,t)}{\partial t} &= a(x,y)\frac{\partial N(x,y,z,t)}{\partial x} \\ &+ \int_{-\infty}^{+\infty} [a(y,z') - a(z',z)]N(x,y,z',t)N(x,z',z,t)\,dz' \end{aligned}$$

with $a(y,z) = a(z,y)$.

We describe now a general approach to integro-differential equations of Barbashin type. Consider the system of ordinary differential equations

$$(19.19) \quad \begin{cases} \dfrac{dx_1}{dt} = f_1(t, x_1, \ldots, x_n) \\[2mm] \dfrac{dx_2}{dt} = f_2(t, x_1, \ldots, x_n) \\[2mm] \qquad \vdots \\[2mm] \dfrac{dx_n}{dt} = f_n(t, x_1, \ldots, x_n). \end{cases}$$

Following VOLTERRA [1959], we replace the discrete index $i \in \{1, \ldots, n\}$ by the continuous variable $s \in [a, b]$ and get an equation of the form

$$(19.20) \qquad \frac{\partial x(t, s)}{\partial t} = F[t, s, \phi(t)]$$

containing the unknown function $x = x(t, s)$; here $\phi(t) = \Phi x(t, \cdot)$ with some functional Φ which acts on $x(t, \cdot)$ for fixed t. If $F(t, s, \cdot)$ is linear, we may write (19.20) in the form

$$(19.21) \qquad \frac{\partial x(t, s)}{\partial t} = \int_a^b f(t, s, \sigma) x(t, \sigma) \, d\sigma;$$

this is a special Barbashin equation which was studied in SCHLESINGER [1914, 1915]. Integrating equation (19.20) over $[t_0, t]$ yields

$$(19.22) \qquad x(t, s) = x_0(s) + \int_{t_0}^t F[\tau, s, x(\tau, s)] \, d\tau$$

with $x_0(s) := x(t_0, s)$. If we suppose that F is continuous in all variables and satisfies a Lipschitz-type condition

$$\sup_{|u-v| \leq \varepsilon} |F(t, s, u) - F(t, s, v)| \leq A\varepsilon$$

with some constant $A > 0$, the solution of (19.22) may be obtained by the usual method of successive approximations with initial approximation $x_0(t, s) \equiv x_0(s)$.

Similarly, consider the following *canonical Hamiltonian system*:

$$(19.23) \quad \begin{cases} \dfrac{dq_1}{dt} = \dfrac{\partial H(t, q_1, \ldots, q_n, p_1, \ldots, p_n)}{\partial p_1} \\[2mm] \dfrac{dq_2}{dt} = \dfrac{\partial H(t, q_1, \ldots, q_n, p_1, \ldots, p_n)}{\partial p_2} \\[2mm] \quad\quad\quad \vdots \\[2mm] \dfrac{dq_n}{dt} = \dfrac{\partial H(t, q_1, \ldots, q_n, p_1, \ldots, p_n)}{\partial p_n} \\[2mm] \dfrac{dp_1}{dt} = -\dfrac{\partial H(t, q_1, \ldots, q_n, p_1, \ldots, p_n)}{\partial q_1} \\[2mm] \dfrac{dp_2}{dt} = -\dfrac{\partial H(t, q_1, \ldots, q_n, p_1, \ldots, p_n)}{\partial q_2} \\[2mm] \quad\quad\quad \vdots \\[2mm] \dfrac{dp_n}{dt} = -\dfrac{\partial H(t, q_1, \ldots, q_n, p_1, \ldots, p_n)}{\partial q_n}. \end{cases}$$

Replacing again the discrete index $i \in \{1, \ldots, n\}$ by the continuous variable $s \in [a, b]$, we may transform this system in the same way into the integro-differential equation (see VOLTERRA [1959])

$$(19.24) \quad \begin{cases} \dfrac{\partial q(t, s)}{dt} = \dfrac{\partial F(t, s, \phi(t), \psi(t))}{\partial p} \\[3mm] \dfrac{\partial p(t, s)}{dt} = -\dfrac{\partial F(t, s, \phi(t), \psi(t))}{\partial q}, \end{cases}$$

where ϕ and ψ are defined similarly as ϕ in (19.20). We point out that other *nonlinear integro-differential equations of Barbashin type* arise in the *theory of permutable functions* which have important applications to integral equations (VOLTERRA [1959]).

Consider now the *matrix differential equation*

$$(19.25) \quad \frac{dV(\tau)}{d\tau} = CV + LV + VM + F,$$

where C, V, L, M, and F are $n \times n$-matrices with C containing $c_1, \ldots, c_n$ on the diagonal. Such equations arise, in particular, as evolution equations for the transition matrix in *inverse problems for nonlinear Schrödinger equations* (TAKHTADZHJAN-FADEEV [1986]). Passing (formally) in the matrix equation (19.25) to the limit $n \to \infty$ gives the equation

(19.26)
$$\frac{\partial v(\tau, s, t)}{\partial \tau} = c(s, t)v(\tau, s, t)$$
$$+ \int_{-\infty}^{+\infty} l(\tau, \sigma, t)v(\tau, s, \sigma) \, d\sigma$$
$$+ \int_{-\infty}^{+\infty} m(\tau, s, \sigma)v(\tau, \sigma, t) \, d\sigma + f(\tau, s, t).$$

This is a generalized Barbashin equation of the form we have studied in Subsection 16.1.

19.3. Systems with substantially distributed parameters

As we have shown in the previous subsection, integro-differential equations of Barbashin type, both linear and nonlinear, naturally arise from systems of n differential equations after passing to the limit $n \to \infty$ or, what is essentially the same, after replacing discrete indices by continuous variables. While this way of obtaining Barbashin equations may seem rather "academic", it has in fact a reasonable physical background which we will discuss in this subsection (see BRACK [1985]).

Consider, for instance, the system

(19.27)
$$\begin{cases} \dfrac{dq}{dt} &= Aq(t) + Bu(t) \\ x(t) &= Cq(t) + Du(t), \end{cases}$$

where the values of A, B, C, and D are $m \times m$-matrices, and the values of q, x, and u are $m \times n$-matrices. The system (19.27) represents the nonstationary balance equations for certain quantities (called *substances*) over given balance spaces, and the state vector q represents a *distribution density function* for these substances. Physically, n denotes the *number of independent balance spaces*, while m is the *number of independent substances*. Here passing to the limit $n \to \infty$ or $m \to \infty$ has a natural physical meaning.

In fact, $n \to \infty$ means that no more discrete balance spaces can be distinguished which occurs in *systems with spatially distributed parameters*. The mathematical model leads here to a system of m partial differential equations.

On the other hand, $m \to \infty$ means that one has to deal with an arbitrarily large number of substances; the corresponding systems are then called *systems with essentially distributed parameters* (BRACK [1985]). In this case one is led to the system

(19.28)
$$\frac{\partial q(s,t)}{\partial t} = a(s)q(s,t)+$$
$$\int_a^b A(s,\sigma)q(\sigma,t)\, d\sigma + \int_a^b B(s,\sigma)u(\sigma,t)\, d\sigma$$

which consists of n integro-differential equations of Barbashin type. If B is a diagonal matrix, the system (19.28) may be written in the form

(19.29)
$$\frac{\partial q(s,t)}{\partial t} = a(s)q(s,t)+$$
$$\int_a^b A(s,\sigma)q(\sigma,t)\, d\sigma + b(s)u(s,t).$$

The variable $s \in [a,b]$ in (19.28) or (19.29) may have different meanings, according to the physical setting. For example, s may be the *number of C-atoms in chemical compouds*, the *chain length in polymers*, or the *boiling temperature of a multicomponent mixture* (see e.g. ROTH [1972], RÄTZSCH-KEHLEN [1984], GILLES-KNÖPP [1967]). For fixed t, the function $q(\cdot,t)$ may be regarded as *density function* of the distribution of the material.

If the matrices A and B in (19.27) depend on t as well, and B is a diagonal matrix, the system (19.29) has to be replaced by the system

(19.30)
$$\frac{\partial q(s,t)}{\partial t} = a(s,t)q(s,t)+$$
$$\int_a^b A(s,\sigma,t)q(\sigma,t)\,d\sigma + b(s,t)u(s,t).$$

This is a Barbashin equation of the type we have studied in Chapters I and II.

We remark that some applications of Barbashin equations to problems in *control theory* have been given in BISJARINA [1964], KHOTEEV [1976, 1984], BEESACK [1987], and KURZWEIL [1957].

19.4. The Kimura continuum-of-alleles model

In 1965, M. Kimura proposed a model for describing mutations as a source for genetic variability in quantitative characters (KIMURA [1965], see also BÜRGER [1986]). The crucial assumption in this model is that new alleles are produced continuously through genetic mutation, and that the expression of a metric trait is determined by a continuum of possible alleles. Assuming that the selection mechanism in the system is constant, this model leads to certain integro-differential equations as we shall show now.

In Kimura's model, one assumes that a quantitative character is under control of a single gene locus at which every mutation may produce a new allele different from the previous ones. The effect of a new allele on a character is assumed to be only slightly different from that of the parent allele.

Let us denote by s the effect of the action of an allele on the quantitative character. It is assumed that an allele having the effect s mutates to an allele with effect σ with *probability density* $u(s - \sigma)$. By $p(t, s)$ we denote the relative frequency of the allele s at time t in the population under consideration, i.e. $p(t, \cdot)$ is a probability distribution. Let $m(s)$ denote the Malthusian fitness of the population, and $\overline{m}(t)$

its average at time t, i.e.

$$\overline{m}(t) = \int_{-\infty}^{+\infty} m(s)p(t,s)\,ds.$$

In a population of competeting generations, the evolution of allelic effects is described by the nonlinear integro-differential equation of Barbashin type

(19.31)
$$\frac{\partial p(t,s)}{\partial t} = [m(s) - \overline{m}(t)]p(t,s)$$
$$+\mu \int_{-\infty}^{+\infty} u(s-\sigma)p(t,\sigma)\,d\sigma - \mu p(t,s),$$

where μ is the *mutation rate*. Based on biological considerations, KI-MURA [1965] proposes the choice

$$m(s) = -as^2, \qquad u(s) = \frac{e^{-s^2/2\gamma^2}}{\sqrt{2\pi\gamma^2}}$$

which leads to the Barbashin equation

(19.32)
$$\frac{\partial p(t,s)}{\partial t} = -as^2 p(t,s) + \mu \int_{-\infty}^{+\infty} \frac{e^{-(s-\sigma)^2/2\gamma^2}}{\sqrt{2\pi\gamma^2}} p(t,\sigma)\,d\sigma$$
$$+ap(t,s)\int_{-\infty}^{+\infty} \sigma^2 p(t,\sigma)\,d\sigma - \mu p(t,s).$$

Here the *equilibria* of equation (19.32) are particularly important, i.e. solutions of the form $p(t,s) \equiv p(s)$. From (19.32) it follows that every equilibrium must staisfy the relation

$$s^2 p(s) - \frac{\mu}{a}\int_{-\infty}^{+\infty} u(s-\sigma)p(\sigma)\,d\sigma = \left(\lambda - \frac{\mu}{a}\right)p(s).$$

Integrating this relation leads to

$$\lambda = \int_{-\infty}^{+\infty} s^2 p(s)\,ds.$$

In BÜRGER [1986] it is shown that equation (19.32) has a unique equilibrium solution $p = p(s)$; this solution is strictly positive, symmetric, and globally asymptotically stable.

We shall prove now that the equation (19.32) with initial condition

$$(19.33) \qquad\qquad p(t_0, s) = p_0(s)$$

has, under appropriate conditions on $\dot{p}_0$, always a local solution:

Lemma 19.1. *Let*

$$(19.34) \qquad\qquad g(t, s) := p_0(s)e^{-as^2(t-t_0)},$$

and suppose that the function $g(\cdot, s)$ is integrable for any s. Let $f(\cdot, s)$ be continuous for all s. Then the initial value problem

$$(19.35) \qquad \begin{cases} \dfrac{\partial p(t, s)}{\partial t} = -as^2 p(t, s) + f(t, s) \quad (t > t_0), \\[2mm] p(t_0, s) = p_0(s) \end{cases}$$

$(s \in \mathbb{R})$ *has the unique solution*

$$(19.36) \qquad p(t, s) = g(t, s) + \int_{t_0}^{t} e^{-as^2(t-\tau)} f(\tau, s) \, d\tau.$$

☐ For fixed s, the problem (19.35) is a linear initial value problem which can be solved by the classical "variation of constants" formula. ∎

Consider the operator

$$(19.37) \qquad \begin{aligned} Kx(t, s) &= \mu \int_{t_0}^{t} \int_{-\infty}^{+\infty} \frac{e^{-(s-\sigma)^2/2\gamma^2}}{\sqrt{2\pi\gamma^2}} x(\tau, \sigma) \, d\sigma \, d\tau \\ &+ a \int_{t_0}^{t} x(\tau, s) \int_{-\infty}^{+\infty} \sigma^2 x(\tau, \sigma) \, d\sigma \, d\tau - \mu \int_{t_0}^{t} x(\tau, s) \, d\tau. \end{aligned}$$

Let $w : \mathbb{R} \to (0, \infty)$ be bounded, continuous, and integrable, and denote by $C_{T,w}$ the space of continuous functions $x : [t_0 - T, t_0 + T] \times$

$\mathbb{R} \to \mathbb{R}$, equipped with the norm

$$||x||_{T,w} :=$$

(19.38)

$$\sup \left\{ \frac{(1+s^2)|x(t,s)|}{w(s)} : |t - t_0| \le T, s \in \mathbb{R} \right\}.$$

We assume that w is chosen such that the initial function p_0 belongs to $C_{T,w}$ for some (and thus all) $T > 0$. The next lemma gives a local Lipschitz condition for the operator (19.37) between $C_{T,w}$ and C:

Lemma 19.2. *For each $\varepsilon > 0$ and each $R > 0$ there is some $T > 0$ such that*

(19.39) $||Kx - Ky||_C \le \varepsilon ||x - y||_{T,w}$

for $||x||_{T,w}, ||y||_{T,w} \le R$.

□ By the definition of the norm (19.38) we have

$$|x(t,s) - y(t,s)| \le \frac{||x - y||_{T,w} w(s)}{1 + s^2},$$

$$|x(t,s)| \le Rw(s), \quad s^2|y(t,s)| \le Rw(s).$$

Consequently, we get

$$|Kx(t,s) - Ky(t,s)| \leq$$

$$\mu \left| \int_{t_0}^{t} \int_{-\infty}^{+\infty} \frac{e^{-(s-\sigma)^2/2\gamma^2}}{\sqrt{2\pi\gamma^2}} |x(\tau,\sigma) - y(\tau,\sigma)| \, d\sigma \, d\tau \right|$$

$$+a \left| \int_{t_0}^{t} \int_{-\infty}^{+\infty} \sigma^2 |x(\tau,s)[x(\tau,\sigma - y(\tau,\sigma)] \right.$$

$$+[x(\tau,s) - y(\tau,s)]y(\tau,\sigma| \, d\sigma \, d\tau|$$

$$+\mu \int_{t_0}^{t} |x(\tau,s) - y(\tau,s)| \, d\tau$$

$$\leq \mu \left| \int_{t_0}^{t} \int_{-\infty}^{+\infty} \frac{e^{-(s-\sigma)^2/2\gamma^2}}{\sqrt{2\pi\gamma^2}} ||x-y||_{T,w} w(\sigma)(1+\sigma^2)^{-1} \, d\sigma \right|$$

$$+a \left| \int_{t_0}^{t} \int_{-\infty}^{+\infty} \sigma^2 Rw(s) ||x-y||_{T,w} w(\sigma)(1+\sigma^2)^{-1} \, d\sigma \, d\tau \right|$$

$$+a \left| \int_{t_0}^{t} \int_{-\infty}^{+\infty} ||x-y||_{T,w} w(s)(1+s^2)^{-1} Rw(\sigma) \, d\sigma \, d\tau \right|$$

$$+\mu \int_{t_0}^{t} ||x-y||_{T,w} w(s)(1+s^2)^{-1} \, d\tau$$

$$\leq C |t-t_0| \, ||x-y||_{T,w} [\mu + aRw(s) + \mu w(s)]$$

with some constant $C > 0$ depending only on w. Now the statement is obvious. ∎

Theorem 19.2. *Let $w : \mathbb{R} \to (0,\infty)$ be bounded, continuous, and integrable. Assume that $p_0 : \mathbb{R} \to \mathbb{R}$ is continuous with*

$$(19.40) \qquad \sup_{-\infty < s < \infty} \frac{(1+s^2)p_0(s)}{w(s)} < \infty.$$

Then for each sufficiently large $R > 0$ there is some $T > 0$ such that the initial value problem (19.32)/(19.33) has a unique solution $x \in C_{T,w}$ with $||x||_{T,w} \leq R$.

□   We show first that the initial value problem (19.32)/(19.33) is equivalent to the fixed point equation

(19.41) $x = GKx + g,$

where g is given by (19.34) and $G : C \to C_{T,w}$ is defined by

(19.42) $Gy(t,s) = \int_{t_0}^{t} e^{-as^2(t-\tau)} y(\tau, s)\, d\tau.$

Indeed, if x solves (19.41), then Lemma 19.1, applied to $f = Kx$, implies that x solves the initial value problem (19.32)/(19.33). Conversely, if $x \in C_{T,w}$ is a solution of the initial value problem (19.32)/(19.33), then again Lemma 19.1 implies that x satisfies (19.41).

Without loss of generality, let $R > ||g||_{T,w}$. Fix

$$\varepsilon \le \frac{R - ||g||_{T,w}}{R||G||},$$

and choose T as in Lemma 19.2. To prove existence and uniqueness of a solution of (19.41), we apply the Banach-Caccioppoli fixed point principle to the operator $Ax = GKx + g$ on the closed ball $B := \{x : x \in C_{T,w}, ||x||_{T,w} \le R\}$. In fact, A maps B into itself since

$$||Ax|| \le ||GKx|| + ||g|| \le ||G||\,\varepsilon\,||x|| + ||g|| \le R \quad (||x|| \le R).$$

From Lemma 19.2 we conclude that $||Ax - Ay|| \le ||G||\,\varepsilon\,||x - y||$ with $||G||\varepsilon < 1$. Thus, A has a unique fixed point in B.   ■

In THIEME [1980, 1986] one may find other problems from population dynamics which lead to nonlinear integro-differential equations of Barbashin type. We also mention the nonlinear Barbashin equation studied in HADELER [1986], which is some kind of continuous analogue to the ordinary differential equations occurring in Fisher's model of n alleles at a locus (HOFBAUER-SIGMUND [1984]) or in evolutionary game dynamics (HOFBAUER-SIGMUND [1984], SEEMAN [1980]). The paper ARINO-KIMMEL [1987] provides an analysis of the asymptotic behaviour of a cell cycle kinetic model, based on the assumption of unequal RNA division during cytokinesis.

19.5. An application to a radiation problem

We are now going to apply the existence and uniqueness theorems for boundary value problems obtained in Subsection 9.5 to other physical and mechanical "real life" problems. Consider the integro-differential equation

$$(19.43) \qquad s\frac{\partial x(t,s)}{\partial t} + x(t,s) = \int_{-1}^{1} k(s,\sigma)x(t,\sigma)\,d\sigma + f(t,s)$$

with boundary conditions

$$(19.44) \qquad \begin{cases} x(0,s) = \phi(s) & (0 < s \le 1), \\ x(T,s) = \psi(s) & (-1 \le s < 0). \end{cases}$$

Problems of this type arise in the mathematical modelling of the *propagation of radiation* through the atmosphere of planets and stars (CASE-ZWEIFEL [1967], MININ [1988], SOBOLEV [1956, 1972, 1985]); here f describes the *interior radiation*, k the *scattering properties of the atmosphere*, and x the (unknown) *radiation density*. In another physical interpretation, the above problem arises in the modelling of *transfer of neutrons* through thin plates and membranes in nuclear reactors (VAN DER MEE [1983]); here f describes the *interior neutron emission*, k the *diffraction properties* of the plates and membranes, and x the (unknown) *angular density* of neutrons. The problem of isotropic scattering of neutrons in an infinite homogeneous medium consiederd in MARCHUK [1961] also leads to equation (19.43).

We shall study the problem (19.43)/(19.44) in the space $X = L_p([-1,1])$, i.e. we take $\phi \in X_+ = L_p([0,1])$ and $\psi \in X_- = L_p([-1,0])$ ($1 < p < \infty$). By $X(s), X_+(s)$, and $X_-(s)$ we denote the weighted spaces with norm (9.43) for the special weight function $w(s) = s$ (which is of course suggested by the form of equation (19.43)). Obviously, we have $X \subseteq X(s)$ and $X_\pm \subseteq X_\pm(s)$ (continuous imbeddings with imbedding constants 1).

Observe that the equation (19.43) exhibits a singularity, since the weight function $w(s) = s$ is not bounded away from zero. We have to pay a price for this, inasmuch as we must require more regularity for the right-hand side f in the next lemma:

Lemma 19.3. *Let* $X = L_p([-1,1])$ *and* $Y = L_r([-1,1])$ *with* $p < r < \infty$, *and let* $Q = [0,T] \times [-1,1]$. *For any* $\phi \in X_+, \psi \in X_-$, *and* $f \in C_t(Y)$, *the problem*

$$
(19.45) \qquad
\begin{cases}
s\dfrac{\partial x(t,s)}{\partial t} + x(t,s) = f(t,s) & \text{if } (t,s) \in Q, \\[2mm]
x(0,s) = \phi(s) & \text{if } 0 < s \le 1, \\[2mm]
x(T,s) = \psi(s) & \text{if } -1 \le s < 0
\end{cases}
$$

has then a unique solution $x \in C_t(X)$ *with* $\partial x/\partial t \in C_t(X(s))$; *this solution is given, for almost all* $(t,s) \in Q$, *by*

$$
x(t,s) =
\begin{cases}
\dfrac{1}{s} \displaystyle\int_0^t e^{-(t-\tau)/s} f(\tau,s)\, d\tau + \phi(s)e^{-t/s} & (0 < s \le 1), \\[4mm]
\dfrac{1}{s} \displaystyle\int_T^t e^{-(t-\tau)/s} f(\tau,s)\, d\tau + \psi(s)e^{-(t-T)/s} & (-1 \le s < 0).
\end{cases}
$$

□ We only have to prove that $x \in C_t(X)$; the fact that $\partial x/\partial t \in C_t(X(s))$ follows then from $f \in C_t(Y)$ and the structure of equation (19.43). For fixed $t_0 \in [0,T]$ we have

$$
\|x(t,\cdot) - x(t_0,\cdot)\|_X \le \left\{ \int_0^1 \left| \frac{1}{s}\int_0^t e^{-(t-\tau)/s} f(\tau,s)\, d\tau \right. \right.
$$

$$
\left. \left. - \int_0^{t_0} e^{-(t_0-\tau)/s} f(\tau,s)\, d\tau \right|^p ds \right\}^{1/p}
$$

$$
+ \left\{ \int_0^1 |\phi(s)[e^{-t/s} - e^{-t_0/s}]|^p\, ds \right\}^{1/p}
$$

$$
+ \left\{ \int_0^1 |\psi(s)[e^{-(t-T)/s} - e^{-(t_0-T)/s}]|^p\, ds \right\}^{1/p}
$$

$$
+ \left\{ \int_{-1}^0 \left| \frac{1}{s}\int_T^t e^{-(t-\tau)/s} f(\tau,s)\, d\tau - \int_T^{t_0} e^{-(t-\tau)/s} f(\tau,s)\, d\tau \right|^p ds \right\}^{1/p}.
$$

For definiteness, let $t > t_0$. We only prove that the first term in (19.45) tends to zero as $t \to t_0$; the proof for the other terms is similar. By the Hölder inequality we have

$$\left\{ \int_0^1 \left| \frac{1}{s} \int_0^t e^{-(t-\tau)/s} f(\tau, s) \, d\tau - \int_0^{t_0} e^{-(t_0-\tau)/s} f(\tau, s) \, d\tau \right|^p ds \right\}^{1/p}$$

$$\leq \left\{ \int_0^1 \left| \frac{1}{s} \int_{t_0}^t e^{-(t-\tau)/s} f(\tau, s) \, d\tau \right|^p ds \right\}^{1/p}$$

$$+ \left\{ \int_0^1 \left| \frac{1}{s} \int_0^{t_0} [e^{-(t-\tau)/s} - e^{-(t_0-\tau)/s}] f(\tau, s) \, d\tau \right|^p ds \right\}^{1/p}$$

$$\leq \int_{t_0}^t \left\{ \int_0^1 \left| \frac{1}{s} e^{-(t-\tau)/s} \right|^q ds \right\}^{1/q} \left\{ \int_0^1 |f(\tau, s)|^r \, ds \right\}^{1/r} d\tau$$

$$+ \int_0^{t_0} \left\{ \int_0^1 \left| \frac{1}{s} [e^{-(t-\tau)/s} - e^{-(t_0-\tau)/s}] \right|^q ds \right\}^{1/q} \times$$

$$\times \left\{ \int_0^1 |f(\tau, s)|^r \, ds \right\}^{1/r} d\tau,$$

where $\frac{1}{p} = \frac{1}{r} + \frac{1}{q}$. Thus, it suffices to show that

$$A_q(t) = \int_{t_0}^t \left\{ \int_0^1 \left| \frac{1}{s} e^{-(t-\tau)/s} \right|^q ds \right\}^{1/q} d\tau \to 0$$

and

$$B_q(t) = \int_0^{t_0} \left\{ \int_0^1 \left| \frac{1}{s} [e^{-(t-\tau)/s} - e^{-(t_0-\tau)/s}] \right|^q ds \right\}^{1/q} d\tau \to 0$$

as $t \to t_0$. Suppose first that $q = 2$. Then

$$A_2(t) = \int_{t_0}^{t} \left\{ \int_0^1 \frac{1}{s^2} e^{-2(t-\tau)/s} \, ds \right\}^{1/2} d\tau$$

$$= \int_{t_0}^{t} \left\{ \frac{e^{2(t-\tau)} - 1}{2(t - \tau)} \right\}^{1/2} d\tau \leq \frac{1}{\sqrt{2}} \int_{t_0}^{t} \frac{e^{t-\tau}}{\sqrt{t - \tau}} \, d\tau \leq \sqrt{2} e^T \sqrt{t - t_0},$$

hence $A_2(t) \to 0$ as $t \to t_0$. If $1 < q < 2$, we get $A_q(t) \to 0$, as $t \to t_0$, by the continuity of the imbedding $L_2 \subseteq L_q$. It remains to analyze the case $2 < q < \infty$. Let first $q = 3$. Integrating by parts and observing that $(a + b)^\alpha \leq a^\alpha + b^\alpha$ for $0 < \alpha < 1$ yields

$$A_3(t) = \int_{t_0}^{t} \left\{ \int_0^1 \frac{1}{s^3} e^{-3(t-\tau)/s} \, ds \right\}^{1/3} d\tau$$

$$= \int_{t_0}^{t} e^{-(t-\tau)} \left[\frac{1}{3(t - \tau)} - \frac{1}{9(t - \tau)^2} \right]^{1/3} d\tau$$

$$\leq \frac{1}{\sqrt[3]{3}} \int_{t_0}^{t} \frac{d\tau}{\sqrt[3]{t - \tau}} + \frac{1}{\sqrt[3]{9}} \int_{t_0}^{t} \frac{d\tau}{\sqrt[3]{(t - \tau)^2}}$$

$$= \frac{\sqrt[3]{9}}{2} \sqrt[3]{(t - t_0)^2} + \sqrt[3]{3} \sqrt[3]{t - t_0},$$

hence $A_3(t) \to 0$ as $t \to t_0$. If $q \in \mathbb{N}$ with $q > 3$, the proof is the same as for $q = 3$. Finally, if $q \in (2, \infty)$ is arbitrary, the assertion follows from the continuity of the imbedding $L_{[q]+1} \subseteq L_q$.

We have shown that $A_q(t) \to 0$, as $t \to t_0$, for any $q \in (1, \infty)$. To see that $B_q(t) \to 0$ as well is easy. In fact,

$$B_q(t) = \int_0^{t_0} \left\{ \int_0^1 \left| \frac{1}{s} e^{-(t_0-\tau)/s} \left[1 - e^{-(t-t_0)/s} \right] \right|^q ds \right\}^{1/q} d\tau$$

$$\leq \int_0^{t_0} \left\{ \int_0^1 \left| \frac{1}{s} e^{-(t_0-\tau)/s} \right|^{2q} ds \right\}^{1/2q} \times$$

$$\times \left\{ \int_0^1 \left| 1 - e^{-(t-t_0)/s} \right|^{2q} ds \right\}^{1/2q} d\tau.$$

Since the first integral remains bounded, and $e^{-(t-t_0)/s} \to 1$ as $t \to t_0$, by Lebesgue's dominated convergence theorem we conclude that $B_q(t) \to 0$ for any $q \in (1, \infty)$. ■

Lemma 19.3 is the basic tool to obtain existence and uniqueness results for the problem (19.43)/(19.44) building on the preceding abstract theorems. For the sake of completeness, we summarize these results with the following two theorems.

Theorem 19.3. *Suppose that the integral operator defined by the kernel function k is regular from $X = L_p([-1, 1])$ into $Y = L_r([-1, 1])$, where $p < r < \infty$, and $f \in C_t(Y)$. Then every solution of the problem (19.43)/(19.44) solves the equation (9.47), where*

$$(19.46) \qquad Lx(t, s) = \begin{cases} \dfrac{1}{s} \displaystyle\int_0^t e^{-(t-\tau)/s} \int_{-1}^1 k(s, \sigma) x(\tau, \sigma) \, d\sigma \, d\tau, \\[2ex] \dfrac{1}{s} \displaystyle\int_T^t e^{-(t-\tau)/s} \int_{-1}^1 k(s, \sigma) x(\tau, \sigma) \, d\sigma \, d\tau, \end{cases}$$

and

$$(19.47) \qquad g(t, s) = \begin{cases} \dfrac{1}{s} \displaystyle\int_0^t e^{-(t-\tau)/s} f(\tau, s) \, d\tau + \phi(s) e^{-t/s}, \\[2ex] \dfrac{1}{s} \displaystyle\int_T^t e^{-(t-\tau)/s} f(\tau, s) \, d\tau + \psi(s) e^{-(t-T)/s} \end{cases}$$

for $0 < s \leq 1$ and $-1 \leq s < 0$, respectively. Conversely, every solution $x \in C_t(X)$ of (9.47), with L given by (19.46) and g given by (19.47), is a solution of (19.43)/(19.44), and the partial derivative $\partial x/\partial t$ belongs to $C_t(X(s))$.

Theorem 19.4. *Suppose that the integral operator defined by the kernel function k is regular from $X = L_p([-1,1])$ into $Y = L_r([-1,1])$, where $p < r < \infty$. Assume that the estimates (9.36) hold, where the functions $a, b, c,$ and d are defined by*

$$\begin{cases} a(t,\tau,s,\sigma) = \frac{1}{s}e^{-(t-\tau)/s}k(s,\sigma), \\[2mm] b(t,\tau,s,\sigma) = \frac{1}{s}e^{-(t-\tau)/s}k(s,-\sigma), \\[2mm] c(t,\tau,s,\sigma) = -\frac{1}{s}e^{(t-\tau)/s}k(-s,\sigma), \\[2mm] d(t,\tau,s,\sigma) = -\frac{1}{s}e^{(t-\tau)/s}k(-s,-\sigma). \end{cases}$$

Suppose, moreover, that one of the four conditions (a), (b), (c) *or* (d) *of Theorem 9.7 is satisfied. Then the problem (19.43)/(19.44) has a unique solution $x \in C_t(X)$, with $\partial x/\partial t \in C_t(X(s))$, for any $\phi \in X_+, \psi \in X_-,$ and $f \in C_t(Y)$.*

19.6. Applications to astrophysics

The following approach to the Kolmogorov-Feller equations of a purely discontinuous Markov process is taken from AGEKJAN [1974]. A Markov process is called *purely discontinuous* if the the corresponding random function which takes the value x at the moment t, remains unchanged in the time interval Δt with probability $1 - P(t,x)\Delta t + o(\Delta t)$, and changes with probability $P(t,x)\Delta t + O(\Delta t)$. Let $\xi(t,x,y)\,dy$ denote the probability with which this function takes values from the interval $[y, y+dy]$, and $\varphi(t,x;\tau,y)\,dy$ the probability which this function takes values from this interval at time τ. Then

$$\varphi(t, x; t + \Delta t, y)\, dy =$$

(19.48)
$$[1 - P(t, x)\Delta t + o(\Delta t)]\delta(y - x)\, dy$$

$$+[P(t, x)\Delta t + o(\Delta t)]\xi(t, x; y)\, dy,$$

where δ denotes the delta distribution. Putting (19.48) into the Kolmogorov-Chapman equation

$$\varphi(t, x; \tau, z) = \int_{-\infty}^{+\infty} \varphi(t, x; t + \Delta t, y)\varphi(t + \Delta t, y; \tau, z)\, dy$$

we get

(19.49)
$$\varphi(t, x; \tau, z) = \int_{-\infty}^{\infty} \Big[(1 - P(t, x)\Delta t + o(\Delta t))\delta(y - x)$$

$$+(P(t, x)\Delta t + o(\Delta t))\xi(t, x; y)\Big]\varphi(t + \Delta t, y; \tau, z)\, dy.$$

Since

$$\int_{-\infty}^{+\infty} \varphi(t + \Delta t, y; \tau, z)\delta(y - x)\, dy = \varphi(t + \Delta t, x; \tau, z),$$

from (19.49) we further get

$$\frac{\varphi(t + \Delta t, x; \tau, z) - \varphi(t, x; \tau, z)}{\Delta t}$$

(19.50)
$$= \left[P(t, x) + \frac{o(\Delta t)}{\Delta t}\right]\varphi(t + \Delta t, x; \tau, z) -$$

$$\left[P(t, x) + \frac{o(\Delta t)}{\Delta t}\right]\int_{-\infty}^{\infty} \xi(t, x; y)\varphi(t + \Delta t, y; \tau, z)\, dy.$$

Taking the limit $\Delta t \to 0$ in (19.50), we obtain the first Kolmogorov-Feller equation

$$\frac{\partial \varphi(t, x; \tau, z)}{\partial t} = P(t, x)(\varphi(t, x; \tau, z) - \int_{-\infty}^{\infty} \xi(t, x; y)\varphi(t, y; \tau, z)\, dy).$$

Furthermore, using the identity

$$\varphi(\tau, y; \tau + \Delta\tau, z)\, dz = (1 - P(\tau, y)\Delta\tau + o(\Delta\tau))\delta(z - y)$$

$$+(P(\tau, y)\Delta\tau + o(\Delta\tau))\xi(\tau, y; z)\, dz$$

and well-known properties of the delta distribution, we get

$$\varphi(t, x; \tau + \Delta\tau, z) = \varphi(t, x; \tau, z)$$

(19.51) $$-\varphi(t, x; \tau, z)(P(\tau, x)\Delta\tau + o(\Delta\tau))$$

$$+ \int_{-\infty}^{+\infty} \varphi(t, x; \tau, y)(P(\tau, y)\Delta\tau + o(\Delta\tau))\xi(\tau, y, z)\, dy.$$

Putting now $\varphi(t, x; \tau, z)$ in the left hand side of (21.12), dividing by $\Delta\tau$, and taking the limit $\Delta\tau \to 0$, we end up with the second Kolmogorov-Feller equation

$$\frac{\partial\varphi(t, x; \tau, z)}{\partial\tau} = -P(\tau, z)\varphi(t, x; \tau, z)$$

$$+ \int_{-\infty}^{\infty} P(\tau, y)\xi(\tau, y; z)\varphi(t, x; \tau, y)\, dy.$$

Let $L(t, x; z)\, dt\, dz$ be the probability with which the corresponding random function which takes the value x at the moment t, makes a jump after time dt and assumes values in the interval $[z, z+dz]$. Then

$$L(t, x; z)\, dt\, dz = P(t, x)\xi(t, x; z)\, dt\, dz.$$

Since

$$\int_{-\infty}^{+\infty} \xi(t, x; z)\, dz = 1,$$

we have

$$\int_{-\infty}^{+\infty} L(t, x; z)\, dz = P(t, x).$$

Consequently, the Kolmogorov-Feller equations for a purely discontinuous random process have the form

(19.52)

$$\frac{\partial \varphi(t,x;\tau,z)}{\partial t} = \varphi(t,x;\tau,z) \int_{-\infty}^{+\infty} L(t,x;y)\,dy$$

$$- \int_{-\infty}^{+\infty} L(t,x,y)\varphi(t,y;\tau,z)\,dy$$

and

(19.53)

$$\frac{\partial \varphi(t,x;\tau,z)}{\partial \tau} = -\varphi(t,x;\tau,z) + \int_{-\infty}^{+\infty} L(\tau,z;y)\,dy$$

$$+ \int_{-\infty}^{+\infty} L(\tau,y;z)\varphi(t,x;\tau,y)\,dy.$$

From (19.52) and (19.53) one sees that the knowledge of the function
$L(t,x,y)$ is indispensable for solving the Kolmogorov-Feller equations.

19.7. Plane-parallel transport problems

In Subsection 19.5 we have studied the boundary value problem
(19.43)/(19.44) for a Barbashin equation related to the theory of
plane-parallel transport problems. This relation is not accidental. In
fact, consider the integro-differential equation of Barbashin type

$$\frac{\partial x(t,s)}{\partial t} + x(t,s) = \int_{-1}^{1} k(t,s,\sigma)x(t,\sigma)\,d\sigma + f(t,s)$$

with boundary conditions

$$x(\tfrac{b}{s},s) = \varphi(s) \ (s > 0), \qquad x(\tfrac{b}{s},s) = \psi(s) \ (s < 0)$$

on the domain

$$D := ([0,b] \times (0,1]) \cup ([-b,0] \times [-1,0))$$

$$\cup \{(t,s) : 0 < s \le \tfrac{b}{t}, \ t \ge b\}$$

$$\cup \{(t,s) : -\tfrac{b}{t} \le s < 0, \ t \le -b\}.$$

After the change of variables $t = \tau/u$ and $s = u$, this equation has the form

$$(19.54) \quad u\frac{\partial y(\tau, u)}{\partial \tau} + y(\tau, u) = \int_{-1}^{1} m(\tau, u, \sigma)y(\tau, \sigma)\, d\sigma + g(\tau, u),$$

with boundary conditions

$$(19.55) \qquad \begin{cases} y(0, u) = \varphi(u) & (u > 0), \\ y(b, u) = \psi(u) & (u < 0). \end{cases}$$

Here (τ, u) belongs to the triangular domain

$$\Delta := \{(\tau, u) : 0 \le \tau \le b,\ -1 \le u \le 1\}.$$

The boundary value problem (19.54)/(19.55) is one of the classical problems describing plane-parallel transport.

We point out that several integro-differential equations may be transformed into Barbashin equations by means of Fourier transforms. For instance, applying to the non-stationary transport equation with isotropic scattering (CASE-ZWEIFEL [1967])

$$\frac{\partial x}{\partial t} + s\frac{\partial x}{\partial r} + x = \frac{c}{2}\int_{-1}^{1} x(t, s, r)\, ds$$

the Fourier transform in r one is led to the integro-differential equation of Barbashin type

$$\frac{\partial x_k(t, s)}{\partial t} + (1 + iks)x_k(t, s) = \frac{c}{2}\int_{-1}^{1} x_k(t, s)\, ds,$$

where

$$x_k(t, s) = \int_{-\infty}^{\infty} x(t, s, r)e^{-ikr}\, dr.$$

Integro-differential equations of the type (19.54) have numerous applications in astrophysics. For example, they describe the radiation transport in the atmospheres of stars, or the absorption lines in star spectra (CHANDRASEKHAR [1950], SOBOLEV [1956, 1972]).

They also occur in the mathematical modelling of wave propagation in media which are bounded by a pair of parallel planes (SOBOLEV [1956, 1972]), in biological experiments (ISHIMARU [1978]), in transport problems for elementary particles in flat media (AGOSHKOV [1988], MASLENNIKOV [1968], CASE-ZWEIFEL [1967], CERCIGNANI [1969], CHANDRASEKHAR [1950]), and in other applications (ENGLAND [1974]).

§ 20. Applications of partial integral equations

In this section we discuss some classes of partial integral operators
and equations which occur in various problems of mathematical physics, such as continuum mechanics, mixed problems of evolutionary
type, axially symmetric contact problems for composite media, frac-
ture mechanics, elasticity and viscoelasticity, creep theory of non-
uniformly aging media, and other fields of solid mechanics. Most of
the material covered in this section is due to ALEKSANDROV-KOVA-
LENKO [1977, 1978, 1980, 1981, 1984, 1986], KOVALENKO [1979, 1981,
1984, 1984a, 1985, 1988, 1989, 1989a, 1990], and MANZHIROV [1983,
1985]. Applying our previous results from § 14 and from the recent
paper KALITVIN [1998] we arrive at several solvability results for such
problems.

Obtaining such solvabiblity results usually amounts to studying the
essential spectrum of the resulting partial integral operator in a suitable Banach space. In fact, this makes it possible to derive existence,
uniqueness, and well-posedness results which are important for the
approximate solution and numerical analysis of the corresponding
equations. Our philosophy is very simple-minded: First we associate
with the given partial integral operator another (simpler) operator
whose spectrum may be calculated explicitly and coincides with the
essential spectrum of the given operator. From a physical reasoning
we may then often conclude that the spectrum of the simpler operator is contained in the positive real half-axis which implies the desired
existence and uniqueness results.

A large part of this section is taken from APPELL-KALITVIN-NASHED
[1999]. For the physical interpretation of the obtained solutions we
refer to the papers cited below.

20.1. Applications to elasticity theory

The following problem from elasticity is studied in VEKUA [1948].
Suppose that a thin elastic plate is clamped in the (x, y)-plane, the
z axis is orthogonal to this plate, and $\overline{F} = \overline{F}(X, Y, Z)$ denotes the
force acting on a unit area of the plate. Assume further that ano-

ther force is acting on the boundary of the plate in the (x, y)-plane. These forces cause a deformation of the plate which may be described by some displacement vector (u, v, w) and which is the sum of the vertical bending and the horizontal deformation of the plate. The mathematical modelling of the horizontal deformation is a classical problem of two-dimensional elasticity theory (see e.g. VEKUA [1948]). The problem of finding the normal displacement leads in turn to the equation

$$(20.1) \quad D\Delta^2 w = Z + X_x \frac{\partial^2 w}{\partial x^2} + Y_y \frac{\partial^2 w}{\partial y^2} + 2X_y \frac{\partial^2 w}{\partial y^2} - X_y \frac{\partial w}{\partial x} - Y \frac{\partial w}{\partial y};$$

here Δ is the Laplace operator, X_x, X_y, and Y_y are the known components of the stress tensor, and $D = Eh^3/12(1-\mu^2)$ is the cylindrical rigidity, where h denotes the thickness of the plate, E the Young module, and μ the Poisson coefficient.

If the orthogonal force $\overline{F}$ to the plate and the horizontal deformation are negligible, then $X = Y = 0$ and $X_x \approx X_y \approx Y_y \approx 0$. In this case (20.1) becomes the classical equation of a bending plate

$$(20.2) \qquad\qquad D\Delta^2 w = Z.$$

If the plate lies on an elastic medium, one has to replace the equations (20.1) and (20.2) by the equations

$$(20.3) \qquad\qquad D\Delta^2 w + kw = Aw$$

and

$$(20.4) \qquad\qquad D\Delta^2 w + kw = Z,$$

where A is the operator defined by the right hand side of (20.1), and k is some positive constant.

In complex coordinates, equation (20.3) may be written in the form

$$(20.5) \quad \begin{aligned} &\frac{\partial^4 w}{\partial z^2 \partial \zeta^2} + a(z, \zeta) \frac{\partial^2 w}{\partial z^2} + 2b(z, \zeta) \frac{\partial^2 w}{\partial z \partial \zeta} + a_1(z, \zeta) \frac{\partial^2 w}{\partial \zeta^2} \\ &+ c(z, \zeta) \frac{\partial w}{\partial z} + c_1(z, \zeta) \frac{\partial w}{\partial \zeta} + \lambda w = f(z, \zeta), \end{aligned}$$

where $z = x + iy$, $\zeta = x - iy$, and

$$a(z, \zeta) = \tfrac{1}{16D}(-X_x + Y_y - 2iX_y),$$

$$a_1(z, \zeta) = \tfrac{1}{16D}(-X_x + Y_y + 2iX_y),$$

(20.6)

$$b(z, \zeta) = -\tfrac{1}{16D}(X_x + Y_y), \quad c(z, \zeta) = \tfrac{1}{16D}(X + iY),$$

$$c_1(z, \zeta) = \tfrac{1}{16D}(X - iY), \quad \lambda = \tfrac{k}{16D}.$$

Equation (20.5) in turn may be rewritten as a partial integral equation of Volterra type

(20.7)
$$
\begin{aligned}
w(z, \zeta) &+ \int_0^z \bar{l}(z, \zeta, t)w(t, \zeta)\, dt \\
&+ \int_0^\zeta \bar{m}(z, \zeta, \tau)w(z, \tau)\, d\tau \\
&+ \int_0^z \int_0^\zeta \bar{n}(z, \zeta, t, \tau)w(t, \tau)\, d\tau\, dt = g(z, \zeta),
\end{aligned}
$$

where we have put

$$\bar{l}(z, \zeta, t) = (z - t)a_1(t, \zeta),$$

$$\bar{m}(z, \zeta, \tau) = (\zeta - \tau)a(z, \tau),$$

$$\bar{n}(z, \zeta, t, \tau) = 2b(t, \tau) + (\zeta - \tau)\bar{c}(t, \tau) + (z - t)\bar{c}_1(t, \tau)$$

(20.8)
$$+ (z - t)(\zeta - \tau)d(t, \tau),$$

$$\bar{c} = -2\frac{\partial a}{\partial z} - 2\frac{\partial b}{\partial \zeta} + c, \quad \bar{c}_1 = -2\frac{\partial b}{\partial z} - 2\frac{\partial a_1}{\partial \zeta} + c_1,$$

$$d = \frac{\partial^2 a}{\partial z^2} + 2\frac{\partial^2 b}{\partial z \partial \zeta} + \frac{\partial^2 a_1}{\partial \zeta^2} - \frac{\partial c}{\partial z} - \frac{\partial c_1}{\partial \zeta} + \lambda,$$

and

$$
\begin{aligned}
g(z, \zeta) &= \int_0^z d\tau \int_0^\zeta (z - t)(\zeta - \tau)f(t, \tau)\, d\tau \\
&+ z\varphi_0(\zeta) + \zeta\psi_0(z) + \varphi_1(\zeta) + \psi_1(z).
\end{aligned}
$$

Here φ_0, φ_1, ψ_0, and ψ_1 are arbitrary holomorphic functions. The equation (20.7) with holomorphic functions (20.6) and f is a special case of the equation

$$(20.9) \qquad w(z,\zeta) + Kw(z,\zeta) = g(z,\zeta),$$

where K is the partial integral operator of Volterra type

$$
\begin{aligned}
Kw(z,\zeta) &= \int_0^z l(z,\zeta,t)w(t,\zeta)\,dt \\
(20.10) \qquad\qquad &+ \int_0^\zeta m(z,\zeta,\tau)w(z,\tau)\,d\tau \\
&+ \int_0^z \int_0^\zeta n(z,\zeta,t,\tau)w(t,\tau)\,d\tau\,dt.
\end{aligned}
$$

In VEKUA [1948] it was shown that equation (20.9) has a holomorphic solution if all data l, m, n, and f are holomorphic. However, this equation may have a unique solution also for more general data. The reason for this is that, under reasonable conditions on the kernels l, m, and n, the spectral radius of the operator (20.10) is zero in many natural function spaces.

In fact, let $\Gamma_1 = \{z : z = \gamma_1(u) : 0 \le u \le a\}$ and $\Gamma_2 = \{\zeta : \zeta = \gamma_2(v) : 0 \le v \le b\}$ be two smooth simple curves in the complex z-plane and ζ-plane, respectively, and let $\Gamma = \Gamma_1 \times \Gamma_2$. The integrals in (20.10) are understood as contour integrals over Γ_1 joining 0 and z, and over Γ_2 joining 0 and ζ, respectively. If the data are not holomorphic, these integrals, and hence also the operator K, depend of course on the contours Γ_1 and Γ_2. Nevertheless, for fixed Γ_1 and Γ_2 the spectral radius of K is still zero. Put

$$l_1(u,v,\tilde{u}) = l(\gamma_1(u),\gamma_2(v),\gamma_1(\tilde{u}))\gamma_1'(\tilde{u}),$$

$$m_1(u,v,\tilde{v}) = l(\gamma_1(u),\gamma_2(v),\gamma_2(\tilde{v}))\gamma_2'(\tilde{v}),$$

$$n_1(u,v,\tilde{u},\tilde{v}) = n(\gamma_1(u),\gamma_2(v),\gamma_1(\tilde{u}),\gamma_2(\tilde{v}))\gamma_1'(u)\gamma_2'(v),$$

$$w_1(u, v) = w(\gamma_1(u), \gamma_2(v)),$$

and

$$g_1(u, v) = g(\gamma_1(u), \gamma_2(v)).$$

Then the operator (20.10) takes the form

(20.11)

$$K_1 w_1(u, v) = \int_0^u l_1(u, v, \tilde{u}) w_1(\tilde{u}, v) \, d\tilde{u}$$

$$+ \int_0^v m_1(u, v, \tilde{v}) w_1(u, \tilde{v}) \, d\tilde{v}$$

$$+ \int_0^u \int_0^v n_1(u, v, \tilde{u}, \tilde{v}) w_1(\tilde{u}, \tilde{v}) \, d\tilde{v} \, d\tilde{u},$$

and the equation (20.9) becomes

(20.12) $$w_1(u, v) + K_1 w_1(u, v) = g_1(u, v).$$

Now we may apply the results of KALITVIN [1998] to this equation. For instance, the spectral radius of the operator (20.11) is zero in the spaces $C(D)$ and $L_p(D)$ ($D = [0, a] \times [0, b]$, $1 \leq p \leq \infty$), provided that the kernels l, m, and n are bounded and integrable. Consequently, equation (20.12) has a unique solution in $C(D)$ or $L_p(D)$, and thus equation (20.9) is uniquely solvable for any function g from $C(D)$ or $L_p(D)$, repsectively.

Analogous results hold for systems of differential equations occuring in the modelling of thin elastic hulls (VEKUA [1948]). Here the kernels depend only on the form of the hull, and the solution determines the stress function. A solution (in a generalized sense) also exists for non-holomorphic entries of the stress tensor and for non-smooth surfaces.

Several conditions for the solvability, or even unique solvability, of equation (20.12), as well as formulas for the resolvent kernel, may be found in KALITVIN [1998]. Unfortunately, only in quite exceptional cases it is possible to construct the resolvent kernel exlicitly. It is therefore useful to have methods for the approximate computation of the resolvent. We point out that also the equations (20.9) and (20.12) themselves are approximations.

To obtain an approximate solution for equation (20.12), one has to replace this equation by the equation

$$\omega(u,v) = \int_0^u \tilde{l}(u,v,\tau)\omega(\tau,v)\,d\tau$$

(20.13)
$$+ \int_0^v \tilde{m}(u,v,\sigma)\omega(u,\sigma)\,d\sigma$$

$$+ \int_0^u \int_0^v \tilde{n}(u,v,\tau,\sigma)\omega(\tau,\sigma)\,d\sigma\,d\tau + \tilde{f}(u,v).$$

In what follows, we denote the partial integral operator defined by the right hand side of (20.13) by $\tilde{K}$. A crucial problem is here to get estimates for the norm of the difference between the solutions and resolvents of the equations (20.12) and (20.13).

Suppose that l, m, n, f, $\tilde{l}$, $\tilde{m}$, $\tilde{n}$, and $\tilde{f}$ are continuous functions. Then the equations (20.12) and (20.13) both are uniquely solvable in the space $C(D)$, where $D = [0,a] \times [0,b]$). Denote by $(I - K_1)^{-1}$ and $(I - \tilde{K})^{-1}$ the corresponding resolvent operators, and by ω and $\tilde{\omega}$, respectively, the corresponding solutions. We get then the obvious estimates

$$\|(I - \tilde{K})^{-1} - (I - K_1)^{-1}\|$$

(20.14)
$$\leq \|(I - \tilde{K})^{-1}\|\,\|(I - K_1)^{-1}\|\,\|K_1 - \tilde{K}\|$$

$$\leq c\,\|K_1 - \tilde{K}\|,$$

where c is some constant which will be specified below. The estimate

(20.14) in turn implies that

$$\|(I - \tilde{K})^{-1} - (I - K_1)^{-1}\|$$

(20.15)

$$\leq c \sup_{(u,v)\in D} \left[\int_0^u |l_1(u, v, \tau) - \tilde{l}(u, v, \tau)| \, d\tau \right.$$

$$+ \int_0^v |m_1(u, v, \sigma) - \tilde{m}(u, v, \sigma)| \, d\sigma$$

$$\left. + \int_0^u \int_0^v |n_1(u, v, \tau, \sigma) - \tilde{n}(u, v, \tau, \sigma)| \, d\sigma \, d\tau \right].$$

Now, the smallness condition

$$(20.16) \qquad \max\{|l_1 - \tilde{l}|, |m_1 - \tilde{m}|, |n_1 - \tilde{n}|\} < \frac{\varepsilon}{c(a + b + ab)}$$

trivially implies that $\|(I - \tilde{K})^{-1} - (I - K_1)^{-1}\| < \varepsilon$, by (20.14). This means that the resolvent operators of the equations (20.12) and (20.13) do not differ very much if the kernels in (20.13) are sufficiently "close" to those in (20.11).

From (20.14) and the obvious inequality

$$\|\omega - \tilde{\omega}\| \leq \|(I - K_1)^{-1}\| \left[\|(I - \tilde{K})^{-1}\| \|K_1 - \tilde{K}\| \|\tilde{f}\| + \|f - \tilde{f}\| \right]$$

it follows that

$$(20.17) \qquad \|\omega - \tilde{\omega}\| \leq c \|K_1 - \tilde{K}\| \|\tilde{f}\| + \|(I - K_1)^{-1}\| \|f - \tilde{f}\|;$$

this shows that also the difference $\|\omega - \tilde{\omega}\|$ of the solutions is small if (20.16) holds and $\tilde{f}$ is chosen sufficiently "close" to f.

The problem of calculating the constant c in (20.14) is related to estimates for the norms of the operators $(I - \tilde{K})^{-1}$ and $(I - K_1)^{-1}$. Since the kernels occurring in (20.12) and (20.13) are continuous, by assumption, they are bounded by a common constant c_0, say. Therefore we have

$$\max\{\|(I - \tilde{K})^{-1}\|, \|(I - K_1)^{-1}\|\} \leq q_0 := \left\| \sum_{n=0}^{\infty} Q_0^n \right\|,$$

where the operator Q_0 is defined by

$$Q_0 x(t, s) = c_0 \int_0^t x(\tau, s)\, d\tau + c_0 \int_0^s x(t, \sigma)\, d\sigma$$

$$+ c_0 \int_0^t \int_0^s x(\tau, \sigma)\, d\sigma\, d\tau$$

for $(t, s) \in D$. To calculate the number q_0 one may use the methods discussed in MÜNTZ [1934]. However, this leads to very cumbersome calculations. Therefore we propose here another method for estimating the constant q_0.

Let $c_1 = \max\{1, c_0\}$, and define an operator Q_1 by

$$Q_1 x(t, s) = c_1 \int_0^t x(\tau, s)\, d\tau$$

$$+ c_1 \int_0^s x(t, \sigma)\, d\sigma + c_1^2 \int_0^t \int_0^s x(\tau, \sigma)\, d\sigma\, d\tau.$$

Since $Q_0 x \leq Q_1 x$ for all positive functions x, we have $q_0 \leq \|(I - Q_1)^{-1}\| =: q_1$. It suffices therefore to estimate the constant q_1. Put

$$L_1 x(t) = c_1 \int_0^t x(\tau)\, d\tau \quad (0 \leq t \leq a)$$

and

$$M_1 x(s) = c_1 \int_0^s x(\sigma)\, d\sigma \quad (0 \leq s \leq b).$$

Then $I - Q_1 = I \overline{\otimes} I - L_1 \overline{\otimes} I - I \overline{\otimes} M_1 - L_1 \overline{\otimes} M_1$. Using the classical tensor product formula (ICHINOSE [1978, 1978a])

$$(I - Q_1)^{-1} = -\frac{1}{4\pi^2} \int_{|\eta|=r} \int_{|\xi|=r} \frac{(\xi I - L_1)^{-1} \overline{\otimes} (\eta I - M_1)^{-1}}{1 - \xi - \eta - \xi\eta}\, d\xi\, d\eta,$$

we get

$$q_1 \leq \frac{r^2}{4\pi^2} \int_0^{2\pi} \int_0^{2\pi} \frac{\|(re^{i\tau} I - L_1)^{-1}\|\, \|(re^{i\sigma} I - M_1)\|}{|1 - r(e^{i\tau} + e^{i\sigma} + re^{i(\tau+\sigma)})|}\, d\sigma\, d\tau.$$

For estimating the last integral it suffices of course to estimate all factors in the numerator and denominator. The evident inequality

$$(I - \rho L_1)^{-1} x(t) = x(t) + \rho \int_0^t e^{\rho(t-\tau)} x(\tau) \, d\tau$$

implies, for the special choice $\rho = \frac{1}{r} e^{-i\tau}$, that

$$\|(re^{i\tau} I - L_1)^{-1}\| = \frac{1}{r} \|(I - \rho L_1)^{-1}\|$$

$$= \frac{1}{r} \left(1 + \sup_{0 \leq t \leq a} \left| \rho \int_0^t e^{\rho(t-\tau)} \, d\tau \right| \right) =: \varphi(r)$$

$$\leq \frac{2e^{a/r} + r + 1}{r^2}.$$

Analogously,

$$\|(re^{i\sigma} I - M_1)^{-1}\| =: \psi(r) \leq \frac{2e^{b/r} + r + 1}{r^2}.$$

Now, for $r = \frac{1}{6}$, say, we have $|1 - r(e^{i\tau} + e^{i\sigma} + re^{i(\tau+\sigma)})| \geq \frac{1}{2}$, and thus $q_0 \leq q_1 \leq \frac{1}{18} \varphi(\frac{1}{6}) \psi(\frac{1}{6})$. Other effective upper estimates for the norm of $(I - Q_1)^{-1}$ may be obtained through other choices of r. Here is yet another method for estimating the norm $\|(I - K_1)^{-1} - (I - \tilde{K})^{-1}\|$. For any two operators A and B we have

$$\|(A + B)^{-1} - A^{-1}\| \leq$$

$$\|(A + B)^{-1} - A^{-1} + A^{-1} B A^{-1}\| + \|A^{-1} B A^{-1}\|$$

$$\leq 2\|A^{-1}\|^3 \|B\|^2 + \|A^{-1}\|^2 \|B\|$$

$$= \|A^{-1}\|^2 \left(2\|A^{-1}\| \|B\| + 1 \right) \|B\|.$$

Choosing, in particular, $A = I - \tilde{K}$ and $B = \tilde{K} - K_1$ yields

$$\|(I - K_1)^{-1} - (I - \tilde{K})^{-1}\|$$

(20.18) $$\leq \|(I - \tilde{K})^{-1}\|^2 (2\|(I - K_1)^{-1}\| \times$$

$$\times \|K_1 - \tilde{K}\| + 1) \|K_1 - \tilde{K}\|.$$

Combining this with the obvious estimate

$$\|K_1 - \tilde{K}\| = \sup_{(u,v)\in D} \left[\int_0^u |l_1(u,v,\tau) - \tilde{l}(u,v,\tau)|\,d\tau \right.$$

(20.19)
$$+ \int_0^v |m_1(u,v,\sigma) - \tilde{m}(u,v,\sigma)|\,d\sigma$$

$$\left. + \int_0^u \int_0^v |n_1(u,v,\tau,\sigma) - \tilde{n}(u,v,\tau,\sigma)|\,d\sigma\,d\tau \right],$$

we see that it suffices to know the norm of the operator $(I - \tilde{K})^{-1}$ in order to estimate $\|(I - K_1)^{-1} - (I - \tilde{K})^{-1}\|$. Finally, using (20.18) and (20.19) yields estimates for the error $\|\omega - \tilde{\omega}\|$ between the two solutions ω and $\tilde{\omega}$.

In the preceding discussion we supposed throughout that the operators K_1 and $\tilde{K}$ act in the space $C(D)$ and the functions f and $\tilde{f}$ are continuous on D. For numerical studies equation (20.12) is often replaced by (20.13), where $\tilde{f}(t_i, s_j)$ and $\tilde{\omega}(t_i, s_j)$ $(i = 1,\ldots,n;\ j = 1,\ldots,m)$ are $n \times m$ matrices; in this case $\tilde{K}$ has to be considered in the matrix space $\mathbb{R}^{n\times m}$. To apply the estimates for the difference $\|\omega - \tilde{\omega}\|$ we may then identify this matrix space with an appropriate subspace of $C(D)$.

20.2. Applications to mechanics of continuous media

The mathematical modelling of some mixed problems of elasticity theory, viscoelasticity, and hydrodynamics (ALEKSANDROV-KOVA-LENKO [1978, 1980], KOVALENKO [1981, 1985]) leads to the equation of Volterra-Fredholm type

$$(20.20) \qquad \lambda x(t,s) + Kx(t,s) = g(t,s)$$

$((t,s) \in D = [0,a] \times [-1,1],\ \lambda > 0)$ involving a partial integral operator

$$(20.21) \quad Kx(t,s) = \int_0^t l(t,\tau)x(\tau,s)\,d\tau + \int_{-1}^1 m(s-\sigma)x(t,\sigma)\,d\sigma,$$

with a kernel m of the form

(20.22) $$m(\xi) = \frac{1}{2} \int_{-\infty}^{\infty} \tilde{m}(x) e^{i\theta x \xi} \, dx \quad (\theta > 0).$$

Here $\tilde{m} = \tilde{m}(x)$ is a positive continuous even function on $\mathbb{R}$ satisfying

$$\tilde{m}(x) = A + O(x^2) \ (x \to 0), \quad |x|\tilde{m}(x) = B + O(x^{-1}) \ (x \to \infty),$$

the kernel l is continuous or weakly singular, and the function g has the form

$$g(t, s) = g_1(t) + s g_2(t) + f(s),$$

with $g_1, g_2 \in C([0, a])$ and $f \in L_2([-1, 1])$.

The equation (20.20) is usually studied in mechanics (ALEKSAN-DROV-KOVALENKO [1980], KOVALENKO [1981, 1985]), where either the functions g_1 and g_2 are given, or the functions

$$p(t) := \int_{-1}^{1} x(t, \sigma) \, d\sigma, \qquad q(t) := \int_{-1}^{1} \sigma x(t, \sigma) \, d\sigma.$$

A solution of (20.20) is then a function x such that $x(t, \cdot) \in L_2([-1, 1])$ and $x(\cdot, s) \in C([0, a])$. Thus, an appriopriate space for studying this equation is the tensor product of $L_2([-1, 1])$ and $C([0, a])$ with uniform crossnorm and, in particular, the space $C_t(L_2)$ we introduced in Subsection 1.1. It is clear that the solvability results for (20.20) are closely related to the properties of the operator (20.21) in these spaces. In KALITVIN-ZABREJKO [1991] it was shown that the spectrum of the operator (20.21) in $L_p(D)$ $(1 \le p < \infty)$ coincides with the spectrum of the compact operator $-\tilde{M}$, where

(20.23) $$\tilde{M}x(s) = \int_{-1}^{1} m(s - \sigma) x(\sigma) \, d\sigma$$

maps $L_p([-1, 1])$ $(1 \le p < \infty)$ into $C([-1, 1])$ and is selfadjoint and positive definite in $L_2([-1, 1])$. This implies that for any positive number $\lambda \notin \sigma(K)$ and each function $g \in L_p(D)$, equation (20.20) has a unique solution $x \in L_p(D)$.

We remark that, for $t = 0$, (20.20) gives a classical equation of the theory of mixed problems (ALEKSANDROV-KOVALENKO [1980], KO-VALENKO [1981]).

Let us now consider the operator (20.21) in $C(D)$ and in $C_t(L_p)$ $(1 \leq p < \infty)$. By $\tilde{L}$ we denote the operator

$$(20.24) \qquad \tilde{L}x(t) = \int_0^t l(t, \tau)x(\tau)\, d\tau,$$

which in the case of a continuous or weakly singular kernel l is compact in the space $C([0, a])$ and has zero spectral radius. Under our assumptions on $\tilde{M}$ and $\tilde{L}$, the operator (20.21) is bounded both in $C_t(L_p)$ $(1 \leq p < \infty)$ and in $C(D)$, we have $\sigma(\tilde{L}) = \{0\}$, and $\sigma(\tilde{M})$ consists of zero and a countable (possibly finite) set of positive eigenvalues $\lambda_1, \lambda_2, \lambda_3, \ldots$ with corresponding continuous eigenfunctions. Since the spaces $C_t(L_p)$ and $C(D)$ are isometric to the weak tensor product $C \overline{\otimes} L_p$ and $C([0, a]) \overline{\otimes} C([-1, 1])$, respectively (GRO-THENDIECK [1955]), and the weak cross-norm is uniform (SCHATTEN [1950]), one may apply the results of ICHINOSE [1978, 1978a]. In particular, the equality

$$(20.25) \qquad \begin{aligned} \sigma(K) &= \sigma_{es}(K) = \sigma_{ew}(K) = \sigma_{eb}(K) \\ &= \sigma_+(K) = \sigma_-(K) = \sigma_l(K) = \sigma_d(K) = \sigma(\tilde{M}) \end{aligned}$$

is true, where we use the notation of Subsection 14.1. Consequently, the operator K in (20.20) is invertible (Fredholm, Fredholm of index zero, n-normal, or d-normal) if and only if $-\lambda \notin \sigma(\tilde{M})$. In particular, equation (20.20) has a unique solution for any positive λ. We summarize with the following

Theorem 20.1. *Suppose that the integral operator*

$$(20.26) \qquad \tilde{M}x(s) = \int_{-1}^1 m(s - \sigma)x(\sigma)\, d\sigma$$

is bounded in $L_p([-1, 1])$ $(1 \leq p < \infty)$ or in $C([-1, 1])$, and the operator (20.24) is bounded and weakly compact in $C([0, a])$. Then the operator

$$(20.27) \qquad Kx(t, s) = \int_0^t l(t, \tau)x(\tau, s)\, d\tau + \int_{-1}^1 m(s - \sigma)x(t, \sigma)\, d\sigma$$

is bounded in $C(L_p)$ $(1 \leq p < \infty)$ or in $C(D)$, respectively, and satisfies the equalities

$$\sigma_l(K) = \sigma_l(\tilde{M}), \qquad \sigma_d(\dot{K}) = \sigma_d(\tilde{M})$$

and

$$\sigma(K) = \sigma_{es}(K) = \sigma_{ew}(K) = \sigma_{eb}(K) = \sigma_+(K) = \sigma_-(K) = \sigma(\tilde{M}).$$

Consequently, the following four statements are equivalent:

(a) *The operator $\lambda I + K$ is Fredholm;*

(b) *the operator $\lambda I + K$ is Fredholm of index zero;*

(c) *the operator $\lambda I + K$ is invertible;*

(d) $-\lambda \notin \sigma(\tilde{M})$.

In particular, equation (20.20) has a unique solution for any $\lambda > 0$.

□ For the proof it suffices to remark that $\sigma(\tilde{L}) = \sigma_{es}(\tilde{L}) = \sigma_{ew}(\tilde{L}) = \sigma_{eb}(\tilde{L}) = \sigma_+(\tilde{L}) = \sigma_-(\tilde{L}) = \sigma_l(\tilde{L}) = \{0\}$, and to apply Theorem 14.1 and Theorem 14.2. ■

Consider the equation of continuous media

$$(20.28) \qquad\qquad x + Kx = f,$$

where K is the operator (20.27). Under the hypotheses of Theorem 20.1, the invertibility, the Fredholmness, and the Fredholmness with index zero of $I + K$ in $C_t(L_p)$ $(1 \leq p < \infty)$ or in $C(D)$ are all equivalent to the condition $-1 \notin \sigma(\tilde{M})$. But this condition is clearly satisfied, since the kernel $m = m(s, \sigma)$ arising in the equations of continuous media generates a selfadjoint positive definite compact operator $\tilde{M}$ in $L_2([-1, 1])$. Moreover, the unique solution of equation (20.28) has the integral representation

$$(20.29) \qquad x = -\frac{1}{4\pi^2} \left(\int_{\Gamma_1} \int_{\Gamma_2} \frac{(\xi I - \tilde{L})^{-1} \overline{\otimes} (\eta I - \tilde{M})^{-1}}{1 - \xi - \eta} \, d\xi \, d\eta \right) f,$$

where $\tilde{L}$ is the operator (20.24), $\tilde{M}$ is the operator (20.23), and Γ_1 and Γ_2 are simple closed contours surrounding $\sigma(\tilde{L})$ and $\sigma(\tilde{M})$, respectively. In this way, (20.29) gives an explicit representation for the contact stress x.

20.3. Mixed problems of evolutionary type

Some contact problems of elasticity which involve abrasion effects between the rough surfaces of two bodies lead to partial integral equations of Volterra-Fredholm type. Further examples arise in some mixed problems describing visco-elastic multilayer ground, when the relative thickness and rigidity are small, or in the study of so-called refined equations of elastic plates (see e.g. ALEKSANDROV-KOVALENKO [1977, 1980, 1986], ALEKSANDROV-KOVALENKO-MKHITARJAN [1982], ALEKSANDROV-SMETANIN [1985], and KOVALENKO [1981, 1989, 1990]).

The equation in question has the form

$$(20.30) \qquad x(t,s) + Kx(t,s) + Nx(t,s) = g(t,s)$$

$((t,s) \in D = [0,a] \times [-1,1])$, where K is the operator (20.21), and N is the Volterra-Fredholm operator

$$(20.31) \qquad Nx(t,s) = \int_0^t \int_{-1}^1 n(t,\tau)m(s-\sigma)x(\tau,\sigma)\,d\sigma\,d\tau.$$

The given functions l, m, and g satisfy here the same conditions as in Subsection 20.2, and the kernel $n = n(t,\tau)$ is supposed to be continuous or weakly singular. An important particular case of (20.30), namely

$$m(s-\sigma) = -\log|s-\sigma|+d, \quad n(t,\tau) = l(t,\tau) = e^{-\alpha(t-\tau)},$$

and

$$g(t,s) = g_1(t) + f(s),$$

has been studied in ALEKSANDROV-KOVALENKO [1986].

Conditions for the operator $I + K + N$ to be invertible or Fredholm in the space $C_t(L_p)$ $(1 \le p \le \infty)$ may be found in KALITVIN [1994, 1997, 1997a, 1997b]. Let

$$(20.32) \qquad \tilde{N}x(t) = \int_0^t n(t,\tau)x(\tau)\,d\tau.$$

Theorem 20.2. *Suppose that the integral operator* (20.26) *is compact in* $L_p([-1,1])$ $(1 \le p < \infty)$ *or in* $C([-1,1])$, *and the operators* (20.24) *and* (20.32) *are compact in* $C([0,a])$. *Then the operator* $K + N$ *acts in* $C_t(L_p)$ *or* $C(D)$, *respectively, and the equalities*

$$\sigma(K+N) = \sigma_{es}(K+N)$$
$$(20.33) \qquad = \sigma_{ew}(K+N) = \sigma_+(K+N) = \sigma_-(K+N)$$
$$= \sigma_l(K+N) = \sigma_d(K+N) = \sigma(\tilde{M})$$

are true.

☐   The fact that the operator $K + N$ acts in $C_t(D)$ or $C(D)$ is obvious. The compactness of N is a consequence of the compactness of the operators (20.24) and (20.32). The equalities

$$\sigma(K+N) = \sigma_{es}(K+N) = \sigma_{ew}(K+N) = \sigma_+(K+N)$$
$$= \sigma_-(K+N) = \sigma_l(K+N) = \sigma_d(K+N)$$

follow from Theorem 20.1 and the stability of Fredholmness and the index with respect to compact perturbations.

It remains to prove that $\sigma(K+N) = \sigma(\tilde{M})$. The inclusion $\sigma(\tilde{M}) \subseteq \sigma(K+N)$ is obvious. Let $\lambda \notin \sigma(\tilde{M})$. Then the equation $\lambda x - (K+N)x = f$ has the form

$$(I - \tfrac{1}{\lambda}\tilde{L})\overline{\otimes}(\lambda I - \tilde{M})x = (\tfrac{1}{\lambda}\tilde{L} - \tilde{N})\overline{\otimes}\tilde{M}x + f.$$

Since the operator $(I - \tfrac{1}{\lambda}\tilde{L})\overline{\otimes}(\lambda I - \tilde{M})$ is invertible, this equation is equivalent to the equation $x = Ax + g$, where

$$A = (I - \tfrac{1}{\lambda}\tilde{L})^{-1}(\tfrac{1}{\lambda}\tilde{L} - \tilde{N})\overline{\otimes}(\lambda I - \tilde{M})^{-1}M$$

and $g = (I - \frac{1}{\lambda}\tilde{L})^{-1}\overline{\otimes}(\lambda I - \tilde{M})^{-1}f$. Moreover, since $(I - \frac{1}{\lambda}\tilde{L})^{-1}(\frac{1}{\lambda}\tilde{L} - \tilde{N})$ is a compact integral operator of Volterra type in $C([0, a])$, its spectral radius is zero. Consequently, A has also spectral radius zero, and thus the equation $x = Ax + g$ admits a unique solution. But then the equation $\lambda x - (K + N)x = f$ has a unique solution as well which means that $\lambda \notin \sigma(K + N)$. ∎

Theorem 20.2 implies, in particular, that the operator $\lambda I + K + N$ is Fredholm of index zero if and only if $-\lambda \notin \sigma(\tilde{M})$. Since $-1 \notin \sigma(\tilde{M})$, the operator $I + K + N$ in equation (20.30) is certainly Fredholm of index zero. Consequently, the Fredholm alternative applies to equation (20.30).

We remark that equation (20.30) occurs also in the mathematical modelling of some axially symmetric contact problems for non-uniformly aging visco-elastic foundations (MANZHIROV [1983, 1985]). The corresponding Volterra-Fredholm equations have the same properties as the operator $K + N$ in Theorem 20.2. However, there are other axially symmetric contact problems which lead to completely different partial integral operators of Volterra-Fredholm type. We will consider such problems in Subsections 20.4 and 20.6.

Again, under the hypotheses of Theorem 20.2 the invertibility, the Fredholmness, and the Fredholmness with index zero of $I + K + N$ in $C_t(L_2)$ are all equivalent to the condition $-1 \notin \sigma(\tilde{M})$. In the important particular case $n(t, \tau) = l(t, \tau)$, this condition also implies the integral representation

$$x = -\frac{1}{4\pi^2} \left(\int_{\Gamma_1} \int_{\Gamma_2} \frac{(\xi I - \tilde{L})^{-1}\overline{\otimes}(\eta I - \tilde{M})^{-1}}{1 - \xi - \eta - \xi\eta} \, d\xi \, d\eta \right) f$$

for the solution x. The condition $-1 \notin \sigma(\tilde{M})$ is again satisfied automatically, since the kernel $m = m(s, \sigma)$ in (20.30) defines a selfadjoint positive definite operator in $L_2([-1, 1])$.

Approximate solutions x_n of equation (20.30) may be constructed in rather the same way as we have done in Subsection 20.1 for the equation (20.12). Let $x(t_i, s)$ denote the value of the solution x of the system (20.30) for $t = t_i$ $(i = 0, 1, \ldots, n - 1)$, and let

$$(20.34) \quad x_n(t, s) = \begin{cases} x(t_{i-1}, s) & \text{if} \quad (t, s) \in [t_{i-1}, t_i) \times [-1, 1], \\ x(t_{n-1}, s) & \text{if} \quad (t, s) \in [t_{n-1}, t_n] \times [-1, 1]. \end{cases}$$

Equation (20.30) may then be approximated by the system of integral equations

$$x_i(s) = -\frac{a}{n} \sum_{k=0}^{i-1} l(t_i, t_k) x_k(s)$$

$$-\frac{a}{n} \sum_{k=0}^{i-1} n(t_i, t_{k-1}) \int_{-1}^{1} m(s, \sigma) x_k(\sigma) \, d\sigma$$

$$-\int_{-1}^{1} m(s, \sigma) x_i(\sigma) \, d\sigma + f(t_i, s) \qquad (i = 0, 1, \ldots, n).$$

Since $-1 \notin \sigma(\tilde{M})$, for each n we get a unique solution $x_0(s), \ldots, x_n(s)$ of this system. Putting this solution into (20.34) we get the approximate solution $(i = 0, 1, \ldots, n - 1)$

$$x_n(t, s) \approx y_n(t, s) = \begin{cases} x_{i-1}(s) & \text{if} \quad (t, s) \in [t_{i-1}, t_i) \times [-1, 1], \\ x_{n-1}(s) & \text{if} \quad (t, s) \in [t_{n-1}, t_n] \times [-1, 1] \end{cases}$$

for equation (20.30). If the kernels l and n in the equations of continuous media and mixed problems of evolutionary type are continuous, the sequences $(x_n)_n$ and $(y_n)_n$ obtained in this way converge in the space with mixed norm $[L_\infty \leftarrow L_2]$ (see Subsection 12.2) to the exact solution.

We point out that our method for the approximate solution of the equations (20.28) and (20.21) is different from that proposed in ALEKSANDROV-KOVALENKO [1980] or KOVALENKO [1981].

For numerical purposes the following approach is suitable. Let $\{t_0, \ldots, t_n\}$ be an equidistant partition of $[0, a]$, and $\{s_0, \ldots, s_m\}$ an equidistant partition of $[-1, 1]$. Replace the original equation by the approximating system

$$x_{ij} = -\frac{a}{n} \sum_{k=0}^{i-1} l(t_i, t_k) x_{kj} + \frac{2}{m} n(t_i, t_k) \sum_{l=0}^{m-1} m(s_j, s_l) x_{kl}$$

$$-\frac{2}{m} \sum_{l=0}^{m-1} m(s_j, s_l) x_{il} = f_{ij} \qquad (i = 0, 1, \ldots, n; \; j = 0, 1, \ldots, m).$$

This system may be solved successively for $i = 0, 1, \ldots, n$, and then the solution is extended by linear interpolation. In this way, one obtains continuous approximate solutions to the original equation which are arbitrarily close to the exact solution.

20.4. Axially symmetric contact problems

The axially symmetric contact problem describing the impression of an annular stamp in an elastic rough surface with abrasion effect leads to the partial integral equation of Volterra-Fredholm type

(20.35) $$\lambda x(t, s) + K_1 x(t, s) = g(t, s)$$

$((t, s) \in D = [0, a] \times [-1, 1], \lambda > 0)$, see ALEKSANDROV-KOVALENKO [1978, 1981, 1984], ALEKSANDROV-SMETANIN [1985], AVILKIN-ALEKSANDROV-KOVALENKO [1985], GALIN-GORJACHEVA [1977], and KOVALENKO [1985, 1988]). Here

(20.36)
$$K_1 x(t, s) = \int_0^t s x(\tau, s) \, d\tau$$

$$+\frac{2}{\pi} \int_c^1 m_1\left(\frac{2\sqrt{s\sigma}}{s+\sigma}\right) \frac{\sigma}{s+\sigma} x(t, \sigma) \, d\sigma,$$

$m_1(k)$ is an elliptic integral of the first kind, and $g(t, s) = f(t) + \overline{f}(s)$, where $\overline{f} \in L_2([c, 1])$ and $f \in C([0, a])$. In the problem described above, f has the form $f(t) = \alpha + \beta t + \gamma(t)$ and characterizes the displacement of the stamp in time, α and β are constants, and $\gamma(t) \to$

0 as $t \to \infty$. The unknown function $x = x(t, s)$ describes the contact pressure, while the function

$$p(t) = 2\pi \int_c^1 \sigma x(t, \sigma) \, d\sigma$$

is the force which the stamp imposes on the surface.

In this connection, two problems are of particular interest (ALEKSAN-DROV-KOVALENKO [1978], GALIN-GORJACHEVA [1977], KOVALENKO [1985]): First, how to find the contact pressure x and the contact force p for given f; second, how to find the contact pressure x and the displacement γ for given p. These problems have been studied in KOVALENKO [1985], however, without referring to the spectrum of the operator K_1. Since the operator K_1 is not of the form (14.1), the results established in § 14 do not apply. Nevertheless, it is possible to give a fairly complete description of the spectrum. Define two operators L_1 and M_1 by

$$L_1 x(t) = \int_0^t x(\tau) \, d\tau \qquad M_1 x(s) = \int_c^1 \sigma \tilde{m}(s, \sigma) x(\sigma) \, d\sigma,$$

where we have put

$$\tilde{m}(s, \sigma) = \frac{2}{\pi} \frac{m_1(\sqrt{s\sigma}/(s + \sigma))}{s + \sigma}.$$

Theorem 20.3. *Under the above hypotheses, the equalities*

$$\sigma(K_1) = \sigma_{es}(K_1) = \sigma_{ew}(K_1) = \sigma_{eb}(K_1)$$

(20.37)

$$= \sigma_+(K_1) = \sigma_-(K_1) = \sigma_l(K_1) = \sigma_d(K_1) = \sigma(M_1)$$

are true. Consequently, the following four statements are equivalent:
(a) *The operator $\lambda I + K_1$ is Fredholm;*
(b) *The operator $\lambda I + K_2$ is Fredholm of index zero;*
(c) *The operator $\lambda I + K_2$ is invertible;*
(d) $-\lambda \notin \sigma(M_1)$.
In particular, equation (20.35) has a unique solution for any $\lambda > 0$.

□ Let $\lambda \in \sigma(M_1)$ and $\lambda \neq 0$. Then λ is an eigenvalue of the operator M_1 with corresponding normalized eigenfunction $v = v(s)$. Since L_1 is a compact operator in $C([0, a])$, we find a sequence of normalized functions $u_n = u_n(t)$ without convergent subsequence such that $L_1 u_n \to 0$ as $n \to \infty$. We have then

$$(K_1 - \lambda I)u_n(t)v(s) = sv(s)L_1 u_n(t) + u_n(t)(M_1 - \lambda I)v(s)$$

$$= sv(s)L_1 u_n(t) \to 0$$

as $n \to \infty$. Now let $\lambda = 0$. Then we find a sequence of normalized functions $v_n = v_n(s)$ without convergent subsequence such that $M_1 v_n \to 0$ as $n \to \infty$. Here we have $K_1 u_n(t)v_n(s) \to 0$ as $n \to \infty$. In both cases, $\lambda \in \sigma_a(K_1) \subseteq \sigma(K)$ for a denotes es, ew, eb, $+$, $-$, l, or d; in fact, the sequence $(u_n \otimes v_n)_n$ does not contain a convergent subsequence if $\lambda \neq 0$, and the operator $K_1 - \lambda I$ has either an infinite-dimensional nullspace or a non-closed range.

To prove (20.37), it remains to show that $\sigma(K_1) \subseteq \sigma(M_1)$. Let $\lambda \notin \sigma(M_1)$. Then the equation $(\lambda I - K_1)x = g$, where g is an arbitrary function in $C_t(L_2)$ or $C(D)$, has the form

$$I\overline{\otimes}(\lambda I - M_1)x(t, s) = s(L_1\overline{\otimes}I)x(t, s) + g(t, s).$$

Since the operator $I\overline{\otimes}(\lambda I - M_1)^{-1}$ is inverse to the operator $I\overline{\otimes}(\lambda I - M_1)$, the last equation is equivalent to

(20.38)
$$x(t, s) = L_1\overline{\otimes}(\lambda I - M_1)^{-1}sx(t, s)$$
$$+ I\overline{\otimes}(\lambda I - M_1)^{-1}g(t, s).$$

Let $A = L_1\overline{\otimes}(\lambda I - M_1)^{-1}\Pi$ with $\Pi x(s) = sx(s)$. Since $\sigma(L_1) = \{0\}$, we get $\sigma(A) = \sigma(L_1)\sigma([\lambda I - M_1]^{-1}\Pi) = \{0\}$ as well (ICHINOSE [1978]). Consequently, equation (20.38) has a unique solution, and hence also the equation $(\lambda I - K_1)x = g$. This shows that the operator $\lambda I - K_1$ is invertible, hence $\lambda \notin \sigma(K_1)$. In this way we have proved that $\sigma(K_1) \subseteq \sigma(M_1)$.

We show now that no number $\lambda < 0$ may belong to the spectrum $\sigma(M_1)$. In fact, since the integral operator with kernel $\tilde{m}$ is compact,

self-adjoint, and positive definite in $L_2([-1,1])$, the same is true for
the integral operator $\hat{M}$ with kernel $\hat{m}(s,\sigma) = s\sigma\tilde{m}(s,\sigma)$. Therefore
M_1 is a compact, self-adjoint, and positive definite operator in the
space $L_2([-1,1];w)$ with weight function $w(s) = s$, and thus its spectrum in this space contains only nonnegative real numbers. Since the
spaces $L_2([-1,1];w)$ and $L_2([-1,1])$ are isomorphic, the spectrum of
M_1 in $L_2([-1,1])$ also contains only nonnegative real numbers. From
this and (20.37) it follows that the operator $\lambda I + K_1$ is invertible for
positive λ. Consequently, for $\lambda > 0$ equation (20.35) has a unique solution in $C_t(L_2)$ or $C(D)$ if g belongs to $C_t(L_2)$ or $C(D)$, respectively.
■

Define operators $\tilde{L}_1$, $\tilde{M}_1$ and $\tilde{K}_1$ by

$$(20.39) \quad \begin{cases} \tilde{L}_1 x(t) = \displaystyle\int_0^t l_1(t,\tau)x(\tau)\,d\tau, \\[3mm] \tilde{M}_1 x(s) = \displaystyle\int_{-1}^1 m_1(s,\sigma)x(\sigma)\,d\sigma, \end{cases}$$

and

$$(20.40) \quad \begin{aligned} \tilde{K}_1 x(t,s) &= a(s)\int_0^t l_1(t,\tau)x(\tau,s)\,d\tau \\[3mm] &\quad + \int_{-1}^1 m_1(s,\sigma)x(t,\sigma)\,d\sigma, \end{aligned}$$

respectively. In the same way as we have proved (20.37) one may
prove the following slightly more general result (KALITVIN [1997b]):

Theorem 20.4. *Suppose that the integral operator $\tilde{L}_1$ in (20.39) acts
in $C([0,a])$ and is weakly compact, and the integral operator $\tilde{M}_1$ is
compact in $L_p([-1,1])$ $(1 \le p < \infty)$ or in $C([-1,1])$. Assume further
that the function $a = a(s)$ in the definition of $\tilde{K}_1$ is continuous on
$[-1,1]$. Then the operator (20.40) acts in $C_t(L_p)$ and in $C(D)$, and
the equalities*

$$\sigma(\tilde{K}_1) = \sigma_{es}(\tilde{K}_1) = \sigma_{ew}(\tilde{K}_1) = \sigma_{eb}(\tilde{K}_1)$$

$$= \sigma_+(\tilde{K}_1) = \sigma_-(\tilde{K}_1) = \sigma_l(\tilde{K}_1) = \sigma_d(\tilde{K}_1) = \sigma(\tilde{M}_1)$$

are true.

Consider the integral equation (20.35) of the axially symmetric contact problem. In Theorem 20.4 we have proved the unique solvability of this equation for each $\lambda > 0$. For solving equation (20.35) explicitly it is useful to apply the Laplace transform with respect to the variable t. Thus, let $X = X(p,s)$ and $F = F(p,s)$ denote the Laplace transforms of $x = x(t,s)$ and $f = f(t,s)$, respectively. Equation (20.35) becomes then

$$(20.41) \quad \lambda X(p,s) + \frac{p}{p+s} \int_{-1}^{1} \tilde{m}(s,\sigma)X(p,\sigma)\,d\sigma = \frac{p}{p+s}F(p,s),$$

where $\tilde{m}$ is defined through m_1 as above. In this way, we have reduced the problem to the solution of a Fredholm integral equation which depends on a parameter. For $p \geq 0$ it is useful to study this equation in the space $L_2([-1,1];w)$ with weight $w(p,s) = s(p+s)$. In fact, the integral operator with kernel $m(s,\sigma)p/(p+s)$ is obviously compact, selfadjoint and positive definite in this weighted space. Denote by $\lambda_1(p), \lambda_2(p), \ldots$ its eigenvalues, and by $e_1(p,s), e_2(p,s), \ldots$ the corresponding normalized eigenfunctions. We have then

$$\lambda X(p,s) = \sum_{k=1}^{\infty} x_k(p)e_k(p,s)$$

and

$$\frac{p}{p+s}F(p,s) = \sum_{k=1}^{\infty} f_k(p)e_k(p,s).$$

Putting this into equation (20.41) we get

$$(20.42) \quad \sum_{k=1}^{\infty} e_k(p,s)[x_k(p)(\lambda + \lambda_k(p)) - f_k(p)] = 0.$$

Since all eigenvalues $\lambda_k(p)$ are nonegative, from (20.42) we get

$$x_k(p) = \frac{f_k(p)}{\lambda + \lambda_k(p)} \qquad (k = 1, 2, \ldots).$$

Therefore the unique solution of (20.41) may be written in the form

$$X(p,s) = \sum_{k=1}^{\infty} \frac{e_k(p,s)f_k(p)}{\lambda + \lambda_k(p)}.$$

Taking now the inverse Laplace transform of this function yields a solution of equation (20.35).

Analogously, we may solve the equation

(20.43)
$$x(t,s) + a(s) \int_0^t l(t-\tau)x(\tau,s)\,d\tau$$
$$+ \int_{-1}^1 m(s,\sigma)x(t,\sigma)\,d\sigma = f(t,s),$$

where m and f are as above, and the functions $a = a(s)$ and $l = l(t)$ are continuous and positive.

Another type of integral equations arising in axially symmetric contact problems for the creeping of non-uniformly aging bodies is (20.44) below. Under the hypotheses of Subsection 20.5, this equation has a unique solution in $C_t(L_2)$. However, to find this solution explicitly is hardly possible in the general case. Of course, one may again apply approximate and numerical methods; for the equations (20.44), (20.35), and (20.43) this may be done in rather the same way as before. For more general equations we will study this problem in Subsection 20.6 below.

20.5. Creeping of non-uniformly aging bodies

The statement of the following problems may be found in ALEK-SANDROV-ARUTJUNJAN-MANZHIROV [1986] and MANZHIROV [1985], see also ALEKSANDROV-KOVALENKO-MANZHIROV [1984], KOVALEN-KO [1984a] or MANZHIROV [1983]. Consider the axially symmetric plane contact problem of a smooth rigid stamp penetrating into a non-uniformly aging viscous-elastic multilayer surface which consists of an inhomogeneous thin layer, a rod-like intermediate layer, and a homogeneously aging layer of arbitrary thickness. Suppose that an alloy of such layers is in contact with a rigid surface, the boundary of the thin layer is free of tangential stress, and the stress along other directions of the alloy is realized by adhesion or smooth contact.

Problems of this type lead to partial integral equations of Volterra-Fredholm type of the form

(20.44) $x(t, s) - K_2 x(t, s) - N_2 x(t, s) = g(t, s)$

$((t, s) \in D = [1, t] \times [a, b])$, where

$$K_2 x(t, s) = \int_1^t l_2(t, \tau) x(\tau, s) \, d\tau - c(t) \int_a^b m_2(s, \sigma) x(t, \sigma) \, d\sigma,$$

and

(20.45) $N_2 x(t, s) = c(t) \int_1^t \int_a^b n(t, \tau) m_2(s, \sigma) x(\tau, \sigma) \, d\sigma \, d\tau.$

Here $c = c(t)$ is a positive continuous function on $[1, T]$, and the kernels $l_2 = l_2(t, \tau)$ and $n = n(t, \tau)$ are bounded and of Volterra type, i.e. equal to zero for $\tau > t$ and continuous everywhere on $[1, T] \times [1, T]$, possibly except on the diagonal $t = \tau$. We further assume that the operator

$$M_2 x(s) = \int_a^b m_2(s, \sigma) x(\sigma) \, d\sigma$$

with square summable kernel $m_2 = m_2(s, \sigma)$ is self-adjoint and positive definite on $L_2([a, b])$. In this connection, two problems are of particular interest: First, how to find the unknown solution $x \in C_t(L_2)$ and the functions

$$p(t) = \int_a^b x(t, \sigma) \, d\sigma, \quad q(t) = \int_a^b \sigma x(t, \sigma) \, d\sigma$$

for given $g \in C_t(L_2)$; second, how to find the functions x and g if the functions p and q are known.

Given $g \in C_t(L_2)$, let

$$L_2 x(t) = \int_1^t l_2(t, \tau) x(\tau) \, d\tau.$$

The equation (20.44) may then be written in the form

$$[I\overline{\otimes}I + c(t)I\overline{\otimes}M_2][I - L_2)\overline{\otimes}I]x(t,s)$$

$$= [-c(t)L_2\overline{\otimes}M_2 + N_2]x(t,s) + g(t,s).$$

Since $\sigma(c(t)I\overline{\otimes}M_2) = E(c)\sigma(M_2)$ (see ICHINOSE [1978]), where

$$E(c) = \{c(t) : 1 \le t \le T\}$$

is the range of the function c, and the spectrum $\sigma(M_2)$ consists of 0 and a countable (possibly finite) number of positive eigenvalues, the operator $I\overline{\otimes}I + c(t)I\overline{\otimes}M_2$ is invertible. Since $r_\sigma(L_2) = 0$, the operator $(I - L_2)\overline{\otimes}I$ is also invertible. Consequently, we may rewrite the last equation in the form $x = Vx + f$, where

$$f(t,s) = [I + c(t)I\overline{\otimes}M_2]^{-1}g(t,s),$$

and V is the compact integral operator of Volterra type

(20.46)
$$V = [(I - L_2)^{-1}\overline{\otimes}I][I\overline{\otimes}I + c(t)I\overline{\otimes}M_2]^{-1}\times$$

$$\times[-c(t)L_2\overline{\otimes}M_2 + N_2].$$

Evidently, this operator is compact in $L_2(D)$. Consequently, $r_\sigma(V) = 0$, and the equation (20.44) has a unique solution in $C_t(L_2)$ for any $g \in C_t(L_2)$. Moreover, this solution may be obtained by means of the successive approximations $x_{n+1} = Vx_n + f$ $(n = 0,1,2,...)$ for any initial approximation $x_0 \in C_t(L_2)$. Finally, under our hypotheses the equalities

(20.47)
$$\sigma(K_2) = \sigma_{es}(K_2) = \sigma_{ew}(K_2) = \sigma_{eb}(K_2) = \sigma_+(K_2)$$

$$= \sigma_-(K_2) = \sigma_l(K_2) = \sigma_d(K_2) = -E(c)\sigma(M_2)$$

and

(20.48)
$$\sigma(K_2 + N_2) = \sigma_{es}(K_2 + N_2) = \sigma_{ew}(K_2 + N_2)$$

$$= \sigma_{eb}(K_2 + N_2) = \sigma_+(K_2 + N_2)$$

$$= \sigma_-(K_2 + N_2) = -E(c)\sigma(M_2)$$

are true. In fact, from $-\lambda \notin E(c)\sigma(M_2)$ it follows, as in (20.44), that the equations $\lambda x - K_2 x = g$ and $\lambda x - (K_2 + N_2)x = g$ have unique solutions in $C_t(L_2)$ for any $g \in C_t(L_2)$. Consequently, $-\lambda \notin \sigma(K_2) \cup \sigma(K_2 + N_2)$, hence $\sigma(K_2) \subseteq -E(c)\sigma(M_2)$ and $\sigma(K_2 + N_2) \subseteq -E(c)\sigma(M_2)$. Now let $-\lambda \in E(c)\sigma(M_2)$. Then $-\lambda = c(t_0)\gamma$, where t_0 is some point from $[1, T]$, and γ is a nonzero eigenvalue of M_2 or zero. By the compactness of the operators L_2 and N_2, we find a sequence of normalized functions $u_n \in C([1, T])$ without convergent subsequence, and a sequence of normalized functions $v_n \in L_2([a, b])$ such that $L_2 u_n \to 0$ and $M_2 v_n - \gamma v_n \to 0$ for $n \to \infty$. But then the sequence of normalized functions $w_n = u_n \otimes v_n \in C_t(L_2)$ $(n = 1, 2, ...)$ has no convergent subsequence either and satisfies

$$(20.49) \qquad K_2 w_n + \lambda w_n = L_2 w_n - c u_n \otimes (M_2 v_n - \gamma v_n) \to 0$$

as $n \to \infty$. This shows that $-\lambda \in \sigma_a(K_2)$ for $a = es, ew, eb, +, -, l$, or d, and hence (20.47) follows, by the inclusion $\sigma(K_2) \subseteq -E(c)\sigma(M_2)$. The equality (20.48) in turn is a consequence of (20.47), since the Fredholmness, the index, the n-normality and the d-normality are stable under compact perturbations. Thus, we have proved the following

Theorem 20.5. *Suppose that the kernels $l_2 = l_2(t, \tau)$, $m_2 = m_2(s, \sigma)$, and $n = n(t, \tau)$ are square integrable, $c = c(t)$ is a continuous function, and the operator L_2 is compact in $C([1, T])$. Then the operators K_2 and $K_2 + N_2$ act in $C_t(L_2)$, and the equalities (20.47) and (20.48) are true.*

Theorem 20.5 may be generalized in the following way:

Theorem 20.6. *Let $U = L_p([1, T])$, $V = L_{p'}([a, b])$ $(1 < p < \infty, \frac{1}{p} + \frac{1}{p'} = 1)$, $m_2 \in [U \leftarrow V]$, and let $c \in C([1, T])$. Suppose that the operator L_2 acts in $C([1, T])$ and is regular and compact in $L_p([1, T])$. Then the operators K_2 and $K_2 + N_2$ act in $C_t(L_p)$ and the equalities (20.47) and (20.48) are true.*

$\square$ For the proof it suffices to observe that, for fixed $t \in [1, T]$, the resolvent operator for $I + c(t)M_2$ has the form $I + c(t)\tilde{M}_2(t)$, where $\tilde{M}_2(t)$ is a u_0-bounded integral operator (see KRASNOSEL'SKIJ-LIFSHITS-SOBOLEV [1985]). This shows that (20.46) is a compact integral operator of Volterra type in the regular ideal space $[U \leftarrow V]$. Consequently, the spectral radius $r_\sigma(V)$ of the operator (20.46) is zero (ZABREJKO [1967, 1967a]), and the assertion follows in the same way as in Theorem 20.5. ∎

Observe that, if the kernels l_2, m_2 and n are continuous, then the operators K_2 and $K_2 + N_2$ certainly act in $C(D)$ and the equalities (20.47) and (20.48) hold true.

20.6. A unified approach to some equations of mechanics

The operators considered in Subsections 20.2 - 20.5 may be written in the general form

$$Kx(t,s) = a(s) \int_0^t l(t,\tau)x(\tau,s)\,d\tau$$

$$(20.50) \qquad\qquad + c(t) \int_0^b m(s,\sigma)x(t,\sigma)\,d\sigma$$

$$+ d(t,s) \int_0^t \int_0^b n(t,\tau)m(s,\sigma)x(\tau,\sigma)\,d\sigma\,d\tau$$

$((t,s) \in D = [0,T] \times [0,b])$ with continuous functions $a = a(s)$, $c = c(t)$ and $d = d(t,s)$, and measurable kernels $l = l(t,\tau)$, $m = m(s,\sigma)$ and $n(t,\tau)$. The spectral properties of the operator (20.50) may be studied by the methods discussed in the preceding two subsections.

Consider the Volterra integral operators

$$\tilde{L}x(t) = \int_0^t l(t,\tau)x(\tau)\,d\tau, \quad \tilde{N}x(t) = \int_0^t n(t,\tau)x(\tau)\,d\tau$$

and the Fredholm integral operator

$$\tilde{M}x(s) = \int_0^b m(s,\sigma)x(\sigma)\,d\sigma.$$

Put

$$\tilde{K}x(t,s) = a(s) \int_0^t l(t,\tau)x(\tau,s)\,d\tau$$

(20.51)

$$+c(t) \int_0^b m(s,\sigma)x(t,\sigma)\,d\sigma,$$

and $E(c) = \{c(t) : 0 \le t \le T\}$ as before. In the following theorem we use the terminology of Sections 4 and 12.

Theorem 20.7. *Let $U = U([0,T])$ and $V = V([0,b])$ be two regular ideal spaces, and let V' be the Köthe dual of V. Suppose that the operators $\tilde{L}$ and $\tilde{N}$ act in $C([0,T])$ and are compact and regular in U. Then the operators (20.50) and (20.51) act in $C_t(V)$ and satisfy the relations*

$$\sigma(\tilde{K}) = \sigma_{es}(\tilde{K}) = \sigma_{ew}(\tilde{K}) = \sigma_{eb}(\tilde{K})$$

(20.52)

$$= \sigma_+(\tilde{K}) = \sigma_-(\tilde{K}) = \sigma_l(\tilde{K}) = \sigma_d(\tilde{K}) = E(c)\sigma(\tilde{M})$$

and

$$\sigma(K) = \sigma_{es}(K) = \sigma_{ew}(K)$$

(20.53)

$$= \sigma_+(K) = \sigma_-(K) = E(c)\sigma(\tilde{M}).$$

Moreover, if the functions l, m, n, a, c, and d in (20.50) are continuous, then the operators $\tilde{K}$ and K act in $C(D)$ and the relations (20.52) and (20.53) hold.

All partial integral equations arising in the theory of continuous media, in mixed problems of evolutionary type, and in axially symmetric contact problems may be written in the general form

(20.54) $$x(t,s) - Kx(t,s) = f(t,s) \qquad ((t,s) \in D),$$

where K is given by (20.50). Under the above hypotheses, the Fredholm alternative applies to equation (20.54) in the space $C_t(V)$ if and only if $1 \notin E(c)\sigma(\tilde{M})$. In several special cases the solution of this equation may then either be given explicitly, or may be obtained by solving a Fredholm integral equation depending on a parameter. In the general case, however, solving equation (20.54) is a highly non-trivial problem. It is therefore interesting to find algorithms for the approximate or numerical solution of this equation. We describe now one of the simplest of such algorithms.

In the terminology of Theorem 20.7, let $V = L_2([0, b])$, suppose that $1 \notin E(c)\sigma(\tilde{M})$. Then equation (20.54) has a unique solution x in $C_t(L_2)$; in particular, the map $t \mapsto x(t, \cdot)$ is uniformly continuous on $[0, T]$. Let $\{t_0, \ldots, t_n\}$ be an equidistant partition of $[0, T]$. As approximate solutions we choose the functions $(i = 1, 2, \ldots, n - 1)$

$$(20.55) \qquad x_n(t, s) = \begin{cases} x(t_{i-1}, s) & \text{if} \quad (t, s) \in [t_{i-1}, t_i) \times [0, b], \\ x(t_{n-1}, s) & \text{if} \quad (t, s) \in [t_{n-1}, t_n] \times [0, b]. \end{cases}$$

Here $(x(t_1, s), \ldots, x(t_n, s))$ is the solution of the system

$$(20.56) \qquad \begin{aligned} x(t_i, s) &= a(s) \int_0^{t_i} l(t_i, \tau) x(\tau, s)\, d\tau \\ &\quad + c(t_i) \int_0^b m(s, \sigma) x(t_i, \sigma)\, d\sigma \\ &\quad + \int_0^{t_i} \int_0^b d(t_i, s) n(t_i, \tau) m(s, \sigma) x(\tau, \sigma)\, d\sigma\, d\tau \\ &\quad + f(t_i, s) \quad (i = 0, 1, \ldots, n) \end{aligned}$$

which in turn may be approximated by the system of integral equations of the second kind

$$x_i(s) = \frac{T}{n} \sum_{k=0}^{i-1} a(s)l(t_i, t_k)x_k(s)$$

$$+\frac{T}{n} \sum_{k=0}^{i-1} d(t_i, s)n(t_i, t_k) \int_0^b m(s, \sigma)x_k(\sigma)\, d\sigma$$

(20.57)

$$+c(t_i) \int_0^b m(s, \sigma)x_i(\sigma)\, d\sigma$$

$$+f(t_i, s) \quad (i = 0, 1, \ldots, n).$$

Now, each of these equations is uniquely solvable, since $1 \notin c(t_i)\sigma(\tilde{M})$ for $i = 0, 1, \ldots, n$. Using the solutions $x_i = x_i(s)$ of (20.57) we put then $(i = 1, 2, \ldots, n - 1)$

$$x_n(t, s) \approx y_n(t, s) = \begin{cases} x_{i-1}(s) & \text{if } (t, s) \in [t_{i-1}, t_i) \times [0, b], \\ x_{n-1}(s) & \text{if } (t, s) \in [t_{n-1}, t_n] \times [0, b]. \end{cases}$$

The numerical solution is carried out as in Subsection 20.2. Let $\{t_0, \ldots, t_n\}$ be an equidistant partition of $[0, T]$, and $\{s_0, \ldots, s_m\}$ an equidistant partition of $[0, b]$. Replace the original equation by the approximating system

$$x_{ij} = \frac{T}{n} \sum_{k=0}^{i-1} a(s_j)l(t_i, t_k)x_{kj}$$

$$+\frac{T}{n} \sum_{k=0}^{i-1} \frac{b}{m} d(t_i, s_j)n(t_i, t_k) \sum_{l=0}^{m-1} m(s_j, s_l)x_{kl}$$

(20.58)

$$+\frac{b}{m}c(t_i) \sum_{l=0}^{m-1} m(s_j, s_l)x_{il}$$

$$+f(t_i, s_j) \quad (i = 0, 1, \ldots, n; \ j = 0, 1, \ldots, m).$$

This system may be solved successively for $i = 0, 1, \ldots, n$, and then the solution is extended by linear interpolation. In this way, one obtains continuous approximate solutions to the original equation which are arbitrarily close to the exact solution.

20.7. Other applications

In this final subsection we briefly sketch some other applications. We start with a problem from aerodynamics. For the approximate calculation of the aerodynamic characteristics of a thin rectangular wing with finite amplitude and stationary streamline one uses two-dimensional integro-differential equations. In appropriate function classes, these equations may be transformed into operator equations of the form (BELOTSERKOVSKIJ-LIFANOV [1985])   ·

$$(20.59) \qquad x(t,s) - Kx(t,s) = f(t,s)$$

$((t,s) \in D = [-a,a] \times [-b,b])$, where

$$(20.60) \qquad Kx(t,s) = l(t) \int_{-a}^{a} x(\tau,s)\,d\tau + \int_{-b}^{b} m(s,\sigma)x(t,\sigma)\,d\sigma.$$

Here $m = m(s,\sigma)$ and $f = f(t,s)$ are given functions, and $l(t) = -\frac{1}{\pi b}\sqrt{\frac{a-t}{a+t}}$. For numerical calculations, equation (20.59) is usually replaced by an approximating linear algebraic equation. In BELOTSERKOVSKIJ-LIFANOV [1985] one may find some error estimates for this approximation under the assumption that the operator equation (20.59) admits a unique solution. Of, course, this assumption may be verified by studying the spectral properties of the operator (20.60). Define two operators $\tilde{L}$ and $\tilde{M}$ by

$$\tilde{L}x(t) = \int_{-a}^{a} l(t)x(\tau)\,d\tau, \quad \tilde{M}x(s) = \int_{-b}^{b} m(s,\sigma)x(\sigma)\,d\sigma.$$

Observe that now neither the operator $\tilde{L}$ nor the operator $\tilde{M}$ is Volterra, and thus one cannot expect that the spectrum of K may expressed through the spectrum of only one of these operators. Nevertheless, since the operator $\tilde{L}$ is one-dimensional, its spectrum contains just one non-zero eigenvalue.

Let U be a Banach space of functions over $[-a,a]$, V a Banach space of functions over $[-b,b]$, and $X = U\overline{\otimes}V$ their tensor product with respect to a quasi-uniform crossnorm (see Subsection 15.4). The operator $\tilde{L}$ acts in U if and only if $l \in U$ and $U \subseteq L_1([-a,a])$; in this case, $\tilde{L}$ is compact in U and $\sigma(\tilde{L}) = \{0, a/b\}$. On the other hand, if $\tilde{M}$ is a bounded operator in V, then from the results of ICHINOSE [1978, 1978a] we get the equalities (KALITVIN [1994a])

$$\sigma(K) = \sigma(\tilde{M}) \cup [\sigma(\tilde{M}) + \{0, a/b\}],$$

$$\sigma_{eb}(K) = \sigma(\tilde{M}) \cup [\sigma_{eb}(\tilde{M}) + \{a/b\}],$$

$$\sigma_{ew}(K) = \sigma(\tilde{M}) \cup [\sigma_{ew}(\tilde{M}) + \{a/b\}],$$

and

$$\sigma_{es}(K) = \sigma(\tilde{M}) \cup [\sigma_{es}(\tilde{M}) + \{a/b\}].$$

We summarize with the following

Theorem 20.8. *Under the above hypotheses, the operator* $I - K$ *with* K *given by* (20.60) *is:*

(a) *invertible in* X *if and only if* $1 \notin \sigma(\tilde{M}) \cup [\sigma(\tilde{M}) + \{a/b\}]$;

(b) *Fredholm in* X *if and only if* $1 \notin \sigma(\tilde{M}) \cup [\sigma_{ew}(\tilde{M}) + \{a/b\}]$;

(c) *Fredholm of index zero in* X *if and only if* $1 \notin \sigma(\tilde{M}) \cup [\sigma_{es}(\tilde{M}) + \{a/b\}]$.

It is interesting to remark that in the monograph BELOTSERKOVSKIJ-LIFANOV [1985] the solvability of equation (20.59) is proved building on a heuristic physical reasoning. In this connection, one may represent the solution of equation (20.40) in the form (20.29), where $\tilde{L}$ and $\tilde{M}$ are defined as above.

For the numerical solution of (20.59), one has first to regularize the equation; this may be done as described in Subsection 15.1 (see also KALITVIN-ZABREJKO [1991]). After such a regularization one ends up with a two-dimensional Fredholm integral equation of the second kind which may be solved numerically by standard methods. For example, using standard quadrature formulas one is led to a system of linear algebraic equations for x_{ij} ($i = 1, \ldots, n$; $j = 1, \ldots, m$), which is uniquely solvable for n and m sufficiently large. There are also other methods based on Theorem 8.3.1 of the monograph BELOTSERKOVSKIJ-LIFANOV [1985].

We close with a problem which leads to a partial integral equation of the first kind. When calculating a dam by means of the cantilever method, the console load $x = x(t, s)$ satisfies the partial integral equation

$$(20.61) \quad \int_0^a l(t, s, \tau) x(\tau, s) \, d\tau + \int_{-b}^b m(t, s, \sigma) x(t, \sigma) \, d\sigma = f(t, s).$$

The arched load in turn satisfies an analogous equation (MOROZOV [1965]) which may be written in the form

$$(20.62) \quad K x(t, s) = f(t, s) \quad ((t, s) \in D = [0, a] \times [-b, b]),$$

where $f \in L_2(D)$ and

$$(20.63) \quad \begin{aligned} K x(t, s) &= \int_0^a l(t, s, \tau) x(\tau, s) \, d\tau \\ &+ \int_{-b}^b m(t, s, \sigma) x(t, \sigma) \, d\sigma \\ &+ \int_0^a \int_{-b}^b n(t, s, \tau, \sigma) x(\tau, \sigma) \, d\sigma \, d\tau. \end{aligned}$$

The solutions of (20.62) are usually looked for in $W_2^{(1)}$ or in $W_2^{(2)}$, under the assumption that the homogeneous equation $K x = 0$ has only the trivial solution. This means that one has to care for conditions under which the operator (20.63) acts from $W_2^{(1)}$ or $W_2^{(2)}$ into $L_2(D)$ and is bounded. Conditions which are both necessary and sufficient are not known. However, one may find sufficient conditions building on the continuity of the imbeddings $W_2^{(1)}, W_2^{(2)} \subseteq L_2(D)$. In particular, the operator K is bounded from $W_2^{(1)}$ and $W_2^{(2)}$ into $L_2(D)$ if the kernels l, m, and n are measurable and such that the functions $\|l(t, s, \cdot)\|_{L_1}$, $\|m(t, s, \cdot)\|_{L_1}$, and $\|n(t, s, \cdot, \cdot)\|_{L_1}$ belong to $L_2(D)$. This is trivially true, for example, if the kernels l, m, and n are continuous.

To prove that the equation $K x = 0$ has only the trivial solution $x = 0$ one has to show of course that $0 \notin \sigma_p(K)$. In the case of general kernels this is a very hard problem. Nevertheless, in the special case $l(t, s, \tau) = l(t, \tau)$ and $m(t, s, \sigma) = m(s, \sigma)$ the situation is much simpler. If the operators

$$\tilde{L}x(t) = \int_0^a l(t, \tau)x(\tau)\, d\tau, \quad \tilde{M}x(s) = \int_{-b}^b m(s, \sigma)x(\sigma)\, d\sigma$$

are bounded in $L_2([0, a])$ and $L_2([-b, b])$, respectively, and $0 \in \sigma_p(\tilde{L}) + \sigma_p(\tilde{M})$, then $0 \in \sigma_p(K)$. Moreover, if $\tilde{L}$ or $\tilde{M}$ has a pure point spectrum, then $0 \in \sigma_p(K)$ if and only if $\alpha \in \sigma_p(\tilde{L})$ and $-\alpha \in \sigma_p(\tilde{M})$ for some α.

A problem which leads to the solution of a special case of equation (20.63) may be found in SOLODNIKOV [1969] and DIDENKO-KOZLOV [1975]. In that problem the authors give estimates for a choice function $r(t)$ with "white noise". Let $r(t) = m(t) + n(t)$, where $m = m(t)$ is the choice function of a random process with zero mean, finite dispersion and correlation function $f(t, s)$, and $n = n(t)$ is an obstacle with zero mean and correlation function $g(t, s)$. Estimates for $r(t)$ may be obtained through the optimal linear filter with a time parameter. To this end, the impulse transition function $x = x(t, s)$ is chosen in such a way that

$$\min E[(m(t) - \int_0^a u(t, \sigma)r(\sigma)\, d\sigma)^2]$$

$$= E[(m(t) - \int_0^a x(t, \sigma)\, d\sigma)^2]$$

for $0 \leq t \leq a$. The minimality criterion means that u satisfies the Euler equation

$$\int_0^a x(t, \sigma)[f(s, \sigma) + g(s, \sigma)]\, d\sigma = f(t, s)$$

$(0 \leq t \leq a, 0 < s < a)$. If there is no white noise in the system, the last equation is a special case of (20.62). The Euler equation is used in DIDENKO-KOZLOV [1975] for constructing a regularizing solution, under the assumption that this equation is uniquely solvable. Of course, the latter condition is satisfied if and only if $0 \notin \sigma_p(\tilde{M})$ and (20.62) admits a solution, where $\tilde{M}$ is given as before with $m(s, \sigma) = f(s, \sigma) + g(s, \sigma)$.

A particularly important special case of (20.61) is the equation

$$(20.64) \qquad \int_0^t l(\tau)x(\tau, s)\, d\tau + \int_{-1}^1 m(s - \sigma)x(t, \sigma)\, d\sigma = f(t, s)$$

which occurs in the modelling of a contact interaction of overlapping bodies with abrasion (see ALEKSANDROV-KOVALENKO [1984]). Here $l = l(t)$ is a piecewise smooth nonnegative function, and

$$m(s) = \int_0^\infty \tilde{m}(u) \cos \frac{u(s)}{\lambda}\, du$$

with $\lambda > 0$ and $\tilde{m} = \tilde{m}(u)$ being a positive continuous even function satisfying certain growth conditions for $u \to 0$ and $u \to \infty$. The operator K_1 defined by the left hand side of (20.64) is obviously bounded in both $C_t(L_2)$ and $L_2(D)$. Moreover, the operators

$$\tilde{L}_1 x(t) = \int_0^t l(\tau)x(\tau)\, d\tau, \quad \tilde{M}_1 x(s) = \int_{-1}^1 m(s - \sigma)x(\sigma)\, d\sigma$$

are compact in $C([0, a])$ and $L_2([-1, 1])$, respectively. Since $0 \in \sigma_p(\tilde{L}_1)$ and $\tilde{M}_1$ is a positive selfadjoint operator in $L_2([-1, 1])$, we have $0 \in \sigma_p(K_1)$ if and only if $0 \in \sigma_p(\tilde{L}_1)$. Consequently, the unique solvability of (20.64) is equivalent to the condition $0 \notin \sigma_p(\tilde{M}_1)$.

Some initial-boundary value problems for equations describing gravitational gyroscopic waves in the Boussinesq approximation (GABOV-SVESHNIKOV [1990]) may be transformed into the partial integral equation of Volterra-Fredholm type with convolution kernels

$$x(t, s) = \int_0^t l(t - \tau)x(\tau, s)\, d\tau$$

$$(20.65) \qquad - \int_\Gamma m(s - \sigma)x(t, \sigma)\, d\sigma$$

$$+ \int_0^t \int_\Gamma n(t - \tau, s - \sigma)x(\tau, \sigma)\, d\sigma\, d\tau + f(t, s).$$

This equation is uniquely solvable, and the solution may be contructed by the classical method of successive approximations (GABOV-SVESHNIKOV [1990]).

If we apply as above the Laplace transform with respect to t to equation (20.65) we get the parameter-dependent Fredholm integral equation

$$X(p, s)[1 - L(p)] =$$

$$\int_\Gamma [N(p, s - \sigma) - m(s - \sigma)] X(p, \sigma) \, d\sigma + F(p, s),$$

where X, L, N, and F denote the Laplace transforms of x, l, n, and f, respectively.

Let us return to equation (20.62) and its particular cases. Since these equations are of first kind, the existence and uniqueness problem for them is not well-posed. Nevertheless, in MOROZOV [1965] one may find an effective regularization method for equation (20.62) which reduces the problem of solving (20.62) to a uniquely solvable variational problem. The solution of this variational problem is possible by means of the classical Ritz method.

We remark that the solution for some classes of equations of type (20.62) may be found rather effectively. For example, it is quite natural to apply the Laplace transform to the equation

$$\int_0^t l(t - \tau) x(\tau, s) \, d\tau + \int_0^s m(s - \sigma) x(t, \sigma) \, d\sigma$$

$$+ \int_0^t \int_0^s n(t - \tau, s - \sigma) x(\tau, \sigma) \, d\sigma \, d\tau = f(t, s)$$

and to solve then the resulting algebraic equation by standard methods of numerical linear algebra.

Suppose that the kernels l and m in (20.62) are symmetric, the operator $\tilde{L}$ is compact in $L_2([0, a])$, and the operator $\tilde{M}$ is compact in $L_2([-b, b])$. Furthermore, let N be bounded in $L_2(D)$ with eigenfunctions $\varphi_i \otimes \psi_j$ $(i, j = 1, 2, \ldots)$ corresponding to eigenvalues γ_{ij}, where φ_i and ψ_j are the eigenfunctions of $\tilde{L}$ and $\tilde{M}$, respectively, corresponding to eigenvalues α_i and β_j. Equation (20.43) may then be written as

$$\sum_{i,j=1}^\infty \varphi_i(t) \psi_j(s) [x_{ij}(\alpha_i + \beta_j + \gamma_{ij}) - f_{ij}] = 0,$$

where x_{ij} and f_{ij} are the Fourier coefficients of the functions x and f. Evidently, the solvability of this equation is equivalent to the solvability of the system

$$x_{ij}(\alpha_i + \beta_j + \gamma_{ij}) = f_{ij} \qquad (i, j = 1, 2, \ldots).$$

Now, if $\alpha_i + \beta_j + \gamma_{ij} \neq 0$ for all i and j, the unique solution of (20.62) has the form

$$x(t, s) = \sum_{i,j=1}^{\infty} \frac{f_{ij}}{\alpha_i + \beta_j + \gamma_{ij}} \varphi_i(t) \psi_j(s).$$

On the other hand, if $P = \{(i, j) : \alpha_i + \beta_j + \gamma_{ij} = 0\} \neq \emptyset$, a necessary and sufficient condition for the solvability of (20.62) is that $f_{ij} = 0$ for all $(i, j) \in P$. In this case the (non-unique) solution has the form

$$x(t, s) = \sum_{(i,j) \notin P} \frac{f_{ij}}{\alpha_i + \beta_j + \gamma_{ij}} \varphi_i(t) \psi_j(s) + \sum_{(i,j) \in P} \xi_{ij} \varphi_i(t) \psi_j(s),$$

where the scalars ξ_{ij} may be chosen arbitrarily for finite P and must satisfy the condition

$$\sum_{(i,j) \in P} \xi_{ij}^2 < \infty$$

for infinite P.

To conclude, we remark that various problems of mechanics and engineering lead to Fredholm-type partial integral equations of the second kind

$$x(t, s) = \int_T l(t, s, \tau) x(\tau, s) \, d\tau + f(t, s)$$

or of the first kind

$$\int_T l(t, s, \tau) x(\tau, s) \, d\tau = f(t, s),$$

containing a parameter s. In particular, this refers to the Lipman-Schwinger equation (REED-SIMON [1972]), to the homogeneous stationary Schrödinger equation with two-dimensional Laplacian, to the equation for the amplitude function of a regular or singular string

(KATS-KREJN [1958], KATS [1963]), to the state equation for a transversal stroke on a visco-elastic flexible fibre (RAKHMATULLIN-OSOKIN [1977]), to the Euler equation discussed above, to the Gel'fand-Levitan equation (GEL'FAND-LEVITAN [1951]), and to several integral equations coming from inverse problems (LEVITAN-SARGSJAN [1988]). Other problems of this type have been studied in ALEKSANDROV-BRONOVETS-KOVALENKO [1988], ALEKSANDROV-KOVALENKO-MARCHENKO [1983], AVILKIN-KOVALENKO [1982], DIL'MANOV [1982], HALTON-LIGHT [1985], ISHIMARU [1978], KALITVIN [1989], and KOVALENKO-MANZHIROV [1982].

Partial integral operators also occur in applications of control theory (SROCHKO [1984]), as well as in the study of boundary value problems for Navier-Stokes equations (see e.g. the monograph BELONOSOV-CHERNOUS [1985]). Interestingly, there is also a connection with minimal projections in L_p spaces (HALTON-LIGHT [1985]) and with some problems in integral geometry (DIL'MANOV [1982]).

In all these applications it is a useful device to consider the arising equations as partial integral equations, because one has then to study just one equation in a properly chosen function spaces, rather than a whole family of equations.

References

K. D. ABHUANKAR, A. L. FYMAT

[1969] *A new type of multi-dimensional singular linear equations*, J. Math. Mech. **19, 2** (1969), 111-122

J. ABLOWITZ, H. SEGUR

[1981] *Solutions and the Inverse Scattering Transform*, SIAM, Philadelphia 1981

T. A. AGEKJAN

[1974] *Probability Theory for Astronomers and Physicists* [in Russian], Nauka, Moscow 1974

V. I. AGOSHKOV

[1988] *Generalized Solutions of Transport Equations and Their Smoothness Properties* [in Russian], Nauka, Moscow 1988

N. I. AKHIEZER, I. M. GLAZMAN

[1977] *Theory of Linear Operators in Hilbert Spaces* [in Russian], Vyshcha Shkola, Kharkov 1977

R. R. AKHMEROV, M. I. KAMENSKIJ, A. S. POTAPOV, A. E. RODKINA, N. N. SADOVSKIJ

[1986] *Measures of Noncompactness and Condensing Operators* [in Russian], Nauka, Novosibirsk 1986; Engl. transl.: Birkhäuser, Basel 1992

V. M. ALEKSANDROV, N.H. ARUTJUNJAN, A.V. MANZHIROV

[1986] *Contact problems in the theory of creeping of non-uniformly aging bodies* [in Russian], Anal. Vychisl. Met. Resh. Kraev. Zad. Uprug. **1** (1986), 3-13

V. M. ALEKSANDROV, M. A. BRONOVETS, E. V. KOVALENKO

[1988] *Plane contact problems for a linearly deformable foundation with a thin reinforcing covering* [in Russian], Prikl. Mekh. Kiev **24**, 8 (1988), 60-67; Engl. transl.: Sov. Appl. Mech. **24**, 8 (1988), 781-787

V. M. ALEKSANDROV, E. V. KOVALENKO

[1977] *Two effective methods of solving mixed linear problems of mechanics of continuous media* [in Russian], Prikl. Mat. Mekh. **41**, 3 (1977), 688-698; Engl. transl.: J. Appl. Math. Mech. **41**, 3 (1978), 702-713

[1978] *The axially symmetric contact problem for a linearly deformable ground of general type in the presence of abrasion* [in Russian], Izv. Akad. Nauk SSSR Mekh. Tverd. Tela **5** (1978), 58-66

[1980] *On a class of integral equations of mixed problems in continuum mechanics* [in Russian], Dokl. Akad. Nauk SSSR **252**, 2 (1980), 324-328; Engl. transl.: Soviet Phys. Dokl. **25**, 2 (1980), 354-356

[1981] *Stamp motion on the surface of a thin covering of a hydraulic foundation* [in Russian], Prikl. Mat. Mekh. **45**, 4 (1981), 734-744; Engl. transl.: J. Appl. Math. Mech. **45**, 4 (1982), 542-549

[1984] *On the contact interaction of bodies with coatings and abrasion* [in Russian], Dokl. Akad. Nauk SSSR **275**, 4 (1984), 827-830; Engl. transl.: Soviet Phys. Dokl. **29**, 4 (1984), 340-342

[1986] *Problems of Continuum Mechanics with Mixed Boundary Conditions* [in Russian], Nauka, Moscow 1986

V. M. ALEKSANDROV, E. V. KOVALENKO, A. V. MANZHIROV

[1984] *Some mixed problems of creep theory of inhomogeneously aging media* [in Russian], Izv. Akad. Nauk Arm. SSR Mekh. **37**, 2 (1984), 12-25

V. M. ALEKSANDROV, E. V. KOVALENKO, S. M. MARCHENKO

[1983] *Two contact problems of the theory of elasticity for a layer with a Winkler-type coating* [in Russian], Prikl. Mekh. Kiev **19, 10** (1983), 47-54; Engl. transl.: Sov. Appl. Mech. **19, 10** (1983), 871-878

V. M. ALEKSANDROV, E. V. KOVALENKO, S. M. MKHITARYAN

[1982] *On a method of obtaining spectral relationships for integral operators of mixed in the mechanics of continuous media* [in Russian], Prikl. Mat. Mekh. **46, 6** (1982), 1023-1031; Engl. transl.: J. Appl. Math. Mech. **46, 6** (1983), 825-832

V. M. ALEKSANDROV, B. I. SMETANIN

[1985] *On symmetric and nonsymmetric contact problems of elasticity theory* [in Russian], Prikl. Mat. Mekh. **49, 1** (1985), 136-141; Engl. transl.: J. Appl. Math. Mech. **49, 1** (1985), 100-105

J. APPELL

[1981] *Implicit functions, nonlinear integral equations, and the measure of noncompactness of the superposition operator*, J. Math. Anal. Appl. **83, 1** (1981), 251-263

J. APPELL, E. DE PASCALE, O. W. DIALLO

[1994] *La fonction de Green pour des équations intégro-différentielles de type Barbachine*, Rend. Accad. Sci. Lomb. **A-127** (1994), 149-159

J. APPELL, E. DE PASCALE, A. S. KALITVIN, P. P. ZABREJKO

[1996] *On the application of the Newton-Kantorovich method to nonlinear partial integral equations*, Zeitschr. Anal. Anw. **15, 2** (1996), 397-418

J. APPELL, E. DE PASCALE, P. P. ZABREJKO

[1991] *On the application of the Newton-Kantorovich method to non-linear integral equations of Uryson type*, Numer. Funct. Anal. Optim. **12** (1991), 271-283

J. APPELL, O. W. DIALLO

[1995] *Problèmes aux limites pour des équations intégro-différentielles de type Barbachine*, Zeitschr. Anal. Anw. **14, 1** (1995), 95-104

J. APPELL, O. W. DIALLO, P. P. ZABREJKO

[1988] *On linear integro-differential equations of Barbashin type in spaces of continuous and measurable functions*, J. Integral Equ. Appl. **1, 2** (1988), 227-247

J. APPELL, E. V. FROLOVA, A. S. KALITVIN, P. P. ZABREJKO

[1997] *Partial integral operators on $C([a, b] \times [c, d])$*, Integral Equ. Oper. Theory **27** (1997), 125-140

J. APPELL, A. S. KALITVIN, M. Z. NASHED

[1999] *On some partial integral equations arising in the mechanics in the mechanics of solids*, ZAMM **79, 10** (1999), 703-713

J. APPELL, A. S. KALITVIN, P. P. ZABREJKO

[1994] *Boundary value problems for integro-differential equations of Barbashin type*, J. Integral Equ. Appl. **6, 1** (1994), 1-30

[1996] *Fixed point theorems in K-normed spaces and boundary value problems for nonlinear integro-differential equations of Barbashin type*, Rivista Mat. Pura Appl. **18** (1996), 31-44

[1998] *Partial integral operators in Orlicz spaces with mixed norm*, Coll. Math. **78** (1998), 293-306

J. APPELL, I. MASSABÒ, A. VIGNOLI, P. P. ZABREJKO

[1988] *Lipschitz and Darbo conditions for the superposition operator in ideal spaces*, Annali Mat. Pura Appl. **152** (1988), 123-137

J. APPELL, P. P. ZABREJKO

[1990] *Nonlinear Superposition Operators*, Cambridge Univ. Press, Cambridge 1990

O. ARINO, M. KIMMEL

[1987] *Asymptotic analysis of a cell cycle model based on unequal division*, SIAM J. Appl. Math. **47, 1** (1987), 128-145

EH. R. ATAMANOV

[1977] *An inverse initial problem for an integro-differential equation* [in Russian], Issled. Integro-Diff. Uravn. Kirg. Frunze **2** (1977), 251-257

V. I. AVILKIN, V. M. ALEKSANDROV, E. V. KOVALENKO

[1985] *Using the more accurate equations of thin coatings in the theory of axially symmetric contact problems for composite foundations* [in Russian], Prikl. Mat. Mekh. **49, 6** (1985), 1010-1018; Engl. transl.: J. Appl. Math. Mech. **49, 6** (1985), 770-777

V. I. AVILKIN, E. V. KOVALENKO

[1982] *On a dynamical contact problem for composite foundations* [in Russian], Prikl. Mat. Mekh. **46, 3** (1982), 847-856; Engl. transl.: J. Appl. Math. Mech. **46, 3** (1983), 679-686

S. BANACH

[1932] *Théorie des Opérations Linéaires*, Ed. Acad. Sci. Polon., Warszawa 1932

E. A. BARBASHIN

[1957] *On conditions for the conservation of stability of solutions to integro-differential equations* [in Russian], Izv. VUZov Mat. **1** (1957), 25-34

[1958] *On the stability of solutions of integro-differential equations* [in Russian], Trudy Ural. Polit. Inst. Mat. Sverdlovsk (1958), 91-98

[1967] *Introduction to Stability Theory* [in Russian], Nauka, Moscow 1967

[1970] *Ljapunov Functions* [in Russian], Nauka, Moscow 1970

E. A. BARBASHIN, L. P. BISJARINA

[1963] *On the stability of solutions of integro-differential equations* [in Russian], Izv. VUZov Mat. **3** (1963), 3-14

E. A. BARBASHIN, L. KH. LIBERMAN

[1958] *On the stability of solutions of systems of integro-differential equations* [in Russian], Nauchn. Dokl. Vyssh. Shkoly Fiz.-Mat. Nauk **3** (1958), 18-22

J.A. BARKER, M. R. HOARE, S. RAVAL

[1981] *A numerical study of the general Rayleigh's piston model*, J. Phys. Math. Gen. **14, 2** (1981), 423-438

E. A. BARKOVA, P. P. ZABREJKO

[1991] *The Cauchy problem for higher order differential equations with deteriorating operators* [in Russian], Diff. Uravn **27, 3** (1991), 472-478

P. R. BEESACK

[1987] *On some variation of parameter methods for integro-differential equations and quasilinear partial integro-differential equations*, Appl. Math. Comp. **22, 2-3** (1987), 189-215

S. M. BELONOSOV, K. A. CHERNOUS

[1985] *Boundary Value Problems for Navier-Stokes Equations* [in Russian], Nauka, Moscow 1985

O. M. BELOTSERKOVSKIJ, I. K. LIFANOV

[1985] *Numerical Methods for Singular Integral Equations* [in Russian], Nauka, Moscow 1985

A. BENEDEK, R. PANZONE

[1961] *The spaces L^P with mixed norm*, Duke Math. J. **28, 3** (1961), 301- 324

J. BERGH, J. LÖFSTRÖM

[1976] *Interpolation Spaces - an Introduction*, Springer, Berlin 1976

E. A. BIBERDORF, M. VÄTH

[1999] *On the spectrum of orthomorphisms and Barbashin operators*, Zeitschr. Anal. Anw. (to appear)

L. P. BISJARINA

[1963] *On the existence of periodic solutions of a certain integro-differential equation* [in Russian], Mat. Zap. Ural. Univ. **4, 2** (1963), 18-27

[1964] *On the stability of solutions of integro-differential equations with partial derivatives* [in Russian], Izv. VUZov Mat. **2** (1964), 31-40

[1964a] *The problem of perturbation accumulation for systems which are described by integro-differential equations* [in Russian], Mat. Zap. Ural. Univ. **4** (1964), 12-22

[1967] *The Cauchy problem and stability conditions for solutions of a certain integro-differential equation* [in Russian], Diff. Uravn. **3, 7** (1967), 1127-1134

N. N. BOGOLJUBOV

[1945] *On some Statistical Methods in Mathematical Physics* [in Russian], Ed. Acad. Sci., Kiev 1945

N. N. BOGOLJUBOV, JU. A. MITROPOL'SKIJ

[1963] *Asymptotic Methods in the Theory of Nonlinear Oscillations* [in Russian], Fizmatgiz, Moscow 1963

I. V. BOJKOV

[1985] *The approximate solution of multisingular integral equations* [in Russian], Dokl. Semin. Inst. Prikl. Math. I. N. Vekua (Tbilisi) **1**, 1 (1985), 51-54

V. V. BOLTJANSKIJ

[1979] *On a class of operators in Hilbert space* [in Russian], Funct. Anal. (Ul'janovsk) **13** (1979), 69-78

[1980] *A theorem on power series* [in Russian], Funct. Anal. (Ul'janovsk) **15** (1980), 24-32

[1981] *On the solvability of a partial integral equation with a kernel which depends on three variables* [in Russian], Diff. Uravn. (Rjazan') (1981), 3-14

V. V. BOLTJANSKIJ, L. M. LIKHTARNIKOV

[1982] *On a class of linear integral equations with partial integrals* [in Russian], Diff. Uravn. **18**, 11 (1982), 1939-1950; Engl. transl.: Diff. Equ. **18**, 11 (1982), 1395-1404

N. K. BOSE

[1977] *Problems and progress in multidimensional system theory*, Proc. IEEE **65** (1977), 824-840

A. BÖTTCHER

[1982] *On some two-dimensional Wiener-Hopf operators with vanishing symbols* [in Russian], Math. Nachr. **109** (1982), 195-213

[1983] *On the Noether property and reduction of two-dimensional Wiener-Hopf operators with piecewise continuous symbols* [in Russian], Dokl. Akad. Nauk SSSR **279, 6** (1983), 1298-1300

A. BÖTTCHER, A. EH. PASENCHUK

[1982] *On the invertibility of Wiener-Hopf operators in a quadrant of the plane with kernels whose supports are contained in a half-plane* [in Russian], Ehlista **1** (1982), 9-19

G. BRACK

[1985] *Systems with substantially distributed parameters*, Math. Res. **27** (1985), 421-424

F. E. BROWDER

[1960] *On the spectral theory of elliptic differential operators I*, Math. Ann. **142** (1960), 22-130

[1968] *Nonlinear maximal monotone mappings in Banach spaces*, Math. Ann. **175** (1968), 81-113

F. E. BROWDER, D. G. DE FIGUEIREDO, C. P. GUPTA

[1970] *Maximal monotone operators and nonlinear integral equations of Hammerstein type*, Bull. Amer. Math. Soc. **76** (1970), 700-705

F. E. BROWDER, C. P. GUPTA

[1969] *Monotone operators and nonlinear integral equations of Hammerstein type*, Bull. Amer. Math. Soc. **75** (1969), 1347-1353

A. V. BUKHVALOV

[1983] *An application of methods of the theory of order-bounded operators to operator theory in L_p* [in Russian], Uspekhi Mat. Nauk **6, 1** (1983), 37-83

A. V. BUKHVALOV, V. B. KOROTKOV, A. G. KUSRAEV, S. S. KUTATELADZE, B. M. MAKAROV

[1992] *Vector Lattices and Integral Operators* [in Russian], Nauka, Novosibirsk 1992; Engl. transl.: Kluwer, Dordrecht 1996

R. BÜRGER

[1986] *On the maintenance of genetic variation: global analysis of Kimmura's continuum-of-alleles model*, J. Math. Biol. **24, 3** (1986), 341-351

D. C. J. BURGESS

[1954] *Abstract moment problems with applications to l^p and L^p spaces*, Proc. London Math. Soc. 4 (1954), 107-128

JA. B. BYKOV

[1953] *On the theory of linear integro-differential equations* [in Russian], Dokl. Akad. Nauk Uzb. SSR **6** (1953), 3-6

[1953a] *On some class of integro-differential equations* [in Russian], Dokl. Akad. Nauk Uzb. SSR **8** (1953), 3-6

[1961] *On some questions in the qualitative theory of integro-differential equations* [in Russian], Issled. Integro-Diff. Uravn. Kirg. Frunze **1** (1961), 3-54

[1962] *On the stability of solutions of a certain class of integro-differential equations* [in Russian], Issled. Integro-Diff. Uravn. Kirg. Frunze **2** (1962), 41-56

[1962a] *On the analytical theory of linear integro-differential equations of Barbashin type* [in Russian], Issled. Integro-Diff. Uravn. Kirg. Frunze **2** (1962), 57-78

[1962b] *On bounded solutions of a certain class of integro-differential equations* [in Russian], Issled. Integro-Diff. Uravn. Kirg. Frunze **2** (1962), 79-89

[1962c] *On the stability of solutions of integro-differential equation and integro-difference equations of Barbashin type I* [in Russian], Issled. Integro-Diff. Uravn. Kirg. Frunze **2** (1962), 121-143

[1983] *Oscillations of solutions of a first order functional-integro-differential equation of Barbashin type* [in Russian], Issled. Integro-Diff. Uravn. Kirg. Frunze **16** (1983), 76-89

JA. V. BYKOV, T. C. KULTAEV

[1984] *Oscillatory properties of solutions of some class of integro-differential equations involving partial derivatives* [in Russian], Issled. Integro-Diff. Uravn. Kirg. Frunze **17** (1984), 15-33

JA. V. BYKOV, T. A. TANKIEV

[1982] *On a generalized boundary value problem for a countable system of iterated equations* [in Russian], Issled. Integro-Diff. Uravn. Kirg. Frunze **15** (1982), 44-62

EH. JA. BYKOVA

[1967] *Approximate solutions of Barbashin type equations* [in Russian], Izv. Akad. Nauk Kirg. SSR **6** (1967), 36-41

[1968] *Approximate solutions of a certain class of integro-differential equations and integro-difference equations* [in Russian], Issled. Integro-Diff. Uravn. Kirg. Frunze **5** (1968), 156-164

[1971] *Some variants of functional corrections for integro-differential equations of Barbashin type* [in Russian], Izv. Akad. Nauk Kirg. SSR **5** (1971), 31-36

EH. JA. BYKOVA, L. JA. BYKOVA

[1969] *On the existence of periodic solutions with algebraic singularity for integro-differential equations of Barbashin type* [in Russian], Issled. Integro-Diff. Uravn. Kirg. Frunze **6** (1969), 205-225

H. CARTAN

[1967] *Calcul Différentiel*, Hermann, Paris 1967

K. M. CASE, P. F. ZWEIFEL

[1967] *Linear Transport Theory*, Addison-Wesley, Reading Mass. 1967

C. CERCIGNANI

[1969] *Mathematical Methods in Kinetic Theory*, Macmillan, New York 1969

S. CHANDRASEKHAR

[1950] *Radiative Transfer*, Oxford Univ. Press, Oxford 1950

C.-J. CHEN

[1995] *On a generalized integro-differential equation of Barbashin type*, Zeitschr. Anal. Anw. **14, 4** (1995), 899-912

[1996] *Nonlinear integro-differential equations of Barbashin type: topological and monotonicity methods*, J. Integral Equ. Appl. **8, 3** (1996), 287-305

[1997] *Barbashin Equations and Partial Integral Operators*, Ph.D. thesis, Univ. Würzburg 1997

C.-J. CHEN, A. S. KALITVIN

[1997] *Nonlinear operators with partial integrals in spaces of summable functions*, Nonlin. Anal. TMA (to appear)

[1998] *On Uryson operators with partial integrals in Lebesgue spaces with mixed norm*, Zeitschr. Anal. Anw. **17, 1** (1998), 3-8

C.-J. CHEN, M. VÄTH

[1997] *On the L-characteristic of the superposition operator in Lebesgue spaces with mixed norm*, Zeitschr. Anal. Anwn. **16, 2** (1997), 377-386

Ju. L. Daletskij, M. G. Krejn

[1970] *The Stability of Solutions of Differential Equations in Banach Spaces* [in Russian], Nauka, Moscow 1970; Engl. transl.: Transl. Math. Monogr. Amer. Math. Soc. **43**, Providence R.I. 1974

Ju. L. Daletskij, N. D. Tsvintarnaja

[1982] *Random diffusion functions of multi-dimensional time* [in Russian], Ukrain. Mat. Zh. **34**, 1 (1982), 20-24

G. Darbo

[1955] *Punti uniti in trasformazioni a codominio non compatto*, Rend. Sem. Mat. Univ. Padova **24** (1955), 84-92

I. K. Daugavet

[1963] *On some property of completely continuous operators in the space C*, Uspekhi Mat. Nauk **18** (1963), 157-158

O. W. Diallo

[1988] *Linear integro-differential equations of Barbashin type in function spaces* [in Russian], Kach. Pribl. Metody Issled. Oper. Uravn. **13** (1988), 67-76

[1988a] *On the theory of linear integro-differential equations of Barbashin type in Lebesgue spaces* [in Russian], VINITI No. 1013-V88, Minsk 1988

[1988b] *On the theory of linear integro-differential equations in the space of continuous functions* [in Russian], Vestnik Belor. Gos. Univ. Ser. Fiz.-Mat. Nauk **2** (1988), 40-43

[1989] *Methods of Functional Analysis in the Theory of Barbashin Integro-Differential Equations* [in Russian], Kand. Diss., Univ. Minsk 1989

O. W. Diallo, P. P. Zabrejko

[1987] *On the Daugavet-Krasnosel'skij theorem* [in Russian], Vestnik Akad. Nauk. BSSR Ser. Fiz.-Mat. Nauk **3** (1987), 26-31

[1987a] *The Bogoljubov averaging method in the problem of bounded solutions of integro-differential equations of Barbashin type* [in Russian], Proc. 11th Int. Conf. Nonlin. Osc., Ed. Janos Bolyai Math. Soc., Budapest 1987

[1990] *Criteria for asymptotic stability for the solutions of integro-differential equations of Barbashin type* [in Russian], Dokl. Akad. Nauk BSSR **34, 2** (1990), 101-104

V. P. DIDENKO, N. N. KOZLOV

[1975] *A regularized method for the solution of message estimation problems* [in Russian], Dokl. Akad. Nauk SSSR **222, 5** (1975), 1045-1048

T. B. DIL'MANOV

[1982] *On a class of Volterra operator equations in the plane* [in Russian], Vopr. Korr. Obratn. Zad. (Novosibirsk) (1982), 77-84

X. DING

[1984] *On global random solutions for random integral and differential equations in Banach spaces*, Appl. Math. (Novy Sad) **14, 2** (1984), 101-109

N. DUNFORD, J. T. SCHWARTZ

[1963] *Linear Operators I*, Int. Publ., Leyden 1963

V. K. DZJADYK, L. A. OSTROVETSKIJ

[1981] *An approximate method for the Goursat problem solution for linear hyperbolic equations with polynomial coefficients* [in Russian], Issled. Teor. Appr. Funk. (Kiev) (1981), 20-37

A. W. ENGLAND

[1974] *Thermal microwave emission from a half-space containing scatterers*, Radio Sci. **9, 4** (1974), 447-454

E. A. ERMOLOVA

[1995] *Ljapunov-Bohl-Exponent und Greensche Funktion für eine Klasse von Integro-Differentialgleichungen*, Zeitschr. Anal. Anw. **14, 4** (1995), 881-898

N. P. ERUGIN

[1963] *Linear Systems of Ordinary Differential Equations with Periodic and Quasi-Periodic Coefficients* [in Russian], Izd. Akad. Nauk BSSR, Minsk 1963

N. A. EVKHUTA, P. P. ZABREJKO

[1985] *On the A. M. Samojlenko method for finding periodic solutions of quasilinear differential equations in Banach spaces* [in Russian], Ukrain. Mat. Zhurn. **37, 2** (1985), 162-168

S. FENYÖ

[1955] *Beiträge zur Theorie der linearen partiellen Integralgleichungen*, Publ. Math. **4, 1** (1955), 98-103

E. V. FROLOVA

[1995] *On the algebra of partial integral operators in $\mathfrak{L}(C)$* [in Russian], Oper. Uravn. Chastn. Integr. Integro-Diff. Uravn. (Lipetsk) (1995), 13

[1996] *Acting criteria for operators with partial integrals in the space of continuous functions*, Oper. Uravn. Chastn. Integr. (Lipetsk) (1996), 3-12

E. V. FROLOVA, A. S. KALITVIN, P. P. ZABREJKO

[1997] *Operator functions with partial integrals on C and L^p* (to appear)

C. J. FUNK

[1973] *Multiple scattering calculations of light propagation in ocean water*, Appl. Optim. **12, 2** (1973), 301-313

B. G. GABDULKHAEV

[1980] *Finite-dimensional approximations of singular integrals and direct solution methods for singular and integro-differential equations* [in Russian], Itogi Nauki Tekh. **18** (1989), 251-307

S. A. GABOV, A. G. SVESHNIKOV

[1990] *Linear Problems for Non-Stationary Intrinsic Waves* [in Russian], Nauka, Moscow 1990

H. GACKY

[1986] *On the existence of solutions of a certain random integral equation*, Fasc. Math. **16** (1986), 89-93

F. D. GAKHOV

[1977] *Boundary Value Problems* [in Russian], Nauka, Moscow 1977

F. D. GAKHOV, V. A. KAKICHEV

[1967] *Cases of degeneracy of two-dimensional singular integral equations with Cauchy kernels and constant coefficients* [in Russian], Funkt. Anal. Teor. Funkt. (Kazan) 4 (1967), 91-96

L. A. GALIN, I. G. GORJACHEVA

[1977] *An axially symmetric contact problem of elasticity theory in the presence of abrasion* [in Russian], Prikl. Mat. Mekh. **41, 5** (1977), 807-812; Engl. transl.: J. Appl. Math. Mech. **41, 5** (1979), 826-831

I. M. GEL'FAND, B. M. LEVITAN

[1951] *On the determination of a differential equation through its spectral function* [in Russian], Izv. Akad. Nauk SSSR **15** (1951), 309-360

E. D. GILLES, U. KNÖPP

[1967] *Die Dynamik von Rührkesselreaktoren bei Polymerisations-reaktoren*, Regelungstechnik **15** (1967), 199-203

V. I. GLIVENKO

[1936] *The Stieltjes Integral*, ONTI, Moscow-Leningrad 1936

B. V. GNEDENKO

[1988] *A Course in Probability Theory* [in Russian], Nauka, Moscow 1988; German transl.: Akademie-Verlag, Berlin 1989

I. TS. GOKHBERG, M. G. KREJN

[1965] *Introduction to the Theory of Linear Non-Selfadjoint Operators* [in Russian], Nauka, Moscow 1965; Engl. transl.: Math. Monogr. Amer. Math. Soc., Providence 1970

E. GOURSAT

[1943] *Cours d'analyse mathématique*, Gauthier-Villars, Paris 1943

A. A. GOVORUKHINA, N. V. KOVALENKO, I. A. PARADOKSOVA

[1985] *Two-dimensional integral equations with partial integrals on the plane and half-plane* [in Russian], Ehlista **4** (1985), 23-32

[1986] *Two-dimensional integral equations with partial integrals and polynomial coefficients on the plane* [in Russian], VINITI No. 3747-V86, Rostov-na-Donu 1986

[1987] *The discrete analogue to a two-dimensional integral equation with partial integrals* [in Russian], VINITI No. 6583-V87, Rostov-na-Donu 1987

[1987a] *Multidimensional integral operators with partial integrals in a half-space* [in Russian], Int. Operat. Uravn. (Krasnodar) (1987), 41-49

[1989] *Integral equations with partial integrals in a half-strip* [in Russian], Proc. Vol. I, Novgorod 1989

W. GREENBERG, C. V. M. VAN DER MEE, V. PROTOPOPESCU

[1987] *Boundary Value Problems in Abstract Kinetic Theory*, Birkhäuser, Basel 1987

A. GROTHENDIECK

[1955] *Résumé des résultats essentiels dans la théorie des produits tensoriels topologiques et des espaces nucléaires*, Ann. Inst. Fourier Grenoble **4** (1955), 73-112

K. P. HADELER

[1986] *The hypercycle, travelling waves, and Wright's equation*, J. Math. Biol. **24, 5** (1986), 473-477

HALBERG, TAYLOR

[1956] *On the spectra of linked operators*, Pacific J. Math. **6, 6** (1956), 283-290

P. HALMOS, V. SUNDER

[1978] *Bounded Integral Operators on L^2 Spaces*, Springer, Berlin 1978

E. J. HALTON, W. A. LIGHT

[1985] *Minimal projections in L_p spaces*, Math. Proc. Cambridge Phil. Soc. **97, 1** (1985), 127-136

J. V. HEROD

[1983] *Series solutions for nonlinear Boltzmann equations*, Nonlin. Anal. TMA **7, 12** (1983), 1373-1387

E. HILLE, R. PHILLIPS

[1957] *Functional Analysis and Semigroups*, Coll. Publ. Amer. Math. Soc., Providence 1957

E. HILLE, J. D. TAMARKIN

[1930] *On the theory of linear integral equations*, Ann. Math. **31** (1930), 479-528

T. ICHINOSE

[1978] *Spectral properties of tensor products of linear operators I*, Trans. Amer. Math. Soc. **235** (1978), 75-113

[1978a] *Spectral properties of tensor products of linear operators II*, Trans. Amer. Math. Soc. **237** (1978), 223-254

M. IMANALIEV, M. DZHURAEV, P. S. PANKOV

[1974] *On an inverse initial value problem for integro-differential equations with partial derivatives* [in Russian], Issled. Integro-Diff. Uravn. Kirg. Frunze **10** (1974), 3-15

A. ISHIMARU

[1978] *Wave Propagation and Scattering in Random Media*, Academic Press, New York 1978

V. G. IVANITSKIJ

[1971] *Approximate solution of a Barbashin type equation by means of the method of averaging of functional corrections* [in Russian], Pribl. Kach. Metody Teor. Diff. Int. Uravn. Kiev (1971), 183-194

[1973] *Numerical realization of the method of averaging of functional corrections for Barbashin type equations* [in Russian], Pribl. Kach. Metody Teor. Diff. Integr. Uravn. Kiev (1973), 76-88

M. C. JOSHI

[1987] *Weak compactness of solution measures of nonlinear approximative random operator equations*, Stochast. Anal. Appl. **5, 2** (1987), 151-166

V. A. KAKICHEV

[1967] *On the regularization of singular integral equations with Cauchy kernels for bicylindric domains* [in Russian], Izv. VUZov Mat. **7** (1967), 54-64

[1968] *Degenerate two-dimensional singular integral equations with Cauchy kernels for bicylindric domains I* [in Russian], Teor. Funkts. Funk. Anal. Pril. **7** (1968), 13-19

[1969] *Degenerate two-dimensional singular integral equations with Cauchy kernels for bicylindric domains II* [in Russian], Teor. Funkts. Funk. Anal. Pril. **8** (1969), 25-28

[1978] *Solution Methods for some Boundary Value Problems for Analytic Functions of Two Complex Variables* [in Russian], Izdat. Tjumens. Univ., Tjumen 1978

V. A. KAKICHEV, N. V. KOVALENKO

[1973] *On the theory of two-dimensional integral equations with partial integrals* [in Russian], Ukr. Mat. Zhurnal **25, 3** (1973), 302-312

S. O. KALIEV

[1974] *On the existence of a solution for one class of integro-differential equations* [in Russian], Inst. Mat. Mekh. Akad. Nauk Kaz. SSR Alma-Ata (1974), 116-121

A. S. KALITVIN

[1983] *On the solvability of a partial integral equation of V. I. Romanovskij type with a degenerate kernel* [in Russian], VINITI No. 3462-83, Leningrad 1983

[1983a] *On partial integral operators* [in Russian], VINITI No. 3461-83, Leningrad 1983

[1984] *On the spectrum and eigenfunctions of a partial integral operator of V. I. Romanovskij type* [in Russian], Funct. Anal. (Ul'janovsk) **22** (1984), 35-45

[1984a] *The superposition operator in spaces with mixed quasinorm* [in Russian], VINITI No. 6883, Leningrad 1984

[1985] *On the spectrum of some classes of partial integral operators* [in Russian], Oper. Prilozh. Pribl. Funk., Leningrad 1985, 27-35

[1985a] *On the multispectrum of linear operators* [in Russian], Oper. Prilozh. Pribl. Funk., Leningrad 1985, 91-99

[1986] *On the spectrum of a partial integral operator in a space with mixed norm* [in Russian], Diff. Uravn. Chastn. Proizv., Leningrad 1986, 128-131

[1986a] *Investigations on Partial Integral Operators* [in Russian], Kand. Diss., Pedag. Inst. Leningrad 1986

[1987] *On continuity and regularity of partial integral operators* [in Russian], VINITI No. 504-V87, Lipetsk 1987

[1987a] *Partial integral equations of V. I. Romanovskij type* [in Russian], VINITI No. 503-V87, Lipetsk 1987

[1987b] *Partial integral equations of V. I. Romanovskij type II* [in Russian], VINITI No. 502-V87, Lipetsk 1987

[1988] *Partial integral equations of V. I. Romanovskij type III* [in Russian], VINITI No. 3229-V88, Lipetsk 1988

[1988a] *On the spectrum of linear partial integral operators with positive kernels* [in Russian], Pribl. Funk. Spektr. Teor., Leningrad 1988, 43-50

[1988b] *On positive eigenvalues of operators with partial integrals* [in Russian], Mezhduvuz. Konf. Molod. Uchen., Lipetsk 1988, 101

[1989] *On the solvability of some classes of integral equations with partial integral operators* [in Russian], Funkts. Anal (Ul'janovsk) **29** (1989), 68-73

[1989b] *On one class of integral equations arising in the mechanics of continuous media* [in Russian], Nelin. Probl. Diff. Uravn. Mat. Fiz. (Ternopol') (1989), 181-182

[1991] *On an equation with partial integrals in function spaces* [in Russian], Vsesojuzn. Shkola Teor. Oper. Prostr. Funkts. (Nizhny Novgorod) (1991), 90

[1991a] *On generalized solutions of integro-differential equations of Barbashin type* [in Russian], Vestnik Cheljabinsk. Univ. Ser. Math. Meth. **1** (1991), 147-148

[1992] *On a boundary value problem for the Barbashin integro-differential equation* [in Russian], Sovrem. Metody Teor. Gran. Probl., Voronezh 1992, 51

[1993] *On the solvability of the boundary value problem for the Barbashin integro-differential equation* [in Russian], Sbornik Trudov, Voronezh 1993, 90

[1993a] *A boundary value problem for the Barbashin integro-differential equation* [in Russian], 2nd Mezhdunar. Seminar Negladk. Razryvn. Zad. Uravn. Optim. Cheljabinsk 1993, 65-66

[1994] *On partial integral operators in contact problems of elasticity* [in Russian], Sbornik Trudov, Voronezh 1994, 54

[1994a] *On some class of partial integral equations in aerodynamics* [in Russian], Sost. Persp. Razv. Nauchn.-Tekhn. Pot. Lipetsk. Obl. (Lipetsk) (1994), 210-212

[1995] *On one class the spectrum of Barbashin integro-differential equations* [in Russian], Oper. Uravn. Chastn. Integr. Integro-Diff. Uravn., Lipetsk 1995, 3-8

[1995a] *The norm of partial integral operators in L^∞ and L^1* [in Russian], Sovr. Metody Nelin. Anal. (Voronezh) (1995), 46-47

[1996] *On the spectral radius of operators with partial integrals.* Voron. Vesen. Shkola, Voronezh 1996, 90

[1996a] *Integro-differential equations of Barbashin type with partial integrals* [in Russian], Intern. Conf. Bound. Value Probl. Special Funct. Fract. Calculus, Minsk 1996, 137

[1996b] *Criteria for compactness and weak compactness for operators with partial integrals in the space of continuous functions* [in Russian], Oper. Chastn. Int., Lipetsk 1996, 13-17

[1996c] *On the theory of Romanovskij type operators* [in Russian], Kraev. Zad. Spets. Funk. Drobn. Ischisl., Minsk 1996, 93-97

[1997] *Nonlinear operators with partial integrals,* Nonlin. Anal. TMA **30, 1** (1997), 521-526

[1997a] *On the spectral radius of Volterra operators with partial integrals* [in Russian], Sovrem. Metody Teor. Gran. Problem, Voronezh 1997, 66

[1997b] *On a Volterra-Fredholm equation with partial integrals arising in axially symmetric contact problems* [in Russian], Funkt. Anal. Uravn. Mat. Fiz., Voronezh 1997, 31

[1998] *Spectral properties of partial integral operators of Volterra-Fredholm type,* Zeitschr. Anal. Anw. **17, 2** (1998), 297-309

A. S. KALITVIN, O. N. DEMANOVA

[1995] *On an operator with partial integrals in spaces with weight* [in Russian], Mat. Itog. Konf. Oper. Uravn. Chastn. Int. Integro-Diff. Uravn., Lipetsk 1995, 9

A. S. KALITVIN, E. V. FROLOVA

[1995] *Operator functions with partial integrals in the space of continuous functions* [in Russian], Sovr. Metody Nelin. Anal., Voronezh 1995, 46-47

A. S. KALITVIN, O. JU. GALITSKAJA, T. V. SOKOLOVA

[1988] *Spectral properties of operators with partial integrals in the space of continuous functions* [in Russian], Mezhduvuz. Konf. Molod. Uchen., Lipetsk 1988, 59

A. S. KALITVIN, S. N. GLOTOV

[1996] *Uryson operators with partial integrals in the space of continuous functions*, Oper. Chastn. Int., Lipetsk 1996, 18-22

A. S. KALITVIN, E. V. JANKELEVICH

[1993] *Operators with partial integrals in the space of continuous functions* [in Russian], Vestnik Cheljabinsk. Univ. **2** (1993), 61-68

[1994] *Operators with partial integrals in a space of vector functions* [in Russian], Trudy Voron. Zimn. Shkoly (Voronezh) (1994), 55

A. S. KALITVIN, I. M. KOLESNIKOVA

[1995] *On the continuity of operators with partial integrals in the space of functions of bounded variation* [in Russian], Mat. Itog. Konf. Oper. Uravn. Chastn. Int. Integro-Diff. Uravn., Lipetsk 1995, 12

A. S. KALITVIN, G. M. KORENCHUK, I. V. EVTJUKHINA

[1993] *On the differentiability of a nonlinear operator with partial integrals* [in Russian], Konf. Molod. Uchen., Lipetsk 1993, 87-88

A. S. KALITVIN, O. A. LAVROVA

[1993] *On the generalized spectral radius of an operator with partial integrals in the space C* [in Russian], Konf. Molod. Uchen., Lipetsk 1993, 92-94

A. S. KALITVIN, S. P. MILOVIDOV

[1981] *An interpolation theorem for a partial integral operator* [in Russian], Funkts. Anal (Ul'janovsk) **21** (1981), 76-81

A. S. KALITVIN, S. N. NASONOV

[1996] *On operators with partial integrals in a Hölder space of functions of two variables* [in Russian], Oper. Chastn. Int., Lipetsk 1996, 23-31

A. S. KALITVIN, S. V. PROVOTOROVA

[1995] *Hardy-Littlewood type operators with partial integrals* [in Russian], Mat. Itog. Konf. Oper. Uravn. Chastn. Int. Integro-Diff. Uravn., Lipetsk 1995, 15

A. S. KALITVIN, L. M. ROMANOVA

[1993] *On the integral equation of V.I.Romanovskij type with partial integrals* [in Russian], Konf. Molod. Uchen., Lipetsk 1993, 89-91

A. S. KALITVIN, P. P. ZABREJKO

[1991] *On the theory of partial integral operators*, J. Integral Equ. Appl. **3, 3** (1991), 351-382

H. KAMOWITZ

[1984] *A property of compact operators*, Proc. Amer. Math. Soc. **91, 2** (1984), 231-236

T. KAMYTOV

[1976] *The approximate solution of an initial value problem for an integro-differential equation of Barbashin type by means of spline functions* [in Russian], Trudy Kirg. Univ. Mat. **11** (1976), 46-50

L. V. KANTOROVICH

[1956] *On integral operators* [in Russian], Uspekhi Mat. Nauk **11, 2** (1956), 3-29

L. V. KANTOROVICH, G. P. AKILOV

[1977] *Functional Analysis* [in Russian], Nauka, Moscow 1977; Engl. transl.: Pergamon Press, Oxford 1982

L. V. KANTOROVICH, B. Z. VULIKH, A. G. PINSKER

[1950] *Functional Analysis in Semi-Ordered Spaces* [in Russian], Gostekhizdat, Moscow 1950

S. KANTOROVITZ

[1957] *A note on partial linear integral equations*, Bull. Res. Council Israel **7, 4** (1957), 181-186

[1958] *On the integral equation* $\varphi(x,y) - \lambda a(x,y) \int \varphi(x,y)\, dx - \mu b(x,y) \int \varphi(x,y)\, dy = c(x,y)$ [in Hebrew], Riveon le Matematika **12** (1958), 24-26

T. KATO

[1966] *Perturbation Theory for Linear Operators*, Springer, Berlin 1966

I. S. KATS

[1963] *Behaviour of the solutions of a second order linear differential equation* [in Russian], Matem. Sbornik **62, 4** (1963), 476-495

I. S. KATS, M. G. KREJN

[1958] *Criteria for the discreteness of the spectrum of a singular string* [in Russian], Izv. VUZov Mat. **2/3** (1958), 136-153

C. T. KELLEY

[1980] *The Chandrasekhar H-equation for radiative transfer through inhomogeneous media*, J. Integral Equ. **2, 2** (1980), 155-165

Z. I. KHALILOV

[1956] *On the investigation of the asymptotic stability of solutions of boundary value problems for equations with partial derivatives* [in Russian], Dokl. Akad. Nauk Azerb. SSR **12, 6** (1956), 375-378

[1961] *On the stability of solutions of differential equations in Banach spaces* [in Russian], Dokl. Akad. Nauk Azerb. SSR **17, 5** (1961), 367-370

J. A. KHOLUB

[1987] *Daugavet's equation and operators on $L^1(\mu)$*, Proc. Amer. Math. Soc. **100, 2** (1987), 295-300

A. L. KHOTEEV

[1976] *An optimal control problem for integro-differential equations of Barbashin type* [in Russian], Probl. Upravl. Optim. Minsk (1976), 74-87

[1984] *Necessary multi-point conditions for singular controls for integro-differential equations of Barbashin type* [in Russian], Vestnik Akad. Nauk. BSSR **2** (1984), 20-25

T. KIFFE

[1980] *A perturbation of an abstract Volterra equation*, SIAM J. Math. Anal. **11, 6** (1980), 1036-1046

M. KIMURA

[1965] *A stochastic model concerning the maintenance of genetic variability in quantitative characters*, Proc. Nat. Acad. Sci. USA **54** (1965), 731-736

JU. L. KLIMONTOVICH

[1982] *Statistical Physics* [in Russian], Nauka, Moscow 1982

W. KOHL

[2000] *On a class of parabolic integro-differential equations*, Zeitschr. Anal. Anw. (to appear)

A. N. KOLMOGOROV

[1986] *Probability Theory and Mathematical Statistics* [in Russian], Nauka, Moscow 1986

V. B. KOROTKOV

[1980] *On some properties of integral and partial integral operators* [in Russian], Sib. Mat. Zh. **21** (1980), 98-105; Engl. transl.: Sib. Math. J. **21** (1980), 73-78

[1983] *Integral Operators* [in Russian], Nauka, Novosibirsk 1983

E. V. KOVALENKO

[1979] *An effective method of solving contact problems for a linear-deformable basis with a thin reinforcing covering* [in Russian], Izv. Akad. Nauk Arm. SSR Mekh. **32, 2** (1979), 76-82

[1981] *An approximate solution of a certain type of integral equations of elasticity and mathematical physics* [in Russian], Izv. Akad. Nauk Arm. SSR Mekh. **34, 5** (1981), 14-26

[1984] *On an integral equation of contact problems of elasticity theory in the presence of abrasive wear* [in Russian], Prikl. Mat. Mekh. **48, 5** (1984), 868-873; Engl. transl.: J. Appl. Math. Mech. **48, 5** (1984), 630-634

[1984a] *The solution of contact problems of creep theory for combined aging foundations* [in Russian], Prikl. Mat. Mekh. **48, 7** (1984), 1006-1014; Engl. transl.: J. Appl. Math. Mech. **48, 7** (1984), 739-745

[1985] *Investigation of an axially symmetric contact problem of the wear of a pair consisting of an annular stamp and a rough halfspace* [in Russian], Prikl. Mat. Mekh. **49, 5** (1985), 836-843; Engl. transl.: J. Appl. Math. Mech. **49, 5** (1985), 641-647

[1988] *Contact problems for bodies with coatings* [in Russian], Izv. Akad. Nauk Arm. SSR Mekh. **41, 1** (1988), 40-50

[1989] *Some approximate methods of solving integral equations of mixed problems* [in Russian], Prikl. Mat. Mekh. **53, 1** (1989), 107-114; Engl. transl.: J. Appl. Math. Mech. **53, 1** (1989), 85-92

[1989a] *Refined equations of the deformation of elastic plates* [in Russian], Prikl. Mekh. Kiev **25, 10** (1989), 111-116; Engl. transl.: Sov. Appl. Mech. **25, 10** (1989), 1053-1058

[1990] *On the analysis of thin porous coatings* [in Russian], Prikl. Mat. Mekh. **54, 3** (1990), 469-473; Engl. transl.: J. Appl. Math. Mech. **54, 3** (1990), 388-392

E. V. KOVALENKO, A. V. MANZHIROV

[1982] *A contact problem for a two-layer aging viscoelastic foundation* [in Russian], Prikl. Mat. Mekh. **46, 3** (1982), 674-682; Engl. transl.: J. Appl. Math. Mech. **46, 3** (1983), 536-542

N. V. KOVALENKO

[1968] *On the solution of a two-dimensional integral equation with partial integrals in the space L_2* [in Russian], Proc. Conf. Rostov. Mat. Obshch. (1968), 41-49

[1971] *On the theory of two-dimensional integral equations with partial integrals* [in Russian], Proc. Conf. Mat. Belor., Vol. I (1975), 38-40

[1972] *On a homogeneous integral equation with partial integrals* [in Russian], Fiz.-Mat. Issledov., Rostov 1972, 3-7

[1975] *Two-dimensional integral equations of convolution type with partial integrals which are solvable in closed form* [in Russian], Izv. VUZov **1** (1975), 111-114

[1981] *On the approximate solution of one type of integral equations of elasticity theory and mathematical physics* [in Russian], Izv. Akad. Arm. SSR **34, 5** (1981), 14-26

M. A. KRASNOSEL'SKIJ

[1956] *Topological Methods in the Theory of Nonlinear Integral Equations* [in Russian], Gostekhizdat, Moscow 1956; Engl. transl.: Macmillan, New York 1964

[1962] *Positive Solutions of Operator Equations* [in Russian], Fizmatgiz, Moscow 1962; Engl. transl.: Noordhoff, Groningen 1964

[1966] *The Shift Operator along the Trajectories of Differential Equations* [in Russian], Nauka, Moscow 1966; Engl. transl.: Math. Monogr. Amer. Math. Soc., Providence 1968

[1967] *On a class of nonlinear operators in the space of abstract continuous functions* [in Russian], Mat. Zametki **2, 6** (1967), 599-604

M. A. KRASNOSEL'SKIJ, S. G. KREJN

[1955] *On the averaging principle in nonlinear mechanics* [in Russian], Uspekhi Mat. Nauk **10, 3** (1955), 147-153

[1956] *On the theory of ordinary differential equations in Banach spaces* [in Russian], Trudy Sem. Funk. Anal. **2** (1956), 3-23

M. A. KRASNOSEL'SKIJ, L. A. LADYZHENSKIJ

[1954] *Conditions for the complete continuity of the P. S. Uryson operator* [in Russian], Trudy Mosk. Mat. Obshch. **3** (1954), 307-320

M. A. KRASNOSEL'SKIJ, E. A. LIFSHITS, A. V. SOBOLEV

[1985] *Positive Linear Systems* [in Russian], Nauka, Moscow 1985; Engl. transl.: Heldermann, Berlin 1989

M. A. KRASNOSEL'SKIJ, JA. B. RUTITSKIJ

[1958] *Convex Functions and Orlicz Spaces* [in Russian], Fizmatgiz, Moscow 1958; Engl. transl.: Noordhoff, Groningen 1961

M. A. KRASNOSEL'SKIJ, G. M. VAJNIKKO, P. P. ZABREJKO, JA. B. RUTITSKIJ, V. JA. STETSENKO

[1969] *Approximate Solutions of Operator Equations* [in Russian], Nauka, Moscow 1969; Engl. transl.: Noordhoff, Groningen 1972

M. A. KRASNOSEL'SKIJ, P. P. ZABREJKO, E. I. PUSTYL'NIK, P. E. SOBOLEVSKIJ

[1966] *Integral Operators in Spaces of Summable Functions* [in Russian], Nauka, Moscow 1966; Engl. transl.: Noordhoff, Leyden 1976

M. G. KREJN

[1971] *Linear Equations in Banach Spaces* [in Russian], Nauka, Moscow 1971

M. G. KREJN, M. RUTMAN

[1948] *Linear operators leaving invariant a cone in a Banach space* [in Russian], Uspekhi Mat. Nauk **3** (1948), 3-95

S. G. KREJN, JU. I. PETUNIN, E. M. SEMENOV

[1978] *Interpolation of Linear Operators* [in Russian], Nauka, Moscow 1978; Engl. transl.: Math. Monogr. Amer. Math. Soc., Providence 1982

L. E. KRIVOSHEJN, EH. JA. BYKOVA

[1974] *Approximate solution of a boundary value problem for one class of integro-differential equations of Barbashin type* [in Russian], Trudy Kirg. Univ. Ser. Mat. **9** (1974), 26-32

J. KURZWEIL

[1957] *Generalized ordinary differential equations and continuous dependence on a parameter*, Czech. Math. J. **7** (1957), 418-449

I. B. LEDOVSKAJA

[1987] *The averaging principle for integro-differential equations* [in Russian], VINITI No. 1892-1987, Voronezh 1987

U. LENHARDT

[1986] *Numerische Lösung linearer Volterrascher partieller Integralgleichungen zweiter Art unter Verwendung der Quadratur-Kubaturformelmethode*, Wiss. Zeitschr. Päd. Hochschule "Liselotte Hermann" Güstrow **24**, **1** (1986), 67-80

A. JU. LEVIN

[1967] *Passing to the limit in nonsingular systems $\dot{x} = A_n(t)x$* [in Russian], Dokl. Akad. Nauk SSSR **176** (1967), 774-777

V. L. LEVIN

[1969] *Tensor products and functors in categories of Banach spaces which are defined by KV-lattices* [in Russian], Trudy Moskov. Mat. Obshch. **20** (1969), 43-82

B. M. LEVITAN, I. S. SARGSJAN

[1988] *Sturm-Liouville and Dirac Operators* [in Russian], Nauka, Moscow 1988; Engl. transl.: Kluwer, Dordrecht 1990

L. KH. LIBERMAN

[1958] *On the stability of integro-differential equations* [in Russian], Izv. VUZov Mat. **3** (1958), 142-151

L. M. LIKHTARNIKOV

[1974] *On an operator equation with two parameters in Hilbert space* [in Russian], Funkts. Anal. (Ul'janovsk) **3** (1974), 92-95

[1975] *On the spectrum of a class of linear integral equations with two parameters* [in Russian], Diff. Uravn. **11, 6** (1975), 1108-1117; Engl. transl.: Diff. Equ. **11, 6** (1975), 833-840

[1976] *On the spectrum of a partial integral equation of V. I. Romanovskij type* [in Russian], Funkts. Anal. (Ul'janovsk) **7** (1976), 95-105

[1980] *On the spectrum of a linear partial integral operator of V. I. Romanovskij type with a completely continuous operator* [in Russian], Diff. Uravn (Rjazan) **11** (1980), 42-51

[1981] *On the solvability of a linear partial integral equation of V. I. Romanovskij type with a completely continuous operator* [in Russian], Diff. Uravn (Rjazan) **12** (1981), 64-72

[1987] *On the spectrum of an integral equation of V. I. Romanovskij type in the self-adjoint case* [in Russian], Proc. Conf., Tambov 1987, 128

L. M. LIKHTARNIKOV, A. S. KALITVIN

[1989] *On equations of V. I. Romanovskij type* [in Russian], Proc. Conf. Vol. II, Novgorod 1989, 57

L. M. LIKHTARNIKOV, L. L. MOROZOVA

[1983] *On a method for the investigation of partial integral equations* [in Russian], Funkts. Anal. (Ul'janovsk) **21** (1983), 108-112

[1988] *On a class of equations of Romanovskij type in Hilbert space* [in Russian], Funkts. Anal. (Ul'janovsk) **28** (1988), 68-81

L. M. LIKHTARNIKOV, L. V. SPEVAK

[1976] *On the solvability of a partial integral equation of V. I. Romanovskij type* [in Russian], Funkts. Anal. (Ul'janovsk) **7** (1976), 106-115

[1976a] *A linear partial integral equation of V. I. Romanovskij type with two parameters* [in Russian], Diff. Uravn (Rjazan) **7** (1976), 165-176

L. M. LIKHTARNIKOV, L. Z. VITOVA

[1975] *On the spectrum of an integral equation with partial integrals* [in Russian], Litov. Mat. Sbornik **15** (1975), 41-47; Engl. transl.: Lith. Math. J. **15** (1975), 228-233

[1976] *On the solvability of a linear integral equation with partial integrals* [in Russian], Ukr. Mat. Zh. **28**, 1 (1976), 83-87; Engl. transl.: Ukrain. Math. J. **28**, 1 (1976), 64-67

A. A. LOBUZOV

[1982] *Investigation of boundary value problems for the characteristic operators of diffusion processes in Banach spaces* [in Russian], Vestnik Mosk. Gos. Univ. **1**, **3** (1982), 22-26

M. E. LORD

[1980] *Existence, uniqueness, and validation of parameter imbedding equations for nonlinear Fredholm integral equations*, Nonlin. Anal. TMA **4**, **5** (1980), 1001-1009

G. JA. LOZANOVSKIJ

[1966] *On quasi-integral operators in KB-spaces* [in Russian], Vestnik Leningr. Gos. Univ. **2** (1966), 35-44

W. A. J. LUXEMBURG, A. C. ZAANEN

[1971] *Riesz Spaces I*, North-Holland Publ., Amsterdam 1971

V. A. MALYSHEV

[1970] *Random walks, Wiener-Hopf equations in a quadrant, Galois automorphisms* [in Russian], Preprint Univ. Moscow 1970

D. MAMYTOV

[1985] *An estimate for the solutions of one class of integro-differential equations* [in Russian], Trudy Inst. Avtom. Akad. Nauk Kirg. SSR Frunze (1985), 3-9

A. V. MANZHIROV

[1983] *Axialsymmetric contact problems for non-uniformly aging viscoelastic foundations with layer rheology* [in Russian], Prikl. Mat. Mekh. **47, 4** (1983), 684-693; Engl. transl.: J. Appl. Math. Mech. **47, 4** (1984), 558-556

[1985] *On a method for solving two-dimensional integral equations for axialsymmetric contact problems for bodies with layer rheology* [in Russian], Prikl. Mat. Mekh. **49, 6** (1985), 1019-1025; Engl. transl.: J. Appl. Math. Mech. **49, 6** (1985), 777-782

V. A. MARCHENKO

[1977] *Sturm-Liouville Operators and Applications* [in Russian], Naukova Dumka, Kiev 1977; Engl. transl.: Birkhäuser, Basel 1986

G. I. MARCHUK

[1961] *The Method of Calculation for Nuclear Reactors* [in Russian], Atomizdat, Moscow 1961

M. V. MASLENNIKOV

[1968] *The Milne problem with anisotropic scattering* [in Russian], Trudy MIAN **97** (1968), 3-133

P. MAURO

[1976] *Su un'equazione integrale lineare di tipo non ancora considerato*, Rend. Accad. Naz. Detta XL **5, 1** (1976), 55-59

I. N. MININ

[1988] *Theory of Radiation Transfer in the Atmospheres of Planets* [in Russian], Nauka, Moscow 1988

G. MINTY

[1962] *Monotone operators in Hilbert space*, Duke Math. J. **29** (1962), 341-346

Ju. A. MITROPOL'SKIJ

[1971] *The Averaging Method in Nonlinear Mechanics* [in Russian], Naukova Dumka, Kiev 1971

N. I. MORARU

[1969] *Wiener-Hopf equations in a quadrant* [in Russian], Diff. Uravn. **5, 5** (1969), 1445-1457

[1975] *On a two-dimensional integral equation of convolution type* [in Russian], Trudy Kishin. Politekh. Inst. **1** (1975), 111-114

V. A. MOROZOV

[1965] *Application of the regularization method to the solution of an ill-posed problem* [in Russian], Vestnik Mosk. Gos. Univ. **1, 4** (1965), 13-25

L. L. MOROZOVA

[1984] *Approximate solution of a partial integral equation of V. I. Romanovskij type by the momentum method* [in Russian], Diff. Uravn. Reshen. Prikl. Zad. (1984), 164-165

[1984a] *On an application of bi-cubic splines to the solution of linear partial integral equations of V. I. Romanovskij type* [in Russian], VINITI No. 4568-84, Leningrad 1984

[1984b] *On the averaging method for functional corrections for the approximate solution of a partial integral equation of V. I. Romanovskij type* [in Russian], VINITI No. 3484-84, Leningrad 1984

[1986] *The structure of eigenfunctions and eigenvalues of a partial integral operator of V. I. Romanovskij type* [in Russian], Diff. Uravn. Chast. Proizv., Leningrad 1986, 58-63

G. MÜNTZ

[1934] *Integral Equations* [in Russian], ONTI, Moscow 1934

V. MURESAN

[1982] *Bezüglich eines Integraloperators vom Typ Volterra-Sobolev,* Studia Univ. Babes-Bolyai Math. **27** (1982), 68-72

[1984] *Die Methode der sukzessiven Approximationen für die Integralgleichung vom Typ Volterra-Sobolev,* Mathematica (RSR) **26, 2** (1984), 129-136

I. P. NATANSON

[1950] *Theory of Functions of Real Variables* [in Russian], GITTL, Moscow 1950; German transl.: Akademie-Verlag, Berlin 1969

S. M. NIKOL'SKIJ

[1969] *Approximation of Functions of Several Variables and Imbedding Theorems* [in Russian], Nauka, Moscow 1969

T. NUREKENOV

[1966] *On properties of the shift operator along the trajectories of systems of integro-differential equations* [in Russian], Izv. Akad. Nauk Kaz. SSR Ser. Fiz.-Mat. Nauk 1 (1966), 43-47

T. NUREKENOV, S. KALIEV

[1974] *An existence theorem for solutions of integro-differential equations in the classical sense* [in Russian], Inst. Mat. Mekh. Akad. Nauk Kaz. SSR Alma-Ata (1974), 280-282

[1974a] *An existence theorem for solutions of the Cauchy problem for integro-differential equations in the classical sense* [in Russian], VINITI No. 2244-74, Alma-Ata 1974

O. P. OKOLELOV

[1967] *On the theory of two-dimensional integral equations with partial integrals* [in Russian], Diff. Integr. Uravn. (Khabarovsk) **3** (1967), 142-149

[1967a] *On two-dimensional equations with partial integrals*, Uch. Zapiski Irkutsk. Ped. Inst. **34** (1967), 65-84

[1967b] *Investigation of Equations with Partial Integrals* [in Russian], Kand. Diss., Univ. Irkutsk 1967

[1968] *The analogue of some Fredholm theorems for integral equations with multiple and partial integrals* [in Russian], Trudy Irkutsk. Univ. **3** (1968), 17-24

V. I. OVCHINNIKOV

[1983] *An exact interpolation theorem in L_p spaces* [in Russian], Dokl. Akad. Nauk SSSR **272** (1983), 300-303; Engl. transl.: Soviet Math. Dokl. **28** (1983), 381-385

B. G. PACHPATTE

[1981] *On nonlinear coupled parabolic integro-differential equations*, J. Math. Phys. Sci. **21**, **1** (1981), 39-49

[1983] *On some application of the Ważewski method for multiple Volterra integral equations*, Ananele Stiin. Univ. A. L. I. Cuza Iasi **29** (1983), 75-83

[1984] *On a nonlinear functional integral equation in two independent variables*, Ananele Stiin. Univ. A. L. I. Cuza Iasi **30** (1984), 31-38

[1986] *On mixed Volterra-Fredholm type integral equations*, Indian J. Pure Appl. Math. **17**, **4** (1986), 488-496

V. C. PHAN

[1984] *Existence of solutions for random multivalued Volterra integral equations I*, J. Integral Equ. **7**, **1** (1984), 143-173

[1984a] *Existence of solutions for random multivalued Volterra integral equations II*, J. Integral Equ. **7, 2** (1984), 175-185

V. S. PILIDI

[1971] *The local method in the theory of operator equations for bi-singular integral equations* [in Russian], Math. Anal. Prilozh. (Rostov-na-Donu) **3** (1971), 81-105

[1971a] *On multidimensional bisingular operators* [in Russian], Dokl. Akad. Nauk SSSR **201**, 4 (1971), 787-789

[1982] *Necessary conditions for the Fredholmness of characteristic bi-singular integral operators with measurable coefficients* [in Russian]), Mat. Zametki **31, 1** (1982), 53-59

[1988] *A proof of the method of excision of singularities for bisingular integral operators with continuous coefficients on closed contours* [in Russian], VINITI No. 6753-B88, Rostov-na-Donu 1988

[1989] *On the method of excision of singularities for bisingular integral operators with continuous coefficients* [in Russian], Funkt. Anal. Prilozh. **23, 1** (1989), 82-83

V. S. PILIDI, L. I. SAZONOV

[1983] *Bi-singular integral operators in spaces of Hölder functions* [in Russian], VINITI No. 6771-83, Rostov-na-Donu 1983

A. I. POVOLOTSKIJ, A. S. KALITVIN

[1983] *Interpolation of a partial integral operator in spaces with mixed norms* [in Russian], Oper. Prilozh., Leningrad 1983, 67-75

[1985] *On nonlinear operators with partial integrals* [in Russian], Funkt. Anal. (Ul'janovsk) **25** (1985), 104-115

[1985a] *On Hammerstein operators with partial integrals* [in Russian],
        Oper. Uravn. Funk. Mnozhestv (Syktyvkar) (1985), 100-107

[1986] *On a functor related with a partial integral operator* [in Rus-
        sian], Kachestv. Teor. Slozhn. Sistem, Leningrad 1986, 50-53

[1987] *Solvability of a Hammerstein type equation with partial inte-
        grals* [in Russian], Pribl. Funk. Spets. Klass. Oper. (Vologda)
        (1987), 141-146

[1989] *On the solution of Hammerstein equations with partial inte-
        grals* [in Russian], Oper. Uravn. Funk. Mnozhestv (Syktyv-
        kar) (1989), 8-15

[1989a] *On the solvability of Hammerstein equations with partial in-
        tegrals* [in Russian], Vsesojuzn. Shkola Teor. Oper. Prostr.
        Funk. (Novgorod) (1989), 2-3

[1991] *Nonlinear Operators with Partial Integrals* [in Russian], Izd.
        Ross. Gos. Pedag. Univ., St. Petersburg 1991

[1994] *Solvability of Uryson equations with partial integrals* [in Rus-
        sian], Vestnik Cheljabinsk. Univ. Ser. Mat. Mech. 1 (1994),
        96-103

JU. V. PROKOFIEV, JU. A. ROZANOV

[1987] *Probability Theory: Basic Notions, Limit Theorems, Random
        Processes* [in Russian], Nauka, Moscow 1987

I. O. RADON

[1919] *Über lineare Funktionaltransformationen und Funktionalglei-
        chungen*, Sitzungsber. Akad. Wiss. Wien **128** (1919), 1083-
        1121

X. A. RAKHMATULLIN, A. E. OSOKIN

[1977] *On the transversal stroke on a viscoelastic flexible fibre* [in Russian], Vestnik Mosk. Gos. Univ. **4** (1977), 90-95

V. S. H. RAO

[1983] *On random solutions of Volterra-Fredholm integral equations*, Pacific J. Math. **108, 2** (1983), 397-405

M. M. RAO, Z. D. REN

[1991] *Theory of Orlicz Spaces*, M. Dekker, New York 1991

T. M. RÄTZSCH, H. KEHLEN

[1984] *Kontinuierliche Thermodynamik komplexer Vielstoffsysteme*, Wiss. Zeitschr. TH Leuna-Merseburg **26, 3** (1984), 391-399

M. REED, B. SIMON

[1972] *Methods of Modern Mathematical Physics*, Acad. Press, New York 1972

R. T. ROCKAFELLAR

[1970] *On the maximal monotonicity of sums of nonlinear monotone operators*, Trans. Amer. Math. Soc. **149** (1970), 75-88.

V. ROMANOVSKIJ

[1932] *Sur une class d'équations intégrales linéaires*, Acta Math. **59** (1932), 99-208

L. R. F. ROSE

[1982] *Asymptotic analysis of the Cook-Noble integral equation*, Math. Proc. Cambridge Phil. Soc. **92, 2** (1982), 293-306

H. ROTH

[1972] *Ein Modell und seine Anwendungen für stoffliche und verfahrenstechnische Berechnungen der Erdölverarbeitung*, Diss. B, TH Leuna-Merseburg 1972

H. L. ROYDEN

[1963] *Real Analysis*, Macmillan, New York 1963

W. RUDIN

[1973] *Functional Analysis*, McGraw-Hill, New York 1973

JA. B. RUTITSKIJ

[1962] *New criteria for the continuity and complete continuity of integral operators in Orlicz spaces* [in Russian], Izvest. VUZov **5** (1962), 87-100

B. N. SADOVSKIJ

[1972] *Limit-compact and condensing operators* [in Russian], Uspekhi Mat. Nauk **27, 1** (1972), 81-146; Engl. transl.: Russian Math. Surveys **27, 1** (1972), 85-155

KH. M. SALGAPAROV

[1962] *On the stability of integro-differential equations of Barbashin type* [in Russian], Issled. Integro-Diff. Uravn. Kirg. Frunze **2** (1962), 117-120

S. A. SAMEDOVA

[1958] *Asymptotic stability of solutions of a certain integro-differential equation* [in Russian], Dokl. Akad. Nauk Azerb. SSR **14, 6** (1958), 419-423

J. SARVAS

[1982] *Quasiconformal semiflows*, Ann. Acad. Scient. Fenn. Math. **7** (1982), 197-219

H. H. SCHAEFER

[1971] *Topological Vector Spaces*, Springer, Berlin 1971

R. SCHATTEN

[1950] *A Theory of Cross-Spaces*, Princeton Univ. Press, Princeton 1950

M. SCHECHTER

[1965] *Invariance of the essential spectrum*, Bull. Amer. Math. Soc. **71** (1965), 365-367

L. SCHLESINGER

[1914] *Sur les équations intégro-différentielles*, C. R. Acad. Sci. Paris **158** (1914), 1872-1875

[1915] *Zur Theorie der linearen Integrodifferentialgleichungen*, Jahresber. Dtsch. Math. Vgg. **24** (1915), 84-123

V. A. SHCHELKUNOV

[1972] *Integral equations whose kernels depend on three variables I* [in Russian], Sb. Nauchn. Trudov Kaf. Vyssh. Mat. Tul'sk. Politekh. Inst. **1** (1972), 34-38

[1974] *Integral equations whose kernels depend on three variables II* [in Russian], Sb. Nauchn. Trudov Kaf. Vyssh. Mat. Tul'sk. Politekh. Inst. **2** (1974), 45-51

I. B. SIMONENKO

[1968] *On higher-dimensional discrete convolutions* [in Russian], Mat. Issled. (Kishinev) **3, 1** (1968), 108-122

[1971] *On the solvability of bi-singular and poly-singular equations* [in Russian], Funk. Anal. Prilozh. **5, 1** (1971), 93-94

A. V. SKORIKOV

[1980] *Bi-singular convolution operators and bi-potentials in spaces with mixed norm* [in Russian], Izv. VUZov Mat. **220, 9** (1980), 85-88

V. V. SOBOLEV

[1956] *The Transfer of Radiation Energy in the Atmosphere of Stars and Planets* [in Russian], Gostekhizdat, Moscow 1956; Engl. transl.: Van Nostrand, Princeton 1963

[1972] *Light Scattering in the Atmosphere of Planets* [in Russian], Nauka, Moscow 1972; Engl. transl.: Pergamon Press, Oxford 1975

[1985] *A Course in Theoretical Astrophysics* [in Russian], Nauka, Moscow 1985

V. V. SOLODNIKOV

[1969] *Theory of Automatic Regulation* [in Russian], Nauka, Moscow 1969

V. A. SROCHKO

[1984] *Optimality conditions of maximum principle type in Goursat-Darboux systems* [in Russian], Sibir. Mat. Zh. **25, 1** (1984), 126-132

J. SYNNATZSCHKE

[1984] *Über eine additive Normgleichheit für Operatoren in Banach-verbänden*, Math. Nachr. **117** (1984), 175-180

D. SZYNAL, S. WEDRYCHOWICZ

[1987] *On existence and asymptotic behavior of random solutions of a class of stochastic functional-integral equations*, Coll. Math. **51** (1987), 349-364

L. A. TAKHTADZHJAN, L. D. FADEEV

[1986] *The Hamilton Approach in Soliton Theory* [in Russian], Nauka, Moscow 1986

H. R. THIEME

[1980] *On the boundedness and the asymptotic behaviour of the non-negative solutions of Volterra-Hammerstein integral equations*, Manuscr. Math. **31, 4** (1980), 379-412

[1986] *A differential-integral equation modelling the dynamics of populations with a rank structure*, Lect. Notes Biomath. **68** (1986), 496-511

V. I. TIVONCHUK

[1971] *On the solution of two-dimensional linear integral equation with variable limits by the method of averaging of functional corrections* [in Russian], Pribl. Kachestv. Metody Teor. Diff. Integr. Uravn. (Kiev) (1971), 142-155

S. K. TOKILASHVILI

[1985] *A regular linear integral equation of general type* [in Russian], Nauchn. Trudy Gruz. Politekh. Inst. **3** (1985), 112-119

[1987] *Basic theorems on regular linear integral equations of general form* [in Russian], Nauchn. Trudy Grusin. Polit. Inst. **2** (1987), 107-109

M. VÄTH

[1997] *Ideal Spaces*, Lecture Notes Math. **1664**, Springer, Berlin 1997

M. M. VAJNBERG

[1969] *Variational Methods and the Method of Monotone Operators in the Theory of Nonlinear Equations* [in Russian], Nauka, Moscow 1969; Engl. transl.: Halsted Press, Jerusalem 1973

G. K. VALEEV, O. A. ZHAUTYKOV

[1974] *Infinite Systems of Differential Equations* [in Russian], Nauka, Alma-Ata 1974

C. V. M. VAN DER MEE

[1983] *Transport theory in L_p spaces*, Integral Equ. Oper. Theory **6**
        (1983), 405-443

A. B. VASIL'EVA, N. A. TIKHONOV

[1989] *Integral Equations* [in Russian], Izdat. Mosk. Gos. Univ., Mos-
        cow 1989

I. N. VEKUA

[1948] *New Methods for the Solution of Elliptic Equations* [in Russi-
        an], Gostekhizdat, Moscow 1948; Engl. transl.: Wiley-Inter-
        science, New York 1967

L. Z. VITOVA

[1975] *The form of the eigenvalues and eigenfunctions of an inte-
        gral operator with partial integrals and degenerate kernel* [in
        Russian], Funct. Anal. (Ul'janovsk) **6** (1975), 34-43

[1976] *Associated functions of an integral operator with partial inte-
        grals and degenerate kernels* [in Russian], Funct. Anal. (Ul'-
        janovsk) **7** (1976), 35-44

[1976a] *Solvability of an integral equation with partial integrals and
        degenerate kernels* [in Russian], Funct. Anal. (Ul'janovsk) **8**
        (1976), 41-52

[1977] *On the Theory of Linear Integral Equation with Partial In-
        tegrals* [in Russian], Kand. Diss., Univ. Novgorod 1977

[1984] *The form of the eigenvalues and eigenfunctions of an integral
        operator with partial integrals and Jordan kernel* [in Russian],
        Funct. Anal. (Ul'janovsk) **23** (1984), 52-61

[1988] *Solvability of linear integral equations with partial integrals
        and Jordan kernels* [in Russian], VINITI No. 1280-1388, Nov-
        gorod 1988

V. S. VLADIMIROV

[1961] *Mathematical problems of uniform-speed particle transport theory* [in Russian], Trudy Mat. Inst. Steklova Akad. Nauk SSSR **61** (1961), 1-158

V. VOLTERRA

[1959] *Theory of Functionals and of Integral and Integro-Differential Equations*, Dover Publ., New York 1959

B. Z. VULIKH

[1961] *Introduction to the Theory of Semi-Ordered Spaces* [in Russian], Fizmatgiz, Moscow 1961

[1977] *Introduction to the Theory of Cones in Normed Spaces* [in Russian], Izdat. Kalinin. Univ., Kalinin 1977

F. WOLF

[1959] *On the invariance of the essential spectrum under a change of boundary conditions of partial differential boundary operators*, Indag. Math. **21** (1959), 142-147

A. C. ZAANEN

[1953] *Linear Analysis*, North Holland Publ., Amsterdam 1953

[1967] *Integration*, North Holland Publ., Amsterdam 1967

[1983] *Riesz Spaces II*, North Holland Publ., Amsterdam 1983

P. P. ZABREJKO

[1966] *Nonlinear integral operators* [in Russian] Trudy Sem. Funk. Anal. Voronezh **8** (1966), 1-148

[1967] *On Volterra integral operators* [in Russian], Uspekhi Mat. Nauk **22, 1** (1967), 167-168

[1967a] *On the spectral radius of Volterra integral operators* [in Russian], Lit. Mat. Sbornik **7, 2** (1967), 281-287

[1968] *Investigations in the Theory of Integral Operators* [in Russian], Doct. Diss., Univ. Voronezh 1968

[1974] *Ideal function spaces* [in Russian], Vestnik Jarosl. Univ. **8** (1974), 12-52

[1976] *On the spectrum of linear operators which act in different Banach spaces* [in Russian], Kach. Pribl. Metody Issled. Oper. Uravn. Jarosl. Univ. **1** (1976), 39-47

[1987] *Bounded solutions of functional and ordinary differential equations* [in Russian], Proc. Conf. Funct. Diff. Equ., Dushanbe 1987

[1989] *Existence and uniqueness theorems for solutions of the Cauchy problem for differential equations with deteriorating operators* [in Russian], Dokl. Akad. Nauk BSSR **33, 12** (1989), 1068-1071

[1990] *The contraction mapping principle in K-metric and locally convex spaces* [in Russian], Dokl. Akad. Nauk BSSR **34, 12** (1990), 1065-1068

[1992] *Iteration methods for the solution of operator equations and their application to ordinary and partial differential equations*, Rend. Mat. Univ. Roma **12** (1992), 381-397

P. P. ZABREJKO, A. S. KALITVIN

[1997] *A boundary value problem for the Barbashin integro-differential equation with variable operator* [in Russian], submitted

P. P. ZABREJKO, JU. S. KOLESOV, M. A. KRASNOSEL'SKIJ

[1969] *Implicit functions and the Bogoljubov-Krylov averaging principle* [in Russian], Dokl. Akad. Nauk SSSR **184, 3** (1969), 526-529

P. P. ZABREJKO, A. I. KOSHELEV, M. A. KRASNOSEL'SKIJ, S. G. MIKHLIN, L. S. RAKOVSHCHIK, V. JA. STETSENKO

[1968] *Integral Equations* [in Russian], Nauka, Moscow 1968; Engl. transl.: Noordhoff, Leyden 1975

P. P. ZABREJKO, A. N. LOMAKOVICH

[1987] *On a generalization of the Volterra theorem* [in Russian], Ukrain. Mat. Zh. **39, 5** (1987), 648-651

[1990] *Volterra integral operators in spaces of functions of two variables* [in Russian], Ukrain. Mat. Zh. **42, 9** (1990), 1187-1191; Engl. transl.: Ukrain. Math. J. **42, 9** (1990), 1055-1058

P. P. ZABREJKO, T. A. MAKAREVICH

[1987] *On some generalization of the Banach-Caccioppoli principle to operators in pseudometric spaces* [in Russian], Diff. Uravn. **23, 9** (1987), 1497-1504

[1987a] *A fixed point principle and a theorem of L. V. Ovsjannikov* [in Russian], Vestnik Belgos-Univ. **3** (1987), 53-55

P. P. ZABREJKO, M. KH. MAZEL', L. G. TRETJAKOVA

[1984] *The Green function in the problem on bounded solutions for generalized ordinary differential equations* [in Russian], Dokl. Akad. Nauk BSSR **28, 1** (1984), 18-20

P. P. ZABREJKO, N. F. NGUYEN

[1987] *The majorant method in the theory of Newton-Kantorovich approximations and the Pták estimates*, Numer. Funct. Anal. Optim. **9** (1987), 671-684

[1989] *The Pták estimates in the Newton-Kantorovich method for operator equations* [in Russian], Vestnik Akad. Nauk Belor. SSR **3** (1989), 8-13

P. P. ZABREJKO, T. NUREKENOV

[1966] *On the existence of nonnegative periodic solutions of systems of integro-differential equations* [in Russian], Vestnik Akad. Nauk Kaz. SSR **5** (1966), 32-36

P. P. ZABREJKO, O. M. PETROVA

[1980] *Theorem on the continuation of bounded solutions for differential equations and the averaging principle* [in Russian], Sibir. Math. Zh. **21**, 4 (1980), 50-61

P. P. ZABREJKO, S. V. SMITSKIKH

[1979] *On a theorem by M. G. Krejn and M. A. Rutman* [in Russian], Funkts. Anal. Prilozh. **13, 3** (1979), 81-82

P. P. ZABREJKO, P. P. ZLEPKO

[1983] *On majorants of Uryson integral operators* [in Russian], Kach. Pribl. Metody Issled. Oper. Uravn. **8** (1983), 67-76

List of Symbols

$|A|$ (operator module)   58
$]A[$ (special operator)   58
A' (associate operator)   61
A^* (adjoint operator)   61
$A^{\#}$ (special operator)   61
$\alpha(s)$ (special function)   34
$\alpha(t,s)$ (special function)   46
$\mathfrak{A}$ (σ-algebra)   85
$\mathfrak{A}(T) \otimes \mathfrak{A}(S)$ (product algebra)   214
$A \otimes B$ (tensor product)   290
$B(\mathbb{R})$ (Bogoljubov space)   157
$BC(\mathbb{R})$ (function space)   140
$BC^1(\mathbb{R})$ (function space)   157
$BV([a,b])$ (function space)   35
$\beta(z,s)$ (special function)   34
$\beta(z,t,s)$ (special function)   46
C (multiplication operator)   3
$C(t)$ (multiplication operator)   2
$C([a,b])$ (Chebyshev space)   19
$C(J)$ (function space)   140
$C(J,X)$ (function space)   92
$C^1(J,X)$ (function space)   92
$C_t(X)$ (function space)   20
$C_t^1(X)$ (function space)   20
$\hat{C}_t(X)$ (function space)   198
$\hat{C}_t^1(X)$ (function space)   198
$\check{C}_t(X)$ (function space)   198

539

Index